A comprehensive book about the future modern power grid with renewable sources – gives a lot of insight into the challenges and possibilities.
Frede Blaabjerg, *Professor, Power Electronics, Aalborg University, Denmark*

This great book focusses the inner core of the ongoing energy transition with the power grid and its most essential components, the supply of solar and wind electricity, the role of batteries and how power electronics enable the system combined with the practical investigation of microgrids.
Christian Breyer, *Professor for Solar Economy, LUT University, Finland*

Provides a good introduction to the wide range of technologies that are transforming the power system. Offers a comprehensive overview of the disruptive technologies that are driving the energy transition.
Babu Chalamala, *Senior Scientist, Grid Modernization and Energy Storage, Sandia National Laboratories, USA*

This is a critical resource that navigates the challenges and opportunities lying ahead of us as we transition to a more sustainable and resilient energy future.
Mark Z. Jacobson, *Professor, Stanford University, USA*

Dr. Ertuğrul has achieved an exceptional result in this textbook. The book covers some of the most essential verities associated with the current modernization of electric power grids around the world. It will surely serve as an essential text for both students and practicing engineers in the power and energy sector. The book provides excellent coverage of both the physical concepts associated with energy conversion for modern inverter based resources (IBRs) and the opportunities and challenge ahead in the continued global push for greater deployment of IBR technologies.
Pouyan Pourbeik, *President, Power and Energy, Analysis, Consulting and Education, Texas, USA*

As renewable energy, especially solar and wind energy, becomes cheaper and more widely deployed, it has become urgent to shift in the basic paradigm that has governed electricity planning for close to a century to integrate these variable sources of electricity. This detailed examination of the technical elements of an evolving power grid will help with transitioning to a new paradigm.
M. V. Ramana, *Professor and Simons Chair, School of Public Policy and Global Affairs, University of British Columbia, Vancouver, Canada*

Dr. Ertuğrul's book is required reading for anyone interested in the decarbonisation of the power industry. At last, there is a readable, comprehensive engineering explanation on how the power system works today and how it will need to work when fossil plants are retired and the grid is more distributed through the introduction of electrification and distributed energy.
Audrey Zibelman, *Former CEO, Australian Energy Market Operator (AEMO)*

Reinventing the Power Grid

In this comprehensive guide for practicing engineers and students, Ertuğrul explains the field of renewable energy and distributed generation technologies and describes the transformation occurring in power grids due to the rise of renewable energy sources and emerging technologies.

This book covers key areas such as the status of grid transformation, photovoltaic (PV) solar energy, wind energy systems, distributed energy resources, microgrids, grid-scale and domestic battery storage systems, e-mobility and emerging distributed energy technologies. The text presents an equilibrium between theoretical concepts and practical applications, with each chapter emphasizing both theory and practical application. Each chapter commences with a lucid explanation of the subject matter, which is then succeeded by an investigation into its real-world applications and implications. Supplementary material is also provided, such as real wind data files, PV data files and Matrix Laboratory (MATLAB®) and Excel codes. This includes a sample real data set from grid-scale autonomous microgrid test platforms and household, distribution and transmission-level power system data. The book also incorporates a section consisting of problems, quizzes and solutions. This element prompts the reader to put the theoretical knowledge to use in addressing real-world challenges, thereby cultivating a more in-depth grasp of the topic. Through this in-depth approach, readers will be able to apply their comprehensive knowledge and practical understanding to decision-making regarding future challenges in the energy industry.

This book is an invaluable guide for professionals working in the field, particularly those who aim to stay updated on the latest technologies and trends. Undergraduate and postgraduate students will also benefit from the book's comprehensive approach and inclusion of real-world data and problems to solve, which will build their expertise and give them a solid foundation for their future careers.

Nesimi Ertuğrul is an Associate Professor in the School of Electrical and Electronic Engineering at the University of Adelaide. He earned a BSc and MSc at Istanbul Technical University and a PhD in electrical engineering at Newcastle University, UK, in 1993. He is a senior member of IEEE.

Reinventing the Power Grid
Renewable Energy, Storage, and Grid Modernization

Nesimi Ertuğrul

CRC Press is an imprint of the
Taylor & Francis Group, an **informa** business

Designed cover image: Components of modernized power system infrastructure around Australia: Wind turbine, solar PV array with 2-axis tracking system, lithium battery modules, power lines and servers. Nesimi Ertuğrul

First edition published 2025
by CRC Press
2385 NW Executive Center Drive, Suite 320, Boca Raton FL 33431

and by CRC Press
4 Park Square, Milton Park, Abingdon, Oxon, OX14 4RN

CRC Press is an imprint of Taylor & Francis Group, LLC

© 2025 Nesimi Ertuğrul

Reasonable efforts have been made to publish reliable data and information, but the author and publisher cannot assume responsibility for the validity of all materials or the consequences of their use. The authors and publishers have attempted to trace the copyright holders of all material reproduced in this publication and apologize to copyright holders if permission to publish in this form has not been obtained. If any copyright material has not been acknowledged, please write and let us know so we may rectify in any future reprint.

Except as permitted under U.S. Copyright Law, no part of this book may be reprinted, reproduced, transmitted or utilized in any form by any electronic, mechanical or other means, now known or hereafter invented, including photocopying, microfilming and recording, or in any information storage or retrieval system, without written permission from the publishers.

For permission to photocopy or use material electronically from this work, access www.copyright.com or contact the Copyright Clearance Center, Inc. (CCC), 222 Rosewood Drive, Danvers, MA 01923, 978-750-8400. For works that are not available on CCC, please contact mpkbookspermissions@tandf.co.uk

Trademark notice: Product or corporate names may be trademarks or registered trademarks and are used only for identification and explanation without intent to infringe.

ISBN: 978-1-032-68595-3 (hbk)
ISBN: 978-1-032-69216-6 (pbk)
ISBN: 978-1-032-69217-3 (ebk)

DOI: 10.1201/9781032692173

Typeset in Minion
by codeMantra

To Olivia and Atacan

Science advances one funeral at a time.

Max Planck

Contents

Foreword, xvii

Preface, xx

CHAPTER 1 ▪ The Power Grid: Evolution, Limitations and with Snapshot of the Current Status and Future 1

  1.1 HISTORICAL CONTENTS AND TURNING POINTS 1

  1.2 A GLOBAL OVERVIEW OF ELECTRICITY GENERATION AND POWER GRID 5

      1.2.1 Basic Components of Power Grid 5

      1.2.2 Electricity Costs and Components 6

      1.2.3 Major Blackouts and Consequences 8

      1.2.4 The Communication Structure and Analysis of Smart Energy Meters 9

      1.2.5 Transmission/Distribution Losses and Loss Ownership 12

  1.3 CHANGING ENERGY LANDSCAPE: LCOE, GLOBAL WARMING AND TOWARDS CLEANER ENERGY SOURCES AND JOB OPPORTUNITIES 13

      1.3.1 Levelized Cost of Energy 14

      1.3.2 Global Warming and Prioritization, and Towards Cleaner Energy Sources 15

      1.3.3 Job Opportunities 16

  1.4 GRID MODERNIZATION TOWARDS FULL GRID TRANSFORMATION 18

      1.4.1 Intermediate (Hybrid) Steps of Transitions 19

1.5	ELECTRIFICATION		22
	1.5.1	Distributed Energy Resources	24
	1.5.2	e-Mobility	25
	1.5.3	Smart Houses	27
	1.5.4	Smart Appliances	29
	1.5.5	Mining Electrification	30
	1.5.6	Growing Needs for Data Centres	35
	1.5.7	Changing Appliance Landscape	40
1.6	CHALLENGES AND SOLUTIONS IN POWER SYSTEMS		47
	1.6.1	Variability, Intermittency and Rate of Change of Power	47
	1.6.2	Efficiency	51
	1.6.3	Power Quality	56
	1.6.4	Power System Inertia and Fault Levels	58
	1.6.5	Reverse Power Flow (Back Feeding)	64
	1.6.6	Supply/Demand Imbalance and Load Shedding/Curtailment	66
	1.6.7	Domestic Wiring Issues in Rooftop PV Systems	70
	1.6.8	Environmental Events and Reliability	71
1.7	POWER SYSTEM SECURITY AND CYBERSECURITY		73
1.8	THE ROLES OF STANDARDS AND REGULATIONS		77
BIBLIOGRAPHY			82

CHAPTER 2 ▪ Photovoltaic Solar Energy   85

2.1	INTRODUCTION		85
2.2	SOLAR RESOURCES, DEFINITIONS AND MEASUREMENTS		88
	2.2.1	Types of Solar Radiation	88
	2.2.2	Solar Energy Potential and Availability	90
	2.2.3	Air Mass (AM) Ratio, Solar Spectrum and Solar Irradiation	95
	2.2.4	Solar Irradiation Measurements	97
2.3	PV MATERIALS		99

2.4	OPERATION PRINCIPLES OF SOLAR PV CELLS AND PRACTICAL STRUCTURE		101
2.5	IMPACTS OF TILTING AND TRACKING IN SOLAR PV PANELS		103
2.6	ELECTRICAL EQUIVALENT CIRCUIT OF SOLAR PV CELLS		107
	2.6.1	Blackbox Testing and Identification of Equivalent Circuit Parameters	108
	2.6.2	Analysis of Equivalent Circuits for PV Cells	112
2.7	PV CELLS TO MODULES TO ARRAYS		115
	2.7.1	The Current–Voltage (I-V) and Power–Voltage (P-V) Characteristics of PV Panels (Building Blocks) under Standard Test Conditions	122
	2.7.2	Impacts of Irradiance and Temperature on I-V Characteristics	124
	2.7.3	Nominal Operating Cell Temperature (NOCT)	125
	2.7.4	Series- and Parallel-Connected Solar PV Cells and Impacts of Shading	126
	2.7.5	Panel and Array Level Shading Scenarios	130
2.8	PV SYSTEMS AND PRINCIPLES OF MPPT		135
	2.8.1	Typical Electrical Loads for Solar PV Panels	137
	2.8.2	Principles of MPPT	137
	2.8.3	MPPT Methods	140
	2.8.4	The Choice of PV Panel Configurations for Applications	144
2.9	PERFORMANCE RATIO (PR)		144
2.10	DESIGN, INSTALLATION GUIDELINES AND STANDARDS		147
2.11	ISSUES AND FAILURES IN SOLAR PV SYSTEMS		149
BIBLIOGRAPHY			152

CHAPTER 3 ▪ Wind Energy Systems 153

3.1	INTRODUCTION	153
3.2	TYPES OF WIND TURBINE SYSTEMS	156

		3.2.1	Shadow Effect in Turbine Types	161
3.3	MAJOR COMPONENTS OF MODERN WIND TURBINES			162
		3.3.1	Rotor Blades	165
		3.3.2	Tower and Foundation	167
		3.3.3	Nacelle and Hub	170
		3.3.4	Generator Types and Choice	171
3.4	PHYSICS OF WIND ENERGY/POWER IN THE WIND			174
		3.4.1	Power in the Wind	174
		3.4.2	Temperature and Altitude Correction for Air Density	175
		3.4.3	Impact of Tower Height and Terrain on Wind Energy Production	177
3.5	CHARACTERISTIC FEATURES OF WIND TURBINES			180
		3.5.1	Axial Momentum Theory and Maximum Efficiency of the Turbine (Betz Limit)	180
		3.5.2	Tip Speed Ratio (TSR)	189
		3.5.3	Turbine/Blade Arrangements, Solidity	191
		3.5.4	Yaw Angle	193
		3.5.5	Pitch Angle, Power versus the Rotor Speed and the Cp versus TSR Characteristics	194
3.6	THE EQUATION OF MOTION, BLADE AND GENERATOR TESTING AND MODELLING			196
		3.6.1	The Equation of Motion	196
		3.6.2	Generator Testing and Modelling	198
		3.6.3	The Blade Testing	205
3.7	WIND CHARACTERISTICS, RESOURCES AND ANALYSIS OF WIND REGIMES			207
		3.7.1	Wind Quality and Measurements	209
		3.7.2	Analysis of Wind Resources	212
3.8	WIND POWER TO ELECTRICITY: CONTROL ISSUES WITH SENSORS			219
		3.8.1	Blade Design and Regulation	219

Contents ■ xiii

	3.8.2	Modelling of Wind Turbines	225
	3.8.3	Control of Large Wind Turbines	229
3.9	WIND FARM LAYOUTS, CONNECTION AND SUBSTATIONS		237
3.10	CAPACITY FACTOR (CF)		239
3.11	WIND TURBINE SYSTEM STANDARDS AND WIND FARM INCIDENT CATEGORIES		240
BIBLIOGRAPHY			243

CHAPTER 4 ■ Power Electronics and Control of Power System Components — 246

4.1	POWER ELECTRONICS (PE) AND WIDE BANDGAP DEVICES (WBG)		246
	4.1.1	Wide Bandgap Devices Status and Application Benefits	247
4.2	CLASSIFICATIONS AND SELECTION OF CONVERTERS AND SWITCH CONTROL PRINCIPLES		250
	4.2.1	DC-DC Converters	253
	4.2.2	Typical Common Converters: AC-DC (Rectifiers) and DC-AC (Inverters)	254
	4.2.3	Types of Switch Control and Modulation Techniques	258
4.3	STEP-DOWN (BUCK) CONVERTER: ANALYSIS AND OPERATION		262
4.4	STEP-UP (BOOST) CONVERTER: ANALYSIS AND OPERATION		265
4.5	SYNCHRONOUS BUCK-BOOST CONVERTER: ANALYSIS AND BENEFITS		268
4.6	POWER FLOW ANALOGY: VARIABLE SPEED MOTOR DRIVES VERSUS CONVERTERS USED IN WIND TURBINE GENERATORS		272
	4.6.1	Utilization of Axes Transformations in Three-Phase Converter Control	274

xiv ■ Contents

- 4.7 OPERATION PRINCIPLES OF BPMG, THREE-PHASE CONVERTERS OPTIONS IN WIND TURBINE SYSTEMS AND THEIR CONTROL 280
  - 4.7.1 Common Features of Generator-Side Converters/Rectifiers 282
  - 4.7.2 Control of Generator-Side PWM Rectifiers 286
  - 4.7.3 Control Principles of Grid-Side Inverters 291
- 4.8 COMMON INVERTER CHARACTERISTICS, CLASSIFICATIONS, OTHER CONTROL METHODS AND THE FUTURE 298
  - 4.8.1 Droop Control 306
- 4.9 FOUR-QUADRANT CONTROL CONCEPT OF CONVERTERS/INVERTERS 313
  - 4.9.1 Converters in Battery Storage, Electric Vehicles and Others 317
- 4.10 FAULTS, FAILURES AND RELIABILITY 323
- BIBLIOGRAPHY 326

Chapter 5 ■ Batteries and Fuel Cells in Energy Storage 328

- 5.1 INTRODUCTION 328
- 5.2 BATTERY CELL STRUCTURES AND BASIC OPERATION 331
- 5.3 COMMON DEFINITIONS USED IN BATTERIES 336
  - 5.3.1 Battery Voltage 336
  - 5.3.2 State of Charge (SOC) 340
  - 5.3.3 Battery Efficiency 343
  - 5.3.4 Battery Life 346
- 5.4 ELECTRICAL EQUIVALENT CIRCUITS OF BATTERY CELLS 349
- 5.5 BATTERY CONNECTIONS, PACKAGING CONFIGURATIONS AND TEMPERATURE CONTROL 352
  - 5.5.1 Battery Temperature Control: Cooling/Heating 359
- 5.6 COMPARISON OF BATTERIES FOR SELECTION 369
- 5.7 CHARACTERISTIC CURVES OF BATTERIES, DISCHARGING AND CHARGING 378

5.8	BATTERY BALANCING AND BATTERY MANAGEMENT	393
5.9	SENSORS USED IN BATTERY SYSTEMS	401
5.10	FUEL CELLS AND SUPERCAPACITORS: DESCRIPTIONS, MODELLING AND OPERATION	405
	5.10.1 Equivalent Circuit, Electrical Characteristics and Definitions	411
5.11	PROTECTION, SAFETY AND BATTERY TESTING METHODS	416
	BIBLIOGRAPHY	423

CHAPTER 6 ▪ Microgrids with Distributed Energy Resources: Design, Operation and Case Studies — 425

6.1	BACKGROUND	425
6.2	VOLTAGE REGULATION IN POWER GRID, ROLES OF INVERTER-BASED RESOURCES (IBRS) AND IMPACTS ON TRADING	429
6.3	ANALYSIS OF REVERSE CURRENT FOR VOLTAGE REGULATION	433
6.4	POWER QUADRANTS, POWER FACTOR AND HARMONIC POWER	435
	6.4.1 Harmonic Power	444
6.5	GENERAL DESIGN CRITERIA OF BATTERY STORAGE SYSTEMS	445
	6.5.1 Reference Applications in BSS Design: LDC, RoCoF and Virtual Inertia	455
6.6	MICROGRIDS	468
6.7	A DETAILED CASE STUDY: AUTONOMOUS MICROGRIDS WITH BSS AND DERS OPTIONS	475
	6.7.1 Mechanical Design Features of a Custom-Built Container	477
	6.7.2 Electrical Design and Characteristic Features	481
6.8	DATA CAPTURE AND POWER QUALITY ANALYSERS	490
	BIBLIOGRAPHY	502

xvi ■ Contents

Chapter 7	■ Comprehensive Results and Analysis of an Autonomous Microgrid with BSS	504
7.1	BASIC OPERATION OF A BATTERY STORAGE SYSTEM (BSS) AND BATTERY	504
7.2	FOUR-QUADRANT POWER, POWER FACTOR CONTROL AND INVERTER IDLE REACTIVE POWER WAVEFORMS	512
7.3	MICROGRID ANTI-ISLANDING, ISLANDING, SEGREGATION AND REINTEGRATION WITH A BSS	518
7.4	POWER QUALITY TESTS WITH NONLINEAR LOAD IN BSS	527
7.5	HARMONIC COMPONENTS AND THD WITH NONLINEAR LOAD IN BSS	537
7.6	VOLTAGE WAVEFORMS AND RMS MEASUREMENTS	541
7.7	COMMUNITY-LEVEL OPERATIONS: DAILY CYCLES OF BSS WITH HIGH PV PENETRATION, P-Q CONTROL AND FAULT RESPONSE	548
	7.7.1 Analysis of Four-Quadrant Control with BSS in Daily Cycles	549
	7.7.2 Longer-Term Operation of the BSS in Grid	551
	7.7.3 Response of the BSS to a Fault	556
	7.7.4 Inverter Efficiency Duration Curve and Thermal Management	561
7.8	MICROGRID WITH FLOW BATTERY IN A REMOTE CAMPUS	563

INDEX, 571

Foreword

As a lifelong researcher and instructor on transitioning the world to a clean, renewable energy future, I am excited to introduce this comprehensive new book on the evolving power grid. This is a critical resource that navigates the challenges and opportunities lying ahead of us as we transition to a more sustainable and resilient energy future.

For over a century, grids have moved electricity to power civilization while remaining largely invisible. However, the traditional grid, heavily reliant on electricity produced centrally and generated by fossil fuels, faces significant and growing questions. Due to aging infrastructure, ever-increasing energy demands and damage due to climate change, all amplified by the integration of renewable energy resources, power grids require a fundamental transformation.

This book is not entirely a chronicle of limitations; it is also a call to action for evaluating our power systems. It dissects the vulnerabilities of the current grid, its dependence on fossil fuels and the environmental consequences of maintaining the status quo. More importantly, it explains the potential benefits of adding vast amounts of clean, renewable and distributed electricity sources.

Within the book's pages, you will find a detailed exploration of the prime clean, renewable electricity-generating technologies that hold the key to a sustainable future. The text explores the large potential of solar and wind electricity, explains the world of battery storage and fuel cells and analyses complex control systems, using helpful analogies that orchestrate these new players within the energy transition.

A particular focus of this book is on microgrids—localized grids supported by distributed energy resources (DERs) that can operate autonomously or be connected to a larger grid. Microgrids offer an insight into the future, providing solutions for remote communities, disaster relief zones and remote military bases. They enhance grid resilience and energy security while inspiring a more decentralized and democratic energy landscape.

The past century has witnessed the transformative power of electricity, yet significant portions of the world remain unserved by a reliable grid or any grid at all. Traditionally, alternating current (AC) electricity dominated, primarily due to its ability to be used in transformers to reduce current and increase voltage, thus transmitting long distance with low line losses. However, this landscape is experiencing a rapid shift driven by advancements in renewable electricity generation and power electronics, which are becoming increasingly complex.

The introduction of power electronics in the past six decades has significantly altered the power system. As discussed in this book, this advancement has enabled the development of variable speed drives, revolutionized converters and inverters and played a critical role in the renewable energy sector and electric vehicles (EVs). While highly beneficial for efficiency and intermittency mitigation, the increasing penetration of renewable electricity sources, like wind and solar, presents a new challenge: reduced power system inertia.

In the past, large, rotating generators provided crucial stability to the grid through inertia. However, with the rise of inverter-based renewable resources and the retirement of aging power stations, this inertia is diminishing. This significantly impacts frequency regulation and overall stability of the conventional power grid. Additionally, the unidirectional flow for which the AC grid was designed is challenged by the bidirectional nature of clean, renewable electricity sources feeding power back into the grid.

As inverter-based resources (IBRs) become the dominant form of generation and interface for inherently direct current (DC) loads (currently disguised within the AC system), this book argues that the need for AC may be re-evaluated in some cases. Advancements in wide bandgap (WBG) devices will further support the operation of IBRs, potentially increasing the use of DC electricity, particularly in microgrids. Additionally, green hydrogen generation and fuel cells offer promising solutions for range and capacity anxieties in long-distance, heavy EVs.

Beyond the grid itself, a significant revolution is quietly unfolding—the widespread "electrification" across various sectors. This shift, discussed herein, is driven by not only efficiency improvements but also the inherent controllability of DC loads currently hidden within the AC system. Imagine a future where transportation, heating and industrial processes are all powered by clean, renewable electricity, significantly reducing our reliance on fossil fuels and air pollution.

The transition to a clean, renewable energy future requires significant changes in the operation of the centralized grid and its market practices. Integrating DC microgrids and decentralized DC power networks will necessitate very fast and secure response mechanisms.

This book hypothesizes that, as the existing hybrid power system transforms into a new form with bi-directional power flow as the norm, power system security will also play a prime role in this future grid. "Reinventing the Power Grid" is a timely exploration of these challenges and technological advancements shaping the power grid of tomorrow.

This well-written book is intended for a broad audience, from renewable energy students and practicing engineers to leading managers and policymakers. It is a crucial and necessary resource for the development of a clean, renewable energy future.

Mark Z. Jacobson
Professor of Civil and Environmental Engineering
Stanford University

Preface

WHILE WIDESPREAD ELECTRICITY USE is a relatively recent phenomenon, with a history of just over a century, it has not reached all corners of the globe. Initially, AC grids dominated due to the ease of generation, transmission at high voltage and distribution at lower voltages. However, this trend is undergoing a rapid transformation within the last two decades.

The introduction of power electronics around 60 years ago significantly altered the power system landscape. Initially used for rectification, power electronics enabled the way for economical and seamless variable speed drives. Further advancements in solid-state switch technology revolutionized the implementation of converters and inverters. These advancements were not limited to the renewable energy sector but also impacted EVs and entire power supply applications. These innovations have mitigated the intermittency in renewable energy sources and augmented the efficiency of battery charging and discharging systems utilized not only in power systems but also in EVs.

As known, the conventional AC grid was designed for unidirectional power flow due to its advantages—low transmission losses, low cost and mass production. However, the rise of renewable energy sources and DERs, facilitated by power electronics, fundamentally changes this structure.

As the penetration of renewable energy has increased, a key challenge has also emerged—the reduction in power system inertia. Traditionally, inertia provided crucial stability to the grid. With the increasing ratio of IBRs (wind, solar and battery technologies including EVs) and the retirement of aging conventional power stations, the contribution of inertia from these sources diminishes. This significantly impacts two critical aspects of AC grid: synchronization and stability. Furthermore, power flow is becoming bidirectional. While unidirectional flow with high inertia worked well for market forces in the past, it presents difficulties as the share of IBRs increases.

As IBRs become the dominant form of generation and interface for DC loads, the need for AC will likely be dramatically questioned. Advancements in WBG devices will further support the operation of IBRs. Green hydrogen generation and fuel cells are potential solutions for range and capacity anxieties, particularly in larger EVs.

An often overlooked but significant revolution is the widespread "electrification" across diverse sectors, from smart appliances to mining electrification. This shift is driven not just by efficiency improvements but also by the controllability of inherently DC loads that are currently disguised within the AC power system.

Furthermore, the operation of the centralized grid and market practices developed around a unidirectional, high-inertia system will likely undergo significant changes. As the grid becomes dominated by IBRs, integrating DC microgrids and decentralized DC power networks will require very fast and secure response mechanisms that cannot be done by human involvement.

This shift toward a DC grid can be seen as a progression through levels 5 and 6 of "electrification", where the existing hybrid system (AC and DC generation with mostly DC loads) transforms into a new form with bi-directional power flow as the norm. Power system security will also be paramount in this future grid.

Therefore, it can be concluded that the limitations of the conventional AC grid are becoming increasingly evident with the rise of renewable energy and distributed generation. Modernization towards a DC grid is not just an option, but an essential step to ensure a stable, secure and efficient power grid for the future. This book aims to address these challenges and explore the technological advancements to explore the way for a more sustainable and resilient power system.

It is intended that this book will not only provide a clear understanding of the connections between sub-technologies and systems for both current renewable energy engineering students and practicing engineers involved in the renewable energy sector. In addition, the leading managers and political decision-makers may obtain a clear direction about the future grid hence to avoid undesirable consequences in terms of reliability, cost and system security.

This book originated from the renewable energy degree program I proposed and led at the University of Adelaide for over 15 years.

The curriculum included a highly customized course titled Distributed Generation Technologies, which explored various interdisciplinary topics relevant to renewable energy.

The curriculum of this course has constantly evolved to reflect the latest advancements in renewable energy technologies. This included insights from industry leaders delivering guest lectures, academic research and my personal experience developing and testing the first autonomous microgrid with battery storage in a distribution network. This project's advanced data acquisition system provided invaluable practical knowledge. My experiences extended beyond the classroom, enriching the course content through conferences I organized on renewable energy topics, microgrid workshops I conducted and consulting work on grid-scale battery storage projects. Encouraged by the publisher, I have transformed these years of experience and knowledge into this book. Instructors seeking to incorporate this book into their curriculum will also have access to supplementary materials, including system-level assignments on wind and solar photovoltaic (PV) systems and foundational laboratory experiment handouts. The book intentionally avoids specific technical standards, focusing on core principles and general considerations due to the dynamic nature of renewable energy technologies. To enhance understanding, complex information is often presented in tables for clear comparison, including equivalent circuits and associated formulas. In addition, circuit diagrams gradually increase in complexity, highlighting similarities among seemingly diverse concepts to simplify explanations and promote a deeper understanding of the material.

In this book, Chapter 1 explores the evolution of power grids, highlighting limitations and future prospects. It examines historical development, provides a global snapshot of electricity generation and discusses the changing energy landscape driven by various concerns. The chapter explores grid modernization, electrification across various sectors and the challenges faced by power systems, including renewable energy variability, efficiency, power quality and security threats. This sets the stage for the book's exploration of the future of power grids and the need for transformation.

Building on the discussion of challenges faced by power grids in Chapters 1 and 2 discusses PV solar energy, a key player in clean energy. It explores solar resources, their measurement and potential availability, laying the groundwork for assessing solar viability in different locations. The operating principles of solar cells and the impact of tilting and tracking on panel efficiency are also discussed. The chapter explains the electrical behaviour of solar cells and how they are connected into modules and arrays, forming the building blocks of larger solar PV systems, which is required for understanding how these systems generate electricity and how to maximize their output. Finally, the chapter covers critical aspects like maximum power point tracking techniques, which ensure optimal power generation, and the concept of performance ratio for evaluating system effectiveness. It concludes with design considerations and standards for installing reliable and efficient solar PV systems.

Chapter 3 discusses another key renewable energy source: wind energy systems. The types of wind turbine systems and their major components, including blades, towers, nacelles and generators, are covered in the chapter. It then explains the physics of wind energy, explaining how wind speed translates into usable power and how factors like temperature, altitude and tower height affect production. The characteristic features of wind turbines are also covered, including the Betz limit for maximum efficiency, tip speed ratio, turbine arrangements and factors like yaw and pitch angles that influence power generation, as understanding these features is vital for optimizing wind turbine performance. Building on this foundation, the chapter explores the equation of motion for wind turbines, along with methods for blade and generator testing and modelling. It then examines wind characteristics, resource assessment techniques and the various control systems employed with wind turbines using sensors to maximize power output and ensure safe operation. Finally, Chapter 3 covers wind farm layouts, connection to the grid through substations and considerations for small-scale wind turbines. It concludes by discussing capacity factors, relevant standards and potential wind farm incident categories.

Chapter 4 bridges the gap between renewable energy sources and grid integration by focusing on power electronics and control of power system components. Understanding how power is converted and controlled is essential for effectively integrating renewable energy sources with variable outputs into the grid. The chapter explains WBG devices, exploring their growing importance and application benefits. It then examines various converter classifications and control principles. The detailed analysis and operation of specific converter types are also discussed. Next, the chapter explores the operating principles of permanent magnet synchronous generators that are leading the generator options, analyses three-phase converter options commonly used in wind turbine systems and discusses their control strategies. This includes examining generator-side converters/

rectifiers, their control using pulse width modulation (PWM) and grid-side inverter control principles. The chapter then expands its focus on inverters, exploring common characteristics, classifications, control methods and future trends. It introduces the concept of four-quadrant control for converters and inverters, a critical aspect for managing power flow in various applications. Finally, the chapter explores the role of power electronics in embedded generation technologies such as battery storage and EVs. It concludes by acknowledging potential faults, failures and considerations for reliability in power electronics systems. This in-depth exploration equips readers with the technical knowledge required to understand how renewable energy sources are integrated and controlled within the power grid.

As renewable energy sources like solar and wind are intermittent, efficient energy storage becomes critical for grid stability and power delivery. Therefore, Chapter 5 focuses on the leading energy storage solutions: batteries and fuel cells. It begins with an exploration of battery cell structures and their basic operation, providing a foundation for understanding how batteries work. It then introduces common battery terminologies, which are essential for effective battery management. The chapter explores electrical equivalent circuits used to model battery behaviour and explains battery connections, packaging configurations and the importance of temperature control for optimal performance and longevity. A comparison of different battery types to aid in selection for specific applications is given, followed by the analysis of characteristic curves for battery discharging and charging, providing insights into their performance under various conditions. To ensure safe and reliable operation, the chapter explores battery balancing techniques and the role of battery management systems. It highlights the importance of sensors used in battery systems for monitoring critical parameters. Recognizing that batteries are not the only energy storage solutions, the chapter concludes by introducing fuel cells and supercapacitors as emerging solutions. It explores their basic descriptions, operating principles and equivalent circuit models. Safety considerations and battery testing methods round out the chapter, providing a comprehensive overview of energy storage technologies essential for integrating renewables into the grid.

Chapter 6 builds upon the knowledge of renewable energy sources, power electronics and energy storage explored in previous chapters. This chapter establishes the background for microgrids, highlighting their growing importance in the evolving power grid landscape. It explains voltage regulation within power grids, exploring the role of IBRs and their impacts on electricity grid. The chapter analyses the concept of reverse current flow within the context of voltage regulation. Understanding power quadrants, power factor and harmonic power becomes crucial for optimizing microgrid performance. A dedicated section explores harmonic power and its potential impacts on the grid. To ensure effective battery storage system design within microgrids, the chapter outlines general design criteria. It explains reference applications for business support system (BSS) design, including load duration curve, rate of change of frequency and the concept of virtual inertia that are all critical aspects for maintaining grid stability. Finally, the chapter introduces microgrids in detail as it lays the groundwork for the following chapter. Therefore, the practical aspects of microgrid design and operation are given to provide a deeper understanding of how the concepts explored in previous chapters come together in a real-world application. The mechanical design features of a custom-built container are explained as they provide insights into the physical infrastructure that houses the microgrid components. It further explores the electrical design and characteristic features of the microgrid, detailing how various DERs and the battery storage system are interconnected and controlled to function as a cohesive unit. To analyse the performance of the microgrid and ensure optimal operation, the data capture and power quality analyser aspects are covered. By examining the data collected by these tools, engineers can gain valuable insights into the microgrid's efficiency, identify potential issues and ensure it delivers reliable power. This comprehensive case study serves as a bridge between theoretical knowledge and practical application, solidifying the reader's understanding of microgrids with DERs and BSS.

The final chapter, Chapter 7, presents the highly detailed and unique results of a real-world microgrid case study with battery storage system integration. While the concepts of microgrids with battery storage system are familiar to industry professionals, the specific details of this case study offer valuable insights not often shared publicly. Sharing highly detailed operational data is uncommon in this competitive industry, possibly due to a desire to limit competition or avoid legal concerns related to power quality exceeding the ideal standards experienced in traditional AC grids. The chapter starts establishing a foundation by reviewing the basic operation of a battery storage system and then explains four-quadrant power control, power factor management, inverter behaviour and the impact of non-linear loads on power quality, including examinations of harmonic components, total harmonic distortion and voltage waveform measurements. In addition, the case study explores the daily operational cycles within a community with high solar power penetration. This analysis covers factors like reverse power flow, solar irradiance, load profiles and how the system utilizes four-quadrant control throughout daily cycles. Additionally, the chapter covers longer-term BSS operation within the grid and its response to potential faults. The impact of environmental factors, particularly temperature, on the microgrid's performance is also explored in this chapter. The final section in the chapter explores the specific case of a microgrid implemented within a university campus, potentially highlighting the use of a flow battery in a microgrid structure.

Finally, I would like to express my gratitude to all my undergraduate and postgraduate students. Their insightful questions over three decades and their research contributions have significantly enriched this book. My appreciation also goes to my colleagues, researchers and the institutions that have supported the contents of this book. Their invaluable contributions include providing supportive information for the manuscript content, as well as their assistance in developing and testing the autonomous microgrid test system with battery storage. These include Carsten Markgraf, Gabriel Haines, Michael Jansen, Graeme Bell, Qiang Gao, Vikram Kenjle, South Australian Power Network and Australian Renewable Energy Agency (ARENA).

I hope this book will be a valuable resource for the greater community and that it will contribute to achieving more reliable, low-cost and highly efficient power sources in the future intelligent world.

Nesimi Ertuğrul
1st April 2024

CHAPTER 1

The Power Grid

Evolution, Limitations and with Snapshot of the Current Status and Future

1.1 HISTORICAL CONTENTS AND TURNING POINTS

Today, energy continues to propel human civilization and remains a crucial component of modern society, powering everything from homes and businesses to transportation and communication networks with an ever-increasing utilization of data centres. Moreover, energy sources define all our daily activities and enable technological advancements. Therefore, the evolution of energy production and consumption can be broken down into several major eras, each marked by unique characteristics and technological advancements:

- Pre-Industrial Era (Pre-1800s): Most energy was sourced from burning wood, charcoal and animal waste.
- Industrial Era (1800s–1950s): The steam engine led to coal becoming the dominant energy source, accelerating industrialization and urban growth.
- Fossil Fuel Era (1950s–2000s): Oil and natural gas became primary energy sources, and nuclear power developed. Energy consumption

surged, and growing concerns about climate change and the finite nature of fossil fuels.

- Renewable Energy Era (2000s–Present): Within the last two decades there has been a growing shift towards renewable energy sources, primarily solar and wind energy.

However, defining turning points in history can be subjective, but several common measures are often used by historians and scholars to identify these moments. These measures include significant and lasting consequences, unpredictable events disrupting the status quo and human actions that substantially influence the course of events. As a result, these turning points form a new reality that helps shape the future. Figure 1.1 is designed to present a broader definition of these turning points.

In this context, it is agreed that turning points typically arise from an interplay of economic, social and governmental elements, the latter being influenced by political and cultural factors. Three critical infrastructure sectors also drive these three elements: energy, mobility and communications, which play a crucial role in modern societies and economies:

- **Energy** is relevant to the production, distribution and consumption of various forms, including coal, oil, natural gas, electricity, renewable energy sources and hydrogen.
- **Mobility** involves the transportation of people and goods, including transportation systems such as trains, cars, planes and ships, and the infrastructure that supports these systems like railways, roads, bridges, ports and airports.
- **Communications** relates to the exchange of information through various means, including telecommunications networks, the Internet and other emerging digital platforms and auxiliary systems.

Even though the evolution of energy, mobility and communication has roots in the prehistoric era, the advent of writing in ancient civilizations has allowed the recording and dissemination of information, enabling us to define a consensus of opinions. The evolution of communication and mobility has caused the spread of knowledge, ideas and goods, connecting cultures across vast distances and significantly influencing human history.

The Power Grid ■ 3

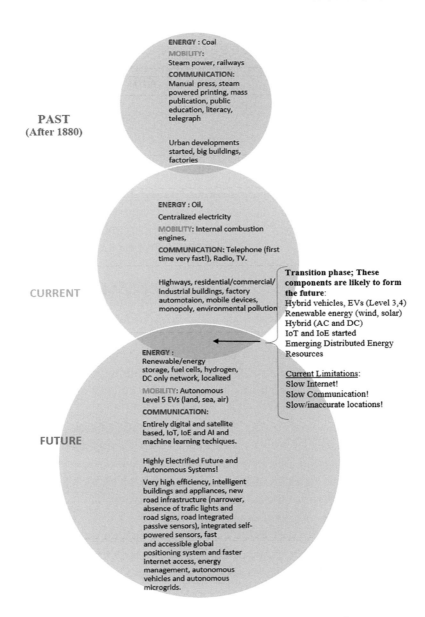

FIGURE 1.1 Three critical infrastructure sectors in human history: Energy, Mobility and Communications, which define the turning points.

Figure 1.1 also provides a brief overview of the evolution over time, along with an estimation of the impacts (indicated by the size of the circles). Moreover, it includes transitional technologies between the overlapping timeframes. A common driving force among these critical infrastructure sectors is the role of "efficiency". It is apparent that as we continuously search for "new energy sources", "new modes of transportation" and "new communication tools", human civilization shifts to a new platform, thus marking significant turning points in history.

Figure 1.1 offers a visual chronicle of critical historical turning points. Potential future trends and developments are expected to focus primarily on electricity generation from renewable energy sources such as solar, wind and wave power. Additionally, advancements in battery technology and energy storage systems are making the use of intermittent renewable energy sources more viable. The deployment of smart grids and other energy management systems contributes to improved efficiency. Furthermore, emerging technologies like hydrogen and nuclear fusion essentially promise a supply of clean energy.

Aviation technology has made notable progress in recent years, enabling faster and more efficient long-distance travel. The integration of electric vehicles (EVs) initially targeted reducing carbon emissions and enhancing safety, but the ultimate aim is to develop various forms of autonomous vehicles. Note that EVs cover a wide range of both domestic and industrial vehicles, such as personal cars, bikes, scooters, mining vehicles, agricultural vehicles, air transport, trams, trains, buses, watercraft and even space vehicles.

Moreover, the advent of the internet and mobile phones in the late Twentieth and early Twenty-first centuries created a new era of instantaneous communication and unprecedented information accessibility. In addition, the rapid increase of 5G and other high-speed mobile networks has enabled faster and more reliable data transmission. Furthermore, the increased adoption of cloud computing and other forms of distributed computing has led to more efficient and flexible use of computing resources. Virtual and augmented reality technologies are already revolutionizing the ways we communicate and interact. Similarly, the rise of the Internet of Things (IoT) and other connected devices is creating new opportunities for automation and data analysis through communication infrastructures.

Power Electronics, as an enabling technology, and distributed energy resources (DERs), which encompass various forms of this enabling technology, are the main elements expected to define future turning points. These elements will have a significant impact on all three critical infrastructure sectors and will be shaped by several factors, including technological advancements, environmental concerns, societal needs and preferences.

Consequently, the future of energy, mobility and communication is likely to be shaped by a complex interplay of technological, environmental and social factors. It is challenging to predict exactly how these trends will evolve, but it is clear that electrical energy technologies will play a central role in these transformations. Moreover, these changes will have profound implications for how people live, work and interact with each other, likely in new urban structures, on intelligent roads and in highly automated workspaces. This book seeks to provide theoretical and practical guidance on the evolution of electrical energy and offers insights, knowledge and training platforms for the future workforce.

1.2 A GLOBAL OVERVIEW OF ELECTRICITY GENERATION AND POWER GRID

1.2.1 Basic Components of Power Grid

A traditional electric grid (or power grid) is a network of electrical generators in power plants, power lines (transmission and distribution), transformers and other equipment (including protection devices, switch gears and substations/connection points) that is used to distribute electricity from a combined power grid to consumers.

The mix of electricity generation (in power plants or in farms) across the globe varies considerably, influenced by factors such as geographical location, availability of natural resources and the energy policies in place. Although the specific contributions of different energy sources change annually, the trend to generate electricity contains a few primary sources: Fossil fuels, renewables (including solar, wind, wave, hydro, geothermal and biomass energy) and nuclear power. Moreover, hydrocarbons, waste-to-energy processes and hydrogen fuel cells are also involved in generation options.

The transmission (primarily including lines, transformers and switch gears as will be discussed later) is responsible for delivering electricity from large power plants to the substations in the power grid by using a high-voltage transmission to minimize energy losses. The distribution system, on the other hand, uses a network of medium and low-voltage power lines to deliver electricity to consumers.

Therefore, the primary duty of any power grid (large, small, thin/wide, short/long) is that it is well-maintained and managed to prevent power outages and disruptions, hence a reliable and stable supply of electricity can be provided to customers.

Overall, the electricity generation landscape around the world is evolving rapidly, with a shift away from fossil fuels towards renewable energy sources and distributed energy resources. However, fossil fuels continue to play a role in many countries, and there are still significant challenges to be overcome in transitioning to a more sustainable, low-carbon energy system. However, some alternative trends and patterns can also be identified.

In the current grid transition phase, the electricity generation methods serve as the prime elements of the modern grid and microgrids, collectively referred to as DERs. The primary target of a DER is to increase energy security, reduce costs and most critically to improve grid reliability. It should be noted here that while energy storage devices such as supercapacitors, batteries, flywheels, pumped hydro and gravitational systems are not "generators" in the traditional sense they act as "virtual generators" in a DER, supplementing or replacing conventional power plants to meet the electricity demand.

1.2.2 Electricity Costs and Components

There are several examples around the world where regulatory changes have been designed to encourage competition and contribute to reducing electricity costs, which include the UK (in 1989), Australia (in 1998, via NEM covering 40,000 km of lines for 9M customers), the EU (since 1990s) and few states in the US that started in 1996. Note that while market liberalization and competition can lead to lower electricity prices, it can lead to price volatility and requires adequate regulatory oversight.

However, so far, electricity prices have generally risen over the past few decades, and it is proven that liberalization and privatization do not lead to lower electricity costs most of the time. This is due to multiple reasons and complex interactions, which include: market concentration and domination, weak regulatory framework and primary focus on profit maximization, external factors (such as fluctuations in global energy prices and exchange rates and pandemics), stranded costs (which were primarily associated with obsolete or inefficient infrastructure that was built under the regulated system passed on to consumers even after deregulation, or even expenses during various components of grid modernization), renewable energy integration due to grid upgrades, customer choice and complexity that leads to ineffective decisions and finally social and political factors (due to political decisions, past or current subsidies or social programs).

Note that although Australia implemented a rapid increase in the installation of renewable energy systems in recent years, the country has one of the highest energy costs in the world. This can be attributed to multiple factors including ageing infrastructure, longer transmission and

TABLE 1.1 Comparison of the Components of Electricity Prices in Three Distinct Regions

Component	Europe	USA	Australia
Supply (Generation)	31%	56%	32% (wholesale cost)
Transmission	28% (network)	13%	45% (network)
Distribution	In Network Costs	31%	In Network Costs
Environmental Costs	Not Specified	Not Specified	10%
Retail Costs	Not Specified	Not Specified	10%
Retail Margins	Not Specified	Not Specified	3%
Taxes & Levies	41% of the bill	Not Specified	Not Specified

distribution lines per capita, decommissioned power plants, investment in renewable energy integration, diminishing demand, focus on profit maximization and regulatory and market structure. Therefore, it is unlikely that electricity costs will be reduced in the near future if alternative measures are implemented to mitigate the above-listed factors.

In addition, it is worth noting that the components of electricity prices vary depending on the country and its market structure. These components can be grouped into network costs, wholesale costs, green energy policies and retail costs. Table 1.1 is given below to compare these components of electricity prices in three distinct regions: Europe, the US and Australia. Note that some of the components were not available in the literature to be included or were classified under a different category.

In the table above, each region has its own unique structure for electricity prices, and thus, the strategies for reducing electricity costs can differ.

For example, in Europe, increasing efficiency in generation through the adoption of advanced technologies and using smart grid technologies, a substantial cost saving can be made. However, the tax reduction needs lobbying and governmental policy changes.

Similarly, generation efficiency improvement should be the prime target in the US followed by the modernizing the distribution network.

Note that transmission and distribution line lengths also vary significantly depending on factors like population density, geography and energy consumption patterns. Densely populated urban areas might have shorter lines due to concentrated demand, while rural regions with scattered populations might require longer lines to reach consumers. A rule of thumb for estimating line lengths based on electricity consumption is about 1 TWh of electricity use is supported by about 225 km of transmission lines, and by about 4,000 km of distribution lines. Therefore, in Australia, network costs make up a significant portion of electricity costs primarily due to the low population density and long transmission and distribution lines per capita which can be addressed by forming renewable-based remote grids.

Technological advancements and increased efficiency in generation can also reduce electricity costs. Moreover, reducing peak demand on warm days can reduce wholesale prices after adapting sufficient energy storage. Furthermore, unlike the other regions, reducing the costs associated with environmental policies and streamlining or minimizing retail operations by effective utilization of smart meters can contribute to reducing overall costs.

1.2.3 Major Blackouts and Consequences

Note that in recent years, as power grids continue to evolve, several major blackouts have occurred worldwide, one of which was the September 2016 blackout in South Australia. This blackout was due to a series of events and technical limitations which lasted several hours, affecting more than 1.7 million residents and leading to significant economic and infrastructure losses. The incident brought to light questions about the dependability of renewable energy sources and the stability of the grid, leading the South Australian government to introduce a range of measures to improve the resilience and reliability of their electricity network, as part of an expensive energy plan.

These strategies included the deployment of large grid-scale battery storage, the commissioning of additional gas-fired power plants to cater for peak demand periods, establishing more interconnections with the national grid to increase power flow and installing synchronous condensers to aid in grid stability. Moreover, new regulations were implemented for wind farms to increase their grid-stability technologies.

Although some near-blackout events occurred in consecutive years, these measures have generally been successful in improving grid reliability, the financial impact on the public and on energy prices remains unclear. Furthermore, as more traditional power plants are decommissioned, the ongoing reduction in power system inertia in the national grid still poses a vulnerability to sudden changes in frequency in the AC grid.

However, with a balanced blend of innovative technologies and policies, it is feasible to maintain the AC grid's stability in the face of future traditional power plant decommissioning. The lessons learned from South Australia's experience have informed both policy and technological innovation, contributing to the enhancement of resilience and reliability in power systems across the globe.

Another major blackout including the Northeast blackout of 2003, which affected the north-eastern US and parts of Canada, affecting about 55 million people for up to 4 days, was caused by a combination of human error and equipment failures.

Moreover, in an event in 2012, India experienced one of the world's largest blackouts, which left over 600 million people without power for several hours, which was caused by excessive demand for power and insufficient infrastructure to meet the demand.

In 2021, Texas, US also experienced a severe winter storm that caused power outages and blackouts throughout the state, which was due to a surge in demand for electricity as consumers turned up their heating while many power generation facilities were offline due to the extreme conditions. In this particular event, there was also a loss of lives and serious injuries and illnesses.

The consequence of these blackouts often extends beyond inconvenience, resulting in economic losses, disruption of essential services and negative impacts on public health and safety. These outcomes highlight the necessity of maintaining a balance between generation and demand, the importance of reliable and robust infrastructure, and the need for rapid response in the highly complex hybrid AC grid.

Therefore, it is crucial to emphasize that as the utilization of AC power increases, along with greater penetration of renewable energy sources and changes to load characteristics (mainly due to the introduction of power electronic devices), the complexities and problems in the "hybrid AC grid" have increased significantly. As illustrated in Table 1.2, these complications primarily stem from the inherent characteristics of an AC network.

1.2.4 The Communication Structure and Analysis of Smart Energy Meters

The communication structure of the power grid includes various technologies that allow for real-time monitoring and control of the grid. These technologies are:

- Supervisory Control and Data Acquisition (SCADA) systems that allow grid operators to remotely monitor and control the power grid.

- Energy management systems (EMS) that use data from the SCADA systems to optimize the operation of the power grid.

- Advanced metering infrastructure (AMI) that uses smart meters.

- Demand response systems that allow utilities to incentivize customers to reduce their energy usage during peak periods, which can help prevent grid overload and reduce the need for new power plants, which is done by communicating with the smart meter.

TABLE 1.2 Undesirable Features and Negative Impacts of AC

- Regulating AC voltage can be challenging due to the simultaneous involvement of three parameters: magnitude, frequency and phase, which are primary causes of instability in the power grid.
- To increase power, an AC source can connect to another AC source via the synchronization process. This requires "regulation" of AC voltage using all three parameters listed above. However, this process introduces a significant delay in the connection process and even more delay when reconnected. The synchronization of a grid-connected PV inverter is also heavily affected by the voltage level, especially when it falls outside the voltage settings of the inverter unit.
- AC introduces a unique electric circuit parameter: impedance, which involves two frequency-dependent quantities, capacitance and inductance.
- Impedances in AC circuits can cause resonance, potentially leading to oscillations in electrical energy and generating undesirable over-voltage and over-current situations.
- Impedances result in AC (complex) powers: apparent, active (real) and reactive. Although the active power performs real work, reactive powers oscillate back and forth at twice the frequency of the supply. This results in additional loading and losses in the power line.
- Impedance and complex power require 4-quadrant control to increase the utilization of a power line, which is performed with reactive power support plants involving AC synchronous machines, or with more modern solutions using static compensators (STATCOMs) and static VAr compensators (SVCs).
- AC grid requires several additional components for power quality improvements. Traditional electromechanical systems (like capacitor banks or voltage regulators at substations) are slow, taking minutes to adjust voltage and often not very accurately. Modern solutions, SVCs, DVRs, active filters and voltage regulators are costly to build and usually integrated into the grid at limited locations.
- AC introduces large currents to power lines when switching on, especially under light load conditions.
- In unbalanced three-phase AC circuits, which are very common due to the nature of the load network, the sequence impedances (positive, negative and zero) are present. Although these components aid in developing techniques for fault current estimation, an unbalanced AC circuit is not desirable in AC transformers and AC rotating machines due to the presence of circulating DC components.
- AC introduces the skin effect, causing higher losses in T/D lines.
- AC causes inductive and capacitive coupling, which prevents complete isolation in various networks concerning harmonics and other disturbances.
- Potentially hazardous voltages in AC networks can occur due to induced voltages from electromagnetic induction and capacitance and Earth Potential Rise (EPR) when large current flows to earth through an earth grid impedance, resulting in step and touch potential.
- As AC diverts from the ideal waveform of a sinewave, it introduces harmonics and poor power factor. Harmonics produce additional power loss (hence heat) in power lines and electromechanical devices, and torque ripples and braking torque in rotating machines.
- AC transformers provide ease of step-up and down conversion but offer limited stepped regulation only using a tapped-winding topology.

(Continued)

TABLE 1.2 (Continued)   Undesirable Features and Negative Impacts of AC

- The AC grid was designed for unidirectional power flow using low-frequency AC transformer technology. Although transformers can perform bi-directional power flow, their utilization in a centralized power grid is unidirectional, as reverse power flow can cause a significant voltage rise at the primary.
- AC transformers have numerous non-ideal characteristics: magnetizing current, losses at no-load (a larger portion of standby power in electronic circuits), in-rush current, voltage drop under load, sensitivity to harmonics/DC offset/load imbalance, lack of overload protection, fire hazard and environmental issues (due to insulation oil), and can contribute to power quality issues.
- Power quality aspects in AC systems cover a much wider spectrum, ranging from voltage sags and swells to transient changes and frequency variations.
- AC sources cause large starting currents and poor starting performance in conventional line-start asynchronous motors.
- AC requires multiple T/D lines that are greatly exposed to environmental factors and are heavier and costlier.
- The AC grid is not ideal for the integration of renewable energy sources that have DC stages.

There are several benefits of smart energy meters, including accurate meter reading and billing, customer engagement and increased energy efficiency when considered timely by consumers, outage detection and response to improve reliability, time-of-use pricing, demand response and load management, integration with renewable energy sources with smart inverters.

Using smart meters in the power grid can significantly reduce costs associated with meter reading and billing and can affect different aspects of retailer costs. However, it does not necessarily eliminate the entire retailer cost. This is due to the fact that there are costs and factors that smart meters do not eliminate, including infrastructure and maintenance costs, wholesale energy costs (which is significant in a de-regulated large market, but not in community level DERs and autonomous microgrids of the future), customer service and administration and regulatory compliance and fees

1.2.5 Transmission/Distribution Losses and Loss Ownership

The average transmission and distribution losses as a percentage of output in the EU were 6.2% as of 2014 (was 9.36% in 1960), in the US were about 5% between 2017 and 2021, and in Australia were 4.8% as of 2014 (was 12.73 in 1973). The reduction in transmission and distribution losses over the given periods can be attributed to several factors including technological advancements, grid modernization and upgrades, energy management systems, regulatory measures and standards, improved maintenance practices and within the last few decades, renewable energy integration and distributed generation near the consumers and load centres, consumer awareness and demand-side management which alters consumption patterns hence losses.

Note that these losses can be higher when the grid is under strain or during high temperatures. In addition, technological improvements, changes in the electricity generation mix and grid modernization can improve the losses.

In a deregulated energy market, the grid is typically divided into separate components, each owned and operated by different entities, including power generation companies, transmission system operators (TSOs), distribution system operators (DSOs) and retail energy providers.

Power and corresponding energy losses occur mainly in the transmission and distribution systems. How these losses are accounted for and who ultimately pays for them may vary based on the regulations in different jurisdictions. Note that TSOs and DSOs estimate the energy losses within their networks, which are used to determine the total amount of electricity to be purchased. However, the cost of this extra electricity is passed on to all grid users, often as part of the transmission and distribution charges in their electricity bills, but eventually to the end-users, the consumers.

In a scenario where prosumers (entities that both produce and consume electricity) generate energy locally, the energy losses in transmission could be significantly reduced. This is due to the fact that electricity does not need to travel as far, specifically if multiple microgrids with DERs are implemented near the load centres. Although there will still be losses in the distribution network, as electricity is delivered to end users, this is significantly low, hence the cost of wasted energy. Note that if the grid connection is still kept for redundancy, the limited cost of energy losses on the grid will be shared.

It can be highlighted that until autonomous microgrids are established, in a more dynamic future energy market, it is possible that prosumers could receive more favourable treatment in terms of how energy losses are allocated. This would be a recognition of the benefits they provide to the grid, such as reducing transmission losses and enhancing grid resilience. The specifics would depend on regulations and market structures, which continue to evolve.

1.3 CHANGING ENERGY LANDSCAPE: LCOE, GLOBAL WARMING AND TOWARDS CLEANER ENERGY SOURCES AND JOB OPPORTUNITIES

The energy landscape around the world is diverse and constantly evolving. The mix of energy sources used to meet the world's energy demands varies by country and region. In recent decades, there has been a shift towards cleaner and renewable energy sources. In addition, energy security is increasingly reliant on a mix of domestic renewable production, imports and various energy storage technologies.

According to the International Energy Agency (IEA) 2021 report, renewable energy accounted for 29% of the world's electricity generation in 2020 and is forecast to rise to 35% in 2025 with hydropower being the largest renewable energy source followed by wind and solar.

However, despite the growing use of renewable energy, fossil fuels still account for a significant portion of the world's energy mix. Interestingly, average energy consumption per capita per year around the world varies drastically, such as 16 kWh in Chad, 1,100 kWh in India, 4,000 kWh in China, 12,000 kWh in the USA and 50,000 kWh in Iceland. Furthermore, it is critical to note that about 760 million people in the world still cannot access electricity.

While about a tenth of the world's population still lacks access to electricity, and a larger portion lacks reliable, affordable energy, limiting economic growth and quality of life, grid modernization clearly offers economic growth and job creation opportunities in a range of energy sectors.

Private companies, investors and governments all have crucial roles to play in this energy revolution, which includes research and development of new technologies and improving energy storage to ensure reliable electricity supply.

However, the changing energy landscape, beyond environmental benefits, requires justifying the utilization of a particular new energy source primarily based on the cost of energy. This is defined by a metric named levelized cost of energy (LCOE).

1.3.1 Levelized Cost of Energy

The LCOE (in $/kWh) is the primary metric for comparing the lifetime unit costs of various technologies at the plant level. It represents the average cost of producing a unit of electricity, including initial capital costs, operating costs, fuel costs (where relevant), capacity factors and any revenue generated from selling electricity or other products. Unlike the financial costs tied to specific projects in individual markets, LCOE reflects the economic costs associated with generic technologies.

LCOE allows for comparing the cost of generating electricity from different sources over the lifetime of a project. This comprehensive view includes all costs, making it possible to identify which renewable technologies may be cost-competitive with conventional ones, both currently and in the future under varying operating assumptions.

Key sensitivities in LCOE analysis include fuel costs, tax subsidies, carbon pricing and costs of capital. However, estimating LCOE can be complex due to factors like the assumed cost of equity capital, and the cost of capital or capital structure for various technologies is open to manipulation.

The specific assumptions for calculating LCOE encompass a broad range, from lifetimes, discount rates and capacity to environmental considerations such as emission costs, carbon price and the long-term societal consequences of various conventional generation technologies. LCOE analysis may also address social and environmental externalities, like social costs and rate consequences for those who cannot afford distributed generation solutions.

Geographical trends and technological advances continuously affect LCOE. For example, the LCOE of solar photovoltaic (PV) technology and wind (both onshore and offshore) has been declining rapidly, making it competitive with conventional sources in many parts of the world. The LCOE of energy storage systems is also decreasing, although it varies by technology and application.

In summary, the LCOE of renewable energy and DERs is becoming increasingly competitive with traditional energy sources. This trend is expected to continue, reflecting the rapid decline in the LCOE of renewable technologies such as solar PV and wind, and the ongoing improvements in technology and scaling of deployment.

1.3.2 Global Warming and Prioritization, and Towards Cleaner Energy Sources

Different stakeholders prioritize concerns such as global warming, the cost of electricity and power reliability in varying orders. For a majority of consumers, the priority is on reliability. Disruptions in power can not only cause significant inconveniences but also pose serious safety threats. Once the reliability is established, the cost typically becomes the next consideration. Historically, even though environmental concerns have been on the rise, they have generally taken a backseat to both reliability and cost. Yet, with the surge in environmental consciousness and the decreasing costs of renewable energy, this hierarchy is witnessing a shift.

In areas with a high penetration of renewable energy sources and supportive policies, concerns about global warming might even surpass those about costs. Meanwhile, in the commercial and industrial sectors, the nature of the business can sometimes make cost considerations more paramount than even reliability.

The electric power system plays a significant role in increasing global warming by releasing greenhouse gases, predominantly from fossil fuel consumption. These emissions include carbon dioxide, methane and nitrous oxide. In addition, pollutants released from burning fossil fuels contribute to environmental challenges like acid rain and smog. Power system infrastructures can also be implicated in habitat destruction, as they sometimes necessitate the removal of trees and vegetation for their establishment.

The habits in which we use our energy, mobility and communications infrastructure can present both positive and negative impacts on the environment. However, via a combination of environmental concerns and technological advancements, renewable energy sources have become increasingly efficient and cost-effective, and grid modernization shows no sign of slowing down.

Mitigation efforts involve the adoption of alternative energy sources to eliminate greenhouse gas emissions and the rigorous implementation of energy efficiency protocols. Such strategies not only counteract contributions to global warming but also support and accelerate a transition towards a more sustainable energy paradigm.

In the spectrum of alternative energy sources developed and still in the pipeline to improve DERs, hydrogen and micro nuclear reactors stand out for their potential, especially as consistent and robust solutions for high-density load centres. The multifaceted nature of hydrogen, coupled with advances in hydrogen storage and fuel cell technologies position it at the forefront of future energy resources. In the meantime, the inherent modularity of micro nuclear reactors enables them to smoothly adapt to varying power demands, making them more relevant for modern-day urban habitats and high-density settings.

It is worth noting that energy storage options will always be critical in the electrification transition. Their role is not just to mitigate the intermittency in power generation but also to cater to changing demand patterns. While this book will feature a dedicated chapter on battery storage, other notable energy storage options include batteries (like lithium-ion, flow batteries and solid-state variants), pumped hydro storage, tidal storage, thermal storage (both molten salt and ice/water varieties), flywheels, compressed air energy storage, hydrogen storage, superconducting magnetic energy storage, supercapacitors, gravitational potential energy storage and kinetic energy storage. Although each storage method has its unique advantages, constraints and optimal use cases, other factors such as capacity needs, discharge duration, geographical limitations and cost-effectiveness will influence their selection once they are all technically viable.

1.3.3 Job Opportunities

In 2021, the renewable energy industry provided employment for approximately 12.7 million people, both directly and indirectly. This figure represents continued global growth over the past decade, predominantly in the sectors of solar PV, bioenergy, hydropower and wind power. In addition, grid modernizations and transformations also offer a range of new and conventional job opportunities across a variety of disciplines. As modernization continues to grow, the demand for skilled professionals in the disciplines is expected to increase. The job opportunities can be classified under seven groups as given below:

- Engineering and Technical Roles: These are essential in the design, construction and maintenance of renewable energy systems and infrastructure, particularly within solar and wind energy sectors. This includes electrical, electronics, power systems and computer systems engineering roles. Future job growth is projected in areas such as cybersecurity, communications, data processing and AI. Additionally, indirect employment opportunities, not accounted for in the aforementioned figures, have emerged in fields like mining, materials and chemical engineering.

- Research, Development and Production: To be involved in developing new technologies and improving existing ones in areas such as energy storage, solar PV and wind turbines, which involves a range of areas specifically in power electronics. In addition, the basic production of associated technologies requires significant medium-skill employment.

- Project Management: To oversee the development and implementation of renewable energy projects, from planning and design to construction and operation.

- Sales and Marketing: To promote and sell renewable energy products and services to customers, such as solar panels and energy storage systems.

- Installation and Maintenance: To be involved in installing and maintaining renewable energy systems, such as solar panels, wind turbines and heat pumps.

- Policy and Advocacy: To be involved in shaping government policy and public opinion in support of renewable energy, including working with policymakers and other stakeholders.

- Finance and Investment: To be involved in providing capital and financial services to renewable energy projects and companies.

Globally, the number of renewable energy jobs will rise to 38 million by 2030 (more than double the 17.4 million under the planned energy scenarios). In addition, energy efficiency, electric vehicles, power systems/flexibility and hydrogen could employ another 74.2 million people by 2030.

1.4 GRID MODERNIZATION TOWARDS FULL GRID TRANSFORMATION

Over the last two decades, the evolution of the electrical power system towards a two-way power flow model has been rapid and continues to have significant impacts. As the grid moves towards full transition, it is predicted to become more decentralized, flexible and digital, focusing on renewable energy sources and energy storage technologies. Key trends shaping this transformation include decentralization via DERs, digitalization, electrification, resilience and security, energy storage, integration of renewable energy, technology advancements, consumer participation as prosumers, energy efficiency and regulatory and policy changes.

This evolution has made clear that the reliability and safety of electrical load service could be compromised if the transmission and distribution grid become incapable of providing backup for the intermittency issues associated with high-impact renewable power. As a response, various forms of energy storage systems have been developed and integrated into conventional grids. Enhancing the capacity of the power grid to accommodate DERs at higher penetration levels has become critical.

Grid modernization can be described as a process of upgrading the electricity grid to enhance its reliability, resilience and performance. This includes integrating more renewable energy sources, utilizing advanced digital technology to improve grid control and management, implementing smart meters and automation for real-time data and enhancing power quality and security. With the competitiveness of the LCOE for solar PV and wind, grid modernization is accelerating on a greater scale to adapt to legacy grid systems.

The current electrical grid's complexity, with its blend of diverse components, nonstandard or varied network elements and ever-increasing integration of power electronics-based DERs, is further compounded by the presence of multiple disconnected players in the system. The decommissioning of parts of the grid due to aging infrastructure or market forces, often results in reduced synchronizing power, leading to operational and stability issues.

However, the continuous integration of DERs and power electronic devices is reducing the need for traditional synchronizing power. As the grid transitions away from legacy systems towards a power electronics-based architecture, opportunities arise to fundamentally rethink the grid design, such as considering alternative designs like DC grids. With advancements in power electronics, a DC grid could offer a more efficient, flexible and reliable power system that can better accommodate high levels of DER penetration. Nevertheless, further research and development are needed, considering the substantial changes required for such a significant shift in grid architecture.

1.4.1 Intermediate (Hybrid) Steps of Transitions

Transitioning towards a "pure system" like an all-electric vehicle or a pure DC grid composed of smaller DC microgrids requires an intermediate, or hybrid, stage. Various reasons such as technological maturity and adaptation, economic feasibility, energy security and reliability, consumer acceptance as well as regulatory and policy alignment necessitate this stage. The analogies between transitioning to fully electric vehicles and a pure DC grid are supported by Table 1.3.

Just as hybrid vehicles serve as an intermediate step between gasoline (petrol)-powered vehicles and EVs, they present a practical path towards more sustainable transportation. The move to fully electric vehicles represents a significant leap, but this transition depends on a still-developing charging infrastructure and grid adaptation. The landscape of charging infrastructure for EVs differs fundamentally from that for vehicles powered by internal combustion engines, with many more options for charging EVs. As on-board charging technology improves and energy storage becomes widely distributed, the need for public charging will diminish.

Similarly, the existing AC grid with primarily DC loads serves as an intermediate step towards a future fully DC grid. A transition to a DC grid would involve supplying power directly to DC loads, improving efficiency while eliminating numerous limitations of AC as listed in Table 1.3. However, this shift involves substantial technological changes—primarily driven by the evolution of power electronics as well as presents challenges in standardization and safety. Therefore, it is predicted that the grid transition towards a DC network will accelerate more decentralized and accessible infrastructure. Although both hybrid systems have their advantages and challenges. The choice depends on various factors, but the transition towards a DC Network as in EV is inevitable, with hybrid solutions playing an important role.

The power grid of the future will be digitized, autonomous and secure, with a greater diversity of energy sources and a more decentralized structure, and new business models, regulatory frameworks and investment strategies will be required.

TABLE 1.3 Analogy between "Hybrid Electric Vehicle versus All-Electric Vehicle" and "Hybrid AC Grid with Power Electronics and DC Renewable Sources versus DC Only Network"

Hybrid Electric Vehicle (HEV) versus All-Electric Vehicle (EV)	Hybrid AC Grid with Power Electronics (PE) and DC Renewable Sources versus Pure DC Grid and Loads
• HEV has an internal combustion engine (ICE) and an electric drive. • HEV has a more complex architecture that increases the size, weight and maintenance requirements. • HEV has to control both power electronics and engine for energy conversion, and complex drive mechanisms for switching between the two power sources. • HEVs have a longer range due to the energy density of petrol and are quicker to fill (but still in a limited number of petrol stations!). *In contrast:* • EV is relatively simpler, involving fewer components hence requires less maintenance. • EV range is getting better with acceptable range. • EV charging is getting better and will improve further with WBG device electronics and easily accessible charging power (widely distributed to household level!). • EVs can be controlled easily and are highly suitable for the final target, autonomous vehicle.	• Hybrid AC grid has conventional generators with bulky and slow-acting components and ever-increasing PE devices that interface with DC-based renewable energy sources. • Hybrid AC grid is getting increasingly complex with diverse levels of maintenance requirements. • Hybrid AC grid has additional components, which increase system size, complexity, maintenance requirements, and potential points of failure. In addition, such complexity encounters difficulties in standardization due to the diverse nature of power electronics technologies and manufacturers. • Hybrid AC grid ideally has to control both major components but this is not easy. • Hybrid AC Grid appears to have higher power capacity. However, its reliability is getting worse with increased renewable energy penetration, and control and stability issues associated with AC systems. • Hybrid AC Grid is not suitable for the final levels of the "electrified" systems for future autonomous system integration. *In contrast:* • DC grid can eliminate the need for many legacy and PE devices, leading to a reduction in component count and complexity. • DC PE circuits with WBG devices display a promising future to offer fast control from generation to distribution and load levels. • DC network does not have any stability issues and greater regulation capabilities and requires negligible maintenance. • DC network can enable autonomous operation and seamless monitoring and diagnostic options. *(Continued)*

TABLE 1.3 (Continued) Analogy between "Hybrid Electric Vehicle versus All-Electric Vehicle" and "Hybrid AC Grid with Power Electronics and DC Renewable Sources versus DC Only Network"

Hybrid Electric Vehicle (HEV) versus All-Electric Vehicle (EV)	Hybrid AC Grid with Power Electronics (PE) and DC Renewable Sources versus Pure DC Grid and Loads
	• Energy storage is DC and the future large-scale storage options will be DC, which can respond to demand effectively and faster. • DC network allows for direct integration of DC-based renewable energy sources and has the highest round trip and system efficiency.
• In both the hybrid AC grid and HEV cases, delaying the transition towards more DC-oriented solutions (DC grid and EV) would delay the training of the workforce required to maintain and operate these systems. Early adoption, while it may have initial costs, will lead to better long-term outcomes in terms of workforce development, standardization and economies of scale. • However, transitioning to a pure DC network, distribution and utilization with DC loads would be a significant undertaking, requiring considerable changes to existing infrastructure (as highlighted under "Electrification"), standards and regulations.	

The hybrid phase serves as an essential bridge in the transition towards new technologies, balancing various factors. It is guiding the shift from conventional energy sources to modern systems, like transitioning from petrol stations to diverse EV charging options, reflecting changes in infrastructure, technology and user needs. As on-board charging technology advances and the grid moves towards DC, the traditional charging infrastructure may become obsolete.

Both hybrid electric vehicles (HEVs) and Hybrid AC grids present unique advantages and challenges, and the choice between them depends on factors like readiness, cost, policy support and environmental considerations. While the transition towards a DC network is inevitable, hybrid solutions will play a key role until full transformation occurs.

Grid transformation to DC is the specific target of grid modernization. Using DC can eliminate energy losses during repeated conversion between AC and DC. However, a full-scale transformation from AC to DC would be slow but steady while AC and DC systems will coexist for a few decades.

The transformation of the power grid highlights a more interconnected, renewable-focused future, and efficient energy solutions that will reshape the global energy mix, which is illustrated in Figure 1.2 which covers the basic components of a typical power grid from generators to prosumers. The subsequent "Electrification" section provides further insights into this progress. Both hybrid systems (HEVs and Hybrid AC grids) will continue to shape the transition until the legacy technology fully transforms, which is confirmed by the shifting in energy paradigm in recent decades.

1.5 ELECTRIFICATION

In essence, electrification, the process of converting systems to operate on electricity, is a diverse trend that offers a range of benefits. However, it requires a careful approach for a seamless transition while minimizing destructive impacts. Its success depends on clean, renewable energy to realize various benefits across major sectors such as transportation, heating, industrial processes, agriculture, shipping, aviation and rail, all of which align well with power grid modernization.

From a broader perspective, electrification is primarily driven by six major factors. These include societal impacts (primarily targeting noise reduction and improved air quality), technological advancements (resulting from improvements in battery technology, appliances and grid infrastructure), demand management (which requires grid-scale storage and smart grid technologies), new job markets in renewable energy and manufacturing (necessitating training and reskilling initiatives), economic

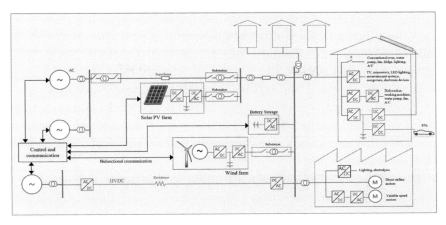

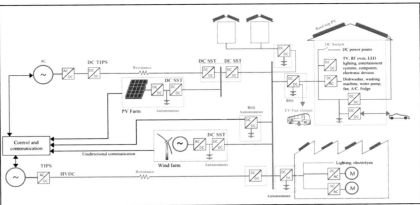

FIGURE 1.2 Basic components and structure of the current (top) and future (bottom) power grid.

considerations (where higher upfront costs can be offset by lower operational expenses) and finally, policy and regulation (which evolves steadily to support grid modernization).

It is worth noting that there is a need for a structured framework in these sectors, which will be summarized using a tiered system level. This system categorizes technological fields and guides timelines, goals and innovation in electrification, details of which are covered in the following paragraphs. In addition, note that these classifications of the characteristics of electrification primarily target efficiency improvement, emission reduction, reduced costs, longer lifespan, control and precision, utilizing various power and energy sources that were not available in the legacy grid.

1.5.1 Distributed Energy Resources

The transition to DERs within the power grid is primarily driven by the following key factors: reliability, the ever-rising cost of electricity, growing environmental awareness, the benefits of electrification with DER (such as economic, practical and technological in a digitized world), efficiency improvements and energy savings and sustainability. In addition, there is a need to explore the differences between electrical and other forms of energy in driving various applications, to justify why electrification with DER is the prime choice.

The shift to electrification with DERs is a complex and structured approach that is likely to support advancements in reliability, energy efficiency as well as the integration of renewable sources. Looking ahead, continued technological advancements are essential for ensuring the power systems' security and optimal performance of the electrification infrastructure. Table 1.4 summarizes the levels of progression of the grid's capacity to accommodate and effectively utilize DERs, outlining different

TABLE 1.4 The Progression of the Grid's Capacity to Utilize DERs

Levels	Remarks
Level 1 (DER-agnostic grid)	The traditional centralized grid was not initially designed to accommodate DERs.
Level 2 (DER-aware grid)	This level acknowledges the presence of DERs but does not fully exploit their potential. Power electronics are utilized to interface DERs with the grid, enabling their integration to some extent. It involves numerous AC/DC and DC/AC conversion stages.
Level 3 (DER-leveraging grid) (Partially started in 2022)	At this level, DERs are actively utilized to optimize grid operations. Advanced power electronics and control systems are employed to leverage the capabilities of DERs and enhance the overall performance of the grid. Both AC/DC and DC/AC conversions also co-exist at this level.
Level 4 (DER-dependent grid)	DERs become an integral part of grid operations at this stage and have the potential to replace centralized generation to a significant extent. The grid may support sector coupling, enabling integration with various sectors such as transportation and heating. The DC grid will form the backbone of the power system.
Level 5 (DER-dependent grid)	DERs form a DC grid with multiple autonomous microgrids. At this stage, all DERs become essential components of grid operations, potentially replacing centralized generation to a large degree. The grid can support sector coupling, enabling integration with various sectors like transportation and heating. This level signifies a paradigm shift towards a highly distributed, decentralized and DER-dominated grid.

levels of grid readiness and functionality. Note that, the ultimate goal is full integration, relying on advanced power electronics, control systems, communication infrastructure and robust cybersecurity measures. The primary structure and grid integration of DERs under a microgrid platform will be covered in detail in Chapter 6.

1.5.2 e-Mobility

Table 1.5 presents a roadmap towards the development of fully autonomous EVs, beginning with internal combustion engine (ICE) vehicles equipped

TABLE 1.5 The Driving Levels in EVs and Current Levels Achieved

Levels	Remarks
Level 0 (no automation)	The driver performs all driving tasks, even when assisted by warning or intervention systems.
Level 1 (driver assistance)	The vehicle can assist with some functions, but the driver must be actively engaged in driving at all times. Examples include adaptive cruise control or parking assistance.
Level 2 (partial automation)	The vehicle can control both steering and acceleration/deceleration in certain modes, but the driver must remain alert and ready to take control at any moment. An example could be a car with both adaptive cruise control and lane centring. AC charging.
Level 3 (conditional automation) (2024)	The vehicle can manage all aspects of driving in certain conditions without human intervention, but the driver must be prepared to retake control when the system requests. Examples include highway autopilot systems with limitations. Charging: AC Grid, DC fast charging, and potentially V2G (Vehicle-to-Grid), V2V (Vehicle-to-Vehicle) and V2B (Vehicle-to-Building) charging capabilities depending on the system and infrastructure availability.
Level 4 (high automation)	The vehicle can perform all driving tasks and monitor the driving environment in specific designated areas or under certain conditions. In those modes, the driver need not pay attention. This level is not yet commercially available (as of 2024). Charging: Primarily DC fast charging due to higher power requirements for extended autonomous operation.
Level 5 (full automation)	The vehicle can perform all driving tasks, under all conditions that a human driver could. There is no need for a human driver in a Level 5 vehicle. This level is likely to be achieved when all vehicles are electric and autonomous with new positioning systems and road infrastructures. Charging: Likely primarily DC fast charging due to the power demands of fully autonomous vehicles.

FIGURE 1.3 A typical skate-board chase design in EVs illustrating significantly reduced component size and simplification in prime drive and drive train, hence reduced maintenance cost.

with basic driver-assist technologies and culminating in self-driving capabilities that can handle any situation.

In addition, the transition from ICE to electric drivetrains often results in a reduction in the total number of components. ICE vehicles have many moving parts required for the combustion process and the related systems, while EVs benefit from the inherent simplicity of electric motors and fewer moving parts (see Figure 1.3). This reduced complexity can lead to advantages in manufacturing, maintenance and reliability.

Conventional ICE vehicles typically convert only a portion of the energy stored in the fuel into useful power at the wheels, with efficiency ranging between 17% and 21%. In contrast, EVs can achieve a significantly higher efficiency, typically ranging between 59% and 62%. It is essential to recognize that the efficiency calculations for ICE vehicles and EVs are distinct. For ICE vehicles, efficiency is measured from the energy content of the fuel at the well to the power delivered at the wheels. For EVs, the efficiency is measured from the electrical energy consumed from the grid (excluding the efficiency of the generator system, including the prime mover) to the power delivered at the wheels. This distinction underscores the vital role local renewable energy sources play in enhancing the overall system efficiency. By employing renewable energy in DERs to generate the electricity needed for EVs, the energy conversion process becomes both more environmentally friendly and more efficient.

The ultimate goal, as detailed in the driving levels in Table 1.5, is to develop fully autonomous vehicles capable of navigating any condition without human intervention (Level 5). This tiered progression supports the development of safer and more efficient transportation systems, applicable to e-mobility. This broad category includes all electrically powered vehicles used in domestic or industrial settings, such as cars, trains, buses, trucks (for transportation or mining), ships, boats, bicycles and even aircraft. Importantly, this electrification initiative is closely linked with advanced electrification technologies utilized in various areas, such as DERs, smart homes and appliances and mining electrification, which all integrate advanced communication tools in fully digitized platforms.

Furthermore, it should be emphasized here that EVs, referred to as e-vehicles, are likely to be seen and widely used as "mobile generators" or "batteries" in microgrids and/or remote consumers. For example, consider the size of an e-bus or e-truck, which could be deployed as an emergency power source during environmental events. The size of the battery in such vehicles (around 500 kW and kWh) can easily provide power to small settlements for multiple hours, which can be deployed fast and seamlessly, which also highlights the potential impact of this technology.

1.5.3 Smart Houses

As in other sectors, smart homes can significantly benefit from electrification. This includes improvements in space heating and cooling, water heating and the integration of smart appliances, all the while optimizing overall energy usage and reliability. Some estimates suggest that a fully optimized smart house could use up to 10%–30% less energy than a conventional home.

In addition, a fully optimized smart house with smart appliances has the potential to achieve a much flatter demand profile compared to a conventional home with standard appliances. This optimization can minimize or even eliminate peaks in energy consumption, contributing to efficiency.

Currently, a range of smart home capabilities are available, each with increasing levels of automation and integration. Partially smart houses incorporate basic automation features or remote control for specific components. Connected houses have internet-connected devices that can be controlled remotely but lack full integration. Integrated smart houses feature interconnected systems and a central control interface, often including home-integrated heat pumps and/or electric water heaters for energy storage and utilization. Advancing further, autonomously smart houses will utilize AI capabilities for optimization, operating on a DC power network. At the highest level, fully autonomous smart houses will represent the pinnacle of automation, with AI systems handling complex tasks.

It is worth noting that, in the literature, there is no universally agreed-upon set of levels specifically defined for smart houses. However, Table 1.6 can be derived from the earlier concept of levels for categorizing the progression of smart home capabilities. This classification may also include specific considerations for DC-powered houses, which could signify a move towards more efficient and innovative energy solutions in residential settings, but most importantly the transformation of the domestic applications from power circuits to control and communication.

TABLE 1.6 A Proposed Framework to Classify the Increasing Levels of Automation and Integration in "Smart Houses"

Levels	Remarks
Level 0 (traditional house)	Operates solely on an AC power network and no smart technology, all rely on manual control.
Level 1 (partially smart house)	Some systems, like thermostats or lighting, have basic automation features but overall, the house requires significant manual control and operates on an AC network.
Level 2 (Internet-connected house) (2024)	The house has internet-connected devices that can be controlled remotely but are not fully integrated with each other. The devices include appliances with limited smartness, security systems, heat pumps and entertainment systems that are also powered by an AC network, and powered partially from a DC supply generated directly from renewable sources or rectified from AC.
Level 3 (integrated smart house)	All devices and systems in the house can communicate with each other and be controlled through a central interface. With limited AI capabilities, such as learning routines. Human input is required for setup and decision-making. Powered by an AC/DC or direct DC network. The generation and demand balance are still in question.
Level 4 (autonomously smart house)	The house has fully integrated smart systems and AI capabilities that allow it to learn and adapt to the occupants' behaviour over time. It can make decisions to optimize comfort, energy efficiency and security. The occupants can manually override the system. All systems are on a DC power network. Renewable energy production, energy storage and advanced energy management are common features. The generation and demand balance are mostly realized.
Level 5 (fully autonomous smart house)	The house is capable of complete autonomy. Complex AI systems handle unforeseen situations or perform complex tasks without human intervention, and all under a DC network. The generation and demand balance are fully realized.

1.5.4 Smart Appliances

A possible framework to define the capabilities of smart home appliances is categorized as shown in Table 1.7. As appliances become smarter, their efficiency will increase, notably through the utilization of high-efficiency motor drives and power electronics converters, specifically designed for each application. This can optimize system efficiency, potentially leading to energy savings of up to 20%–30%.

TABLE 1.7 A Proposed Framework to Classify the Evolution of "Smart Appliances"

Levels	Remarks
Level 0 (non-smart)	The standard appliances that require full manual control, accepting AC input only.
Level 1 (basic-smart)	These appliances have basic features such as timers or pre-sets for specific tasks requiring significant human input, and accepting AC input.
Level 2 (intermediate-smart) (2024)	They are capable of connecting to networks via Wi-Fi or Bluetooth for remote control but still require human supervision for tasks.
Level 3 (conditional-smart)	Appliances at this level can perform multiple tasks independently based on sensor input, and can integrate with other smart devices, but may require human intervention for unexpected situations. For example, cooking appliances at this level can handle entire cooking processes on their own once programmed correctly but may require human intervention in case of unexpected situations or complex recipes. They are powered via AC/DC and DC/DC converters.
Level 4 (high-smart)	These appliances can adapt to user patterns and preferences over time through machine learning or AI, and offer a high degree of autonomy while retaining the option for manual control. They will be capable of performing complex tasks, such as adjusting recipes based on the ingredients present, and can independently handle most situations. However, they may still provide the option for human control if desired. They will be powered from a DC power network only.
Level 5 (fully-smart)	At this level, appliances are fully autonomous, incorporating features like DC power, wireless charging, complex energy management and complete integration with the smart home ecosystem as described previously. These devices will perform tasks independently, optimize energy use and require no human intervention. For example, at this level, cooking appliances will independently perform any cooking task, from recipe selection to ingredient adjustment and cooking, without any human intervention.

Smart appliances can optimize energy usage through sensors and machine learning algorithms. For instance, a smart refrigerator may adjust its energy consumption based on usage patterns, and a smart washing machine could optimize cycles based on the load, although some brands have partial capabilities.

As illustrated in the table, the highest level of intelligence and automation of smart appliances signifies the peak of technological evolution in home appliances. The industry is likely to evolve significantly and develop new standards.

In Level 5, smart appliances will reach the following characteristic features: autonomy, learning and adaptation, energy efficiency, advanced sensors and data processing, predictive maintenance, system integration (communication with other home systems), complex task execution (such as smart ovens deciding meals based on available ingredients), user interaction, security (including encryption and biometric authentication), over-the-air updates for continuous improvement and added efficiency features such as brushless motors, higher efficiency inverters, wireless energy management and wide bandgap (WBG) devices. In the domain of smart home appliances, this form of electrification reflects the transition from conventional devices to fully smart, energy-efficient, DC-powered, wirelessly charged appliances, fully integrated into the smart home ecosystem. Therefore, it can be envisaged that this evolution will also integrate with the operation of developed DERs.

1.5.5 Mining Electrification

Mine electrification is emerging as one of the most consequential and transformative developments in the mining industry. In 2021, the State of Play organization disseminated a comprehensive report centred around mine electrification, providing an in-depth examination of the associated challenges and opportunities. The report was meticulously curated, leveraging over 450 individual surveys, high-level interviews with prominent mining executives, webinars and workshops conducted with a select group of mining companies and their leaders. The survey revealed that a staggering 89% of global mine executives anticipate that existing mines will be entirely electrified within the next two decades, foreseeing an era of fully electric next-generation mines. Furthermore, the report underscored the significant economic, health and environmental advantages presented by mine electrification while also acknowledging the inherent challenges that accompany this transformative process.

The framework for mining electrification, given in Table 1.8, offers a systematic approach to track and facilitate the progress towards fully autonomous and sustainable mining operations. These levels help clarify the increasing complexity and capabilities as the industry moves from a "hybrid" system to "pure" electrification. Note that actual capabilities at each level can vary significantly based on the specific technologies and mining practices used. In addition, each level will contain a degree of integration, autonomy

TABLE 1.8 A Proposed Framework to Classify "Mining Electrification"

Driving Levels	Remarks
Level 0 (traditional mining)	Mining operations rely solely on diesel or other fuels for power, and equipment, vehicles and facility operations have no electric components. The power grid is AC.
Level 1 (partial electrification)	Some aspects of the mining operations, such as lighting or specific machinery (including the first USSR electric quarry excavator driven by thyristor-controlled converters) are powered by electricity, while the primary sources of energy remain fossil fuels and with an AC electricity network.
Level 2 (hybrid electrification) (Started around 2020 and likely to continue next decade)	Electricity is mainly sourced from the AC grid, with charging infrastructure in place for electric equipment. Renewable energy sources are also used. Numerous electric/battery-powered mining machinery is in operation. In both open pit and underground settings, gravity energy storage, ventilation, air conditioning, water heating and heat pumps are considered. AC and DC networks started to co-exist.
Level 3 (full electrification)	All mining operations are powered by electricity. The electricity is primarily sourced from the grid; and AC and DC, AC/DC, DC/AC options may all co-exist. Smart and IoT technologies will start to integrate.
Level 4 (smart electrification)	DC Grid only and mining operations are powered by locally produced renewable energy sources. The grid power (from AC/DC converters or from a DC network) is used as a backup or supplementary source, and there are capabilities for energy storage and demand response.
Level 5 (fully autonomous electrification)	Smart and autonomous grid technologies are fully integrated to optimize energy use and ensure reliability which can include AI. The operations are powered by a self-regulating, fully renewable and autonomous energy system that maximizes efficiency, minimizes environmental impact, and can operate independently of the grid by demand forecasts and real-time data. Note that autonomy will be driven by power systems and cyberspace security.

and smartness, offering a comprehensive understanding of the current state and future directions. Therefore, more sustainable progress will be made towards efficient and autonomous mining systems of the future.

Electrification in the mining industry is driven by significant potential energy savings and the desire for autonomy and enhanced power systems and cyberspace security. Some keynote studies and updates for machine ratings suggest that electrifying mining equipment can lead to substantial energy savings. These savings are key reasons for the ongoing shift towards electrification, making operations more economically viable and environmentally sustainable.

Note that the demand for minerals and metals is projected to increase significantly in the coming decades due to the growth of renewable energy, electric vehicles and other sustainable technologies. As a result, the mining industry will need to continue to find ways to increase productivity and efficiency while reducing its carbon footprint.

Electrification, including the deployment of commercially available electrically/battery-powered mining machines and supporting devices such as rope shovels, load-haul-dump trucks, drills, service EVs, electric crushers, scoops, smaller truck models, electric locomotives, explosive chargers, rock bolting rigs, LED light towers and conveyors, is seen as a key solution to help achieve these goals.

Advances in battery technology and renewable energy sources, such as solar and wind, are enabling mines to reduce their reliance on fossil fuels and operate more sustainably. Furthermore, governments and international organizations are pushing for greater sustainability in the mining sector, setting climate targets and driving further innovation and investment in electrification and other clean technologies for mining.

In the next decade, "Mine electrification" will become a more common term, alongside concepts like autonomous microgrid, e-mobility, e-mining, digitization and power system security.

Similar to the previous electrification trend, the framework outlined in Table 1.8 provides a systematic approach to transitioning towards fully autonomous and sustainable mining operations. The table also highlights the growing complexity from a "hybrid" system to "pure" electrification. Note that each level's actual capabilities may vary significantly, depending on specific technologies and mining practices. Furthermore, each stage integrates a level of autonomy and intelligence, offering insight into present states and future trajectories.

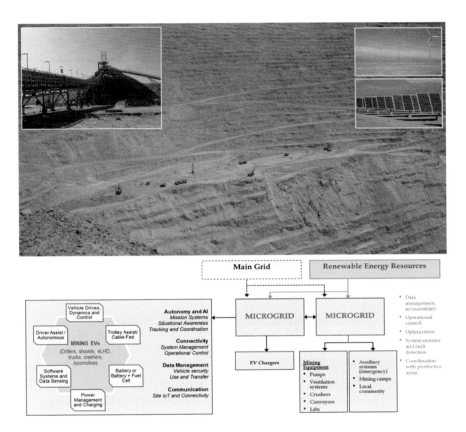

FIGURE 1.4 A large open-pit mining sites with local renewable energy sources and the components of the future mining electrification system.

Figure 1.4 illustrates the concept of a modern open-pit mining site that is integrated with local renewable energy sources to create a more sustainable and efficient operation. This is indicative of e-mining, or electronic mining, where there is a heavy reliance on EVs and other electric-powered equipment.

In the image, there are two main components:

1. **Open-pit mining site:** The large photograph shows an open-pit mine with various levels and heavy machinery at work. This is typical of large-scale mining operations where minerals or other geological materials are extracted from a large open pit in the ground.

2. **Renewable energy sources:** In the inset images at the top-right corner, there are wind turbines and solar panels, representing renewable energy sources that can be used to power mining operations, reducing reliance on fossil fuels and decreasing the environmental footprint.

3. **Mining electrification system:** At the bottom of the figure, there is a diagram showing various components of a mining electrification system, including EVs, battery or fuel cell systems and other technologies like vehicle drives, dynamics and control systems. These systems are integrated into a microgrid that can operate independently from the main power grid, using energy generated from renewable sources.

Regarding e-mining in undersea environments:

- E-mining could potentially be applied to undersea mining, where EVs designed for underwater use—such as submersible drones or remotely operated vehicles (ROVs)—would extract resources from the ocean floor.

- The use of mining EVs underwater can minimize the environmental impact compared to traditional methods, as they can be powered by clean energy and designed to reduce disturbance to the sea bed.

- Integration with renewable energy sources, as depicted in the figure, could be more challenging in undersea environments due to the difficulty of transmitting power. However, technologies like energy storage in batteries or using renewable energy sources located on support ships or platforms could be solutions.

- The future of undersea e-mining would rely heavily on advanced autonomy and AI, as shown in the diagram, to navigate and operate in the complex and often hazardous underwater environment.

In summary, the figure presents an advanced vision of mining operations that are both more sustainable through the use of renewable energy and more technologically advanced through the integration of electric vehicles and smart systems. Applying this to undersea mining could revolutionize the way marine resources are extracted with a focus on sustainability and minimal environmental impact.

1.5.6 Growing Needs for Data Centres

Energy efficiency improvement lists in the future will focus on products that consume the most electric power. This list includes data centres (which employ unidirectional power flow), e-mobility vehicles (with bidirectional power flow) and variable-speed motor drives (encompassing high-efficiency motors and inverters for both domestic and industrial applications).

In a typical data centre, a multitude of electrical equipment is interconnected in complex ways. Environmental tools, such as heating, ventilation and air conditioning (HVAC), and electrical power management systems (comprising renewable sources, grid-connected solutions and uninterruptible power supplies) work collaboratively. The smooth operation of these components is crucial, especially as electricity demands and HVAC needs are ever-evolving. Elements like data processing and storage, internal and external network devices, information security tools, voice-over IP systems and physical security mechanisms all interact at varying levels. Such interactions alter power demands, produce losses and consequently affect the overall efficiency of data centres.

Several factors highlight the increasing importance of data centres and power network infrastructures in the future. These factors include exponential data growth driven by IoT, the surge in video content, online activities and digitization. There is also a rising dependency on cloud services (such as software, platform and infrastructure) and an increasing reliance on remote servers. The emergence of edge computing, which demands real-time data processing on-site for applications like autonomous vehicles or augmented reality, requires the presence of micro or mini data centres. Furthermore, the ongoing electrification and decentralization of the power grid, along with the increased electricity demand and the need for DERs, reliability and cybersecurity, emphasize the significance of data centres and power networks.

Historically, data centres operated primarily on AC power, given that the conventional utility grid supplies AC and most traditional appliances and devices were designed for AC usage. However, many data centre-specific devices, including servers, storage systems and networking equipment, inherently use DC. This means that any AC-to-DC conversion introduces energy losses and generates heat, leading to inefficiencies.

The proposition of power data centres using DC for increased efficiency dates back to the early 2000s. Organizations, such as the Lawrence

Berkeley National Laboratory (LBNL), lead this cause. Advantages of transitioning to DC-powered data centres include diminished conversion losses, reduced cooling demands due to lower heat generation and potential savings in infrastructure costs owing to simpler power distribution units and fewer transformers.

However, the transition to DC in data centres has not been rapid. Challenges include substantial investments in existing AC infrastructure, a lack of standardization for DC equipment, device compatibility concerns and a general reluctance to move away from traditional practices. Interestingly, these are the same challenges that hinder the broader adoption of a DC power network in the conventional power network.

By the 2010s, with an intensified focus on energy efficiency and the increasing utilization of renewable energy sources (primarily DC), there was a notable uptrend in the adoption of DC power solutions. This trend was particularly evident in newer data centres or those undergoing significant upgrades. While high energy consumption remains a concern in data centres, the use of an internal DC power network has become standard. Concurrently, investigations and trials are ongoing to explore more energy-efficient systems, such as the incorporation of WBG devices in power electronic converters.

Note that data centres represent various electrical and electronic systems and equipment, their interactions define the electricity demand and electrical efficiency. Although Table 1.9 is given to describe the elements that have an impact on efficiency and potential future directions which are likely to define the future growth as well as the electricity demand,

TABLE 1.9 The Elements of Data Centres Which Have an Impact on Efficiency and Potential Future Directions

Elements	Description and Challenges	Potential Future Directions and Solutions
Data processing	High power demand with varying intensities based on data traffic. Requires reliable and efficient power supply systems.	Optimization of processing infrastructure using energy-efficient hardware, optimized algorithms and server virtualization.
Data storage	Power demand changes based on read/write operations. High temperatures can reduce the lifespan of the hardware.	Transition to energy-efficient storage solutions. Advanced storage hierarchy to minimize active data retrieval times.

(Continued)

TABLE 1.9 (*Continued*)   The Elements of Data Centres Which Have an Impact on Efficiency and Potential Future Directions

Elements	Description and Challenges	Potential Future Directions and Solutions
Network infrastructure	Bandwidth use and data traffic influence power consumption. Equipment must maintain reliable connections.	Energy-efficient routing algorithms, AI-based traffic prediction, and adaptive network hardware that adjusts based on load.
Information security equipment	Security appliances, firewalls and intrusion detection systems add to power demand.	Integrated, multi-layered security hardware that is energy-optimized. Use of software-based virtual appliances.
Voice over IP equipment	Constant power for uptime, but traffic can vary.	Power-saving modes during low traffic. Transition to more efficient codec and compression techniques.
Physical Security Equipment	Cameras, biometric systems and alarms require constant power.	Adoption of energy-efficient surveillance equipment, motion-triggered systems, and AI-driven adaptive security measures.
HVAC	Environmental equipment uses significant power, especially cooling for data centres. As stated in Figure 1.5a direct relationship exists between IT equipment heat generation and HVAC load.	Advanced cooling techniques such as free cooling, liquid cooling and hot/cold aisle containment. Adaptive HVAC systems that adjust based on real-time heat output from IT equipment.
Power management equipment	Ensuring uninterrupted power supply can be challenging with varying demands from information technology and HVAC. DERs with renewable and energy storage are critical approaches. Noting that the DC power network is already a norm.	Smart grid integration, AI-driven power distribution systems that adapt based on demand, and efficient energy storage systems for balancing renewable input. However, the key is to achieve an autonomous and reliable microgrid achieved by WBG device-based power electronics.
Operational automation	With varying power demand and HVAC requirements, real-time monitoring and adjustments are crucial for efficiency and uptime.	AI-driven predictive analytics for equipment failure, automated load balancing and adaptive power management systems that adjust equipment performance based on real-time needs.
Data storage	Power demand changes based on read/write operations. High temperatures can reduce the lifespan of hardware.	Transition to energy-efficient storage solutions. Advanced storage hierarchy to minimize active data retrieval times.

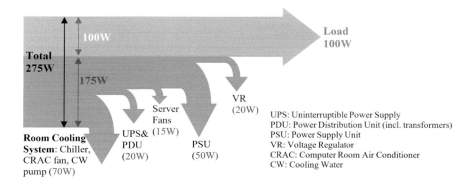

FIGURE 1.5 Typical power consumption in a 100 W system load in a data center.

Figure 1.5 is provided to highlight the typical power consumption in a 100 W system load. Note that the efficiency of the components of the power delivery system varies but provides insights about the potential energy-saving opportunities. For example, the efficiency range for different sections are UPS (88%–92%), PDS (98%–99%), PSU (68%–72% when with AC input, 90%–92% when high-efficiency DC and DC/DC (also known as VR, 78%–85%).

It is worth noting that the worldwide energy demands in data centres within the period of 2015–2021 have reduced in Traditional Data Centres by 66% (from 97 to 33 TWh), and have increased in Cloud (Non-Hyperscale) Data Centres by 16% (from 62 to 72 TWh) and in Hyperscale Data Centres by 181% (from 31 to 87 TWh).

Estimates and forecasts of data centre energy consumption reinforce the importance of using energy-efficient equipment in data centres. For example, in 2014, data centres in the US consumed about 1.8% of total electricity consumption. These changes are primarily due to the efficiency improvement measures and new DC power supply technologies involving emerging power electronics switching devices (GaN, SiC, WBG devices). However, operational system efficiency also offers a potential for energy efficiency.

Note that air conditioning in data centres is highly linked to efficiency improvements as temperature, humidity and heat load define air conditioning needs. It is important to note that internal temperature rise in a data centre is also a by-product of power supply losses that are directly related to the power electronics technology utilized.

It can be concluded that heat generated from the losses in the components of data centres requires additional cooling power, hence "cooling" can be considered as the backbone of energy savings. Note that processor racks in data centres cooled by water take much less space than the

TABLE 1.10 A Guide to the Effectiveness and Cost of Cooling Methods

Resultant Energy Requirements for Data Centres at a Cooling Capacity of 35.17 kW

Heat Transfer Medium	Fluid Flow Rate (litre/min)	Conduit Size, Round (cm)	Power (kW)
Forced air	260,996.4	86	2.7
Water pump	75.71	5.08	0.19

An Emerging System: AirJet® at a Cooling Capacity of 4.25 W, Suitable for Processors at a 29 dBA Noise Level (Very Quiet)

Air flow	5.95	Dimensions: 27.5 × 41.5 × 2.8 mm	0.001

air-cooled racks. Table 1.10 summarizes a critical guide about the effectiveness and cost of different cooling methods, which also includes an emerging cooling method, AirJet, which may drastically alter the existing cooling practices both in data processing hardware and power electronics converters. AirJet® cooling technology employs a thin membrane that oscillates at ultrasonic frequencies. This holds significant potential for future processor applications due to its multiple advantages: substantial space savings, reduced power consumption and quietness.

Today, as a backbone of the digital economy, to store, process and distribute vast amounts of data, the localized data centres (or "edge computing") are having significant implications on electrification and grid transformation. Such centres already offer a number of benefits including timely real-time decision-making, effective bandwidth utilization, improved local data privacy and security, improved reliability by minimizing external connectivity and facilitation of effective digitization for AI, machine learning and IoT in power systems' operations, which are all related to the power system security as well.

Therefore, data centres are among the highest consumers of electric power and have a growing trend with greater energy-saving opportunities. The emerging cloud (hyperscale) data centres are likely to drive the development of future integrated power electronic solutions, which will integrate alternative energy sources (such as renewables and hydrogen), energy storage and power supply equipment. In addition, they contain support systems such as air conditioning, security and lighting as well as internal transportation vehicles of some form.

1.5.7 Changing Appliance Landscape

The modernized grid system and the increasing emphasis on energy efficiency and renewable energy integration have altered the appliances' landscape and several emerging electrical appliances are already available and growing, which support and represent the electrification trend. Note that all new appliances have commonalities including high efficiency, potential to operate from DC sources and ease of integration into smart grid systems. Some of the major appliances include DC-powered (direct solar PV-powered) air conditioners, RF solid-state cooking ovens and electric space heating/cooling and water heating devices, heat pumps. Note that environmentally responsible refrigerants are also the common characteristics of air conditioning appliances.

1.5.7.1 Solar PV-Based DC-Powered Air Conditioners

Although they are still in the early adopter stage, standalone solar-powered air conditioners are emerging at the forefront of sustainable cooling/heating. These units can operate entirely on solar power during daylight hours, offering significant cost savings while increasing efficiency. However, it is essential to be aware of their initial installation costs, space requirements for solar panels and limitations during prolonged cloudy periods or at night, which can be addressed by battery storage or by using minimal grid power, preferably during off-peak times. Key features of these appliances include inherently high efficiency (using direct DC and brushless PM motors in compressors), operation over a broad range (from $-10°C$ to $+50°C$) and integrated digital communication for full control, timers and consumption tracking. These systems also incorporate built-in Maximum Power Point Tracking (MPPT) circuits directly connected to the DC bus of the inverter driving the compressor motor. By eliminating the power losses associated with AC/DC rectification, they offer higher system efficiency. Due to their reliance on rooftop PV systems, they are especially suitable for remote areas and grid-independent setups.

1.5.7.2 Electric Water Heaters

There are several primary types of water heaters, including electric resistance, air-source heat pumps, ground-sourced heat pumps (geothermal heat pump), solar heat pumps (which combine a heat pump with a rooftop solar collector), instantaneous electric, solar electric, instantaneous gas, gas storage, solar gas and other varieties that utilize sources like wood, coal and oil.

TABLE 1.11 Comparison of the Capital and Running Costs of Different Water Heating Methods

Water Heating Methods	Capital Cost (US$)	Running Cost	Remarks
Electric Resistance Heater	$150–$800	12–15 cents/kWh	Costly, if not renewable
Air-source Heat Pump (ASHP)	$4,000–$8,000	5–8 cents/kWh	Equivalent cost considering efficiency
Ground-Sourced Heat Pump (Geothermal Heat Pump	$10,000–$30,000	3–5 cents/kWh equivalent	Equivalent cost considering efficiency
Solar Heat Pump	$6,000–$12,000	3 cents/kWh	Combining both (heat pump and a rooftop solar collector), only on sun exposure
Instantaneous Electric Heater	$500–$1,500	10–12 cents/kWh	Requiring periodic maintenance, limited lifetime
Instantaneous Gas Heater	$1,000–$3,000	$10–$15/month	For average use
Gas Storage Heater	$900–$2,000	$20–$30/month	For average use
Solar Electric Heater	$10,000–$15,000	Very low after the payback period	Potentially just maintenance and replacement of minor parts
Solar Gas Heater	$3,000–$6,000	$10/month	Can be much less with sufficient solar contribution

In terms of efficiency, electric resistance, heat pumps and instantaneous gas heaters stand out. Electric resistance-based water heaters, for instance, boast an efficiency close to 100%. This is mainly because they can function as a form of energy storage, drawing power directly from a DC source, such as a solar PV system.

Water heating technologies have significant variations in their capital and operating costs. Table 1.11 offers approximate values in the US and in US$, though these can vary based on factors like region, brand and local incentives.

Electric resistance water heaters are among the most affordable to install but, currently, the priciest to run (unless paired with rooftop solar or under very concessional controlled load tariffs). They also provide the highest flexible demand capacity. These heaters fall into the flexible demand (FD) category, enabling activation when surplus solar is available or during off-peak periods. While manual switches and timers can be employed with these heaters, they can also be controlled through smart meters, solar relays, solar diverters or digitally via energy management platforms.

1.5.7.3 Heat Pumps: Space Heating, Cooling and Water Heating

Heat pumps work by transferring heat from one place to another. In heating mode, they extract heat from the outside (even when it is cold) and move it inside. In cooling mode, they work in reverse, pulling heat out of a room and discharging it outside (as in a fridge), thereby acting like an air conditioner.

This dual functionality makes heat pumps a highly efficient and versatile solution for both heating and cooling needs, which could potentially replace both air conditioners and water heaters. Heat pump water heaters, for instance, use electricity not to heat water directly but to move heat from the air or ground to heat the water, which can be significantly more energy-efficient than traditional electric water heaters.

However, while heat pumps can operate effectively in most climates, their efficiency tends to decrease in regions with extremely cold temperatures. In such cases, a secondary heat source might still be necessary for the coldest periods, meaning heat pumps might not entirely replace traditional heating systems in those areas.

Also, replacing existing systems with heat pumps involves upfront costs and may require modifications to the infrastructure, which could be a barrier for some consumers. Therefore, while heat pumps represent an important technology for reducing emissions from buildings, their adoption will depend on factors like improving technology, reducing costs and policy support.

In conclusion, while heat pumps have the potential to replace air conditioners and electric water heaters to a large extent, whether they can do so entirely will depend on a variety of factors. It is important to consider individual circumstances, including local climate and the specific needs of the building, when deciding on the best heating and cooling solutions.

The motivations for the transition of this power system load can be defined under three groups: economic considerations, demand stabilization and revenue maintenance and promotion of electrification. While this transition offers clear benefits, if EV penetration also increases, it may require upgrades to the electricity grid to handle increased demand, particularly during peak times. However, this may easily be addressed by energy management due to the "energy storage" characteristics of these applications.

Combining heat pumps with direct solar PV in domestic applications offers multiple benefits as summarized in Table 1.12.

TABLE 1.12   The Benefits of Heat Pumps

Benefit	Description
Renewable energy use	Reducing dependence on the grid and carbon footprint.
Cost savings	Eliminating associated electricity bills due to self-generation from solar PV.
High efficiency	Since electricity is primarily used for heat transfer.
Scalability	Solar PV systems can be expanded based on energy needs and available space.
Low operating costs	After initial installation, operating costs are typically lower than conventional heating/cooling systems.
Grid independence	Operate off-grid (although it may be connected to the grid as well)
Triple functionality	They can provide heating, cooling and water heating.
Integration with storage	Can be combined with battery storage for continuous operation.
Low maintenance	System components generally have low maintenance needs once installed.

It should be emphasized here that the term "*efficiency*" implies a ratio of output to input, hence it should not exceed 100%! However, for heat pumps, this efficiency definition is somewhat stretched because they move heat rather than directly convert electrical energy to heat. Therefore, the Coefficient of Performance (COP) is defined which can be greater than 1 (or 100%) due to the larger heat transfer.

There are alternative ways to frame or describe the performance of heat pumps without using the term "*efficiency*" or by using efficiency in a different context. Table 1.13 provides a summary of various metrics used to describe the performance of heat pumps in different contexts and applications.

Note that if it is preferred to use the term "efficiency" to be capped at 100%, then it will be better to refer to the "*electrical conversion efficiency*" of components within the heat pump. Table 1.14 presents the overall electrical losses in heat pumps.

Therefore, given the loss ranges listed in Table 1.14, it can be concluded that the overall electrical efficiency of heat pumps typically falls between 65% and 90%, with the remaining percentage attributed to above given losses. In addition, it can be concluded that efficiency improvements in heat pumps are still possible if these individual losses and "operational system losses" are improved as in, where a heat pump system is optimized directly without a compressor. Therefore, the improvements in energy

TABLE 1.13 The Alternative Ways to Describe the "*Performance*" of Heat Pumps

Metric	Description	Application
Coefficient of performance (COP)	Represents the ratio of heating or cooling provided to the energy consumed.	General measure for heat pump efficiency.
Energy factor (EF)	Energy delivered as hot water divided by the energy input to the heater over a 24-hour period.	Heat pump water heaters.
Heating seasonal performance factor (HSPF)	Total space heating during the heating season divided by the electrical energy consumed by the heat pump during the same season.	Air-source heat pumps (heating mode).
Seasonal energy efficiency ratio (SEER)	Total heat removed during the cooling season divided by the electrical energy consumed by the heat pump during the same season.	Air-source heat pumps (cooling mode).
Energy efficiency ratio (EER)	Ratio of heat output to energy input at a specific condition.	Geothermal heat pumps (cooling mode).
Energy consumption efficiency	Amount of electrical energy consumed by the heat pump to operate.	Descriptive metric to denote energy use, usually in kWh/month or similar units.

TABLE 1.14 A Summary of the Losses in the Electrical System of Heat Pumps

Loss Type	Description	Typical Electrical Losses
Converter losses	Losses from power electronic converters, in variable-speed systems.	2%–5% of input power
Compressor losses	Mechanical and electrical losses from the compression of refrigerant.	5%–10%
Control and ancillary circuit losses	Losses from controlling electronics and auxiliary systems.	1%–3%
Fan motor losses (if air-sourced)	Losses associated with driving the fans in air-source heat pumps.	5%–20%
Pump motor losses (if water-sourced)	Losses in motors of circulation pumps in ground-source or water-source systems.	10%–15%
Distribution losses	Losses in the electrical distribution system, including transformers and conductors.	1%–5%
Start-up losses	Transient losses during heat pump start-ups, significant in frequently cycling systems.	Variable; can affect overall efficiency by several points

performance are achieved: 1.52% and 3.58% for cooling and heating system COPs respectively, and 0.76% and 0.81% for the heat pump COP.

Note that, air-source heat pump water heaters boast energy efficiencies that are three to five times greater than those of resistance water heaters. While their installation costs are on the higher end, their superior efficiency renders them more economical to operate, subsequently reducing associated emissions. Instantaneous gas water heaters are compact and relatively inexpensive to produce. They have effectively supplied hot water (and space heating) for several decades. However, gas utilization has shifted dramatically in recent years due to factors like power generation and export opportunities. The fluctuating price of gas and strategic decisions have made them dependent on external variables. Furthermore, digital control and consumption tracking in gas heaters are challenging (not fitting well into the digitized world), and they have a substantial emission footprint.

Electric resistance water heaters offer high flexibility but are less efficient, while heat pumps are more efficient but less flexible. To balance efficiency and flexibility in the electricity system, different technologies can be applied based on specific needs. Heat pumps should be used where outdoor space is available for optimal efficiency. If not, electric resistance units can be preferred for better flexibility. When rooftop solar is present, self-consumption can be initiated using solar diverters or timers. In other cases, controlled flexibility is achieved using a Demand Response Enabling Device (DRED) or smart meter. Note that heat pumps can efficiently heat as well as cool spaces as they are becoming more popular as a replacement for traditional systems.

The radar chart (also known as a spider, star, polar or Kiviat diagram) in Figure 1.6 presents a comparison of three primary water heater types, which illustrates the unique correlation of six quantities. When used in conjunction with a customized version of Table 1.12 for specific regions or countries, it can help determine the optimal water heating method. It is worth noting that a similar approach can be applied for space heating and cooling as well. Moreover, suitable heat pumps, often used for space heating, cooling and water heating, can be combined with solar PV systems. This combination can lead to significant energy savings and incorporates digital control and communication features. However, the initial capital cost will be higher considering the existing market price and competition.

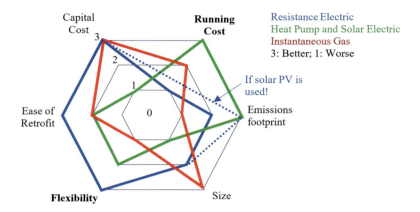

FIGURE 1.6 The radar chart of three main water heater types.

It is important to note that space cooling and heating systems can function as "energy storage" by shifting and varying peak demand. This smoother demand profile can better accommodate future EV charging needs. Additionally, future homes may integrate heat pump systems that also handle cooling functions typically provided by refrigerators. For clarification, heat pumps and reverse-cycle air conditioners are often used interchangeably, particularly in some regions. This is because reverse-cycle air conditioners utilize the same principles as heat pumps. However, there is a key distinction: while all reverse-cycle air conditioners are heat pumps, not all heat pumps are reverse-cycle air conditioners.

1.5.7.4 Induction Cooktops

Induction cooktops are a type of stovetop that heats cookware through induction rather than thermal conduction from a flame or an electrical heating element that with a significant air gap to reach the cooking substance. They utilize electromagnetic fields to heat pans directly, making them more efficient than traditional electric stoves (85%–90% efficiency compared to 70%–75%).

Within these cooktops, there is an electronically controlled coil that operates at high frequencies. The heat is generated through eddy currents induced in the cookware's magnetic base material, such as iron or magnetic stainless-steel bottom, a process known as magnetic induction.

Therefore, contrary to traditional cooking methods where heat is transferred from the burner to the cookware, induction cooktops transmit heat directly to the vessel. This direct heat transfer ensures a larger portion of the energy used is channelled straight into the bottom of the cookware, speeding up the heating process and minimizing energy wastage as well as significantly eliminating nearby fire risk. Consequently, they achieve high efficiency. Furthermore, the consistent heat, modulated using high-frequency switching with power electronics, offers precise cooking temperatures. It can be summarized that when such cooktops are utilized up to 20% energy saving is possible, which can also be adapted to the IoT.

1.5.7.5 Solid-State RF Cooking Ovens

With the rise of solid-state radio frequency (RF) technology, the landscape of microwave oven systems is undergoing a transformative shift. Traditional magnetrons, once the backbone of microwave generation, are now being sidelined in favour of more advanced solid-state devices, which leverage WBG devices such as gallium nitride (GaN). These new devices, including the likes of Si LDMOS, offer enhanced power and frequency control, ensuring uniform cooking without the need for turntables. The integration of such technology, notably the GaN HEMT, is redefining microwave ovens' efficiency and precision. Given their superior performance, cost-effectiveness and reliability, solid-state RF ovens are likely to become the new benchmark in microwave cooking at higher efficiency. These ovens use high-frequency electromagnetic waves to cook food, similar to microwaves, but with potential advantages in terms of speed and cooking uniformity.

1.6 CHALLENGES AND SOLUTIONS IN POWER SYSTEMS

1.6.1 Variability, Intermittency and Rate of Change of Power

"Variability" and "intermittency" are characteristics commonly associated with wind, solar and wave power. These characteristics have significant implications for the control systems and the reliable operation of the power grids to which they are connected. In this book, "variability" defines fluctuations in the power output of renewable energy sources over longer time frames, ranging from minutes to hours as in sun rise and set. In contrast, "intermittency" indicates faster changes, varying from brief interruptions to periods of seconds to minutes. In addition, the "rate of change" refers to the speed at which these power sources can ramp up or down based on conditions or demands. This concept will be briefly discussed below in the context of grid stability and the integration of renewable energies.

1.6.1.1 Solar PV

In solar PV systems, variability is shaped by natural day/night cycles and seasonal changes, leading to fluctuations in solar power production. These fluctuations continue over extended periods and do not influence the design of power electronics or their control. Additionally, solar eclipses can introduce significant changes in solar power production for their duration, presenting a temporary variability as well as intermittency that may affect the power grid, depending on its relative power contribution. Note that the timing, duration and location of solar eclipses can be predicted with great accuracy.

The variability in solar PV's rate of change spans hours, progressing sinusoidally from zero at dawn, peaking at solar noon and returning to zero by dusk. Taking a 12-hour from sunrise to sunset as an example, irradiance may rise by approximately 8%–10% per hour (see Figure 1.7a). The exact rate, however, depends on both latitude and season.

Conversely, intermittency in solar PV arises from rapidly changing elements, such as transient clouds, tree shading or the cloud-edge effect (also known as scattering and "enhanced edge illumination") (see Figure 1.7b). In this phenomenon, cloud edges can function as lenses, concentrating sunlight, which sometimes results in solar panels capturing more sunlight than they would under clear sky conditions. These intermittent events can result in significant effects on solar energy production, influencing both power generation and the design of maximum power point tracking control. Rate changes of 50%–80% reductions within seconds to minutes are possible during dense cloud cover.

When analysing real irradiance variations in solar PV systems, the practical characteristics of intermittency can be defined by the maximum possible rate of change in power ($\Delta P/\Delta t = 700\text{--}1{,}200$ W/s) and current ($\Delta I/\Delta t = 2\text{--}5$ A/s). These characteristics are crucial for designing associated power electronics and integrating them into the power grid as energy sources.

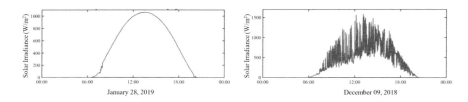

FIGURE 1.7 Typical variations of irradiances under full sun (a) and under fast-intermittency (with scattering, hence above 1kW/m²) (b).

Furthermore, electricity system operators can predict the impacts of a solar eclipse and take measures to mitigate its effects on the central grid. A solar eclipse can affect areas into complete darkness for a duration ranging from a few seconds to about 7.5 minutes. In addition, it might cause partial darkness, which can span from a few minutes to several hours, depending on the observer's location and the eclipse's specifics. A total eclipse can also cause a phase of partial eclipse lasting several hours. Given that a solar eclipse can affect solar irradiance for around 3 hours, sustaining a consistent electricity supply becomes particularly challenging in grids largely driven by renewable energy. It is worth noting that, during an eclipse, power production from solar PV systems might experience significant reductions, while the demand might also increase in the dark phases. Figure 1.8 provides an overview of the variations in essential environmental parameters (such as solar radiation, temperature, relative humidity and wind speed) during the total solar eclipse that took place in the US on 21 August 2017. This highlights the impacts of a solar eclipse on environmental conditions and on renewable energy generation.

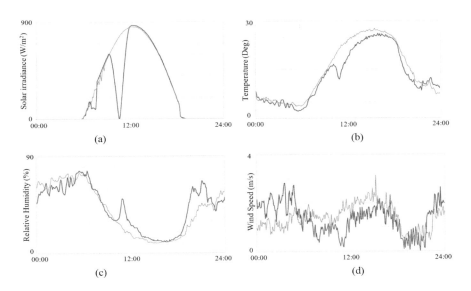

FIGURE 1.8 The impacts of solar eclipse on environmental quantities: (a) Solar irradiance (W/m^2), (b) Temperature (oC), (c) Relative Humidity (%), (d) Wind Speed (m/s). Darker trace: Mean global values. Lighter trace: The average values from the previous and following days.

1.6.1.2 Wind Generation

Wind patterns over onshore areas can change rapidly due to factors such as land topography, vegetation and temperature variations. This results in more frequent and significant fluctuations in power output, especially during storm fronts. The rate of change in wind energy, often referred to as its ramp rate, is vital for grid operators to understand as it can influence the reliability and stability of the electrical grid. Under specific weather conditions, large onshore wind farms have displayed rates of change up to 10% of their capacity per minute.

On the contrary, offshore wind patterns tend to be steadier and more consistent. Large bodies of water moderate wind speed and direction, and the lack of obstructions and minimal influence from local topography contribute to this steadiness. Nonetheless, offshore winds can still experience rapid changes, especially during substantial weather transitions, with typical rates of change being 5%–7% of capacity per minute for offshore farms.

While advancements in wind turbine technology and enhanced forecasting methods are consistently improving our understanding and management of these ramp rates, they significantly influence the control, intermittency of the generation and overall grid stability.

1.6.1.3 Wave Energy

The power output from wave energy converters is primarily determined by wave height and frequency. Since waves originate from distant winds, they offer a degree of short-term predictability. However, similar to wind, weather pattern shifts can lead to fluctuations in wave energy output. Wave energy devices might encounter rapid output changes with evolving sea conditions, but these alterations are generally more gradual than the immediate changes observed with solar PV.

In contrast to solar PV, wind generation displays slower intermittency, largely attributed to the inertia of turbine blades and the generator. Regarding instantaneous power fluctuations, solar PV typically poses the greatest challenge due to its susceptibility to fast-moving cloud cover.

1.6.1.4 The Impacts of Generation and Demand Variability and Their Size

Power and voltage variations in the power network are influenced by both changing generator power and fluctuations in load demand. Both these factors significantly affect stability and the consistency of voltage, which in turn impacts power quality. The nature (whether it is resistive, inductive or capacitive) and size of an electrical load in relation to its source, as well as the robustness of the grid, determine its effect on grid voltage.

Larger grids are generally more adept at handling variability and intermittency than smaller ones due to their greater resource capacity for balancing supply and demand. For instance, in a large and strong grid, if wind power diminishes in one region, the deficit can potentially be offset by wind or solar energy from another area. Moreover, larger grids can distribute the effects of abrupt load changes across a wider array of generators, thus minimizing the strain on any single generator.

The scale of renewable energy sources also plays a role in influencing variability and intermittency. While more substantial renewable energy facilities can yield higher power outputs, they can simultaneously result in more pronounced output fluctuations. For instance, a sizable solar farm can generate substantial power during sunny conditions, but its output can drop rapidly when obscured by clouds. Conversely, smaller installations might experience reduced overall variability, but their relative variability, or the percentage change in output, might be more significant.

Lastly, the size of the load being integrated or removed from the grid is crucial. Connecting a large load suddenly can cause a sharp voltage drop, potentially leading to frequency variation and instability. Similarly, abruptly disconnecting a substantial load can result in a sudden voltage surge. These voltage shifts tend to have a more pronounced effect on smaller grids, which possess fewer resources for equilibrium between supply and demand.

1.6.2 Efficiency

The efficiency of power system components varies based on their design, operational conditions and size. The specific impacts of efficiency in power system components include:

- *Economic impact,* due to lower operational costs and deferring or eliminating the costs associated with upgrades or expansions,
- *Environmental impact,* due to reduced emission and resource conservation,
- *Reliability impact,* due to reduced heat and stress on components,
- *Societal impact,* due to job creation as a result of the retrofitting or upgrading and health benefits as a result of reduced emission,

- *Operational impact*, due to improved power quality that impacts stability as well as lifetime of components, and better flexibility and responsiveness to changing grid conditions.
- *Strategic impact*, due to greater energy security and Regulatory Compliance with efficiency standards.

A general overview of the typical efficiencies of various power system components and potential alternatives to improve the efficiency are summarized in Table 1.15.

The significance of both individual unit efficiency and overall system efficiency is explained through an efficiency comparison presented in Table 1.16. This comparison provides the efficiency of direct electrification in EVs with that of hydrogen-based electrification in EVs. To align with the power transition trend and to eliminate the effects of other power generation methods, it is assumed that both EV types are powered by 100% renewable electricity. As indicated in the table, major components are classified and sub-grouped based on their efficiency range, as well as the anticipated percentage of efficiency improvements by 2050, in line with the NetZero target.

Forecasting the precise efficiencies of these technologies by 2050, especially within the framework of achieving a NetZero carbon emissions target, is challenging. However, considering the potential advancements from power electronics to conversion/transportation/storage, there is a clear potential to enhance the efficiency of power system components, hence making efficiency predictions more feasible.

It can be concluded that with advancements in fuel cell technologies (such as the incorporation of new catalysts and membrane materials), there is potential for achieving higher efficiencies. In addition, although it may look incremental, the efficiency of inverters (in addition to reliability) can be improved using switches based on WBG devices. Note that "flat-efficiency" in WBG devices denotes the converter's capability to sustain high efficiency across a wide range of operating conditions, notably varying load conditions. Unlike many conventional converters where efficiency peaks at specific load conditions and diminishes under light loads, a converter exhibiting flat efficiency consistently maintains elevated efficiency over a broader spectrum of load scenarios.

Furthermore, electric motors stand to achieve enhanced efficiency via the introduction of innovative materials and design modifications.

TABLE 1.15 An Overview of the Typical Efficiencies of Various Power System Components and Potential Alternatives to Improve the Efficiency

Components	Typical Efficiency Range	Remarks about Alternatives and Emerging Components	Potential Efficiency Improvement
Generators			
Coal-fired plant	33%–40%	Integrated gasification combined cycle and with carbon capture and storage.	~45%
Nuclear reactors (>300 MW)	33%–38%	Micro nuclear reactors (<10 MW) Small modular reactors (<300 MW).	~40%
Hydroelectric	90%	Pump hydro storage	70%–80% (round-trip).
Concentrating solar power	30%–40%	Combined with PV and natural gas.	Improvement will be a weighted average.
Wind turbine system	30%–50%	Hybrid with offshore wind (45%–50%) and wave (10%–40%) sharing common infrastructures (substation and MVDC cable).	5%–10% improvements due to shared infrastructures. Sum of the output powers.
Solar PV	20%–24%	Multi-junction solar cells. Perovskite solar cells.	>25%
Batteries (gen and load)	85%–90% (round-trip)	New batteries such as solid state.	Higher than 90% (round-trip).
Transmission and Distribution Network			
AC transmission lines	90%–97%	Improved HVDC and MVDC and with renewables.	~2% improvements.
Large AC transformers	98%–99%	Solid state transformers with renewables and DC.	Limited improvements but added benefits of reactive power control and bidirectional power flow.

(*Continued*)

54 ■ Reinventing the Power Grid

TABLE 1.15 (*Continued*) An Overview of the Typical Efficiencies of Various Power System Components and Potential Alternatives to Improve the Efficiency

Components	Typical Efficiency Range	Remarks about Alternatives and Emerging Components	Potential Efficiency Improvement
AC distribution lines	85%–90%	DC lines and local generation.	90%–98%
Distribution transformers	95%–99%	Solid state transformers with renewables and DC.	
AC substations	94%–99%	Intelligent Substations with renewables and DC.	Reduction in system average interruption duration index: 10%–30%. Reduction in energy loss: 2%–5%. Reduction in maintenance cost: 15%–30%. Renewable integration benefits: 5%–20%. Reduction in operational cost: 5%–15%.
Loads			
Large Industrial Motors	85%–95%	Design and material improvements and integrated WBG power electronics.	96%–98%
Small motors	20%–85%	Design and material improvements and integrated WBG power electronics.	85%–95%
LEDs	15%–20%	WBG LEDs (for high brightness and in water and air purification systems and in medical therapies).	60% (Wall-Plug Efficiency, WPE)
Fridge compressor	200%–250% (for COP 2.0–2.5)	Inverter driven and with brushless PM motor.	30%–40% reduction in energy consumption

(*Continued*)

TABLE 1.15 (*Continued*) An Overview of the Typical Efficiencies of Various Power System Components and Potential Alternatives to Improve the Efficiency

Components	Typical Efficiency Range	Remarks about Alternatives and Emerging Components	Potential Efficiency Improvement
Air Conditioners (Inverter driven and with brushless PM motor)	200%–400% (for COP 2.0–4.0)	DC powered Air conditioners.	10%–15% reduction in energy consumption
Heat Pumps for space heating/cooling and water heating	250%–400% (for COP 2.5–4.0)	Inverter driven and with brushless PM motor, improved heat exchange.	10%–40% improvements in output and 1%–3.5% improvement in COP using advanced control algorithms.
Electric Resistance Heater (water, space)	100%	Not applicable	Not applicable
Electric Oven	70%–80%	Solid-State RF Oven with WBG devices	80%–85%
Microwave Oven	60%–75%		
Electric Cooktop (radiant or coil)	70%–75%	Induction Cooktop	85%–90%
Other Load-Accompanying Components			
Inverters	90–95	WBG Inverters	95%–99%
DC/DC converters and power supplies	70–90	WBG Resonant converters	90%–98%
AC coupled chargers	80%–90%	DC coupled WBG Chargers	90%–98%
EV chargers (slow/fast)	90%–95%	WBG device-based On-board Chargers	97%–99%
As battery systems:	75%–85%	Using new batteries, temperature control and safety measures can be minimized and eliminated.	1%–3% improvements, depends upon the size and external impacts.
EVs	80%–88%		
Community level storage	78%–86%		
Grid level storage			

TABLE 1.16 An Efficiency Comparison for Battery-EVs (BEV) and Hydrogen-Fuel Cell EVs (FCEV) with Potential Efficiency Improvements by 2050, when Powered by 100% Renewables

Process Stage	Direct Electrification (BEV) Efficiency	BEV Efficiency by 2050	Hydrogen-Based Electrification (FCEV) Efficiency	FCEV Efficiency by 2050
Source to Storage				
Electrolysis			70%–76%	76%–85%
Transportation, storage and distribution	95%	97%–99%	~90%	~93%
Charging equipment	90%–95%	97%–99%		
Storage to Wheel				
Battery charger	85%–90%	96%–98%		
H_2 to electricity conversion (fuel cell)			50%–60%	60%–70%
Inverter	90%–95%	95%–98%	90%–95%	95%–98%
Electric motor	90%–95%	96%–98%	90%–95%	96%–98%
Overall Efficiency	59%–73%	82%–92%	26%–37%	39%–53%

Moreover, the efficiency associated with "Transmission, storage and distribution" can benefit if renewable energy is harvested locally as in rooftop solar PV and battery storage solutions.

1.6.3 Power Quality

The integration of renewable energy sources can impact the power quality in an electrical grid. These often appear in the form of voltage variations, frequency variations and harmonics, which need to be maintained within specified limits, particularly due to the intermittency in wind and solar energy, and the inverter technology used to convert their output into AC power.

Note that the smart inverters used in DERs operate in various modes which affect the power quality while enabling the integration of renewable energy resources into the power grid. The primary modes of smart inverters' operation can be listed as grid-tied (grid-following) mode, island (grid-forming) mode and grid-support mode. However, there are further modes that accommodate these three modes with multiple functions as required These are multi-mode (or hybrid) operation (by switching between grid-tied, islanded and grid-support modes), bimodal operation (operating both in grid-tied mode and in islanded mode, and with energy storage), grid-interactive mode (both exporting power to the grid and importing power from the grid), and DC-coupled and AC-coupled modes (to offer efficiency advantages and easier to retrofit storage), which can alter the power quality measures.

1.6.3.1 Voltage Quality
Voltage variability can be a concern because electrical devices are designed to operate within a certain voltage range. For example, in Australia, the standard voltage is 230 V, and a range of +10% to −6% is acceptable. If the voltage is too high or too low, it can cause equipment to malfunction or fail, and it can also reduce the efficiency of power transmission and the distribution network.

As solar PV systems feed electricity back into the grid, the local voltage can rise, sometimes even going beyond the acceptable range. To counteract these issues and maintain grid stability, smart solar PV inverters are equipped with voltage ride-through capabilities, allowing them to continue operating within certain voltage boundaries without disconnecting from the grid (which is equal to unpredictable generator disconnection!).

Therefore, power system operators use various methods to control and manage voltage levels, including transformer tap changers, capacitor banks and voltage regulation devices. However, the actual voltage can fluctuate around the nominal value due to various factors, such as load variations and the integration of distributed generation sources like solar PV. Reactive power control capability in smart inverters can offer reactive power control hence voltage regulation with the cost of reduced active power generation.

1.6.3.2 Harmonics
Power electronic devices, such as inverters used in renewable energy systems, can introduce harmonics into the power system. These harmonics can distort the sinusoidal waveform of the power system's voltage and current, leading to power quality issues. In renewable energy systems, particularly those using inverters (like solar PV systems), managing Total Harmonic Distortions (THD) which can quantify the harmonic contents of electrical quantities is crucial. THD simply refers to the deviation of a pure sinusoidal waveform due to the addition of harmonics. Table 1.17 summarizes the impacts of high THD and the possible mitigation measures used.

TABLE 1.17 Impacts of High THD and Mitigation Methods That Can Be Used in Power Grid

Impacts of High THD	Mitigation Measures
• Equipment malfunction • Overheating of equipment • Increased energy losses • Interference • Reduced power factor • Capacitor bank failures • Grid instability	• Passive filters • Active harmonic filters • Improved inverter design • Use of multi-level inverters • Oversizing neutral conductors • Isolation transformers • Distributed grid architecture • Regular maintenance and monitoring • Grid codes and standards • Dynamic harmonic compensation

1.6.3.3 Frequency Variation

The frequency of a power grid reflects the balance between electrical supply (generation) and demand (load). In an AC power grid, the frequency is primarily determined by the speed at which synchronous generators rotate. Frequency variation in the power grid can be caused by a number of factors including sudden changes in load, loss of generation, variable renewable energy sources (change is generation!), limited power system inertia, network disturbances (such as short-circuits, transmission line faults or equipment failures), large interconnected systems and interconnection dynamics and poor coordination or control failures.

The stability and reliability of a power grid depend on maintaining its frequency close to a specified value (50 Hz in many parts of the world or 60 Hz in North America). With the evolution of the power grid, especially with the integration of more renewables and advanced technologies, frequency control remains an ongoing issue. Frequency Control Ancillary Services (FCAS) are used in power systems to ensure that the frequency remains within a designated range. FCAS ensures that, at all times, supply (generation) matches demand (load) to maintain the system frequency within set limits. The details of frequency variations and their dynamics will be covered in the following paragraphs, under power system inertia and the mitigation methods using FCAS.

1.6.4 Power System Inertia and Fault Levels

As it is known, mechanical inertia in power systems is inherent in traditional generation assets. The combined inertia in a large power grid comes from all the rotating masses, like large generators connected to turbines.

When there is a sudden change in the grid frequency, these rotating masses (large inertia) naturally resist changes due to their kinetic energy and aid in the management of the power grid. Note that the frequency variation can be related to power imbalances (generation/load imbalances). To explain the concept of the power system inertia clearly, an analogy can be made as given below.

Imagine trying to change the speed of a heavy train with multiple engines (generators at different power stations) and carriages (loads) on a track. The heavy weight of the train (representing the power system's inertia) means it will not stop or start immediately, even if the throttles (governor control system) are adjusted. The number and types of carriages (loads of different sizes and characteristics) further influence this inertia. Similarly, in a power system, the collective inertia of all generators and loads determines how the system's frequency (equivalent to the train's speed) responds to changes in power generation or consumption (demand). Note that to keep the train's speed constant (to keep frequency constant!), the total inertia and the status of the loads play critical roles.

In addition, the analogy can be expanded and refined further to explain the stability and other complex interactions in power system operations, which can be grouped under four categories:

1.6.4.1 Transient Instability

Imagine a train where the engine becomes disconnected from the carriages. The engine speeds up because it is no longer pulling weight, potentially leading to a crash (total blackout with serious consequences!).

In power systems, this is similar to the pole slipping of a synchronous generator. Stability can be restored by reconnecting system components gradually. This is known as "black-start" and refers to the process of restoring an electric power station or a part of an electric grid to operation without relying on the external electric power transmission network. The procedures and ability to carry out a black start are a crucial part of a utility's disaster recovery and resilience planning and are very costly.

1.6.4.2 Oscillatory Instability

Two locomotives, when perfectly synchronized, can respond seamlessly to speed changes. If they are not well-synchronized, they may counteract each other, causing a back-and-forth "hunting" oscillation. Similarly, power systems with poorly-tuned generators can oscillate and destabilize.

1.6.4.3 Frequency Control Ancillary Services (FCAS)

- Imagine a long train with multiple powerful locomotives distributed among the carriages. To maintain a constant speed despite challenges and disturbances, each engine's power output must be fine-tuned, just as FCAS ensures a stable frequency by adjusting the output of various generators.

1.6.4.4 Load Shedding and Curtailment (To Be Discussed in Detailed Later)

- Imagine a train powered by two locomotives. If one runs out of fuel, the train slows down, similar to a frequency drop. By disconnecting some carriages (similar to shedding some electrical loads), the remaining locomotive can maintain a more normal speed (constant frequency!).
- If we consider that one of the locomotives is skidding even as the accelerator is pressed further, leading to wasted fuel. This mirrors how energy from PV and wind generators can be wasted (curtailment), though it could be stored in batteries for later use.

In power systems, the relationship between generated power, demand power, system inertia and the synchronous generator's rotor angular velocity can provide the stored energy in the system that equates to kinetic energy, which allows the determination of the power system inertia. For example, power system inertia in Australia, particularly in the National Electricity Market (NEM), has been experiencing a decline, which is primarily due to the transition to renewable energy, the retirement of conventional generators, the spread out of the generation landscape and more reliance on interconnectors between states and regions. Declining power system inertia leads to rapid frequency changes in the grid. The rate of change of frequency (RoCoF) is influenced by both the power imbalance and system inertia as greater inertia leads to a lower RoCoF.

However, both wind turbines and solar PV systems usually employ power electronic converters to interface with the grid. Although such converters process the generated power to match grid requirements, they "decouple" the generation sources from the grid's electrical frequency. Therefore, they do not contribute directly to the grid's mechanical inertia in the traditional sense, but they can provide "synthetic" or "virtual" inertia through advanced control techniques. This synthetic inertia (resistance to the change of speed hence frequency) mimics the behaviour of traditional inertia and can help stabilize the grid.

For example, in a power network with a synchronous generator operating at 50Hz, an abrupt demand causes the system frequency to decrease as shown in Figure 1.9. The figure highlights the RoCoF under both high and low inertia scenarios. Important parameters include the frequency nadir (lowest frequency point) and the eventual steady-state frequency. To stabilize these variations, control actions are implemented over various durations, such as reserve deployments to manage future disturbances by mitigating generation-demand imbalances. Although generators might not immediately compensate for system imbalance, their stored energy helps control excessive frequency drops before primary controls intervene.

A low-inertia, renewable-focused network necessitates rapid FCAS following major frequency disruptions. Within the "RoCoF Area" shown in Figure 1.9, frequency declines faster, posing risks of cascading outages or complete blackouts. Effective network operation demands that RoCoF,

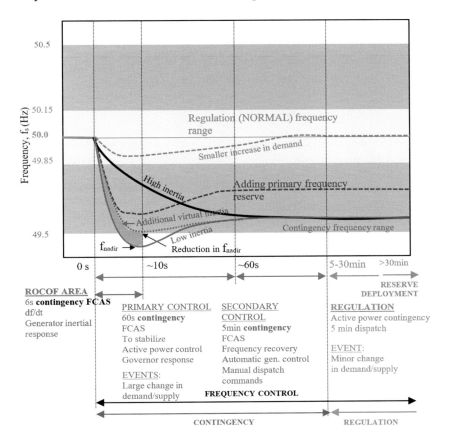

FIGURE 1.9 Impact of load increase and virtual inertia by battery storage on RoCoF in a 50 Hz power grid.

frequency nadir and steady-state frequency be properly managed to prevent issues like brownouts or blackouts. Multiple solutions, such as battery storage system (BSS)-based primary frequency control, can be leveraged to enhance stability. For example, BSS can quickly inject power (offering "virtual inertia"), thus preventing system frequency from dropping to dangerous levels.

Furthermore, the "RoCoF Area" in Figure 1.9 also illustrates the importance of quick responses to disruptions. This response time comprises various factors including the RoCoF detection time, communication delays, decision-making device response time and power ramp-up time.

Note that, energy storage systems, such as grid-scale batteries and Virtual Power Plant (VPP) concepts are integrated into FCAS, which can offer very fast frequency response services, given their ability to rapidly discharge.

Although a visual description is given in the figure, in managing modern power grids, several mechanisms are implemented to maintain system balance and ensure consistent power supply. The overview of power grid balancing and control can be summarized below, which each aims to provide stability, flexibility and efficiency to the energy landscape:

- Demand Response Services:
  - Raising Services: Enhance frequency by upping generation or cutting demand.
  - Lowering Services: Decrease frequency by reducing generation or increasing demand.

- Response Speed:
  - Fast: Primary Frequency Response occurring within seconds.
  - Slow: Secondary Frequency Control, spanning a few minutes.
  - Delayed: Takes several minutes to activate.

- Market Dynamics:
  - Bidding System: Both generators and demand responders bid to offer FCAS as practised in Australia.
  - Dispatch Decisions: The market operator dispatches based on system requirements and bid values.

- Renewable Integration:
    - Enable smooth integration of renewable sources through advanced four-quadrant reactive power control via smart inverters.

- Grid Interconnection:
    - A preventive measure to avoid massive generator outages that might instigate widespread disruptions or blackouts.

- Financial Aspects:
    - The expense tied to reserving capacity for FCAS and its deployment speed dictate FCAS market prices.

It may be considered that inertia and fault levels (short-circuit current levels) address different phenomena: inertia is concerned with frequency stability after disturbances, while fault levels are concerned with the maximum short-circuit currents. However, the two are indirectly related through the types of generators connected to the system. Synchronous generators (like coal, gas or hydro plants) contribute both to system inertia (with their rotating masses) and to fault levels (being capable of delivering high currents during faults).

With the increasing integration of non-synchronous generators (like solar PV, wind turbines and battery storage), system inertia decreases. However, all these generators use power electronic interfaces to connect to the grid. While the power electronics interface dictates much of the grid-interactive behaviour of these resources, the characteristics of the prime source also play a significant role. Therefore, the power electronics dictate the behaviour of these resources during faults, determining how they respond to voltage sags, frequency deviations and other anomalies, which are all classified under the term "smart inverters".

However, the internal characteristics of the prime source (be it a battery's internal resistance and state of charge, a solar panel's current operating point or a wind turbine's instantaneous wind conditions and generator characteristics) will also influence their fault response.

Note that, as was mentioned previously, none of these resources contributes to mechanical inertia in the traditional sense. However, they can be controlled to provide a fast frequency response, which can mimic the stabilizing effects of inertia, which is commonly referred to as "synthetic" or "virtual" inertia. It is also worth noting that utility-scale batteries have a unique

TABLE 1.18 A Structured Understanding of the Significance and Application of Fault Levels in Various Aspects of Power Systems

	Application Area	Use and Implication of Fault Levels
Design considerations	Circuit breaker rating	Determine appropriate ratings to interrupt maximum fault currents.
	Equipment ratings	Design equipment to withstand stresses induced by fault currents.
	Grounding systems	Design grounding systems to safely handle fault conditions.
	Network design and expansion	Analyse fault level changes due to network modifications to ensure safe integration.
Operational aspects	Protection coordination	Set relay pickup settings and time coordination based on fault current magnitudes.
	Power quality	Address voltage sags and other disturbances due to high fault currents.
	Stability studies	Analyse system response during and post-fault for system stability.
Safety implications	Arc flash analysis	Calculate energy released during arc faults to define safety measures.
Renewable integration	Renewable integration	Assess fault current characteristics with increasing renewable penetration.

characteristic due to their bidirectional nature that allows them to provide unique grid services such as voltage support, ramping and energy arbitrage.

It is essential to emphasize that the short-circuit levels under faults are critical parameters for designing and operating power systems with safety and reliability. Table 1.18 offers a summary of the uses and implications of fault levels in power systems. This information is expected to remain relevant until the grid becomes predominantly driven by converter systems and is populated with DC sources and loads.

1.6.5 Reverse Power Flow (Back Feeding)

Traditionally, power flow is from the transmission level to the distribution level, i.e., from the utility to the consumers. However, in recent years, urban areas have also witnessed a significant increase in solar PV hence power is now being fed back into the distribution and then into the transmission network. In addition, large-scale renewable connections and interconnections to the grid introduce more variability in load profiles. Furthermore, network development and asset management teams are faced with challenges in managing changing (decreasing) demand and anticipating how ageing infrastructure will adapt to these shifts.

The primary effects of reverse power flow are the overvoltage and on cyclic ratings of transformers. Research suggests that transformer longevity improves up until reverse power flow becomes prominent, but may differ due to variations in regional commercial activity. Note that the reverse power flow can also cause issues with transformer tap changers, which are designed to maintain a constant voltage level for varying loads but are not typically designed for reverse power flow. If power is flowing in the reverse direction, tap changers may have to operate more frequently, leading to increased wear and tear.

As it is well known, a load duration curve (LDC) represents the load levels on a system, sorted in descending order of magnitude over a specified time period. Therefore, load duration curves in this context provide a visual tool to understand and quantify the impacts of growing reverse power from distributed energy sources over time. When observing LDCs over multiple years (Figure 1.10) with increasing reverse power, certain trends and characteristics emerge. Note that as reverse power increases, the curve shifts downward, indicating that there are periods where the net load on the system is negative (i.e., the system is exporting power rather than consuming it). In addition, with increased distributed generation, peak loads might be reduced, leading to a flatter top portion of the curve. This suggests that distributed sources are offsetting traditional peak demand times.

LDCs indicate increasing variability due to intermittent energy sources such as solar PV as evident in Figure 1.10. The periods of reverse power, signified by parts of the curve below the zero-load line, are illustrated in Figure 1.10c and have expanded over the years. Overlaying LDCs from various years highlights the growing impact of reverse power, with each subsequent year revealing an intensified downward shift influenced by renewable adoption rates. Specifically, in South Australia, several transmission transformers exhibited reverse power flow driven by interconnection flows, large-scale renewable integration and residential rooftop solar PV. The residential transformer, analysed from 2012 to 2018 and illustrated in Figure 1.10c, showed a rising trend in reverse power flow magnitude and duration. By 2018, it reached 9.288 MW for about 60 days, up from 3.439 MW in 2012. Also, zero power flow instances increased from

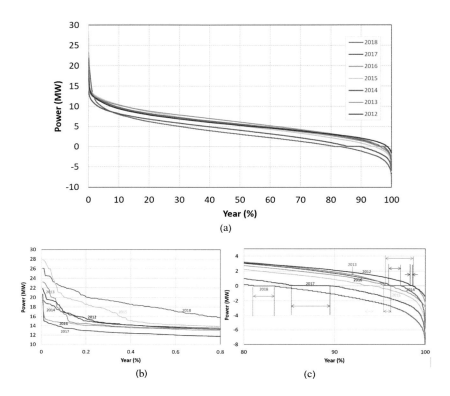

FIGURE 1.10 Load duration curves of a practical residential transformer between 2012 and 2018 demonstrating ever-increasing solar PV penetration and the reverse power flow: (a) entire percentage of LDCs, (b) percentage range where LDCs have the highest power, and (c) percentage range highlighting the ever-increasing reverse power.

none in 2014 to around 9 days in 2018. Annually, peaks in demand during extremely hot days (>40°C), primarily due to heightened air conditioner use, became evident. For instance, 2018 saw a peak demand of 26.08 MW, which exceeded 25 MW only for roughly 2 hours. Evaluating the absolute load offers insight into the transformer's total power flow. The significant power flow values for the transformer are summarized in Table 1.19.

1.6.6 Supply/Demand Imbalance and Load Shedding/Curtailment

Maintaining a supply–demand balance is fundamental for the safe, efficient and reliable operation of electrical power systems.

The characteristics and limitations of the AC grid play a significant role in supply–demand imbalances, which primarily arise from various factors including variability in renewable energy sources, inflexibility of

TABLE 1.19 The Key Values of Power Flow in the Power Transformer Given in Figure 1.10

Year	Magnitude of Forward Power (MW)	Magnitude of Reverse Power (MW)	Magnitude of Reverse Flow Duration (% time)	Zero Power Flow Duration (% time)
2012	21.89	−3.439	1.63	0.00
2013	23.25	−2.579	1.28	0.00
2014	19.31	−6.116	2.64	0.00
2015	27.88	−6.527	5.72	0.03
2016	19.09	−4.162	4.11	0.04
2017	20.73	−6.677	10.20	0.25
2018	26.08	−9.288	16.55	2.39

traditional power plants, transmission constraints, lack of energy storage, demand fluctuations, grid stability and frequency regulation, ageing infrastructure, economic factors and with a degree of regulatory and policy challenges.

In the existing hybrid grid, to address the generation–demand imbalance, a combination of technological, regulatory and market-based solutions is adapted. These include energy storage, demand response, grid modernization and digitalization (including distribution management system), interconnected grids, flexible generation sources, forecasting and predictive analytics, DERs, control of renewable energy resources, infrastructure upgrades, regulatory and market reforms and a very limited level of consumer education and behavioural changes.

In addition, load shedding and curtailment have also been utilized to address supply/demand imbalance, which is intertwined with the history and evolution of electrical power systems, and in recent years become "problems" as more variable renewable energy sources are integrated. However, they are applied in different contexts and have distinct characteristics, which are summarized in Table 1.20.

Primary reasons for curtailment include oversupply, transmission constraints and congestion, high wind ramps, local transmission outages, balancing issues (exhaustion of reserves and oversupply), conventional voltage control challenges (usually by tap changing transformers), wildlife protection measures (e.g., bird migrations, habitats), ice formations on wind turbines, area control error (ACE) violation avoidance (the measures taken to prevent or correct exceeding predefined thresholds), expansion of renewable sources outpacing new transmission infrastructure and maintenance of transmission systems.

TABLE 1.20 Distinct Characteristics of Load Shedding and Curtailment in Power Grid

	Load Shedding (Rolling Blackouts)	**Curtailment (Reduction)**
	"Intentional disconnection of power to prevent total system collapse when demand exceeds supply".	*"Intentional reduction or restriction of power due to oversupply or grid constraints. It is often applied to renewable energy sources".*
Purpose	Prevent a total blackout by reducing demand to match supply.	• To manage oversupply, grid constraints, or maintain grid stability. • Economic curtailment: might be cheaper to curtail renewable energy production than to turn off a coal or gas plant.
Application	Affects end-users, resulting in power outages based on schedule or emergencies.	Applied to power producers, especially renewables, instructing them to reduce output.
Duration	Varies from short to extended periods based on supply-demand imbalance.	Varies based on grid conditions; can be short-lived or extended.
Impact	Affects consumers leading to disruptions, potential economic losses, equipment failures and inconvenience.	Impacts power producers leading to financial losses due to reduced energy sales.
Notification	Users might get advance notice for scheduled outages; little to no warning in emergencies.	Power producers receive notifications from grid operators about the need to reduce output.
Mitigation methods	Infrastructure upgrades, demand side management, DERs, microgrids, diversification of energy sources, energy storage, grid interconnectivity, smart grids and metering.	Energy storage, grid upgrades, demand response, interconnection.

To highlight curtailment events in the real world, Figure 1.11 is provided which also captures the interplay between various energy sources and environmental conditions on a typical day when renewable energy penetration was 125% in South Australia. On 10 September 2023, the demand was low due to pleasant Spring weather. Note that a clear sky ensured that rooftop PV systems were in overdrive around mid-day, signifying peak solar generation. However, this surge leads to notable curtailment events, evident from the clipped solar utility curve. Such overproduction moments,

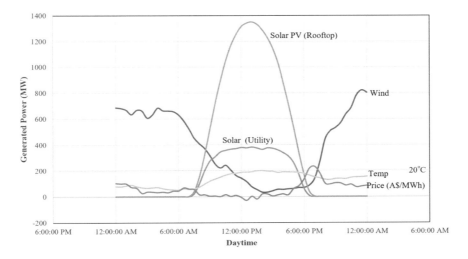

FIGURE 1.11 A sample day when renewable energy penetration was 125% in South Australia.

combined with a lack of immediate demand or storage, manifest in negative pricing intervals. As the sun's intensity reduces towards the evening, a sharp price hike is evident, marking the decrease in solar output and the grid's increased reliance on traditional energy sources. Interestingly, the figure also highlights the complementary nature of wind and solar. On this particular day, as solar dips, wind generation rises, indicating their complementary nature reference to solar PV. However, the temperature trace indicates an increase in daytime temperature, which, while boosting solar, appears to diminish wind output due to air's reduced density. This figure also highlights the intricate balancing act grid operators face, emphasizing the need for robust energy management strategies.

It is important to note that the need for curtailment and the methods used vary based on the specific grid infrastructure, regulatory environment and the penetration of renewable energy in a particular region. As renewable energy becomes a larger portion of the energy mix, effective curtailment strategies will be crucial to ensure grid stability and efficient energy use.

In addition, effective load shedding helps to prevent blackouts and maintain the stability of the power grid. However, it is always a last-resort measure, and utilities aim to minimize using different methods. The possible methods utilized in load shedding and curtailment are listed in Table 1.21.

TABLE 1.21 Types of Load Shedding and Curtailments Used to Support Generation/Load Demand

Types of Load Shedding to Manage Power Distribution	Curtailment Methods to Manage Access Energy Production in Renewable Energy
• Rotational load shedding • Under frequency load shedding • Under voltage load shedding • Manual load shedding • Emergency load shedding • Scheduled load shedding • Thermal load shedding • Industrial load shedding • Brownouts (reducing voltage for a certain period due to overload). • Demand response (or load management)	• Set-point curtailment • Frequency-responsive curtailment • Voltage-responsive curtailment • Economic dispatch and market integration • Priority grid access and dispatch • Storage integration • Advanced forecasting • Grid expansion and interconnection • Dynamic line ratings • Demand-side management

It is worth noting that, as explained in Chapters 6 and 7, autonomous microgrids with advanced features can significantly reduce the frequency and severity of generation and load imbalances. The combination of local generation, storage and intelligent demand management can make these systems far more resilient and adaptable than traditional grids. While they might not eliminate imbalances entirely, they can minimize their impact and enhance overall energy reliability and sustainability.

1.6.7 Domestic Wiring Issues in Rooftop PV Systems

In newly designed, robust electrical connections that are well-maintained, contact resistances remain very low, usually falling between $\mu\Omega$ and $m\Omega$. However, when connections are faulty or have deteriorated, there is a significant increase in contact resistances. This can trigger problems, including overheating, voltage drops and even potential fires. For example, loose connections may have resistances spanning tens of $m\Omega$ to several Ω, and corroded or damaged connections can easily escalate to Ω or even more. In addition, several factors, such as ageing, contamination, oxidization and rusting can compromise the integrity and performance of an electrical connection.

Therefore, while integrating rooftop solar PV systems into traditional domestic electrical configurations, originally designed for unidirectional power flow, may appear simple, it has significant risks. Such integrations can present safety hazards, particularly when connected with legacy systems that are designed for a short duration of unipolar power flow only during the operation of a few appliances.

It is worth noting that a typical house's electrical circuit layout comprises various circuit resistances. These stretch from the distribution network wires right up to the energy meter, covering clamps, service lines, terminals, fuses and AC mains wire resistances. Therefore, even a slight increase in contact resistance can lead to considerable power losses, impacting the system's efficiency. The most critical concern, however, arises with reverse power flow, especially if it is high and for a long duration, as this causes circuit resistances to overheat. This is already experienced by numerous prosumers, which is initially considered as a random electrical fault.

Furthermore, when DC isolators are used on the rooftop in solar PV applications, there are potential risks. This is due to a number of facts including long-term temperature extremes causing premature failure, weathering and UV degradation, moisture ingress, dirt and debris accumulation, mechanical wear and tear, potential for arc faults and operational challenges due to stiffness. Therefore, they should be either positioned under protected areas or integrated inside the power electronics converter.

1.6.8 Environmental Events and Reliability

As weather-related disruptions become more frequent year by year, the push for reliable local energy generation and storage becomes crucial, which will cover both metropolitan areas as well as remote sites. Many countries experienced that tornadoes, strong winds, lightning strikes and heavy rain can cause significant damage to power lines as well as to renewable energy resources leaving thousands without electricity for multiple days, primarily due to the scale of damage and stretched resources. In addition, recovery and repair efforts are also affected by challenging access conditions. Furthermore, the escalating threat of bushfires highlights the vulnerabilities of centralized power grids, which also create hazardous conditions that further impact recovery and repair efforts.

These highlight the importance of local power generation and stand-alone power systems equipped with battery storage since localized energy sources can reduce the impacts of such disruptions in the future. Moreover, the need for medical help, communication as well as mitigation measures (such as pumping water to fight fire) can become possible.

In Figure 1.12, the net load profile in power systems with substantial renewable energy integration graphically illustrates how the load changes throughout the day as renewable energy sources contribute to the grid. This profile (also known as the duck curve) showcases a pronounced trough during mid-day when solar generation is typically at its peak, followed by rapid ramps in the morning and evening and directly relates to the reliability of the grid. Table 1.22 summarizes the impacts of solar PV integration on grid reliability.

72 ■ Reinventing the Power Grid

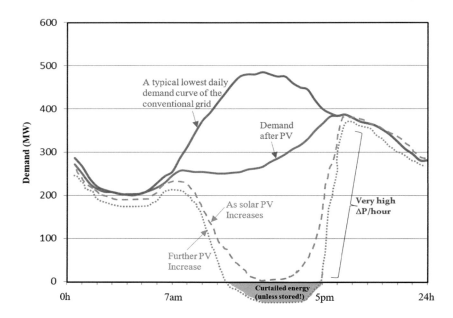

FIGURE 1.12 The net load profiles in power systems with substantial renewable energy integration, also known as duck curve to the outline of curves.

TABLE 1.22 Features of Solar PV Integration and Its Impact on Grid Reliability

Features of Solar PV Integration	Relation to Grid	Impact on Grid Reliability
Over-generation risk (midday)	Potential for excess solar generation during mid-day can change the generation and consumption balance.	The curtailment or shutdown of power sources may be required hence impacting grid stability and reliability.
Rapid ramping requirements	As solar generation decreases during the evening, there is need to quickly ramp up to meet demand using other sources.	If not managed effectively, the rapid ramp-up can threaten grid reliability due to potential imbalances.
Decreased inertia and frequency stability	The widespread adoption of solar PV displaces traditional generators, leading to reduced grid inertia.	As discussed previously, low power system inertia can cause faster frequency deviations, posing a threat to reliability.
Predictability and forecasting challenges	Inherent variability and intermittency in solar PV can lead to deviations from forecasted outputs.	Unpredictability affects supply-demand balance, hence impacting reliability.
Infrastructure strains	Older grid infrastructures with unidirectional power flow can face issues when excess power is sent back to the grid.	Reverse power flow can strain grid components, hence posing potential reliability issues.

The role of the prosumer becomes increasingly important as the challenges associated with variable renewable energy integration, often represented by a deepening in the load profile, intensify. If existing infrastructure remains unchanged, costs could rise. As a response, individuals with solar PV systems might incorporate energy storage, either at the household or community levels. As more energy is stored and self-consumed, the revenue from selling excess energy back to the grid could decline further. Over time, this evolving energy landscape might transform into a much flatter profile. Note that the integration of EVs as well as "electrification" practices are likely to avoid the reduced demand hence the revenue of power system operators in the short term.

1.7 POWER SYSTEM SECURITY AND CYBERSECURITY

Power system security and cybersecurity are closely linked, as the power grid relies on information technology and communication networks to operate efficiently and securely.

Cybersecurity is critical to maintaining power system security, as the power grid is vulnerable to cyber attacks that can disrupt power delivery and cause widespread outages. Cyber threats to the power grid can come from a range of sources, including nation-state actors, hacktivists and cybercriminals.

Power system operators implement robust cybersecurity measures to protect against these threats, including firewalls, intrusion detection systems, access controls and encryption. They also regularly update and patch their systems to address any known vulnerabilities and monitor their networks for any suspicious activity.

Moreover, cybersecurity can also impact power system security in terms of the availability and reliability of the power grid. For example, a cyber attack on a power grid's control systems or communication networks can lead to a loss of situational awareness, which can lead to cascading failures and widespread outages. Therefore, it is essential to address both to ensure the safe, reliable as well as efficient operation of the power grid.

In the context of power system security, autonomous systems can play a critical role in improving the efficiency and reliability of the power grid. Table 1.23 is structured to give an overview of power system security, considering current issues, solutions and potential future directions. However, it is critical to understand that due to the ever-evolving nature of technologies and threats, power system security should be viewed as a continuous process that requires regular updates and innovations. Additionally,

74 ■ Reinventing the Power Grid

TABLE 1.23 An Overview of Power System Security in the Light of Issues and Solutions

		Description/Issues	Solutions	Predictions on Future Developments
Power system stability	Power system dynamics	Maintaining equilibrium during small disturbances.	Advanced control systems, wide-area monitoring.	AI-controlled stabilization mechanisms.
	Fault isolation	Quickly isolating to avoid cascading failures.	Advanced protection relays, automated sectionalizing.	More adaptive and predictive fault response systems.
	System inertia	Resistance to rapid frequency changes.	Flywheels, synchronous condensers, dynamic demand response, virtual inertia	Decline of system inertia with more renewables, mitigation measures will be crucial. Utilization of smart inverters.
Resource management	Distributed generation	Integration of decentralized power sources like solar and wind.	Smart inverters, smart grids and microgrids	Vast increase in prosumers leading to more decentralized generation.
	Energy storage systems	To support supply-demand mismatches and renewable intermittency.	Batteries, pumped storage, flywheels, grid-scale solutions.	Rise in multi-functional storage solutions serving grid needs.
	Demand-side Management	Managing and modifying consumer demand for energy.	Smart meters, homes and appliances, energy management.	AI-involved enhanced consumer control over energy usage and storage.
Cybersecurity and data	Cyber attack prevention	Protecting grid infrastructure from digital threats.	Firewalls, intrusion detection systems, secure coding practices.	Rise in AI-powered threat detection and response systems.

(Continued)

TABLE 1.23 (Continued) An Overview of Power System Security in the Light of Issues and Solutions

		Description/Issues	Solutions	Predictions on Future Developments
	Data Privacy	Ensuring confidentiality and security of grid and user data.	Encryption, access controls, tokenization, data masking.	Enhanced focus on data rights, with stronger regulations.
	Data integrity and availability	Maintaining the accuracy and consistency of data over its life cycle.	Redundant systems, database mirroring, checksum validations.	Cloud-based as well as decentralized data storage and management.
Infrastructure and regulation	Physical infrastructure security	Safeguarding physical assets	Surveillance, physical barriers, security personnel, drones.	Autonomous microgrids, autonomous and one-way communications and monitoring systems with predictive threat analysis.
	Grid codes and standards	Defining technical and operational criteria for grid components and operations.	Regular updates to guidelines, international interconnection standards.	Evolving standards with the integration of new tech and renewables.
	Contingency planning	Preparing for potential large-scale disturbances and failures.	Simulation software, risk assessment tools, emergency response protocols.	AI-enhanced real-time scenario analysis and response.

as previously highlighted, as the grid transitions to become more power electronics-driven, DC-dominant and autonomous, many of the issues listed in the table may become obsolete.

The grid of the future will be dynamic, decentralized and digitized, requiring continuous adaptations in terms of technology, management and regulations. Existing measures will form the foundation, but constant innovation will be necessary to address emerging challenges.

Given the escalating significance of power system security and cybersecurity, it is critical to provide direction for future studies. Major reference materials for power system security involve IEEE Standards (693 and 1547), NERC Reliability Standards (including Critical Infrastructure Protection, CIP), textbooks, industry reports and guidelines. Furthermore, several entities offer guidelines, standards, education and credentials designed to enhance and standardize the proficiency of professionals and institutions within the realm of cybersecurity. These entities can be categorized under the following groups:

- Frameworks and Models
    - Skills Framework for the Information Age (SFIA)
    - Cybersecurity Skills Framework (SPARTA)
    - Information Security Management Systems (ISO 27001)
- Education and Training Initiatives
    - National Initiative for Cybersecurity Education (NIST NICE)
    - Joint Task Force on Cybersecurity Education (CSEC2017)
    - National Centres of Academic Excellence in Cyber Defense (CAE)
- Knowledge Codification
    - Cyber Security Body of Knowledge (CyBOK UK)
- Certifications
    - Certified Information Systems Security Professional (CISSP)

1.8 THE ROLES OF STANDARDS AND REGULATIONS

The development and evolution of standards and regulations play a critical role in facilitating a cost-effective, reliable and robust transition of the power grid. These guidelines cover a wide range of sectors, from e-mobility and smart homes to industries like mining. International bodies establish these standards primarily to ensure safety, interoperability and compatibility on a global scale which is needed for trade while continually adapting to technological advancements and societal needs.

As the power grid evolves to integrate more renewables and distributed energy sources, such standards become indispensable. They address challenges related to grid stability, power quality and data security. In addition, there is a need to maintain reliability and promote cost-effectiveness, while aiding in grid modernization.

Furthermore, as technology advances, standards support innovation and adapt to emerging challenges. This evolution involves regularly updating safety protocols and integrating advancements in energy storage, AI and cybersecurity. Regulatory bodies and organizations, with the assistance of application engineers and legal experts, consistently refine these standards and regulations to ensure seamless integration of renewable resources and their associated components into a hybrid grid.

Therefore, it can be emphasized that standards offer best practice guidelines for the design, safety, manufacture and usage of products in power system applications. Although they are generally voluntary, but demonstrating compliance with certain standards can prove beneficial for marketing purposes, and may even be a contractual or insurance requirement. Standards are established by recognized organizations, as listed in Table 1.24, along with brief explanations.

Standards can be applied to all electrical devices and power system components, covering design, physical construction, integration, assembly, testing, commissioning, safety and maintenance. Therefore, it is essential to identify the appropriate standards for a specific application and its related hardware. This is vital for the understanding of standards associated with the distinct functionalities of technologies, for seamless application from development to deployment.

The classification of standards helps to link the particular subject area or system component they address. For example, in the contents of renewable energy systems, DERs and EVs, standards cover a broad spectrum of

TABLE 1.24 Major Recognized Organizations That Establish Standards

Organizations	Descriptions
IEEE (Institute of Electrical and Electronics Engineers)	It is a professional association for electronic engineering and electrical engineering with its corporate office and operations centre in the US.
IEC (International Electrotechnical Commission)	It is an organization that prepares and publishes international standards for all electrical and electronic technologies.
UL (Underwriters Laboratories)	UL provides safety-related certification, validation, testing, inspection, auditing, advising and training services to a wide range of clients, including manufacturers, retailers, policymakers, regulators, service companies and consumers.
ISO (International Organization for Standardization)	ISO composed of representatives from various national standards organizations, and publishes worldwide technical, industrial and commercial standards.
SAE International (Society of Automotive Engineers)	It is a US-based globally active professional association.
VDE (Verband der Elektrotechnik, Elektronik und Informationstechnik)	VDE is a Germany-based organization that publishes technical standards primarily in the electrical engineering field.
DNV (Det Norske Veritas)	DNV is a Norwegian classification with the objective of "safeguarding life, property and the environment", to identify, assess and advise on risk management.
Renewable UK	Former the British Wind Energy Association. It is the trade association for wind power, wave power and tidal power industries.
AS/NZS (Australian Standard/New Zealand Standard)	Australia/Standards New Zealand are published and sold under a joint imprint, and are called up in both countries' regulations.
OSHA (Occupational Safety and Health Administration)	It is an agency of the US Department of Labour. Its mission is safe and healthy working conditions by setting and enforcing standards and by providing training, outreach, education and assistance.
ITU (International Telecommunication Union)	An agency of the UN, it aims to foster international cooperation and maintain international order to set international standards.

topics. Categorization can be made relevant to power system transition and components under these major groups: photovoltaic systems, wind generation systems, energy management systems, microgrids, electric vehicle charging stations, distributed generation systems, grid-tied energy storage systems, high voltage direct current (HVDC) systems, demand response systems and cybersecurity for power systems.

However, due to the fast pace of grid modernization and electrification activities, it is likely that the existing primary standards will be altered. In addition, a single standard might span multiple categories, and categories might be expended by additional standards to cover specific and emerging elements.

Note that a number of countries, for example, Australia usually have not dedicated local standards for every category. Therefore, in some cases, the local standards also refer to the major international standards or international standards that are used directly when no equivalent standard exists.

However, given the international nature of standardization bodies and the range of technical details they cover, there are potentially hundreds of standards associated with each of these categories and their subsystems. This makes it challenging to generate a comprehensive list of all associated standards for each subsystem within the categories. Moreover, due to the highly digitized nature of hybrid power systems, aspects like "digitization", "communication" and "software" further complicate these categories.,

For example, the "Photovoltaic Systems" category alone includes subsystems like solar cells, modules, inverters, DC/DC converters, mounting systems, tracking systems, electrical wiring and protection (including electrical, mechanical, weather, fire), occupational safety and health, insulation, isolation, earthing and lightning, each with their unique sets of associated standards. Similarly, "Cybersecurity for Power Systems" includes standards related to software, hardware, security risk management, industrial communication networks, substation automation, security for industrial automation and control systems, cybersecurity preparedness, information sharing and incident response.

Furthermore, each of the categories must address the safety, reliability and efficiency of the power system. For example, the standards for "Energy Management Systems" should consider communication protocols, performance metrics and safety. The "Microgrids Standards" would revolve around design, safety, operation, control and integration with the main grid. "Electric Vehicle Charging Stations" would involve standards related to safety, interoperability, energy efficiency, as well as communication between the charger, the vehicle and the grid. "Distributed Generation Systems" would primarily focus on safety, interconnection with the grid and performance. For "Grid-Tied Energy Storage Systems", standards would primarily address safety, performance and interconnection. "High Voltage Direct Current (HVDC) Systems" would cover safety, operation and testing and "Demand Response Systems" would include communication protocols, performance metrics and safety standards.

Therefore, it may be necessary to engage with an expert or a consultant in the specific field or subsystem. Additionally, utilizing the search functionalities offered by each standards organization's website could be considered to find standards relevant to a specific interest. Note also that, as mentioned above, the standards evolve over time, more frequently as the grid modernization and transformation accelerate with technological improvements. Hence, it is crucial to refer to the latest version of the original standards.

It should be emphasized here that the most commonly used IEEE Standards, which also form the basis for standards developed in other regions, provide a highly comprehensive set of guidelines. These standards cover various power system components and distributed generation technologies, such as Batteries, Instrumentation and Measurement, the National Electrical Safety Code (NESC), Power and Energy, Power Electronics, the Smart Grid and Transportation. They can be accessed at *https://standards.ieee.org/standard*.

Note also that a specific set of issues can also be mapped to the categories. As utilized in Chapter 6 in the design of the autonomous microgrid, these include smart inverters, earthing, acoustic noise in air conditioners used in battery management, lighting, enclosure design, protection (fire, extinguishing, lightning), system installations, batteries' installations, weather monitoring systems, monitoring, low voltage, high voltage, auxiliary power, security (for attacks, disruptions and theft). Note that these issues usually overlap, and some categories might require the consideration of more issues, depending on the specific circumstances and requirements. A highly comprehensive list can also be produced for the battery technologies themselves.

Finally, "Regulations" are rules or laws imposed by governmental bodies and are mandatory. They must be obeyed to, and failure to comply can result in varying levels of penalties. Typically, regulations are broader than standards and may reference specific standards as part of their requirements. Table 1.25 provides a possible classification based on

TABLE 1.25 A Possible Classification Based on the Categories and Typical "Regulations"

Categories	Typical Regulations and Guidelines
Photovoltaic systems	*US*: Federal and State Solar Access Laws. *Europe*: EU Renewable Energy Directive. *Australia*: Clean Energy Regulator Rules.
Wind generation systems	*US*: National Environmental Policy Act (NEPA). *Europe*: EU Renewable Energy Directive. *Australia*: Environment Protection and Biodiversity Conservation Act.
Energy management systems	*US*: Energy Policy Act; and guidelines of the Department of Energy (DOE) under Superior Energy Performance (SEP). *Europe*: EU Energy Efficiency Directives. *Australia*: National Greenhouse and Energy Reporting (NGER) Act.
Microgrids	*US*: Public Utility Regulatory Policies Act (PURPA), the Federal Energy Regulatory Commission (FERC). *Europe*: EU Electricity Market Directive. *Australia*: National Electricity Law, the Australian Energy Market Operator (AEMO).
Electric vehicle charging stations	*US*: Fixing America's Surface Transportation (FAST) Act. *Europe*: EU Alternative Fuels Infrastructure Directive. *Australia*: Australian Design Rules, the National Electricity Rules (NER).
Distributed generation systems	*US*: Federal Power Act, the Public Utility Regulatory Policies Act (PURPA). *Europe*: EU Electricity Market Directive. *Australia*: National Electricity Law, the National Electricity Rules (NER).
Grid-tied energy storage systems	*US*: Federal Energy Regulatory Commission Order 841, Public Utility Commissions usually follow IEEE 1547. *Europe*: EU Clean Energy Package. *Australia*: Complies AS 4777, National Energy Retail Law.
High voltage direct current (HVDC) systems	*US*: Federal Energy Regulatory Commission (FERC) Transmission Planning and Cost Allocation regulations, and the North American Electric Reliability Corporation (NERC). *Europe*: EU Guidelines for Trans-European Energy Infrastructure *Australia*: The National Electricity Rules (NER) administered by AEMO.
Demand response systems	*US*: Federal Energy Regulatory Commission Order 745. *Europe*: EU Electricity Market Directive. *Australia*: The National Electricity Rules (NER).
Cybersecurity for power systems	*US*: Federal Information Security Management Act (FISMA), NERC sets mandatory Critical Infrastructure Protection (CIP) standards. *Europe*: EU Network and Information Security Directive. *Australia*: Security of Critical Infrastructure Act, the Australian Energy Sector Cybersecurity Framework (AESCSF).

the above-described categories and associated typical "Regulations". Note that this table is just a general pointer to provide an insight into the type of regulations that might be applied. However, for more precise information, a specialist in regulatory affairs or a legal advisor may need to be consulted.

Note that this chapter has provided an overview of the current hybrid power grid, which is becoming increasingly complex, fraught with ever-increasing power quality problems and rising energy costs, and highly susceptible to escalating environmental conditions. In subsequent chapters, the focus will shift to grid modernization, detailing the operation of key novel power system components. In addition, this chapter aims to shed light on the level and nature of power demand, which is expected to be managed more through technological advances than individual preferences. Therefore, future power system components, as highlighted in "electrification" discussions, will likely be highly autonomous, eliminating legacy demand profiles that are unpredictable or difficult to predict.

While the pace of grid modernization appears swift, leading to expectations of a smarter and more flexible grid, the reality is that it is becoming more complex and is quite far from meeting energy demands in a reliable, cost-effective manner that minimizes environmental impact. Therefore, the subsequent chapter aims to highlight the complexities and limitations in the evolving landscape of electrical power systems.

BIBLIOGRAPHY

K. M. U. Ahmed, M. H. J. Bollen and M. Alvarez, A review of data centers energy consumption and reliability modeling. *IEEE Access*, 9, pp. 152536–152563, 2021, https://doi.org/10.1109/ACCESS.2021.3125092.

Australia – Electric power transmission and distribution losses, https://www.indexmundi.com.

S. Burt, Meteorological impacts of the total solar eclipse of 21 August 2017, available at https://centaur.reading.ac.uk/75670/1/Total%20eclipse%2021%20August%202017%20-%20Stephen%20Burt.pdf.

S. Chae, S. Bae and Y. Nam, Performance improvement of air-source heat pump via optimum control based on artificial neural network. *Energy Reports*, 10, pp. 460–472, 2023, https://doi.org/10.1016/j.egyr.2023.06.051.

Cost of supplying electricity to households at an eight-year low, available at https://www.accc.gov.au/media-release/cost-of-supplying-electricity-to-households-at-an-eight-year-low.

R. Dollinger, Emergency Lateral Control of Autonomous Vehicles using, PhD Thesis, The University of Adelaide, 2023.

Electric power transmission and distribution losses, available at https://www.indexmundi.com.

Electricity explained: Factors affecting electricity prices, available at https://www.eia.gov/energyexplained/electricity/prices-and-factors-affecting-prices.php.

N. Ertugrul and D. Abbott, DC is the future. *Proceedings of the IEEE*, 108(5), pp. 615–624, May 2020.

N. Ertugrul and G. Haines, "Impacts of Thermal Issues/Weather Conditions on the Operation of Grid Scale Battery Storage Systems," *2019 IEEE 13th International Conference on Power Electronics and Drive Systems (PEDS)*, Toulouse, France, 2019, pp. 1–6 https://doi.org/10.1109/PEDS44367.2019.8998808.

N. Ertugrul, Mine electrification and power electronics: The roles of wide-bandgap devices. *IEEE Electrification Magazine*, 12(1), pp. 6–15, March 2024, https://doi.org/10.1109/MELE.2023.3348254.

N. Ertugrul, A. P. Kani, M. Davies, D. Sbarbaro and L. Morán, "Status of Mine Electrification and Future Potentials," *2020 International Conference on Smart Grids and Energy Systems (SGES)*, Perth, Australia, 2020, pp. 151–156, https://doi.org/10.1109/SGES51519.2020.00034.

N. Ertugrul, S.A. Pourmousavi, M. Davies, D. Sbarbaro and L. Moran, "Status of Mine Electrification and Future Potentials", *International Conference on Smart Grids and Energy Systems (SGES)*, November 2020, https://doi.org/10.1109/SGES51519.2020.00034.

Frequently Asked Questions (FAQs) – U.S. Energy Information Administration (EIA), available at https://www.eia.gov.

Q. Gao, B. Ding, N. Ertugrul and Y. Li, Impacts of mechanical energy storage on power generation in wave energy converters for future integration with offshore wind turbine. *Ocean Engineering*, 261, pp. 112136, 1 October 2022.

G. Haines and N. Ertugrul, "Application Sensorless State and Efficiency Estimation for Integrated Motor Systems," *2019 IEEE 13th International Conference on Power Electronics and Drive Systems (PEDS)*, Toulouse, France, 2019, pp. 1–6, https://doi.org/10.1109/PEDS44367.2019.8998868.

G. Haines, N. Ertugrul and W. L. Soong, "Autonomously Obtaining System Efficiency Maps from Motor Drive Systems," *2019 IEEE International Conference on Industrial Technology (ICIT)*, Melbourne, VIC, Australia, 2019, pp. 231–236, https://doi.org/10.1109/ICIT.2019.8755199.

How are EU electricity prices formed and why have they soared?, available at https://www.eurelectric.org/in-detail/electricity_prices_explained/.

Projected Costs of Generating Electricity, 2020 Edition, International Energy Agency, Nuclear Energy Agency Organisation for Economic Co-Operation and Development, available at https://iea.blob.core.windows.net/assets/ae17da3d-e8a5-4163-a3ec-2e6fb0b5677d/Projected-Costs-of-Generating-Electricity-2020.pdf.

Renewable Energy and Jobs Annual Review 2022, available at https://www.ilo.org/wcmsp5/groups/public/---dgreports/---dcomm/documents/publication/wcms_856649.pdf.

M. Salazar and N. Ertugrul, "Potential enhancements for vehicle electrical power management systems in military vehicles," *2013 Australasian Universities Power Engineering Conference (AUPEC)*, Hobart, TAS, Australia, 2013, pp. 1–6, https://doi.org/10.1109/AUPEC.2013.6725360.

A. Shehabi, et al., United States Data Centre Energy Usage Report. Lawrence Berkeley National Laboratory, Berkeley, CA, 2016. LBNL-1005775, Available from: https://www.researchgate.net/publication/305400181_United_States_Data_Centre_Energy_Usage_Report#fullTextFileContent [accessed Dec 29 2022].

J. Stoupis, R. Rodrigues, M. Razeghi-Jahromi, A. Melese and J. I. Xavier, Hierarchical distribution grid intelligence. *IEEE Power and Energy Magazine*, (21(05), pp. 3288598, September/October 2023, https://doi.org/10.1109/MPE.2023.3288596.

E. Thomas, Effect of weather, load and reverse power flow on power transformers, Honours project report, School of Electrical and Electronic Engineering, the University of Adelaide, Australia, 2019.

Water-cooled servers: Common designs, components, and processes, available at https://www.ashrae.org/File%20Library/Technical%20Resources/Bookstore/WhitePaper_TC099-WaterCooledServers.pdf.

S. Zhu, C. J. Kikkert and N. Ertugrul, "A wide bandwidth, on-line impedance measurement method for power systems, based on PLC techniques," *2014 IEEE International Symposium on Circuits and Systems (ISCAS)*, Melbourne, VIC, Australia, 2014, pp. 1167–1170, https://doi.org/10.1109/ISCAS.2014.6865348.

Y. Yao, N. Ertugrul and A. P. Kani, "Investigation of Short-Term Intermittency in Solar Irradiance and Its Impacts on PV Converter Systems," *2022 32nd Australasian Universities Power Engineering Conference (AUPEC)*, Adelaide, Australia, 2022, pp. 1–6, https://doi.org/10.1109/AUPEC58309.2022.10215771.

https://www.statista.com/statistics/186992/global-derived-electricity-consumption-in-data-centres-and-telecoms/.

CHAPTER 2

Photovoltaic Solar Energy

2.1 INTRODUCTION

Solar energy has been an abundant renewable energy source that is going well back human history, and it is harnessed in various ways in the context of their development and complexity:

- Early methods: Solar drying and solar ovens or cookers.

- Mid-twentieth century: Solar water and space heaters, PV systems (PV cells emerged in the 1950s), solar pumps and PV electrolysis for hydrogen production.

- Recent methods: Concentrated solar power (CSP), building-integrated PVs (BIPVs), solar desalination, photoelectrochemical (PEC) water splitting, solar thermochemical hydrogen production and solar-powered transportation.

The above methods are used individually or in combination, depending on the specific needs and conditions of a location. As technology advances and the need for renewable energy sources grows, new methods and improvements on existing methods will likely emerge. However, the importance of PV technology in today's energy landscape is clear.

Photovoltaic, derived from the Greek word "photo" meaning light and "voltaic" from the electrical unit "volt", refers to the conversion of light into electricity. As it is witnessed over the past few decades, the evolution of PV technology has been remarkable, from small-scale applications such as calculators and watches, to power homes, industries, communities and grid scale solutions.

Several factors have contributed to the rapid growth and adoption of PV technology: technological advancements, economic viability, government incentives and environmental awareness. As technology continues to advance and the world shifts towards cleaner energy solutions, PV solar energy will definitely play an even more significant role.

Table 2.1 is provided to compare the energy sources based on their potential and their contribution to global electricity generation. It is important to note that the contribution to global electricity generation is constantly evolving as countries invest in different energy sources and as technology advances. Note also that the table is arranged with renewable and clean energy sources at the top and prioritized by their contribution to global electricity generation.

Considering the prime energy sources, Figure 2.1 can be given that compares power capacity by technology. Note that solar energy, which includes rooftop solar PV installations, solar PV farms and solar thermal power stations, is predicted to experience significant growth alongside offshore wind resources. This growth in solar and wind is expected to continue, driven by decreasing costs, technological innovations and enhanced policy support. By around 2027, it is projected that solar energy will overtake coal, capturing about 22% share, while coal's contribution, influenced by environmental concerns, policy limitations and competition from cleaner energy alternatives, is anticipated to drop below this value. Concurrently, specifically offshore wind energy is on a rise, with projections suggesting that it might surpass natural gas by the early 2030s.

This renewable energy growth is achieved by the potential ascent of green hydrogen, which leverages both solar and wind, and the potential introduction of small modular reactors. Hydropower and bioenergy are likely to remain relatively stable, with their growth potential being constrained by environmental and logistical challenges. Nonetheless, forecasts indicate a decline in both coal and natural gas.

Despite the challenges of initial investments, utility-scale solar PV remains the most cost-effective electricity generation method in many countries. The expansion of distributed rooftop solar PV is being propelled by rising electricity prices and policies focused on increased energy costs for consumers. By 2035, it is estimated that solar might constitute about

TABLE 2.1   Renewable and Clean Energy Sources, Reflecting Their Contribution to Global Electricity Generation as of 2022

Energy Source	Advantages	Limitations	Usage by 2022	Potential
Hydroelectric	Renewable, reliable storage capability.	Environmental impact, limited by geography.	~16%	Limited by river flow and geography: globally ~40,000 TWh/year potential, with ~16,000 TWh/year technically feasible.
Wind	Renewable, decreasing costs.	Intermittent land use, offshore construction difficulties	~5%–6%	Global potential estimated at ~400 TW, but only a fraction is technically and economically feasible to harness.
Solar (PV and thermal)	Renewable, decreasing costs, scalable.	Intermittent, land use, resource use for panels.	~3%–4%	Earth receives ~173,000 TW of solar energy, but only a fraction is technically and economically feasible to harness.
Biomass	Renewable, utilizing waste products.	competing with food sources, emissions, land use.	~2%	Limited by agricultural and forestry residues, dedicated energy crops, and organic waste: ~100–300 EJ/year globally.
Geothermal	Renewable, reliable	Limited by geography, depleting local sources.	<1%	Earth's internal heat: ~100 TW, but only a fraction is technically and economically feasible to harness.
Tidal and wave	Renewable, predictable patterns	Limited by geography, environmental impact.	<1%	Limited by coastal geography: global potential estimated at ~3 TW for tidal and ~10 TW for wave energy.
Nuclear	High energy density, low emissions, reliable.	Nuclear waste, public perception, high capital costs.	~10%	Uranium: decades to centuries, depending on extraction technology and reactor type. Thorium: greater potential than uranium, but less developed technology.
Fossil fuels	High energy density, established infrastructure, reliable, controllable.	Greenhouse gas emissions, finite resource, air pollution.	~60% (coal, natural gas, oil)	Limited by reserves: Coal: ~1 trillion tons (centuries at current rates). Oil: ~1–1.7 trillion barrels (decades at current rates). Natural gas: ~198 trillion cubic meters (decades at current rates).

88 ■ Reinventing the Power Grid

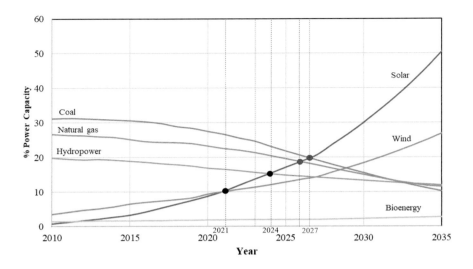

FIGURE 2.1 Comparison of the capacity of the prime energy sources by technology and future projections.

54% and wind might constitute 27% of the global energy mix, signifying a notable shift in energy paradigms. It is interesting to report that the "living renewable energy laboratory" South Australia has already reach to this energy mix with zero coal power stations.

However, it can be concluded that due to the regional constraints, unpredictable events, technological innovations and shifts in policy, these projections are likely to differ.

The primary aim of this chapter is to provide a good understanding of the current status of solar PV technology, its future potential and the challenges it may face. Emphasis will be placed on presenting technical details and real-world applications to illustrate a comprehensive picture of the solar PV landscape, which is set to shape future energy resources as the grid evolves. Subsequent sections will explore the science and engineering behind solar PV cells, modules and systems. Additionally, practical assessments, designs and analyses relevant to both distributed and farm-level applications will be discussed in detail.

2.2 SOLAR RESOURCES, DEFINITIONS AND MEASUREMENTS

2.2.1 Types of Solar Radiation

Figure 2.2 illustrates three types of solar irradiances that come from the sun and reaches to a target (say a solar PV panel):

Photovoltaic Solar Energy ■ 89

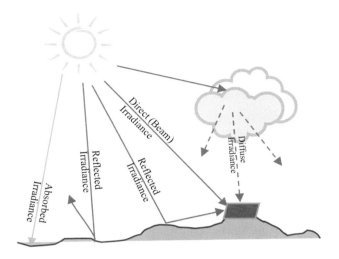

FIGURE 2.2   Types of solar irradiance (radiation).

- Direct (beam) irradiances: This is sunlight that travels straight from the Sun to Earth without any interference, which is the most concentrated form of solar irradiance that can be focused using mirrors or lenses and is best suited for solar concentrators and tracking systems. It can be measured on a surface that is always perpendicular to the sun.

- Diffuse irradiances (scattered lights): The irradiance that interacts with particles gases and water droplets, causing it to scatter in various directions, which reaches the Earth's surface from all directions, which is specifically present on cloudy days.

- Reflected irradiances (albedo): When direct radiation hits a surface, a portion of it will be absorbed by the surface, and a portion will be reflected, depending on the properties of that surface. For example, snow or other bright materials can reflect sunlight, while dark soils or water bodies have a lower albedo and reflect less radiation.

Most of the sunlight that reaches the surface of a PV module comes directly from the sun or reflected from clouds or particles in the atmosphere but a small amount is also reflected from the ground. Since the total solar irradiance received on a surface contains direct, diffuse and reflected irradiances, they can be utilized in "fixed solar panels" that do not track

90 ■ Reinventing the Power Grid

the sun. Note that some of the reflected irradiances can also be captured as in bifacial solar PV cells. It needs to be emphasized that intermittency in solar PV refers to the variability in these solar irradiances due to factors like cloud cover, atmospheric conditions and the angle of the sun.

2.2.2 Solar Energy Potential and Availability

The amount of solar energy a specific location can harness depends on several factors, including its latitude, altitude, topography, climate and the time of year. Figures 2.3 and 2.4 are produced to visually represent these factors.

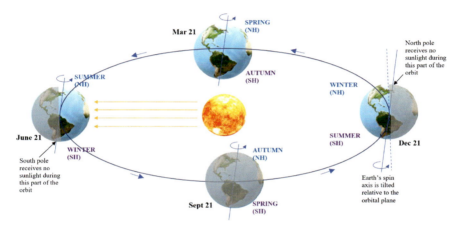

FIGURE 2.3 Sun centred view for Earth's orbit, tilt reference to the orbital plane and seasons. NH: Northern Hemisphere; SH: Southern Hemisphere

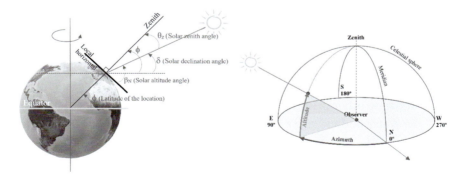

FIGURE 2.4 Earth centred view for representation of solar angles and earth parameters, illustrating the interplay of latitude, azimuth, zenith and solar angles for optimal solar energy utilization.

The Earth orbits the Sun in an elliptical trajectory, taking approximately 365.25 days to complete. As illustrated, there are times when the Earth is slightly closer to the Sun (perihelion) and times when it is a bit farther away (aphelion). Moreover, the Earth's axis of rotation is not perpendicular to its orbital plane (the plane in which the Earth orbits the Sun). Instead, it is tilted at an angle of approximately 23.5°, causing to seasonal variations.

As the Earth orbits the Sun, its axial tilt causes different parts of the planet to receive varying amounts of sunlight throughout the year. When one hemisphere tilts towards the Sun, it experiences summer, marking the longest day of the year. During this time, the Sun reaches its "zenith", resulting in the highest potential for solar energy production. Conversely, when the same hemisphere tilts away from the Sun, it experiences winter, marking the shortest day of the year. The Sun then reaches its lowest point in the sky, correlating with the lowest potential for solar energy production.

It is important to understand that due to the cyclical nature of this process, solar energy production must be estimated over an annual cycle. This estimation significantly influences its levelized cost of electricity (LCOE) and the sizing of energy storage systems. In addition, daily solar variations are considered to ensure a balance between generation and load.

In Figure 2.4, let us assume an observer is located at a specific point on Earth's surface, aiming to harness solar energy for power generation. The Sun, positioned in the sky, emits irradiance that reaches the Earth's surface in the forms discussed in the previous section.

In the figure, "latitude" is defined as the angle ϕ between a line extending from the Earth's centre to the equator and another line from the Earth's centre to the observer's location. This angle signifies the observer's distance north or south from the equator.

"Azimuth" is the angle formed between the true north direction on the observer's local horizontal plane and the line pointing to the Sun's position on that plane, measured clockwise. The figure illustrates the Sun's cardinal direction relative to the observer.

For certain solar calculations, it is convenient to assume an observer at the Earth's centre to streamline geometric relationships. At the observer's actual location on the Earth's surface, a flat plane tangent to the Earth, known as the "local horizontal plane" (an imaginary plane that is perpendicular to the gravitational force), represents the observer's horizon. Directly above the observer, where a line perpendicular to this plane intersects the sky, is the "zenith", the point in the sky directly overhead. Note that opposite the zenith, directly below the observer, is the nadir.

The "solar zenith angle", θ_Z, is the angle between the line pointing to the Sun from the observer and the "zenith", indicating the Sun's distance from being directly overhead.

Furthermore, the "solar declination angle", δ, is the angle between the Sun's rays and the plane of the Earth's equator. As it can be observed, this angle changes throughout the year as the Earth orbits the Sun, signifying the Sun's relative position to the equator.

Therefore, we can define the "solar altitude angle" β_N (also known as solar elevation angle) complementary to the "solar zenith angle", to calculate the Sun's height in the sky, as measured from the observer's horizon.

Although these angles and associated calculations are well known as commonly used in navigations, architecture and urban planning and astronomy, understanding these angles and parameters is critical for maximizing solar energy capture, since solar panels need to be oriented optimally (in addition to the tracking system to follow the sun during daylight). Knowing the sun's position (given by azimuth and solar altitude angles) helps in this orientation. The solar declination angle, which changes throughout the year, helps in understanding seasonal variations in solar energy potential.

Note that in the context of solar energy and solar geometry, the "altitude angle" is important for determining the sun's position in the sky and hence the amount of solar irradiance a location receives. However, "latitude" is a fixed value for a location and plays a role in determining the sun's path across the sky throughout the year.

While the azimuth angle (varies between 0° and 360° over a day) describes the horizontal direction of an object, the altitude angle describes its vertical position (how high it is in the sky). The combination of both azimuth and altitude gives a complete description of the object's position in the sky from a specific location. However, there is a need to know the time as well, which can be represented by "hour angle", h (or H). "Hour angle" is defined as 0° at solar noon and increases by 15° per hour, hence negative before solar noon and positive after solar noon and hence the time of day with respect to the sun's position.

In solar calculations, both the "azimuth" and "hour angles" are vital for determining the sun's position in the sky at any given time and location. Therefore, the following equations can be given using the above definitions and the figures provided.

$$q_Z = 90° - \beta_N \tag{2.1}$$

Since the "solar declination angle", δ, varies throughout the year as the Earth orbits around the Sun, the declination angle is 0° during the equinoxes, +23.45° at the summer solstice, and –23.45° at the winter solstice. There are several methods to calculate the solar declination angle which are listed in Table 2.2, including the most commonly used "approximation".

Similar to the "solar declination angle" described above, the "solar altitude angle", β_N, can also be calculated by various methods considering the required accuracy and the available data. For many practical applications, the solar declination, latitude and hour angle will be sufficient as it provides a good balance between accuracy and computational simplicity as given in Table 2.3.

Furthermore, it can be concluded that as the Earth rotates and seasons change, different parts of the world experience daylight, darkness and varying weather conditions, leading to fluctuations in renewable energy generation. For instance, when solar generation is lower during the winter months in one hemisphere, it could be higher during the summer months in the opposite hemisphere (referred to as "north–south phase shift" or "seasonal phase shift", as shown in Figure 2.5). Similarly, the diurnal phase shift (or east–west phase shift) can help even out solar energy generation throughout the day when there is an interconnected grid covering a broad longitudinal range.

TABLE 2.2 The Methods to Calculate the "Solar Declination Angle", δ

Method	Formula/Description
Approximation	$\delta = 23.45 \sin\left[\dfrac{360}{365}(n+284)\right]$ OR $\delta = 23.45 \sin\left[\dfrac{n-80}{365} 2\pi\right]$ where n is the day of the year (January 1, $n=1$; January 2, $n=2$ ....)
Jean Meeus' method	$\delta = -\arcsin(0.39779 \times \cos(\gamma + 1.914 \times \sin(\gamma)))$ $\gamma = n\, \dfrac{360}{365.24}$
Cooper's method	$\delta = \displaystyle\sum_{i=1}^{24} a_i \cos\left(\dfrac{2\pi\, i\, n}{365}\right)$ where a_i values are coefficients derived from Fourier series approximations of the solar declination.
Solar and moon calculator or astronomical algorithms	Online tools and software packages that use algorithms.

TABLE 2.3 The Methods to Calculate the "Solar Altitude Angle", β_N

Method	Formula/Description
Using solar declination (δ), latitude of location (ϕ) and hour angle (h)	$\sin(\beta_N) = \sin(\delta)\sin(\phi) + \cos(\delta)\cos(\phi)\cos(h)$ $\beta_N = \arcsin[\sin(\delta)\sin(\phi) + \cos(\delta)\cos(\phi)\cos(h)]$
Using solar zenith angle (θ_z)	$\beta_N = 90° - \theta_Z$
Using solar calculation tools or software	Online tools and software packages can be used including the NREL's Solar and Moon Position Algorithm (SAMPA) or the Photovoltaic Geographical Information System (PVGIS) online tool.
Using astronomical algorithms	More complex methods consider the Earth's elliptical orbit and axial tilt.
Using empirical models	There are empirical models based on observational data.

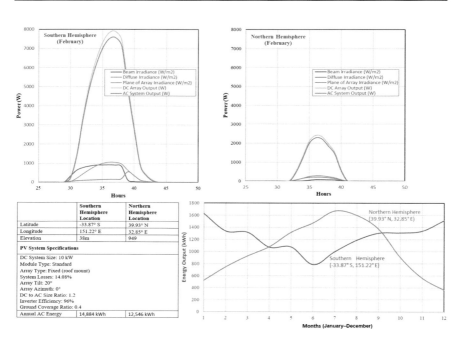

FIGURE 2.5 Seasonal phase-shift in solar generation due to Earth's rotation and changing seasons in two different locations.

While there is not a universally standardized term for this phenomenon, it is often described as "geographic smoothing" or "temporal and spatial diversity" of renewable energy generation. By interconnecting diverse geographic regions, the variability of renewable energy sources can be mitigated. This results in a more stable and reliable energy supply, which not only reduces the generation–demand imbalance but also minimizes the need for energy storage. Although a fully realized global application remains conceptual, there are regional proposals in place, such as connections from North Africa to Europe and from Australia to Singapore using HVDC submarine cables. A notable project is China's HVDC transmission system, designed to transmit power from the renewable-rich west to the high-demand eastern load centres.

2.2.3 Air Mass (AM) Ratio, Solar Spectrum and Solar Irradiation

AM refers to the optical path length through which sunlight travels through the Earth's atmosphere. It is a dimensionless number that represents the ratio of the actual path length of sunlight to the shortest possible path length (when the sun is directly overhead, or at the zenith).

Note that the solar altitude angle is related to the AM ratio. The AM ratio is essentially a measure of the length of the path that sunlight takes through the Earth's atmosphere (see Figure 2.6) standardized ($h_1 = 1$) to its shortest possible length when the sun is directly overhead. The higher the sun is in the sky (greater altitude angle), the shorter the path, and thus, the lower the AM ratio. Conversely, when the sun is low on the horizon, the AM ratio is higher because sunlight has to pass through a greater thickness of the atmosphere.

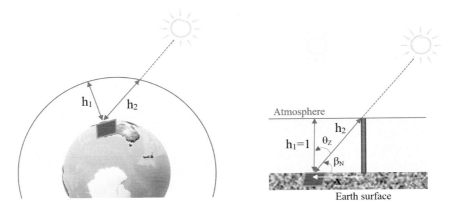

FIGURE 2.6 Visual presentation of the air mass as a function of solar altitude angle.

The AM can be approximated as AM=$h_2/h_1 \approx h_2/(h_2\cos(\theta z))=1/\cos(90°-\beta N)$.

Note that an easy method to find AM is to consider the shadow of a vertical pole as illustrated in Figure 2.6; hence, AM= $\sqrt{(1+(x/h_1)^2)}$, where x is the length of the shadow. However, note that due to atmospheric curvature, these calculations are not exact when the sun is near the horizon.

The AM ratio indicates the atmospheric path length of sunlight, influencing its intensity and spectrum due to scattering and absorption. It is important to know in solar energy for predicting solar panel output and standardizing measurements. Understanding AM is essential for accounting for atmospheric effects on sunlight across various other disciplines, such as atmospheric sciences for studying optical properties and pollutants in astronomy.

For consistent solar cell comparisons, standard spectra and power densities are defined for radiation both outside Earth's atmosphere and on its surface. The standard spectrum on Earth's surface is AM1.5G (global, containing direct and diffuse radiation) or AM1.5D (direct radiation only). The AM1.5D radiation reduces the AM0 (the standard spectrum outside the Earth's atmosphere, in space) spectrum by 28%, resulting in approximately 970 W/m² for AM1.5G. For convenience, the standard AM1.5G spectrum is normalized to 1 kW/m².

The direct sunlight intensity varies daily and can be determined using a specific equation based on the AM. Since the atmospheric layer inconsistencies as well as altitude affects the solar irradiance, solar irradiance I_G at a specific elevation h (in km) can be calculated using the empirical equation given below, which is accurate enough to few kilometres above sea level.

$$I_G = 1.353\left[(1-0.14\,h)\,0.7^{AM^{0.678}} + 0.14\,h\right] \quad (2.2)$$

In the equation, the 1.353 kW/m² value represents the solar constant, with 70% of radiation reaching Earth. The additional 0.678 power term accounts for atmospheric layer inconsistencies. Some typical irradiance values at different locations are given in Table 2.4. Note that this effect is also visible to naked eye, as sunlight intensity rises with altitude, altering the spectral content and making skies appear bluer at higher elevations.

The atmosphere also influences the solar irradiance's spectral distribution, crucial for PV energy conversion, since solar cell efficiencies vary with the incident light's spectrum. Figure 2.7 illustrates the solar spectrum

TABLE 2.4 Sunbeam Normal to Horizon, as at Solar Noon on Summer Solstice at 23.5°N Latitude

Location	Irradiance
Above atmosphere (direct/total)	1.353 kW/m²
Desert sea level (direct)	0.970 kW/m²
Desert sea level (total)	1.050 kW/m²
Standard sea level (direct)	0.930 kW/m²

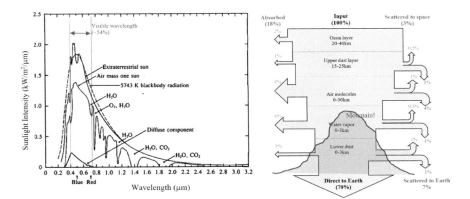

FIGURE 2.7 (a) The solar spectrum, and (b) the impacts of different atmospheric levels, at AM1, clear sky absorption and scattering of incident solar energy.

outside the atmosphere (AM0) and the AMI direct and estimated diffuse radiation spectra at sea level. The spectrum's variations are due to factors like ozone absorption, molecular scattering and absorption by other atmospheric gases.

2.2.4 Solar Irradiation Measurements

Solar irradiance measurements are fundamental to the successful design, deployment and operation of solar energy systems, which can provide insights into the potential energy generation, system performance and efficiency of solar installations.

Different instruments, each with its specific range of measurement and accuracy, are tailored for various applications, from solar thermal systems to large solar PV farms. Table 2.5 provides a detailed overview of these instruments, their utilization in different solar energy applications, their typical accuracy, cost, solar radiation flux density and wavelength range.

TABLE 2.5 Comparison of Solar Irradiance Measurement Instruments

Parameters	Pyranometer	Pyrheliometer	Radiosonde	Sunshine Recorder
Application in solar thermal	Measures total available solar radiation for potential heat generation.	Optimizes orientation of mirrors/lenses in concentrating systems.	Useful in high-altitude or cloudy regions.	
Application in rooftop solar PV	Assesses solar potential of rooftops and monitors system performance.	NA	NA	Tracks effective sunshine hours for monthly energy prediction.
Application in solar PV farms	Measures spatial variations in radiation across the farm.	Used in farms with sun-tracking systems.	Predicts short-term variations in energy generation.	NA
Typical accuracy	<1% error for high-precision models.	<0.5% error.	Specialized for detailed atmospheric data.	Relatively simple device; less detailed information.
Solar radiation flux density (W/m²)	0–2,000 W/m²	0–2,000 W/m²	Varies based on atmospheric conditions.	NA
Wavelength range (nm)	300–2,800 nm	200–4,000 nm	Broad spectrum	NA
Cost considerations	Ranges from a few hundred to several thousand dollars.	Often costs several thousand dollars.	More specialized; used primarily in research or specific applications.	Less expensive than pyranometers and pyrheliometers.

Note that calibration of such instruments is important in ensuring the accuracy and reliability of solar irradiance measurements. In solar thermal systems, calibration ensures precise system sizing and accurate performance prediction. For rooftop solar PV applications, calibration is needed for accurate performance estimation, especially given the smaller scale of these installations. In large solar PV farms however, regular calibration is not just a technical requirement but also a financial one since inaccuracies can significantly impact financial models as well as revenue projections.

Note also that the specific calibration process and frequency can vary based on the instrument in use and the calibration standard to which it is being compared.

Figure 2.8 provides a series of graphs detailing solar irradiance measurements over the course of a month, obtained using a pyranometer as described in a case study in Chapter 6. The measurements span across daylight hours, as indicated by the time of day on the x-axis, and the y-axis provides irradiance values in W/m^2. The bell-shaped curves typical of daily solar irradiance indicate the peak of solar activity around solar noon, while the variability in maximum irradiance indicates the impact of fluctuating weather conditions. Throughout a spring month, November, in Australia, there is a slight change about the average irradiance as well, which reflects seasonal changes in the sun's angle. The data's consistency also demonstrates a well-functioning pyranometer and systematic recording, including sudden meteorological changes and scattering due to clouds.

2.3 PV MATERIALS

As it is known and will be discussed later, PV arrays, comprised of numerous PV cells, are a cornerstone of DER. The PV technology landscape includes a variety of materials, each with unique efficiencies and characteristics tied to their chemical and physical make-up. Table 2.6 summarizes commercially available PV cell types, alongside some research innovations that are near commercialization at the time of preparation of this book.

Since the solar PV cell can be considered as a generator powered by light, our primary interest lies in understanding its operational behaviour and integration into larger energy systems, rather than discussing the intricate molecular details—a subject comprehensively addressed in other studies.

Furthermore, it is important to recognize that solar PV cells and arrays are seldom used in isolation. They are typically paired with compatible power electronics converters. These converters adjust the electrical output to match the load's requirements, aiming for maximum power delivery at peak efficiency. Therefore, the power transfer process relies on multiple components to achieve optimal "system efficiency", with the solar PV cell being just one crucial contributor.

It is worth noting that while new PV cell technologies emerge, the fundamental system configuration will likely to remain unchanged. Therefore, PV cell and converter advancements will likely translate to improved system efficiency and, consequently, increased power output from a given surface area.

100 ■ Reinventing the Power Grid

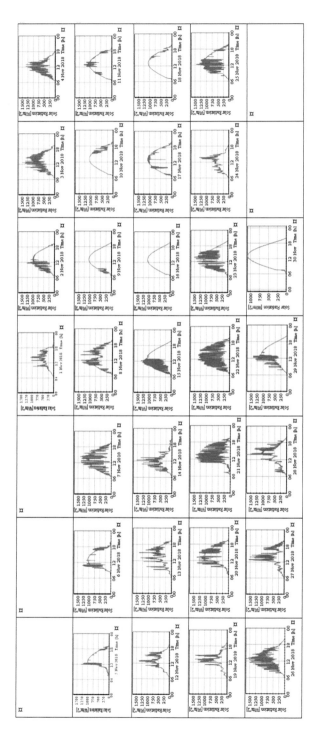

FIGURE 2.8 Monthly solar irradiance measured by a pyranometer (on Nov 2018, South Australia).

TABLE 2.6　A Summary of Commercially Available PV Cell Types

PV Material Type	Typical Efficiency	Remarks
Multi-junction solar cells	30%–40%	Can exceed 40% under concentrated sunlight. Complex and expensive, used in space applications.
Concentrator PVs	30%–40%	Very high efficiencies under concentrated sunlight; and expensive and complex to manufacture.
PSCs	15%–35%	Lab cells have achieved over 33%. Currently issues with long-term stability and sensitivity to air/moisture. It has potential for low-cost production.
Monocrystalline silicon	15%–22%	Highly stable with a long lifespan. Moderate cost due to mature manufacturing processes.
Polycrystalline silicon	13%–18%	Rigid, requires more space. Energy-intensive production.
Copper indium gallium selenide (CIGS)	12%–15%	Part of thin-film category. Can be made semi-transparent for building-integrated applications.
Cadmium telluride (CdTe)	10%–12%	Part of thin-film category. Generally lower cost than crystalline silicon.
Quantum dot solar cells	10%–13%	Potential for high efficiency due to multiple electron–hole pairs from a single photon. Concerns about stability and material toxicity
Dye-sensitized solar cells	8%–12%	Low-cost materials. Suitable for diffused light conditions but concerns about electrolyte leakage
Amorphous silicon (a-Si)	6%–10%	Part of thin-film category. Flexible, lightweight. Can be integrated into buildings.

2.4 OPERATION PRINCIPLES OF SOLAR PV CELLS AND PRACTICAL STRUCTURE

Although diodes, light-emitting diodes (LEDs) and solar PV cells serve different primary functions, they share many similarities (see Table 2.7) because of their foundational reliance on the PN junction and semiconductor materials. Identifying these similarities can help in understanding their commonalities and correlating their electrical characteristics.

It can be summarized that the fundamental behaviour of electrons and holes at the PN junction is crucial to the operation of diodes, LEDs and solar PV cells. All three devices typically possess a layered structure, but solar PV cell is designed to capture more solar irradiance, using large

TABLE 2.7   Comparison of the Primary Functions of Diodes, LEDs and Solar PV Cells

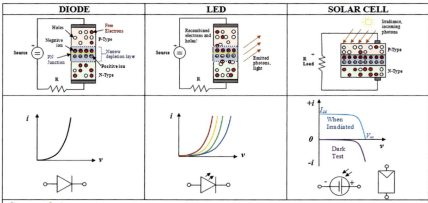

Common features:

N-Type (Negative Type) Semiconductor: Doped with donor impurities that have extra valence electrons. The extra electrons are free to move and are called "majority carriers." There are very few holes (absence of electrons), which are the "minority carriers."

P-Type (Positive Type) Semiconductor: Doped with acceptor impurities that have fewer valence electrons. This creates "holes" which are free to move and are the "majority carriers." There are very few free electrons, which are the "minority carriers."

Depletion Layer (Depletion region):
- It is formed at the PN junction due to the initial movement and recombination of majority carriers across the junction.
- When the PN junction is formed, free electrons from the N-type region move into the P-type region and recombine with holes. Similarly, holes from the P-type region move into the N-type region and recombine with electrons.
- This movement and recombination of carriers leave behind fixed, uncovered positive ions in the N-type region and negative ions in the P-type region creating the depletion layer.
- This region acts as an insulator and establishes an electric field that opposes further movement of majority carriers across the junction.

Diode:	Light Emitting Diode (LED):	Solar Photovoltaic (PV) Cell:
It is a two-terminal device made from a PN junction. When a positive voltage is applied (forward-biased) to the P-type and a negative voltage to the N-type, the depletion layer narrows, allowing current to flow from the P-type to the N-type.	It is a special type of diode that emits light when forward-biased. Electrons from the N-type recombine with holes in the P-type. As they recombine, they release energy in the form of photons (light). The colour of the light depends on the energy gap of the semiconductor material.	Note that a solar cell looks like a large flat diode, which converts sunlight into electricity also using a PN junction. Photons from sunlight hit the solar cell and can give enough energy to electrons to break free from their atoms. This creates electron-hole pairs. Due to the built-in potential of the PN junction, electrons are driven towards the N-type material and holes towards the P-type. This movement creates a DC voltage potential across the terminals hence an electric current which is used by the external circuits.

surface area. Note that the efficiency of these devices is influenced by the purity of the semiconductor materials. Common manufacturing techniques, such as doping, are shared among all. While silicon is currently a prevalent material in all three, compound structures using other materials are also employed. Quantum mechanics plays a pivotal role: in diodes, it governs electron movement; in LEDs, photon emission during recombination; and in solar cells, electron excitation by photons. All three devices exhibit unidirectional current flow, with solar cells also capable of "dark IV tests" that requires reverse current flow. Their performance is temperature-sensitive, with factors like the forward voltage drop in diodes and LEDs or the efficiency of solar cells being affected. Their unique voltage–current characteristics also reflect their specific functionalities in practice.

Solar PV cells operate fundamentally as diodes, p-n junction, designed to absorb photons from the sun and transform them into electrical energy. The term 'irradiance' describes the solar power available on a specific surface area, and it is quantified in watts per square meter. When photons possess adequate energy, they can free an electron within a PV material. If an electric field is in proximity, these electrons can be channelled towards a metallic contact, subsequently generating an electric current.

The bandgap is a fundamental property of semiconductors, critical in determining the efficiency of PV cells. It represents the energy difference between the top of the valence band and the bottom of the conduction band. This gap indicates the minimum energy required to move an electron from its bound state in the valence band to a free state in the conduction band. The bandgap determines the threshold energy required to excite an electron from the valence band to the conduction band. Photons with energy less than the bandgap will not be absorbed, while those with energy significantly greater than the bandgap will be absorbed but lose excess energy as heat. For example, silicon has a bandgap of about 1.1 eV, making it sensitive to a broad range of the solar spectrum. In contrast, gallium arsenide has a bandgap of about 1.43 eV, achieving higher efficiencies than silicon cells under concentrated sunlight. Strategies for optimizing PV efficiency based on bandgap include making tandem cells, using advanced materials like perovskites, introducing an intermediate band and using hot carrier cells.

2.5 IMPACTS OF TILTING AND TRACKING IN SOLAR PV PANELS

The tilt and orientation of solar panels also play a crucial role in their efficiency. While the ideal position is to face the equator at an angle equal to the location's latitude, practical considerations like roof structure and potential shading should be considered. Properly tilted and oriented panels can significantly increase energy generation and the overall efficiency of a solar PV system.

For maximum solar energy capture, panels should ideally be tilted towards the equator. In the Northern Hemisphere, this means facing south, and in the Southern Hemisphere, facing north. The optimal tilt angle is roughly equal to the latitude of the location.

If a roof's slope is already close to the ideal tilt angle for the location, panels can be mounted flush with the roof. If the roof's slope is steep or flat, adjustable mounting systems can be used to achieve the desired tilt.

Deviating from the ideal orientation (e.g., facing east or west instead of south in the Northern Hemisphere or instead of north in Southern Hemisphere) can reduce energy capture. However, east- or west-facing panels are equally beneficial since energy demand peaks in the morning or late afternoon.

Note that the tilt angle can influence how shadows are cast on the panels, especially from nearby obstructions such as trees, chimneys or neighbouring buildings. A steeper tilt might avoid shading from low obstructions but could be more susceptible to shading from taller objects. In addition, a tilt angle optimized for summer might result in more shading during winter when the sun is lower in the sky and vice versa.

As it is well known, the highest energy capture occurs when the sun is at its highest point in the sky known as solar noon. Therefore, panels can be oriented to the sun to maximize energy yield, which can be done by "solar trackers" that can adjust the tilt and orientation of panels throughout the day to face the sun directly, maximizing energy capture.

As it is illustrated in Figure 2.9, in the stationary systems, the panels are set at a fixed angle, ideally equal to the site's latitude, facing the equator. The insolation curve for a fixed tilt system shows a single peak around solar noon. The energy captured is less than tracking systems (Figure 2.10), especially during mornings and evenings. The curve will be symmetrical, reflecting the sun's consistent path across the sky.

In "single-axis horizontal tracking (SAHT)" systems, the panels rotate around a horizontal axis, allowing them to follow the sun's path from east to west (Figure 2.10a). The insolation curve is broader and flatter than the fixed tilt, capturing more energy during mornings and evenings. The total energy obtained throughout the day will be higher than fixed systems.

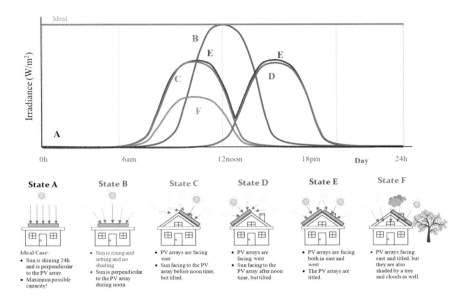

FIGURE 2.9 The fixed tilting scenarios and their impacts on solar irradiances.

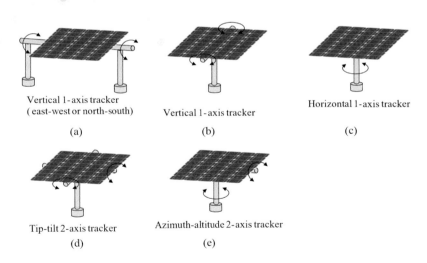

FIGURE 2.10 Principles of tracking methods showing the rotation angles and associated axis.

In "single-axis vertical tracking (SAVT)" systems, the panels rotate around a vertical axis (Figure 2.10b), which is less common than horizontal tracking but can be useful in higher latitudes. The curve will show variations specifically during different seasons; hence, it can capture more energy during seasons when the sun takes a lower path across the sky.

In "dual-axis tracking (DAT)" systems (Figure 2.10d and e), the panels can rotate both horizontally and vertically. This enables them to face the sun directly throughout the day and throughout the year. Therefore, the insolation curve will be the broadest and flattest among all systems, capturing the maximum possible energy throughout the day. Peaks will be more pronounced, reflecting the system's ability to maintain optimal orientation relative to the sun.

Finally, in "tilted single-axis tracking (TSAT)" systems (Figure 2.11c), the axis of rotation is tilted from the horizontal, which is a compromise between SAHT and DAT. The insolation curve will lie between SAHT and DAT, broader than SAHT but not as flat as DAT. This system captures more energy than SAHT, especially during seasons when the sun's elevation changes significantly.

In Figure 2.11, insolation patterns visually represent the energy capture efficiency of each tracking method given in Figure 2.10. A broader curve indicates more energy capture during mornings and evenings, while the height of the curve at solar noon indicates the system's efficiency during peak sunlight. The area under each curve represents the total energy captured throughout the day.

Table 2.8 provides a concise comparison between single-axis and dual-axis tracking systems in terms of cost, energy yield and other factors. As it is summarized in the table while both single-axis and dual-axis tracking systems come with a higher initial investment compared to fixed-tilt

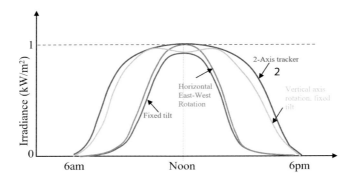

FIGURE 2.11   Typical insolation patterns and impacts of solar tracking during a sunny day.

TABLE 2.8 Comparison between Single-Axis and Dual-Axis Tracking Systems

Factors	Single-Axis Tracking	Dual-Axis Tracking
Cost	10%–25% higher than fixed tilt.	20%–35% higher than fixed tilt.
Energy yield	15%–25% compared to fixed tilt.	25%–40% compared to fixed tilt.
Initial investment	Higher due to tracking system.	Higher due to more complex tracking system.
Maintenance	Moderate (due to moving parts).	Higher (more complex system with more moving parts).
Land use	Requires more spacing between panels, hence needs more land.	Requires more spacing to avoid shading, hence much more land.
System longevity	Potentially extended due to increased energy yield.	Potentially extended due to increased energy yield.
LCOE	Potentially lower due to increased energy yield.	Potentially lower due to increased energy yield.

systems, they offer a significant increase in energy yield. This enhanced energy yield has the potential to reduce LCOE over the system's lifespan. However, dual-axis systems, being more complex, might incur higher maintenance costs and require more land spacing. In solar PV farms, the single-axis tracking system is more commonly used, due to a balance between increased energy yield and cost-effectiveness and reliability.

2.6 ELECTRICAL EQUIVALENT CIRCUIT OF SOLAR PV CELLS

As it is known, solar PV cells, the fundamental building blocks of solar panels, convert sunlight directly into electricity. To understand and model the intrinsic behaviour of such cells, electrical equivalent circuits are often used, which provide a simplified representation of the complex interactions between photons and semiconductor materials within the cell.

If the inherent characteristics of the cell, such as current–voltage relationships, are captured using an equivalent circuit, it can assist in diagnostic evaluations, predictive modelling and tolerance analysis. For instance, if a PV system underperforms, the equivalent circuit can help determine whether the issue arises from the cell's intrinsic properties or external factors. In addition, the ideal performance derived from the equivalent circuit, combined with initial benchmark real data, can be used to identify faults and failures in the PV cell (or module).

In real-world implementations, external factors like shading, temperature variations, soiling and installation discrepancies play a significant role in the system's performance, which can introduce variations and uncertainties that may overshadow the minute details captured by the equivalent circuit. However, such circuits not only provide a conceptual framework for those learning about solar PV technology but also ensure that before external factors come into play, the PV cell or module is inherently optimized.

The equivalent circuit allows for the prediction of the PV cell's behaviour under different conditions, such as varying illumination or temperature. Moreover, by understanding the equivalent circuit, one can optimize the cell's performance. In addition, when integrating PV cells (in the form of PV panels) into larger systems or when designing power electronics that interface with PV cells, the equivalent circuit provides essential information for ensuring efficient operation and to develop control algorithms. Furthermore, the equivalent circuit can highlight potential areas of inefficiency or loss in the PV cell, guiding future improvements.

2.6.1 Blackbox Testing and Identification of Equivalent Circuit Parameters

The electrical equivalent circuit of the solar PV cell or array can be defined after experimentally obtaining its electrical characteristics, voltage and current relationships. Note that there is a parallel to conventional generator tests in solar PV "generators".

An approach that can be used for the development of an equivalent circuit for solar PV cells is illustrated in Figure 2.12. Note that when the terminals of the PV cell are shorted, the current that flows is called the short-circuit current (I_{sc}). This represents the maximum current the cell can produce at zero voltage. Similarly, when the PV cell is not connected to any load, it exhibits an open-circuit voltage (V_{oc}). This is the maximum voltage the cell can produce at zero current, which is highlighted in Figure 2.12a.

At its simplest, the equivalent circuit of an ideal solar PV cell consists of an ideal current source (representing the photocurrent) in parallel with an ideal diode that represents the p-n junction behaviour of the cell. As it is illustrated in Figure 2.12b, such an ideal equivalent circuit has an ideal current source that delivers a constant current (CC) regardless of the voltage across its terminals and an ideal voltage source represented by the diode's potential that provides a constant voltage (CV) irrespective of the current drawn from it.

However, no voltage sources (such as batteries and power supplies) are truly ideal due to their internal resistance, which can cause the voltage

Photovoltaic Solar Energy ■ 109

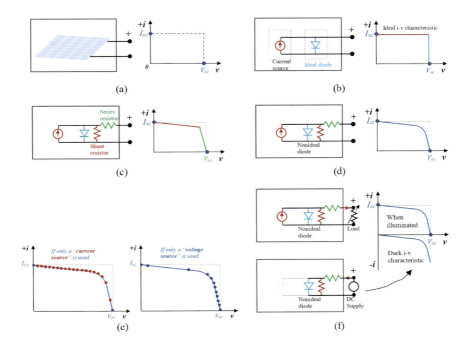

FIGURE 2.12 An approach for the development of an equivalent circuit for solar PV cells.

to drop under high current loads. Similarly, real-world current sources are not ideal. They have limitations in terms of the range of voltages over which they can maintain a CC. Note also that only photodiodes under certain conditions and MOSFET transistors in specific configurations can act as current sources.

Therefore, it can be concluded that the measured load characteristics of a cell are far from an ideal shape. To understand the behaviour of the PV cell under varying loads, a loading test has to be conducted by connecting varying values of resistor loads across the terminals. Since small changes in current can cause large voltage variations up to the "knee point" (maximum power point (MPP)) and small changes in voltage can cause large variations beyond the knee point, it is crucial to vary the load accurately. This requires precise variation of load, which means both low resistance (for high current regions) and high resistance (for high voltage regions) need to be adjusted with precision, which is illustrated in Figure 2.12e.

To represent the diversion from the ideal current and voltage source behaviour, a series and a shunt resistor can be added, which represents the internal resistive losses and the leakage current in the cell, respectively (see Figure 2.12c).

To better match the measured data, especially around the knee point, a more realistic diode model can be introduced. This model accounts for the nonlinear behaviour of the diode, ensuring the curve of the equivalent circuit matches the actual I-V characteristics more closely (Figure 2.12d).

The "dark I-V characteristic" refers to the current and voltage curve of a solar PV cell or module when it is measured in the absence of illumination (see Figure 2.12f), which is highly practical to achieve in practice. This characteristic provides valuable information about the cell's intrinsic properties, such as its diode ideality factor (which indicates how closely the PV cell's behaviour matches that of an ideal diode), series resistance and shunt resistance. Note that in dark I-V measurements, the current is flowing in the opposite direction and the current paths are different, hence causing a lower series resistance in the dark measurements to the light measurements.

It is worth noting that series resistance (R_s) is due to the material of the PV cell, contacts and interconnections and it affects the slope of the I-V curve near the MPP. In addition, parallel (shunt) resistance (R_p) represents the path that the current may take instead of passing through the load, which affects the slope of the I-V curve near the open-circuit voltage as illustrated in Figure 2.12e.

While a single diode model provides a simplified representation of a solar cell's behaviour, multiple diode models offer a more comprehensive and accurate representation, especially for advanced solar cell technologies and detailed performance analysis. Note that the purpose of using multiple diode models in the solar PV equivalent circuit is to more accurately represent the complex behaviour of real solar cells and improve the accuracy of the I-V (current–voltage) characteristic curve. There are various reasons for using multiple diode models: better representation of non-idealities, improved accuracy, to capture different recombination paths (which can be represented by a different diode), versatility (to represent various types of solar cells, from monocrystalline to polycrystalline to thin-film cells, each of which may have different dominant recombination mechanisms) and a better temperature and irradiance variations.

Table 2.9 provides a summary of the different equivalent circuit models used for solar PV cells and panels, along with remarks on their accuracy (based on typical literature) and a brief description.

TABLE 2.9 A Summary of the Different Equivalent Circuit Models

Equivalent Circuit Type	Remarks
Ideal single diode model	Basic representation. Cannot capture all real-world behaviours. Accuracy: ±10%–20% for cells, ±20%–30% for panels.
Single diode with series and shunt resistance model	Captures resistive losses. Widely used for practical applications. Accuracy: ±5%–10% for cells, ±10%–15% for panels
Two (or multiple) diode model	Captures multiple recombination mechanisms. Used for high-efficiency cells. Accuracy: ±2%–5% for cells, ±5%–10% for panels.
Empirical models	Derived from experimental data without a theoretical basis. They fit specific I-V curves but may not generalize well across different conditions or cell types. Accuracy: ±1%–3% for specific cells, ±10%–15% for specific panels.
Distributed resistance models	Accounts for resistances that are distributed throughout the cell or panel, rather than being lumped at one point. Useful for larger cells or panels where resistances can vary across the device. Accuracy: ±5%–8% for larger cells, ±3%–6% for panels
Complex models	Incorporate advanced physical mechanisms or are tailored for specific advanced PV technologies. They can be computationally intensive but offer detailed insights, especially for novel or high-efficiency technologies. Accuracy: ±1%–2% for advanced cells, ±2%–4% for panels.

Although the single diode model is accurate enough to understand the behaviour of basic solar PV cells and their combination behaviour in a panel arrangement, the impedance spectroscopy is an alternative method for modelling further complexities of solar PV cells, which identifies impedance components like resistance, capacitance and inductance. The technique focuses on the interaction between electromagnetic fields and the photoactive materials in the cells. The method is especially suitable in characterizing perovskite solar cells (PSCs) to understand their degradation mechanisms. By integrating electrochemical impedance measurements with MPPT, such cells are comprehensively characterized, explaining the behaviour of charge carriers during operation, which is crucial in optimizing the performance and durability of solar cells. The process typically involves overlaying a small AC voltage onto a DC bias and observing the changing current, which helps determine the cell's impedance at different frequencies. At high frequencies, the impedance matches the series resistance, whereas at lower frequencies, it provides information about the shunt resistance. Moreover, exposing the PV cell to sudden changes in light or load and monitoring its response to identify its internal capacitances and inductances, an accurate modelling of temperature impacts is obtained.

Note that empirical model of a specific solar PV panel derived from experimental data can be as accurate as complex models.

2.6.2 Analysis of Equivalent Circuits for PV Cells

As outlined in Table 2.7, a solar cell fundamentally acts as a diode. Its diode-like behaviour becomes clear when examining its I-V characteristics in the absence of light. Under illumination, the solar cell maintains its diode properties, but with an added light-induced current (photocurrent). This shifts the I-V curve. The intersection of the curve with the vertical axis (zero voltage) represents the short-circuit current (I_{sc}) that is the cell's peak current under light. Where the curve intersects the horizontal axis (zero current) indicates the open-circuit voltage (V_{oc}).

The equivalent circuit given previously in Figure 2.12d can be dissected with increasing detail to emphasize the effects of each circuit component, along with temperature and light factors crucial for real-world operation. These circuits and their corresponding solutions are summarized in Table 2.10.

It is important to mention that a basic equivalent circuit for a PV cell consists of a current source, powered by sunlight, in parallel with an actual diode. From the load's perspective, R side, a real PV cell exhibits two specific states: short circuit and open circuit. The fundamental methodology presented here considers that by defining the terminal current I and the terminal voltage V as a function of I_{sc} and V_{oc} using the real diode characteristic, the I-V characteristic of the solar PV cell can be obtained.

Table 2.10 contains the same approach for two additional circuits: simple equivalent circuit with a parallel resistor and simple equivalent circuit with both parallel and series resistors, which represent a closer match for the measured I-V characteristics.

In solar PV cells, the series resistance, R_s, and parallel resistance, R_p, play significant roles in determining the performance of the cell.

R_p is placed in parallel with the diode to consider imperfections and leakages in the cell. Ideally, R_p would be infinitely large (means no leakage), but in real cells, it has a finite value due to the various reasons including material imperfections, surface recombination of electrons and holes at the surface of the solar cell (hence a leakage current that flows across

TABLE 2.10   The Summary Solutions of the Equivalent Circuits for Solar PV Cell

CIRCUIT NAME	EQUIVALENT CIRCUITS	SOLUTIONS AND REMARKS
The simple equivalent circuit (with a current source and single diode)	In the circuit: I_d is the diode current, V_d is the diode voltage, I is the solar cell's output current, V is the voltage across the solar cell and R is the load resistor. Where the diode equation can be given as $$I_d = I_o\left(e^{\frac{qV_d}{nkT}} - 1\right)$$ I_o is the saturation current in amps that depends on the type, doping density and junction. n is the diode ideality factor q is the electron charge (1.602×10^{-19} C) k is Boltzmann's constant (1.381×10^{-23} J/K), T is the temperature of the cell (K). and for a junction temperature of 25°C $$I_d = I_o(e^{38.9 V_d} - 1)$$	$I = I_{sc} - I_d$ hence $$I = I_{sc} - I_o(e^{38.9 V_d} - 1)$$ $V = V_d$ Note that V_{oc} can be calculated when $I = 0$ and $V_d = V_{oc}$ hence $$V_{oc} = 0.0257 \ln\left(\frac{I_{sc}}{I_o} + 1\right)$$ Since I_{sc} is directly proportional to solar irradiation, I-V curves can be plotted for varying solar irradiance.
The simple circuit with a parallel resistor		$I = I_{sc} - I_d - I_p$ $I = I_{SC} - I_d - \dfrac{V}{R_p}$ $$I = I_{sc} - I_o(e^{38.9 V_d} - 1) - \dfrac{V}{R_p}$$ $V = V_d$ $\Delta I = V/R_p$
The simple circuit with a series resistor.		$I = I_{sc} - I_d = I_{sc} - I_o\left(e^{\frac{qV_d}{kT}} - 1\right)$ $V_d = V + I\,R_s$ $I = I_{sc} - I_o\left(e^{\frac{qV_d}{kT}} - 1\right)$ $\Delta V = I\,R_s$
The simple circuit with a parallel and a series resistors.		$I_{sc} = I + I_d + I_p$ $I = I_{sc} - I_o(e^{38.9 V_d} - 1) - \dfrac{V_d}{R_p}$ $V = V_d - I\,R_s$ $\Delta I = V/R_p$ $\Delta V = I\,R_s$

the cell rather than through the load), edge leakage (if not well-finished or sealed) and partial shading.

A high parallel resistance is desirable to minimize current leakage in the cell, ensuring that most of the generated current is available for extraction. Conversely, a low parallel resistance indicates significant current leakage, which can reduce the cell's efficiency.

The series resistance, R_s, is primarily due to the resistance of the front and back contacts and the resistance of the semiconductor material.

The typical values of R_p and R_s can vary based on the type, quality and manufacturing process of the solar cell.

Typical values for R_s in crystalline silicon solar cells are in the range of 0.1–11 Ω/cm^2. For thin-film solar cells, the R_s can be higher, often in the range of 1–10 Ω/cm^2 or more, depending on the thickness and quality of the material.

A higher R_p value is desirable as it indicates lower leakage currents. For high-quality crystalline silicon solar cells, R_p can be several hundreds to thousands of Ω/cm^2. For thin-film solar cells or lower quality crystalline cells, R_p can be lower, often in the range of 10–1,000 Ω/cm^2.

Furthermore, the I-V characteristics of a solar PV cell have to be calculated for temperatures other than the standard 25°C, which are summarized in Table 2.11.

In addition, to plot I-V characteristics at different temperatures, approximations can also be used by recalculating the saturation current

TABLE 2.11 The Methods to Identify Temperature Effect on the Equivalent Circuit

Parameter and Method	Description and Equations
Temperature Effects on the Equivalent Circuit	
Diode saturation current, I_0	The saturation current increases with temperature due to increased intrinsic carrier concentration in the semiconductor.
Short-circuit current (photocurrent), I_{sc}	Photocurrent increases with temperature because the semiconductor's bandgap decreases, leading to increased photon absorption.
Open-circuit voltage, V_{oc}	V_{OC} decreases with temperature due to the increase in the diode's saturation current.
Series resistance, R_s	Series resistance can increase with temperature due to increased resistivity of the semiconductor material.
Parallel resistance, R_p	Parallel resistance can decrease with temperature due to increased intrinsic carrier concentration, leading to increased leakage currents.

(Continued)

TABLE 2.11 (*Continued*) The Methods to Identify Temperature Effect on the Equivalent Circuit

Parameter and Method	Description and Equations
Calculating I-V Characteristics at Different Temperatures	
Calculating I_{sc} and V_{oc}	$I_{sc}(T) = I_{sc}(25°C) + \beta I_{sc}(T-25)$
	$V_{oc}(T) = V_{oc}(25°C) + \beta V_{oc}(T-25)$
	Temperature coefficients for the short-circuit current (βI_{sc}), open-circuit voltage (βV_{oc}) and maximum power (βP_{max}) are commonly provided by manufactures.
Calculating I_0	$I_0(T) = I_0(25°C)\left(\dfrac{T}{25}\right)^3 e^{-\frac{E}{k}\left(\frac{1}{T} - \frac{1}{25}\right)}$
	Here, E is the bandgap energy and is known for semiconductor materials.
	Note that the bandgap of semiconductors typically decreases with increasing temperature, which can be described by the Varshni equation:
	$E(T) = E(0) - \dfrac{\alpha T^2}{T + \beta}$
	where $E(T)$ is the bandgap at temperature T, $E(0)$ is the bandgap at 0 K, *and* α and β are material-specific constants. By measuring the bandgap at different temperatures and fitting the data to this equation, the constants α and β can be determined, and the bandgap at any temperature can be calculated.
Using the single-diode model	Calculate the I-V characteristics at different temperatures by adjusting the parameters I_0, I_{sc}, R_s and R_p for the new temperature.

for each temperature. This requires the diode equation $I_d = I_0 (\exp(qV_d/kT) -1)$, where I_0 saturation current at 25°C is commonly given and other constants are known.

Note that the respective I-V and power–voltage (P-V) characteristics for each temperature will intersect at the point where V_d and I_d values are equal on both plots. Therefore, we can use the saturation current at 25°C and use ratios to determine the unknown saturation currents at other temperatures.

2.7 PV CELLS TO MODULES TO ARRAYS

A solar PV cell, primarily made from silicon (as detailed in Table 2.6), is the fundamental unit converting sunlight into direct current (DC) through the PV effect. Using individual solar cells directly is not practical due to reasons such as limited power output, fragility, the necessity for a bypass diode (to be discussed later) and installation difficulties. To address these issues, solar panels house multiple cells, increasing power, protection, efficiency and easy to handle. As shown in Figure 2.13, multiple cells are connected in series and parallel (with bypass diodes that will be discussed later) and encapsulated between layers of a clear adhesive, often ethylene-vinyl acetate (EVA). This encapsulant shields the cells from external elements and sandwiches them to the panel's front tempered glass and rear plastic back sheet. An anti-reflective coating enhances sunlight absorption, and the encapsulation safeguards the cell from environmental factors such as moisture. The front tempered glass ensures maximum light penetration while providing resilience against environmental factors such as hail. In addition, the rear plastic back sheet adds insulation and seals the panel from external conditions. An aluminium frame holds the entire assembly and offer structural support and easing mounting. Furthermore, a junction box at the panel's rear contains electrical terminals as well as bypass diodes. The figure of the solar PV panel in Figure 2.13 also highlights that front and back contacts allow current to exit the cell, powering an external circuit via the terminals. Some systems employ concentrating lenses or mirrors to focus sunlight onto a smaller cell, necessitating tracking for sun alignment.

Note also that there are bifacial solar PV cells and associated panels with a transparent back, which involves a double glass structure or a clear back sheet to facilitate the dual-sided absorption. The transparent back layer is made of transparent polymers or glass. The use of high-purity silicon and advanced cell architectures, such as passivated emitter and rear cell (PERC) can further enhance the efficiency of bifacial cells.

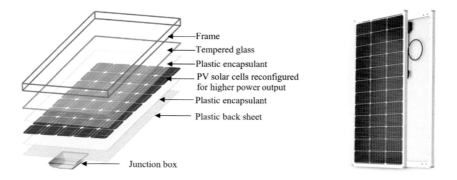

FIGURE 2.13 The components of a "monofacial" solar PV panel structure.

Bifacial solar PV cells generally have a higher efficiency than traditional monofacial cells. This is because they can harness sunlight from both their front and rear sides, capturing direct sunlight and reflected light from the surrounding environment. Depending on the installation and the amount of reflected light available, bifacial modules can produce up to 5%–30% more energy than their monofacial counterparts.

Although bifacial solar PV cells are the most effective, they need to be installed in environments where there is a significant amount of light that can be reflected onto the rear side of the panel. This includes installations over light-coloured or reflective surfaces (such as white rooftops, sand or other light-coloured grounds). Elevated mounting systems are often used to allow sunlight to reach the rear side of the panel. Ground-mounted bifacial systems, especially in large solar farms, can also benefit from the albedo (reflectivity) of the ground.

An equivalent circuit of bifacial solar PV cell is given in Figure 2.14. Note that the electrical equivalent circuit of a bifacial solar PV cell is similar to that of a standard monofacial solar cell but with some modifications to account for the bifacial nature. To determine the photocurrents individually, the short-circuit current of each side should be measured, while the other side is masked or not exposed to light. This gives $I_{sc(front)}$ and $I_{sc(back)}$, which approximate $I_{photo(front)}$ and $I_{photo(back)}$, respectively. When both sides of the solar PV cell are exposed to light, the total I_{sc} of the cell will be approximately the sum of the photocurrents from both sides, that is, $I_{sc} \approx I_{photo(front)} + I_{photo(back)}$. This assumes minimal interaction between the two sides, which is generally a good approximation for bifacial cells.

As illustrated in Figure 2.13, a solar PV panel is formed by assembling multiple solar cells together. The typical sizes of standard square cell wafer size are 156, 166, 182 and 210 mm depending upon the technology used. Therefore, the power ratings of practical cells also vary between 2 and 4.7 W for polycrystalline cells and between 3 and 8 W for monocrystalline

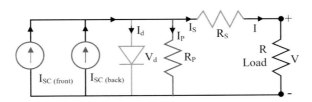

FIGURE 2.14 Single diode equivalent circuit of a bifacial solar PV cell.

cells. Commonly used panel arrangement using individual PV cells are illustrated in Figure 2.15 together with a typical by pass diode arrangement.

However, during the evolution of solar PV technology, a variety of panel sizes and power ratings are emerged, which aimed different applications, from residential rooftops to large utility-scale installations. While specific products from individual manufacturers might have unique specifications, it is possible to generalize the standard sizes and power and voltage ratings. Table 2.12 provides an overview of such products with their primary specifications.

Solar PV array configurations refer to the way the above illustrated solar panels are electrically connected to achieve much higher voltage and power ratings that are critical to match the DC/DC converters' and inverters' inputs. In array arrangements, the two primary configurations are series and parallel and a combination of both, known as a series–parallel configuration. As stated above, the choice of configuration depends on the specific requirements of the installation, whether it is a rooftop or a solar farm as summarized in Table 2.13.

Additional considerations when choosing a solar PV array configuration include the use of microinverters and power optimizers, which maximize

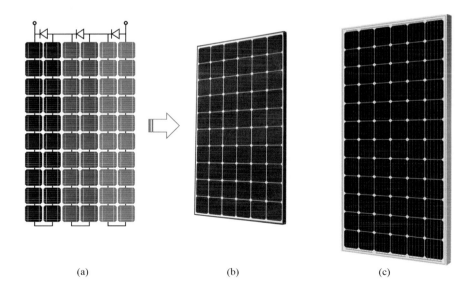

FIGURE 2.15 A commonly used panel arrangement using individual PV cells with three by pass diodes (a), a standard rooftop panel with 60 cells (b) and a standard commercial panel with 72 cells (c).

TABLE 2.12   General Overview of Standard Solar PV Panel Sizes and Power Ratings

Panel Type	Size (Height × Width)	Typical Use	Power Range	Voltage Range (V_{oc})
60 cells	1.65 × 1 m	Residential	250–330 W	30–40 V
88 heterojunction half-cut cells	1.73 × 1.2m	Residential	450–470 W	~65 V
72 cells	2 × 1 m	Commercial and utility	340–450 W	40–50 V
120 half-cut cells	1.65 × 1 m	Residential	280–350 W	30–40 V
144 half-cut cells	2 × 1 m	Commercial and utility	360–480 W	40–50 V
96 cells	Varies	Premium/high efficiency	320–420 W	50–60 V
104 cells	Varies	Premium/high efficiency	>320 W	55–65 V
High power range	Larger sizes	Utility scale	700–900 W	50–70 V

TABLE 2.13   Solar Array Configuration Choices in Residential and Commercial Applications

Configuration	Rooftop Domestic: 1–10 kW Commercial: 10–100 kW	Solar Farm (Solar Parks) Small Scale: 1–10 MW Medium Scale: 10–50 MW Large Scale: 50 MW to over 1 GW
Series connection	• To increase voltage under low current that reduces power losses in cables when the inverter is distant from panels.	• Used when panels are strung together before connecting to a central inverter.
Parallel connection	• To increase current while voltage remains the same, which is effective with shaded locations.	• Less common due to high currents, hence costly cabling.
Combination of series–parallel connection	• Groups of panels in series form strings then connected in parallel. • Used in larger installations to balance benefits, especially for varying orientations or shading issues	• Very common, allowing flexibility, optimizing power output and tailored to inverter requirements as well as the landscape topology.

the output of each panel and are more common in rooftop set-ups, though they can also be found in some solar farms. The chosen configuration primarily depends on the specific voltage and current requirements of the inverter chosen. Furthermore, while centralized systems can prefer simpler configurations to monitor, residential set-ups often aim to maximize the output of individual panels.

Figure 2.16 provides a visual representation of how connecting PV cells in series or parallel affects their collective electrical behaviour in a panel arrangement, specifically in terms of current and voltage outputs. For

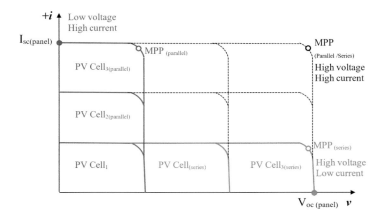

FIGURE 2.16 Connecting PV cells in series or parallel and their effects on electrical quantities.

example, the voltage at the open-circuit point (V_{oc}) for the series combination is cumulative, summing the V_{oc} of each individual cell. In addition, the total short-circuit current (I_{sc}) for the parallel combination in a panel is the sum of the I_{sc} of each individual cell. Therefore, the overall panel (or array) will also have its MPP, as indicated on the graph.

For domestic rooftop applications, localized energy production is a significant advantage as producing energy where it is consumed can lead to reduced transmission losses and lower infrastructure costs. Additionally, many regions offer financial incentives such as tax credits, rebates and feed-in tariffs for homeowners who install solar panels, making the investment financially attractive. On the other hand, solar PV farms benefit from economies of scale, where large-scale production can bring down the cost per unit of electricity. These farms should be strategically located in areas with the highest solar irradiance, ensuring maximum energy production, and preferably on non-arable lands.

As it was mentioned previously, solar PV panels come in various types, each suited to specific applications. Monocrystalline panels, known for their high efficiency and sleek black appearance, are commonly used in residential and commercial installations. Polycrystalline panels, with their distinctive blue hue, are also popular in these settings but are generally less efficient than their monocrystalline counterparts. Thin-film panels, which can be flexible and lightweight, are ideal for portable applications and BIPVs. Emerging technologies, such as bifacial panels, capture sunlight from both sides and are finding unique applications like vertical installations on highways. As the solar industry evolves, a blend of such technologies is applied in innovative ways to maximize energy production across various settings. The common and emerging applications are summarized in Table 2.14, which support both electrification trend and grid transformation.

TABLE 2.14   Common and Emerging Applications of Solar PV Cells, Panels and Arrays

Application Areas	Benefits and Features
Residential rooftop	• Installed on individual homes • Grid-tied or with battery backup
Commercial rooftop	• Larger installations on businesses or factories • Often grid-tied
Solar farms	• Large-scale ground-mounted installations • Feeds into the grid
Solar parking canopies	• Provides shade while generating power • Often found at public parking areas
Portable solar	• Small, flexible panels • Used for camping, hiking or temporary set-ups
BIPV	• Integrated into building materials like roof tiles, windows or facades • Aesthetic and functional
Solar street lights	• Standalone units with integrated battery • Used in public areas, roads and pathways
Solar water pumps	• Used for irrigation in remote areas • Usually directly powered by solar during the day
Off-grid remote systems	• Provides power in remote areas without grid access • Often paired with a battery storage system
Agrivoltaics	• Dual-use of land (agriculture + energy production) • Reduction in water usage due to shading • Improved crop yield in certain conditions due to moderated temperatures • Enhanced PV efficiency due to cooler ambient conditions beneath
Solar PVs in greenhouses	• Consistent energy supply for greenhouse operations • Reduced dependency on external power sources • Allows for controlled environment agriculture with light filtering capabilities
Solar PVs on railways	• Efficient use of space along and between railway tracks • Direct energy supply to train stations and other railway infrastructure • Potential to feed excess energy back into the grid

(*Continued*)

TABLE 2.14 (*Continued*) Common and Emerging Applications of Solar PV Cells, Panels and Arrays

Application Areas	Benefits and Features
Solar PVs in irrigation channels and floating solar	• Prevention of water evaporation • Land conservation • Natural cooling effect for improved PV efficiency • Direct energy for irrigation pumps • Reduction in algae growth
Bifacial vertical panels on highways	• Dual-sided energy harvesting enhancing efficiency • Sound barrier for traffic noise • Efficient use of space along highways • Visual aesthetic enhancement • Potential road lighting or other infrastructure power needs
Solar PVs over bike paths	• Efficient use of space above bike paths • Shading and protection for cyclists from direct sunlight • Potential lighting for paths during low light conditions • Enhanced visual aesthetics and promotion of green transportation

2.7.1 The Current–Voltage (I-V) and Power–Voltage (P-V) Characteristics of PV Panels (Building Blocks) under Standard Test Conditions

In standard test conditions (STC) for solar panels, a controlled light source that closely simulates natural sunlight is used to illuminate the solar panel being tested. The primary method of generating this light source is through the use of specialized equipment known as a solar simulator. Solar simulators are designed to produce light with characteristics that closely match those of sunlight under specific test conditions. They provide uniform illumination across the entire surface of the solar panel to ensure accurate testing results. During the tests, current and voltage sensors are used to measure the electrical characteristics of the solar panel. The simulator also matches the spectrum of the light emitted to match the AM1.5 spectrum which ensures that the testing conditions closely resemble real-world solar radiation. The following light sources are used in solar simulators: xenon arc lamps, halogen lamps and LEDs (with specific wavelengths to achieve the desired spectrum). In addition, filters and reflectors are used to adjust the spectrum and intensity of the light to match the STC conditions. The uniform illumination across the surface of the solar panel is collimated to create parallel rays, similar to sunlight.

Furthermore, the intensity of the light source is adjusted to achieve the irradiance to the desired level (typically 1 kW/m² for STC). To achieve the required cell temperature of 25°C during testing, the solar panel is often placed on a temperature-controlled surface or in a controlled chamber.

Figure 2.17 illustrates I-V and P-V characteristics of a commercial solar PV panel tested using a solar simulator. Note that, in the test, both voltage source and current source have been used to obtain the measurements. In the voltage source region, the voltage is actively controlled by the measurement system, where the current changes significantly with small changes in voltage. In the current source region, the current is also actively controlled by the measurement system, where the voltage changes significantly with small changes in current.

The figure provides the nameplate data as well as mechanical and thermal characteristics of the panel, and calculated values using I-V test results are also given in the figure. In general, these curves are useful for designing

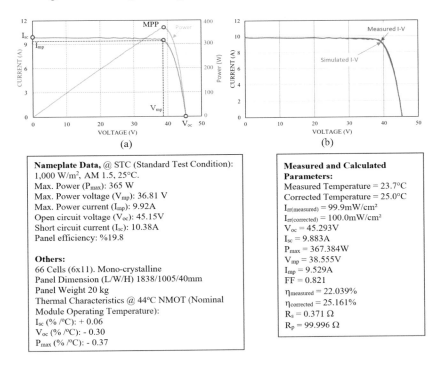

FIGURE 2.17 Measured I-V and P-V characteristics of a commercial solar PV panel using a solar simulator (a); its measured and simulated I-V characteristics (b) and its data sheet and measured/calculated characteristic electrical parameters.

solar energy systems with PV arrays as they can be optimally match with inverters and other system components. Manufacturers and researchers also use this information to evaluate and compare different PV panels and to ensure that the panel meets specified performance standards.

Upon obtaining the I-V curve for a solar panel, various parameters can be determined. These include the open-circuit voltage (V_{oc}), which is observed when the current is zero (representing a no-load case), and the short-circuit current (I_{sc}), which occurs when the terminal voltage is zero. In addition, the MPP and its corresponding voltage and current values (V_{mp} and I_{mp}, respectively) can be identified. From the measured data, the values for series and parallel resistances can be obtained. Another crucial parameter is the fill factor (FF), which serves as a measure of the "squareness" of the I-V curve. This quantity essentially indicates how closely the I-V characteristics of the panel align with those of an ideal solar PV. The FF is calculated by taking the ratio of the actual maximum power from the module to the product of V_{oc} and I_{sc}.

2.7.2 Impacts of Irradiance and Temperature on I-V Characteristics

The I-V characteristic of a solar panel is influenced not just by the amount of sunlight the panel receives but also by the temperature of the solar cells. This behaviour can be observed from the diode characteristics in the panel's equivalent circuit. Figure 2.18 summarizes these characteristic changes, at a constant temperature of 25°C and at a constant irradiance of 1 kW/m².

Note that as irradiance decreases (from 1 to 0.2 kW/m²), both short-circuit current and open-circuit voltage decrease. However, the reduction in Isc is more pronounced compared to Voc. Consequently, the maximum power output also falls significantly.

The characteristic curves at a constant irradiance of 1 kW/m² exhibit unique behaviour across different temperatures (−25°C to 75°C). As the temperature increases, the open-circuit voltage decreases. This is a result of the increased intrinsic carrier concentration in the p-n junction of the cell with temperature. On the contrary, the short-circuit current experiences a slight increase with rising temperature because of augmented carrier mobility. Therefore, it can be concluded that the overall effect of increasing temperature is significant on the efficiency of the solar cell. As expected, the overall power output decreases with increasing temperature, which is evident from the decreasing height of the peaks.

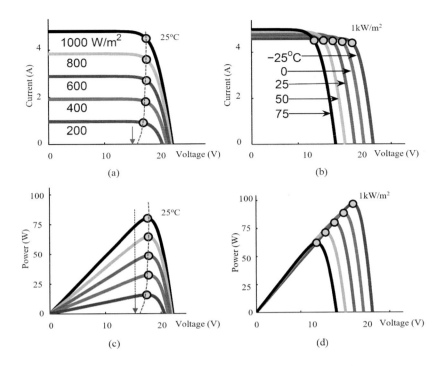

FIGURE 2.18 Characteristics of a solar PV panel: I-V curves under varying irradiances (a) and temperatures (a and b), and corresponding P-V curves (c and d).

As irradiance diminishes from 1 kW/m² to 200 W/m², the short-circuit current proportionally decreases. While the open-circuit voltage also reduces, this reduction is less pronounced than the change in current. In addition, the peak power value experiences a significant decline as irradiance drops.

2.7.3 Nominal Operating Cell Temperature (NOCT)

As it was seen, the temperature of a PV cell is a critical factor that influences its efficiency and performance. The ambient temperature and the insolation—the solar energy incident on the cell—directly affect the cell's temperature. Since only a portion of the incident solar energy is converted into electricity, the remainder is absorbed by the cell, often resulting in an increase in temperature.

The NOCT is a standard metric used to indicate the expected temperature of a PV cell under specific conditions, which are usually an ambient temperature of 20°C, solar irradiation of 0.8 kW/m² and a wind speed of

1 m/s. The NOCT provides a baseline for understanding how a PV cell will perform under these "standard" conditions.

For ambient conditions that differ from those at NOCT, the cell temperature can be calculated using the equation given below:

$$T_{cell} = T_{amb} + \gamma \frac{\text{NOCT} - 20°C}{0.8} S \qquad (2.3)$$

where T_{cell} is the cell temperature in °C, T_{amb} is the ambient temperature, and S is the solar insolation measured in kW/m². Note that the calculation allows for adjustments to the voltage and power output in PV modules and inverter systems based on the varying cell temperatures, which is necessary for system design and energy yield predictions. If the NOCT is not provided, an alternative approach to estimate the cell temperature is to use the following equation:

$$T_{cell} = T_{amb} + \gamma \left(\frac{\text{Insolation}}{1\,\text{kW}/\text{m}^2} \right) \qquad (2.4)$$

Here, γ represents a proportionality factor that is influenced by the wind speed and ventilation conditions around the PV cells. It typically ranges between 25°C and 35°C. This suggests that with an insolation of 1 kW/m² (referred to as "1-sun"), the PV cell temperature will be about 25°C–35°C higher than the ambient temperature.

Both methods described above are essential for assessing and predicting the performance of PV systems under various environmental conditions. The adjustment factors for NOCT and the γ proportionality factor allow for more accurate modelling of PV system outputs, considering the real-world conditions that affect cell temperature and, consequently, the efficiency of generation.

2.7.4 Series- and Parallel-Connected Solar PV Cells and Impacts of Shading

As it is known, series connection of PV cells means that the current flowing through each cell is the same, but the voltage V can vary across each cell.

If three identical cells ($n=3$) are connected in series and are all in the sun, the same current I flow through each of them and if we assume that the top cell as illustrated in Figure 2.19 is shaded at different rates, the behaviour of the series-connected cells will be affected differently.

In the first scenario (Figure 2.19a), if the top cell is 100% shaded, its current source I_{SC} will be reduced to zero. Therefore, the voltage drop across R_P as current flows through it will cause the diode to be reverse-biased, so the diode current is also zero. Then, the entire current flowing through the module will flow through both R_P and R_S in the shaded cell. Therefore, the top cell, instead of adding to the output voltage, actually reduces it. Table 2.15 summarizes the voltage variations under such shaded scenario.

Referring to Figure 2.20. Shading issues are addressed in practice either using bypass and blocking diodes to help current flow around shaded cells or using more complex solutions involving reconfiguration of PV panels and arrays.

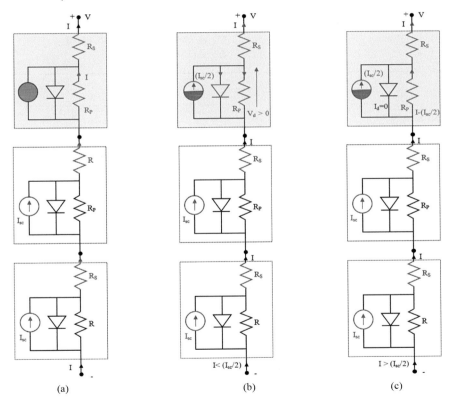

FIGURE 2.19 Equivalent circuit models representing different shading conditions of a PV cell: 100 % shaded cell (a), 50% shaded cells when the current is less than half of the short circuit current $I<I_{sc}/2$ (b) and when $I>I_{sc}/2$ (c).

TABLE 2.15 The Voltage Variations Using the Equivalent Circuit Models Experiencing Different Shading Conditions

Shading Conditions	Voltage Variations
One cell is 100% shaded among the three cells ($n=3$) (see Figure 2.19a).	The output voltage: $V_{shade} = V_{n-1} - I(R_p + R_s)$ where V_{n-1} is the voltage of the bottom $n-1$ cells. Since $V_{n-1} = [(n-1)/n]\ V$, $V_{shade} = [(n-1)/n]\ V - I(R_p + R_s)$ Therefore, the voltage drop ΔV at any given current I: $\Delta V = V - V_{shade} = (V/n) + I(R_p + R_s)$ Since $R_p \gg R_s \quad \Delta V \cong (V/n) + I R_p$
One cell is shade 50% and if the current to the shaded cell, I, from the rest of the module is $< I_{SC}/2$ (see Figure 2.19b).	Some current will still flow through the diode and the cell will still contribute a reduced positive voltage.
One cell is shade 50% and if the current to the shaded cell, I, from the rest of the module is $> I_{SC}/2$ (see Figure 2.19c).	Current equal to the difference between the two will be diverted through the parallel resistance R_P. The diode is reverse-biased, hence $\Delta V \cong (V/n) + [I - (I_{sc}/2)]\ R_p$

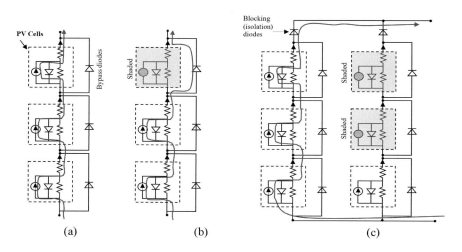

FIGURE 2.20 Schematic representation of solar PV cells/panels illustrating the protective roles of bypass diodes against shading effects and blocking diodes to prevent current backflow.

Schottky diodes can be placed across PV cells, connected in reverse parallel, to create a bypass for the current of the parallel resistance, thus mitigating the impact of shading.

Although placing a bypass diode across every cell in a module is possible, in practice, one bypass diode typically serves a group of cells. For example, three bypass diodes might be used, each covering one-third of the cells in a module as illustrated in the figure.

Bypass diodes can also be situated across each module in a string to decrease its impact on a larger array. The figure illustrates both the circuitry and the I-V characteristics of three PV modules. As shown, after the diversion of the current through the bypass diode, the string retains two-thirds of its potential power output. Without these diodes, three-fourths of the power output would be avoided.

Furthermore, bypass diodes prevent the occurrence of hotspots in individual shaded cells. Such hotspots could arise due to heat generated in the parallel resistance.

When strings of modules are connected in parallel, and they are not generating identical voltages under any operating condition (a highly probable scenario), the strings with lower voltage act as a load for the other strings, thus diminishing the array's output considerably.

Blocking diodes placed at the top of each string prevent reverse current drawn by a shaded string. Figure 2.20 indicates the functions of bypass and blocking (isolation) diodes in solar PV systems, with an emphasis on shade mitigation and protection.

Figure 2.20a shows PV cells or panels arranged in series, with each having a linked bypass diode. The purpose of these diodes is to ensure that shading or underperformance of one or more cells/panels does not affect the entire string. If a PV cell or panel is shaded or underperforming, it can become a load rather than a source, which can result in power dissipation and potential damage.

Should a PV cell or panel become shaded (as denoted by the "shaded" label and the greyed-out box in Figure 2.20b), the connected bypass diode becomes forward-biased, offering an alternative path for the current and thus "bypassing" the shaded cell/panel. This ensures that the other cells/panels in the string continue operating efficiently.

In Figure 2.20c, the top diodes function to prevent the backflow of current into the PV cells/panels, which could occur during shading. They ensure that the generated current travels in the intended direction, that is, towards the load or grid. If certain cells/panels are in shade, the blocking diodes stop the backflow of current into these shaded cells/panels. This feature is particularly important when multiple strings of PV cells/panels are connected in parallel. Absent the blocking diodes, a shaded string could draw current from other functioning strings, which is undesirable. The shading impacts of a panel arrangement with limited number of bypass diodes will be analysed in detail in the following subsection.

Note that parallel connections of PV cells are generally more resilient to shading compared to series connections. In a parallel configuration, shading affects the current produced by the shaded cell, but the overall system voltage remains relatively stable. In addition, the intrinsic passive components within a shaded cell behave like a load. This can lead to localized heating, commonly referred to as hotspots. However, in a parallel set-up, the current passing through these components is reduced, which mitigates the severity of hotspots. Blocking diodes are implemented to further prevent the flow of current through shaded cells, offering protection against these issues. Nevertheless, in systems where cells are connected in parallel, the voltage output of each cell is relatively low. Consequently, if a DC/DC converter is used to step up the voltage to a desired level, it must be capable of achieving a higher voltage gain to compensate for the lower cell voltage. This requirement is essential for ensuring the efficient functioning of PV systems under varying shading conditions.

2.7.5 Panel and Array Level Shading Scenarios

Full shading describes scenarios where sunlight is entirely blocked from reaching the solar panel or array. Such situations can arise during early mornings or late evenings when the sun's position is low or due to obstructions like tall structures. External factors like accumulated debris or snow can also contribute to full shading. While such obstructions significantly diminish power output, solar PV systems employ bypass diodes within the modules.

However, to reduce cost and simplify the assembly process, only a select number of bypass diodes are utilized in commercial PV panels. For example, a standard PV panel for household use includes 66 cells. These cells are organized into three series configurations, with 22 cells connected in series lengthwise. Each of these groups is equipped with a single bypass diode, making for a total of three. While these diodes provide an alternate route for the current in shaded "cell strings", this set-up has unintended consequences on shading that are summarized in Figures 2.21–2.26 under three major groups (diagonal, horizontal and vertical partial shading) and one full-sun (Figure 2.21) and one full-shading (Figure 2.26) conditions.

These shading scenarios are chosen based on the functional design of the cell or panel arrangement, as previously mentioned, and their interaction with light.

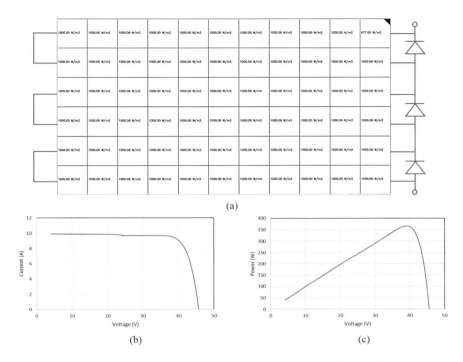

FIGURE 2.21 Full-light performance of the standard 66-cell and 3 bypass diode arranged solar PV panel (a) and its I-V (b) and P-V (c) characteristics.

Diagonal shading (Figures 2.22 and 2.23) occurs in locations where the sun's path is more varied. Horizontal (Figure 2.24) and vertical (Figure 2.25) shadings are present for blocking low-angle morning or evening sun; especially on east- or west-facing facades, shading is often valid in buildings to block high-angle summer sun while allowing low-angle winter sun to penetrate.

As it can be seen in the sample shading patterns, the shading direction does play a significant role. Note that even in shaded conditions in the daytime, there exists some irradiance, approximately 100 W/m². This irradiance is vital in defining the level of current source. Consequently, when shading takes place, neighbouring and partial cells adjust to accommodate this change. A shaded solar cell inherently produces a reduced current. When cells within a module are series-connected, the entire module's current is constrained by the performance of the shaded cell, resulting in a sharp decline in power output.

In horizontal shading, type of shading affects one row of cells at a time. The values of 1,000 W/m² indicate consistent insolation on the

132 ■ Reinventing the Power Grid

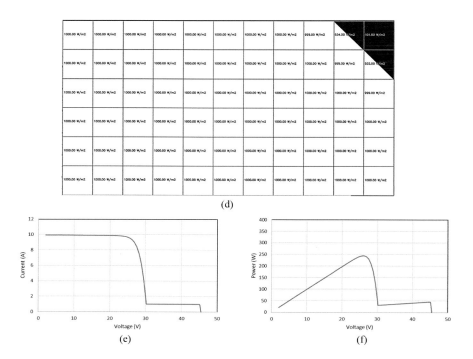

FIGURE 2.22 Diagonally shaded performance of the standard solar PV panel with one string partially shaded (a) and its I-V (b) and two-peaks P-V (c) characteristics.

panels, except the shaded area which has a reduced value indicating partial shading.

The solar panel shown has a diagonal shading pattern in Figure 2.22, affecting multiple cells across different series connections. Since solar cells in a series are dependent on each other for performance, even a single shaded cell can reduce the output of the entire series. With diagonal shading affecting multiple series connections, the overall efficiency and performance of the panel can be significantly impacted. I-V curves show a significant drop in current as the voltage increases, more pronounced than typical IV curves. This indicates that the shaded cells are affecting the panel's performance considerably. The two distinct "knee" points indicate that multiple series of cells are affected by shading. The sharp drop towards the end indicates the maximum voltage point beyond which the current generation drops drastically, making it inefficient to operate beyond this point. This behaviour suggests that the shading is causing a significant reduction in the power output at higher voltages.

Photovoltaic Solar Energy ■ 133

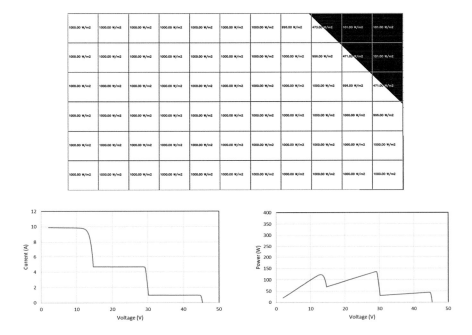

FIGURE 2.23 Diagonally shaded performance of the standard solar PV panel with two strings partially shaded (a) and its I-V (b) and three-peaks P-V (c) characteristics.

In Figure 2.23, the solar panel has a diagonal shading pattern impacting multiple cells across two series connections. The effect of shading is more pronounced compared to the previous case, as the reduction in insolation for the cells is significantly high. In this mode of operation, I-V curve has three distinct "knee" points, indicating that two series of cells have been notably impacted by shading, which also impacts the MPP in the P-V curve.

In Figure 2.24, the solar panel shows an interesting shading pattern. There is a band of cells in the middle of the panel with reduced insolation, surrounded by cells with nearly full insolation of 1,000 W/m² on the top and 978 /m² on the sides. This indicates a strip of shading affecting several cells in series connections of the second strings, spanning the entire width of the panel. Therefore, horizontal shading impacts entire rows. The power curve starts to rise with voltage but then peaks relatively early. The presence of multiple peaks or "knee" points suggests a significant impact across the affected cells.

134 ■ Reinventing the Power Grid

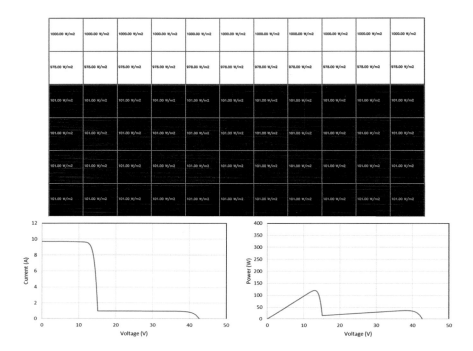

FIGURE 2.24 Horizontally shaded performance of the standard solar PV panel with one string fully-shaded (a) and its I-V (b) and two-peaks P-V (c) characteristics.

In Figure 2.25, the effects of vertical shading on a series of cells are shown. Given that these cells are interconnected in series, the diminished performance of shaded cells drastically reduces the voltage output. Though bypass diodes are employed to offset this decrease, their scattered presence—only one for every group of 22 cells—means that extensive shading can activate all diodes, resulting in a significant voltage output decline. This alteration in power, as visualized in the power curve, represents the collective impact of both voltage and current modifications due to shading.

For the context of earlier shading definitions, it is assumed that the solar PV panels are arranged longitudinally in a north–south direction. This orientation is critical because vertical shading can simulate effects analogous to a full-shading scenario, as evident in Figure 2.26. In this shading situation, the entire panel experiences uniform shading, subjected to a low irradiance level of 100 W/m². This indicates the presence of shading even during daylight hours. As anticipated, with the majority of cells receiving less irradiance due to consistent shading, there is a drop in current output.

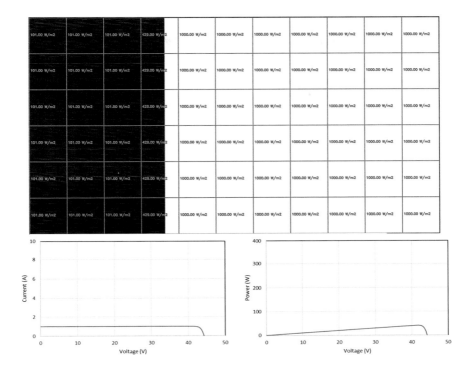

FIGURE 2.25 Vertically shaded performance of the standard solar PV panel with 1/3 shaded area (a) and its I-V (b) and very low P-V (c) characteristics.

Bypass diodes assume a pivotal role in channelling the current away from shaded cells.

The distinct shape of the PV curve, marked by several peaks, emphasizes the profound effect of shading on the efficiency of the panel. Like all solar panels, it is desirable to boost efficiency by minimizing shading. In instances where shading is unavoidable, using microinverters, power optimizers or adopting parallel-connected cell configurations can be beneficial. Nonetheless, it is essential to consider the implications of grouped cells, the scarcity of bypass diodes and the panel's orientation during installations.

2.8 PV SYSTEMS AND PRINCIPLES OF MPPT

Figure 2.27 shows the architecture of a solar PV system designed to transform solar energy into electrical power for either direct consumption or storage and to merge it with an AC grid or network. The DC/DC converter, in conjunction with a control mechanism known as MPPT, ensures

136 ■ Reinventing the Power Grid

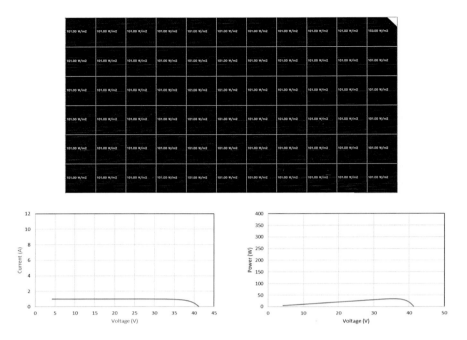

FIGURE 2.26 Fully shaded performance of the standard solar PV panel (a) and its I-V (b) and very low P-V (c) characteristics.

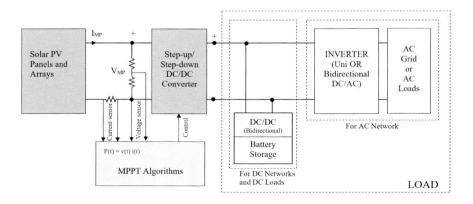

FIGURE 2.27 Generalized solar PV system block diagram.

optimal power extraction. A provision exists for battery storage through a bidirectional DC/DC converter, facilitating the storage and retrieval of energy as needed. Furthermore, an inverter, which can be either unidirectional or bidirectional, converts DC electricity to AC, for supplying to an AC grid or driving AC loads. This system offers flexibility and efficiency, ensuring a smooth integration of solar energy into standard electrical grids.

While solar PV panels can directly power an electrical load, the effective integration of these panels with diverse electrical loads requires the use of power electronic converters, which will be discussed in Chapter 4. These converters primarily regard the solar panel as a power source with distinctive V-I characteristics, aiming to efficiently transmit the available power to the load side.

DC/DC converters are critical for ensuring the voltage derived from solar panels meets system and load demands, after maximizing power extraction from the panels in real-time using MPPT methods. Despite the vast range of DC/DC converter topologies, their operation also centres around MPPT control. The buck, the boost and the buck/boost converters will be explored in Chapter 4, as they illustrate the voltage step-up, step-down and step-up/step-down mechanisms of the solar panel output.

2.8.1 Typical Electrical Loads for Solar PV Panels

Understanding the characteristics of different electrical loads is highly critical for designing and optimizing systems, especially in contexts of solar PV installations, where the efficient conversion and utilization of power is important. Table 2.16 provides a comprehensive overview of the electrical load characteristics, their applications and the major roles of power electronics converters in their operation. Such converters simply act as an intermediate stage between solar PV panels, and these loads primarily manipulate the voltage and current levels while accommodating MPPT algorithms which will be discussed later.

2.8.2 Principles of MPPT

As solar PV systems generate power from sunlight, it is desirable to transfer this active power timely (due to varying light conditions) to a load that may have different voltage and current requirements. During this transfer, it is also desirable to have a minimum power loss.

As illustrated in Figure 2.18, the MPP is the point on the I-V curve where the product of current and voltage is maximized; hence, the solar panel can deliver the maximum power if connected to an electrical load (circuit), while the location of the MPP can change based on environmental conditions. Therefore, operating a solar panel near its MPP ensures the maximum possible power from the panel is obtained, which is why MPPT controllers are needed in solar PV systems.

TABLE 2.16 The Roles of Power Electronics Converters in Various Load Types

Load Type	Voltage–Current Characteristics	Specific Applications	Roles of Power Electronics Converters
CC load	• The V-I curve is a horizontal line.	• LED drivers. • Battery chargers.	• Voltage regulation: while maintaining CC. • Current control.
CV load	• The V-I curve is a vertical line.	• Power supplies. • Final stages of battery charging: maintaining a CV.	• Ensuring a stable voltage across the load. • Protection: preventing overvoltage situations.
Nonlinear load	• Varied V-I curve shapes.	• Rectifiers, variable frequency drives and switch mode power supplies.	• Improving power quality. • Adaptive voltage and current control
Inverter	• Complex V-I characteristics	• A part of solar PV systems	• AC waveform generation • Grid synchronization: regulating frequency, phase and amplitude.
Battery	• Somewhat linear V-I relationship during discharge but nonlinear during charging.	• Lithium-ion and others for a range of application	• Charging control • Optimal voltage conversion during battery discharge. • Overcurrent protection and thermal monitoring.

However, when the P-V characteristic is analysed, it can be observed that operating without MPPT can be a possible option in certain applications, but there are trade-offs to consider:

- If a load is selected that requires a voltage slightly less as highlighted in Figure 2.18.
- The panel will not operate at its MPP during varying conditions, leading to potential inefficiencies, which will be more pronounced at higher irradiance values (see Figure 2.18c)
- Operating without MPPT can simplify the system and reduce costs, such as charging a lead–acid battery. However, if a battery is involved overcharging or deep discharging is possible which also reduces lifetime of the battery.

To determine the voltage level at which a load can operate directly from the solar panel without a MPPT controller, Figure 2.18c can be used which is around 15V (denoted by the dashed line).

In solar PV applications, matching the characteristics of the solar panel with the connected load is of prime importance. Drawing excessive current can lead the panel's voltage to collapse, moving away from its MPP. Alternatively, if the load consumes too little current, the panel might function at a higher voltage, again deviating from its MPP. Therefore, to understand the operation of MPPT controllers, the principle of maximum power transfer will be explained briefly.

The maximum power transfer theorem is a fundamental concept in electrical engineering. It explains that the most power will be channelled from a source to its load when the load resistance equals the internal resistance of the source. For example, if a source has an open-circuit voltage, V, and an internal resistance, R_{source}, the maximum power it can deliver to a load is $V^2/4R_{source}$. This occurs when $R_{load}=R_{source}$.

In AC circuit scenarios, this translates to the complex conjugate of the source's internal impedance. This principle finds practical applications in various domains. For example, to achieve optimal sound quality, the resistance of a speaker should mirror the output resistance of its amplifier. This principle also finds resonance in RF impedance matching, where the objective is to align the impedances of the source and load to ensure efficient power transfer. Similarly, to extract maximum power from a battery, one should connect a load whose resistance matches the battery's internal resistance.

However, it is essential to underline that while the theorem establishes a condition for optimal power transfer, it does not automatically translate to peak efficiency. Hence, the high efficiency of the power electronics converter (MPPT controller) becomes a driving factor.

In the context of solar PV systems, the impedance matching is slightly different, which has the principle diagram illustrated in Figure 2.28. The DC/DC converter (such as a buck, boost or buck/boost converter) adjusts its input impedance based on its duty cycle (or pulse width modulation (PWM)) that will be explained in detail in Chapter 4. The "source impedance" of the solar panel at the MPP can be defined as the derivative of the voltage with respect to the current (dV/dI) at that point. This is analogous to a resistance. By adjusting the duty cycle of the DC/DC converter, the MPPT algorithm ensures that the input impedance of the converter (and thus the combined impedance of the converter and the load) matches the source impedance of the panel at the MPP. This ensures maximum power transfer.

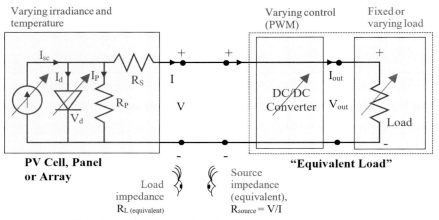

FIGURE 2.28 An illustration of impedance matching between a PV cell and a combined DC/DC converter-load for maximum power transfer.

2.8.3 MPPT Methods

Figure 2.29 illustrates the MPP, P_{max}, occurring at a specific voltage named V_{mpp}, along with several other power values (P_1, P_2, P_3, P_4) and corresponding voltages (V_1, V_2, V_3, V_4). DC/DC converters aim to match source impedance to ensure the solar PV panel or array consistently operates at V_{mpp} regardless of changes in temperature or irradiance.

MPPT algorithms are designed to oversee these DC/DC converters. They monitor temperature and irradiance conditions indirectly through voltage and current readings, and then, they adjust the system's operational

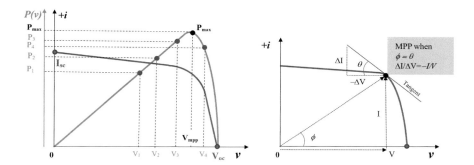

FIGURE 2.29 The P-V characteristic curve of a solar PV panel with several power values (*Pmax*, P1, P2, P3, P4) and corresponding voltages (V_{mpp}, V1, V2, V3, V4) and the illustration of the incremental conductance method.

point to ensure it always remains close to or at the MPP. Various MPPT methods exist to achieve this aim, whether the environmental conditions are consistent or there is shading of some form.

When setting up a PV system, it is important to select an MPPT method that aligns with the system's requirements. Factors to consider in the selection include the complexity of the design, the method's capability to pinpoint the true MPP among possible multiple peaks due to shading (see Figure 2.22); cost implications, especially with the inclusion of voltage and current sensors; its responsiveness to varying irradiance and temperatures; and its speed in adapting to changes, as faster adjustments typically indicate greater sensitivity.

Table 2.17 summarizes MPPT methods based on the above-listed factors and provide insights that can help to select a specific method.

Among these, the incremental conductance method is explained below since it stands out for its precise tracking and ease of implementation.

The underlying principle in the incremental conductance method for MPPT is that the slope of the P-V curve at the MPP is zero. This method utilizes the differential change in power ($P = I\,V$) with respect to voltage (dP/dV) to find the MPP.

Differentiating power with respect to voltage results in

$$\frac{dP}{dV} = I\frac{dV}{dV} + V\frac{dI}{dV} \qquad (2.5)$$

At the MPP, the change in power with respect to voltage is zero ($dP/dV=0$). If the differential change in current with respect to voltage (dI/dV) is approximated to the incremental changes in current and voltage (ΔI and ΔV):

$$0 \approx I + V\frac{\Delta I}{\Delta V} \qquad (2.6)$$

which can lead to the equation at MPP as seen in Figure 2.24b.

$$\frac{\Delta I}{\Delta V} = -\frac{I}{V} \qquad (2.7)$$

This equation justifies that at the MPP, the incremental conductance ($\Delta I/\Delta V$) is approximately equal to the negative value of the instantaneous conductance ($-I/V$). By constantly comparing the incremental

TABLE 2.17 A High-Level Overview of Various MPPT Methods

MPPT Method	Description	Disadvantages	Advantages
Perturb and observe (hill climbing)	• Adjust the PV voltage and observe whether the change made increases or decreases the power delivered.	• Oscillate around the MPP • Not efficient under fast-changing conditions.	• Simple to implement and low cost.
Incremental conductance	• It is based on the fact that the slope of the power versus voltage curve is zero at the MPP and negative after.	• Complexity increases with the number of solar array modules.	• High accuracy • Effective under varying conditions.
CV	• Operates the PV system at a fixed percentage of the open-circuit voltage, which is close to the MPP.	• Less accurate • Not ideal for fast-changing conditions.	• Simple to implement.
CC	• Adjusts the PV system's operation so that its output current remains constant.	• Less accurate under varying irradiance conditions.	• Easy to implement; works well under constant irradiance.
Particle swarm optimization (PSO)	• An optimization technique that iteratively refines potential solutions.	• Can be complex and requires more computational resources.	• Capable of finding global MPP in conditions with multiple peaks.
Differential evolution (DE)	• An optimization technique using the differential of populations.	• Requires higher computational power.	• Efficient in tracking global MPP under varying conditions.
Artificial neural network (ANN)	• Uses trained neural networks to predict the MPP.	• Requires training and a significant data set for accuracy.	• Adaptive • Can be accurate if trained properly.
Fuzzy logic control (FLC)	• Uses fuzzy logic rules to determine the MPP.	• Designing the fuzzy rules can be complex.	• Can be accurate • Adaptable to varying conditions.
Analogue-based MPPT control	• Uses analogue circuits to measure and define the MPP	• The circuit can be complex and may be limited to obtain high voltage gain.	• Can be very fast under varying conditions and accurate.

conductance to the instantaneous conductance, the MPPT algorithm can determine the direction to adjust the operating point to maintain operation at the MPP, even as environmental conditions change.

As shown in the flow chart of Figure 2.30, the method is particularly suitable for digital control systems that utilize sensors to measure voltage and current at discrete sampling instances. At every sampling instant time "n", the digital control system obtains current (I) and voltage (V) measurements from the PV array using sensors. Then, the incremental changes in current (ΔI) and voltage (ΔV) are computed using the difference between the current and previous sampled values.

If $\Delta V=0$, the algorithm returns without any action, as the system is already operating at the MPP. However, if ΔV is not zero, the algorithm checks the value of $\Delta I/\Delta V$ relative to the negative instantaneous conductance ($-I/V$). If $\Delta I/\Delta V$ is greater than $-I/V$, the operating voltage of the array should be increased to approach the MPP which can be done by adjusting the duty cycle of the DC/DC converter used. Conversely, if $\Delta I/\Delta V$ is less

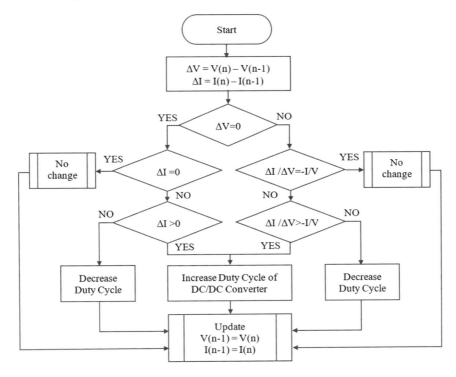

FIGURE 2.30 The principle flow diagram of the incremental conductance algorithm.

than $-I/V$, the operating voltage needs to be decreased to approach the MPP. If $\Delta I = 0$ but ΔV is not zero, the voltage should be increased. After the DC/DC converter makes the required adjustments, the algorithm returns, waiting for the next sampling instant.

Therefore, depending on the decision made in the previous step, the algorithm will adjust the PV array's operating voltage. This adjustment ensures that the system continuously hunts the MPP, under varying temperature and irradiance conditions.

Note that while very rapid algorithm convergence is often considered as a benefit, it is worth noting that for stationary solar applications, such speed is not critical. This is because solar irradiance and temperature do not fluctuate rapidly, allowing a moderate-paced tracking system to be effective and efficient.

However, solar cells on moving objects that experience rapid power fluctuations within milliseconds may require fast MPPT control as in vehicles accommodating solar PV systems. Therefore, in such applications, fast detection of the MPP becomes crucial, requiring control circuits that can compare power within short intervals (tens of milliseconds). While digital methods require high-precision analogue-to-digital converters (ADCs) and fast processors, leading to impractical costs and power consumption for mobile applications, analogue control circuits offer a more efficient solution, despite their complexity. Therefore, analogue-based MPPT control is often considered superior in such scenarios.

2.8.4 The Choice of PV Panel Configurations for Applications

The commonly known system configurations used in the industry for solar PV panels are listed in Table 2.18 with brief descriptions.

These integrated solutions are especially useful in complex installations where panels are subject to varying levels of insolation due to shading, debris or orientation differences. By optimizing power output at the panel level, the methods listed in the last raw of the table can significantly reduce losses compared to systems where MPPT is only done at the string or inverter level.

2.9 PERFORMANCE RATIO (PR)

The diagram in Figure 2.31 illustrates the cascade of power losses in a typical solar PV system from incident sunlight to the final electrical output, either as AC for grid connection or as DC for battery storage. The initial solar irradiance is subject to atmospheric losses (20%–30%) before reaching

Photovoltaic Solar Energy ▪ 145

TABLE 2.18 Commonly Used Solar PV System Configurations

Configurations	Description and Choice	Diagram/Figure Reference
Multi-string Inverter	Ideal for systems with multiple orientations or shading issues, as each string can operate optimally under different conditions.	
Virtual Central Inverter	Provides the benefits of central inverter efficiency with added redundancy and ease of maintenance of string systems.	
Central Inverter	Best for large, commercial installations where all panels have uniform orientation and shading is not an issue.	
String Inverter	Suitable for smaller systems or where each string is expected to have uniform solar exposure, allowing for simpler management.	
Micro Inverter (or AC Modules)	Optimizes performance for each panel individually, ideal for complex roofs with multiple angles and potential shading scenarios.	
Series-Parallel	Combines series and parallel connections to manage voltage and current within desired limits, useful in systems with varying panel numbers.	
Total-Cross-Tied	Enhances resilience to shading and more flexible array design by adding cross-ties to the series-parallel configuration.	
Multi-string	Allows for greater control and optimization of each string, adaptable to varying conditions and system sizes.	
Panel-Level Power Electronics (PLPE) or Smart Modules	Integrated, in the form of: • DC Optimisers • Smart DC Modules	

the solar panel, where panel efficiency limits conversion. In addition, losses due to panel tilt, orientation, temperature and soiling further reduce the available energy. Transmission losses through DC cabling and DC/DC conversion account for 1–3%, while inverter losses when converting to AC power range from 4% to 6%. AC cabling losses to the grid are between 3% and 6%. Consequently, the final output power available to the grid is significantly low, whereas for DC utilization like battery storage is more efficient.

These losses highlight the importance of optimizing each segment of the energy conversion process to maximize the efficiency and efficacy of

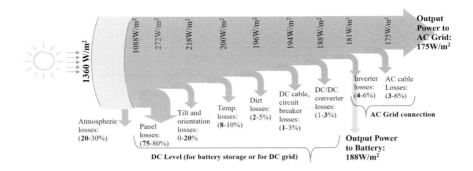

FIGURE 2.31 A comprehensive overview from sunlight to power delivery and typical efficiency breakdown. Solar irradiance is assumed 1360 W/m² outside the atmosphere.

solar PV systems for end use. Understanding and analysing each component of a solar PV system is essential to optimize its performance, increase its reliability and ensure its economic viability. However, it is important to note that these values can vary based on specific conditions (such as seasonal and daytime changes), equipment and other factors.

Considering the entire loss components, the PR of a practical solar PV system is a key metric used to assess the efficiency and health of the PV system. It indicates how well the PV system operates compared to its potential maximum output under ideal conditions. The parameter is defined as follows:

$$\text{Performance Ratio} = \frac{\text{Actual energy output of the PV system in a specific period (day, month or year) in kWh}}{\text{Expected energy output (nominal capacity at as its peak or rated power) under STC in kWh}}$$

(2.8)

To calculate the PR, measure the actual energy output of the PV system over a specific period, such as a day. In addition, the total solar irradiance on the PV module plane during the same period should be measured. This measurement is typically expressed in kWh/m². The next step involves calculating the expected energy output under STCs. This is achieved by multiplying the nominal capacity of the PV system (expressed in kW$_{peak}$) by the solar irradiance (in kWh/m²) received during the period. Finally, dividing the actual energy output by the expected energy output under STC yields the PR.

Regularly calculating the PR allows for tracking the performance and efficiency of a PV system over time, assisting in identifying potential issues or degradation in the system's operation.

Note that the PR is independence from the location and installed capacity of the PV system. Therefore, it can help to compare the performance of different PV systems regardless of their size or location. A PR closer to 1 (or 100%) indicates optimal performance, whereas a lower PR might indicate potential inefficiencies or problems within the system.

The inefficiencies include losses attributable to converters, cables and other components. In calculating the PR, the actual output is the energy delivered to the grid or consumed by the load after accounting for all system losses.

It is worth noting that these losses can arise from several sources: thermal losses in the PV modules as they warm under sunlight; converter/inverter inefficiencies; cable resistance; module mismatch within an array (which is also due to partial shading, dirt accumulation, ageing and degradation, manufacturing imperfections and faults); shading; soiling; system orientation that is not ideal; and system unavailability due to maintenance, faults or curtailment.

Note also that the shape of the I-V curve also indicates the level of energy harvest hence the PR. This is due to FF that is a parameter describing the squareness of the I-V curve. FF is the ratio of the actual maximum power output to the theoretical power output (calculated as the product of open-circuit voltage and short-circuit current). A higher FF indicates a more square-like I-V curve, which in turn suggests higher efficiency and a higher PR. High series resistance flattens the I-V curve at the MPP, while high shunt resistance affects the slope of the curve near the I_{sc}. Optimized resistances ensure minimal power loss, resulting in a better-shaped I-V curve and higher PR.

2.10 DESIGN, INSTALLATION GUIDELINES AND STANDARDS

Design, installation guidelines and standards for solar PV systems are critical for ensuring safety, performance and reliability. These guidelines and standards cover a wide range of topics from the selection of components to the final commissioning of the system.

Table 2.19 summarizes the information into two categories in the design and installation of solar PV systems. In many jurisdictions, certified training programmes are available, and certification is often a requirement for professionals to design and install PV systems legally. Furthermore,

TABLE 2.19 Design and Installation Guidelines in Solar PV Systems

Category	Item	Description
Design guidelines	Site assessment	Evaluation of the geographic location, solar irradiance, shading analysis and local weather patterns.
	System sizing	Determining the capacity of the PV system required to meet the electrical load needs.
	Component selection	Choosing appropriate panels, inverters, mounting systems, batteries (if needed) and other system components.
	System configuration	Deciding on the system type (grid-tied, off-grid or hybrid) and the configuration of the PV array (series, parallel or a combination).
	Energy efficiency considerations	Integrating energy efficiency measures to reduce the overall energy demand before sizing the PV system.
	Electrical design	Proper sizing of cables, overcurrent protection, grounding and meeting voltage drop and maximum current specifications.
	Mechanical design	Ensuring structural integrity of mounts, racks and the PV array against wind, snow and other loads.
Installation guidelines	Mounting systems	Following best practices for installing rooftop or ground-mounted systems, including proper anchoring and alignment.
	Electrical installation	Following the practices of the National Electrical Code (NEC) or local codes for all wiring and electrical connections.
	System grounding	Proper grounding of the system components for safety and lightning protection.
	Commissioning	Verifying the installation is correct, performing initial testing and confirming that the system operates as expected.
	Documentation	Maintaining detailed records of the installation, including schematics, component datasheets and a maintenance schedule.

compliance with the standards is frequently a precondition for obtaining installation permits, for receiving government incentives and for approving system interconnection by utility companies.

As mentioned in the previous chapter, the development and operation of solar PV systems are guided by standards ensuring their safety, reliability and efficiency. The standards, subject to regular updates, are crucial for professionals in the solar industry to follow for compliance with the latest in technological advancements and industry practices. Therefore, it is imperative to stay updated with these evolving standards and any specific local variations that apply. Although international and national standards associated with solar PV systems can vary, they can be classified under a number of groups including PV module standards, inverter and converter standards, balance of system (BoS) components, installation and safety standards, grid interconnection standards, performance testing and energy yield, monitoring/communications/management, environmental testing, BIPV, quality assurance and factory inspection, tracking systems, battery and energy storage standards, electrical installations, building sites, grid connection and other relevant standards.

2.11 ISSUES AND FAILURES IN SOLAR PV SYSTEMS

Table 2.20 provides a summary of common faults and failures in solar PV systems, categorized by their origins. Issues range from production defects and installation errors to environmental challenges and operational wear.

TABLE 2.20   Common Faults and Failures in Solar PV Systems

Category	Fault/Failure Type	Common Reasons
	Cell microcracks	Mechanical stress during installation or transport.
	Hotspots	Shading, soiling, cell defects or poor soldering joints.
Module level	Potential-induced degradation (PID)	High voltage biases between the ground and the voltage generated by the panel, which can reduce the module's MPP and its open-circuit voltage (V_{oc}) along with a reduction in shunt resistance. This can reduce performance and accelerate ageing The choice of glass, encapsulation and diffusion barriers has all been shown to have an impact on PID. It is mostly associated with modules at negative potentials to ground and can often be reversed by applying positive bias to affected modules.
	Delamination and EVA browning	Poor encapsulation, thermal cycling.
	Corrosion	Moisture ingress.
	Electrolytic capacitor degradation	Thermal stress, ageing.

(Continued)

TABLE 2.20 (*Continued*)   Common Faults and Failures in Solar PV Systems

Category	Fault/Failure Type	Common Reasons
Inverter failures	Software bugs	Incorrect MPPT tracking or operational failures.
	Power semiconductor failures	Thermal stress, overvoltage.
	Communication failures	Lack of system monitoring or control.
	Connector and cabling issues	Corrosion, poor connections, UV degradation.
Balance of system failures	DC/AC disconnects	Wear and tear.
	Overvoltage and surge damage	Lightning, grid anomalies.
	Mounting system failures	Corrosion, mechanical failure due to wind loads.
External factors	Weather-related damage	Hail, snow loads, extreme temperatures.
	Soiling and shading	Reduced energy harvest.
	Wildlife interference	Shading, physical damage.
Installation issues	Poor workmanship	Loose connections, incorrect wiring, inadequate sealing.
	Design errors	Incorrect string sizing, inverter selection.
System-level issues	Grid-related problems	Voltage fluctuations, frequency deviations.
	Monitoring system faults	Lack of performance data, incorrect data.
Performance	Thermal cycling	Solder bond failures, cracks.
	UV-induced degradation	Affecting polymers and materials.
Degradation	Wear and tear	Ageing, decrease in efficiency.
	Snail trail contamination	Usually associated with the use of defective front metallization silver paste during production of cells, becomes apparent after a couple of years of production.

Due to the non-uniform nature of these issues, which vary by specific context and environment, precise percentages are omitted.

Fault occurrence is influenced by the quality of PV components, installation practices, environmental conditions, maintenance routines and the system's age. Inverters, for instance, may fail early on due to manufacturing issues, while module degradation, such as PID, may become evident later in the system's life cycle.

To determine exact percentages of each fault type, one would require empirical data from system performance evaluations, maintenance records or research data. Since this table omits specific percentages, it avoids the need for data that must be meticulously tailored to each installation's unique operational and environmental circumstances.

Additionally, "system-level issues" are documented because of their potential effects on PV systems, including grid voltage and frequency disruptions, which, although external, can compromise system integrity.

Estimating the prevalence of faults can be based on general industry insights for demonstration purposes in the absence of empirical data, acknowledging that these figures are estimates and subject to variability. Faults can occur in isolation or concurrently, with some systems experiencing repeated issues or multiple systems encountering similar problems.

In general, every PV module (the backbone of a PV system) experiences product failure in its lifetime under three categories: infant failures, midlife failures and wear-out failures. In addition to these relatively long-term failure-related power degradations, many PV modules also show a light-induced power degradation (LID) just after installation. This is usually considered in the calculation of the rated power. Figure 2.32 shows the distribution of failure types of a PV module at the start of its working life. Note that infant mortality failures occur in the beginning of the working life of a PV module, which are primarily related to flaws in PV modules during manufacturing, and are associated with incorrect transportation and handling.

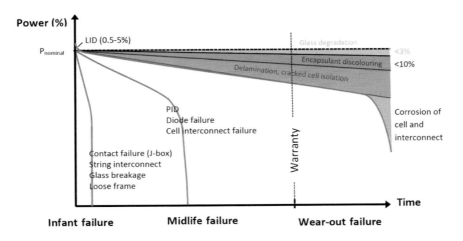

FIGURE 2.32 Three categories of failures of PV modules during its life-time.

BIBLIOGRAPHY

G. Ertasgin, D. M. Whaley, N. Ertugrul and W. L. Soong, "A Current-Source Grid Connected Converter Topology for Photovoltaic Systems," *Proceedings of the Australasian Universities Power Engineering Conference, (AUPEC '06)*, Melbourne, Australia, 10–13 Dec. 2006.

G. Ertasgin, D. M. Whaley, N. Ertugrul and W. L. Soong, "Analysis and Design of Energy Storage for Current-Source 1-ph Grid-Connected PV Inverters," *Proceedings of the IEEE Applied Power Electronics Conference, (APEC '08), 24–28* Feb. 2008, Sydney, Australia, pp. 1229–1234.

G. Ertasgin, D. M. Whaley, N. Ertugrul, W. L. Soong, "Implementation and Performance Evaluation of a Low-Cost Current-Source grid-Connected Inverter for PV Applications," *Proceedings of the IEEE Sustainable Energy Technologies Conference (ICSET '08), 24-27* Nov. 2008, Singapore, pp. 939–944.

N. Ertugrul, *Small-Scale PV Systems Used in Domestic Applications in a Comprehensive Guide to Solar Energy Systems*. Academic Press, New York, pp. 333–350, 2018.

C. Hu and R. M. White, *Solar Cells: From Basic to Advanced Systems (McGraw Hill Series in Electrical and Computer Engineering)*, McGraw-Hill, New York, 1983, pp. 294.

E. G. Laue, The measurement of solar spectral irradiance at different terrestrial elevations. *Solar Energy*, 13, pp. 43–50, 1970.

Y. Yao, N. Ertugrul and A. P. Kani, "Investigation of Short-Term Intermittency in Solar Irradiance and Its Impacts on PV and DC/DC Converters Power Output," *Australasian Universities Power Engineering Conference, AUPEC 2022*, Adelaide, Australia, 2022.

CHAPTER 3

Wind Energy Systems

3.1 INTRODUCTION

Wind energy is a key component of sustainable energy. It is a fundamental source in the mix of renewable energy (see Figure 3.1) as well as a potential source for making green hydrogen in the future. Large wind farms are now common which is achieved by advanced turbine, generator and power electronics technologies. Furthermore, offshore wind is growing because of bigger turbines and new floating platforms that enable the technology to be implemented in deep waters. Although improvements in energy storage and grid connection are making it a reliable energy source, their integration and control are still presenting challenges in a continuously evolving hybrid power grid.

FIGURE 3.1 A combined onshore wind and solar PV farm in South Australia, with large size Horizontal-Axis Wind Turbines (HAWTs).

DOI: 10.1201/9781032692173-3

This chapter will provide a comprehensive overview of the current state of wind energy development and its rapid expansion. Various systems of wind turbines will be explored, alongside an exposition of the primary components that constitute modern turbines. An analysis of wind characteristics and resources, as well as an examination of wind regimes, will be provided. The chapter will discuss the basic physics that allows us to harness wind energy, along with the distinct attributes of wind turbines. The conversion process of wind power to electricity, with particular emphasis on generators and control mechanisms, will be described. The concept of capacity factor, the efficiency of wind farms and the process of testing wind turbines will also be examined.

Onshore wind farms have been the backbone of wind power due to their ease of access and lower initial costs. However, they are limited by available land and wind resources, which can be less consistent than offshore.

On the contrary, offshore wind farms exploit the higher and more consistent wind speeds found over the ocean, which can lead to a higher energy yield per turbine. Despite the higher costs and engineering challenges associated with offshore installations, the trend is moving towards these farms due to their large power generation potential and the reduced impact on human populations. Moreover, technological advancements and experience in the offshore oil and gas industry are being leveraged to lower costs and improve the viability of offshore wind energy.

Table 3.1 provides a comparative table outlining the characteristics and considerations for onshore and offshore wind turbines and farms.

As the wind energy sector advances, there is a clear trend towards larger and more powerful turbines. This progression is anticipated to continue in response to the escalating demand for renewable energy worldwide. Continuous innovations are primarily aimed at enhancing the overall efficiency and output of wind power generation. In addition, it is important to recognize that the actual energy output—and consequently, the capacity factor—of wind turbines is subject to local wind conditions and operational considerations. Although the typical power ratings and blade lengths are available, variations occur across different models and installations. With ongoing technological enhancements, the trend is shifting towards high power, which is expected to yield higher energy with fewer units, particularly in offshore settings.

The push for increasing turbine size and capacity, especially offshore, comes from the goal to lower the levelized cost of electricity. By capturing

TABLE 3.1 Comparison of Onshore and Offshore Wind Turbines and Farms

Feature	Onshore Wind Turbines and Farms	Offshore Wind Turbines and Farms
Location	Situated on land, often in rural or remote areas	Located in bodies of water, typically at sea
Turbine size	Smaller compared to offshore, primarily due to logistical constraints	Larger, benefitting from economies of scale and stronger winds
Power capacity	Typically range from under 1 MW to about 6 MW per turbine	Commonly range from 6 to 14 MW or more per turbine
Wind consistency and speed	Variable, with lower average wind speeds	More consistent and higher average wind speeds
Foundation type	Standard concrete or steel foundations	Requires specialized foundations such as monopiles or jackets
Installation costs	Lower due to easier access and simpler logistics	Higher due to challenging marine operations and foundation costs
Accessibility and maintenance costs	Easier to access for construction, maintenance and repairs, generally has lower maintenance cost.	Access is weather-dependent and can be challenging. Higher maintenance cost, with complex logistics, especially for distant farms.
Energy yield	Lower wind speeds result in lower energy yield	Higher wind speeds and consistent winds increase energy yield
Grid connection	Often easier to connect to existing land-based grid infrastructures and substations	May require undersea cabling and more complex grid integration
Environmental impact	Land use and potential impact on terrestrial habitats	Marine impact, but potential benefits like artificial reefs
Visual impact	More visible, can be a concern for local populations	Less visible from land, reduced visual impact
Noise impact	Noise can be an issue for nearby residents	Noise is less of an issue for human populations
Potential for expansion	Limited by land availability and usage rights	Potentially larger areas available for development

wind more effectively and converting it into electricity with greater efficiency, larger turbines can diminish the number of turbines required to generate a set amount of power in a farm arrangement. This can lead to cost savings in both installation and maintenance for each unit of electricity produced.

3.2 TYPES OF WIND TURBINE SYSTEMS

As one of the prime driving forces of renewable energy, wind turbines can be classified based on several criteria, each serving a purpose for understanding their functionality, design and application. Each category serves different purposes, from individual appliances and remote applications in micro and mini turbines to large-scale energy production in larger sizes. As technology advances, the efficiency and applications of these turbines may expand, offering more versatile solutions for renewable energy needs.

A classification of wind turbines by their size is summarized in Table 3.2, which includes categorizations like micro, small, medium or

TABLE 3.2 Comparison of Wind Turbine Sizes

Size Category	Rated Power	Rotor Diameter	Typical Applications	Tower Height	Remarks
Micro wind turbines	Up to 1 kW	<3 m	Battery charging, small electronics, remote sensing.	Varies, typically <15 m	Portable, urban and remote settings, very small-scale use.
Mini wind turbines	1–10 kW	3–8 m	Cabins, RVs, boats, hybrid systems.	15–25 m	Small-scale use, often in rural or off-grid locations.
Small wind turbines	Up to 100 kW	<18 m	Residential, small businesses, off-grid.	15–30 m	Simple design, on-site installation, minimal infrastructure.
Medium wind turbines	100 kW–1 MW	18–50 m	Small communities, schools, commercial facilities.	30–60 m	Substantial foundations, grid connection required.
Large wind turbines	1 MW–3+ MW	>50 m	Large onshore/offshore wind farms.	>60 m	Major infrastructure, detailed siting, economies of scale.
Ultra-large wind turbines	10 MW and above	Could exceed 200 m	Offshore wind farms in high-wind areas.	Proportionately larger	Advanced technology, maximizing energy yield, on offshore use.

FIGURE 3.2 Small horizonal-axis 3-blade (left) and 5-blade (middle) turbine systems, and a small pole-mount vertical-axis wind turbine with a solar PV panel for street-light post.

large, typically characterized by their rated power, rotor diameter, typical applications and tower height. This size-based classification is also useful for determining the turbine's suitability for specific environments, from residential to industrial scale applications.

Another approach to categorizing wind turbines is by their axis orientation—horizontal-axis wind turbines (HAWTs) and vertical-axis wind turbines (VAWTs) (see Figure 3.2). HAWTs, with the main rotor shaft and electrical generator at the top of a tower and pointed into the wind, are predominant in the industry due to their efficiency in high wind conditions. VAWTs, however, offer advantages in turbulent winds and varied wind directions, making them suitable for areas where wind conditions are less predictable.

These small wind turbines can play a significant role in a diversified approach to renewable energy generation, especially when integrated with other renewable sources like solar power.

Note that Figure 3.2 displays three different small wind turbines, which are designed for a lower power output and are suitable for residential, community or remote applications and sea vessels where a grid connection is not feasible or cost-effective. However, their effectiveness is heavily dependent on local wind conditions, and they have several critical needs, including site assessment, maintenance, zoning and permitting, grid connection and battery storage, operational cost and durability.

Note that in the design of HAWTs, there is a practical relationship between the rotor diameter and the tower height that optimizes energy capture and structural efficiency.

The tower height is typically designed to be greater than the rotor diameter to ensure that the blades are exposed to less turbulent and higher velocity wind, which increases with height due to the wind shear effect. This helps in avoiding ground interference and taking advantage of higher wind speeds available at greater heights, which significantly improves the power output.

A common rule is that the tower height in wind turbines should be at least 1–1.5 times the diameter of the rotor. For example, if a wind turbine has a rotor diameter of 100 m, the tower height might be in the range of 100–150 m. In practice, the exact ratio will depend on detailed site assessments, including wind shear profiles, turbulence intensity and other local conditions. The goal is to balance the additional costs of taller towers against the potential gains in energy production from accessing higher altitude winds.

Table 3.3 offers a simplified overview of the general characteristics of HAWTs and VAWTs. Specific models may have unique features or design optimizations that can influence their performance and suitability for different environments or purposes.

TABLE 3.3 Comparison of Horizontal-Axis Wind Turbines (HAWTs) and Vertical-Axis Wind Turbines (VAWTs)

Feature	Horizontal-Axis Wind Turbines (HAWTs)	Vertical-Axis Wind Turbines (VAWTs)
Axis orientation	Horizontal; parallel to the ground.	Vertical; perpendicular to the ground.
Typical size and capacity	Large; typically, from 2 MW up to 14+ MW for offshore models.	Smaller; typically, less than 100 kW for most designs.
Blade design	Typically, three blades.	Various designs, often with curved or straight blades.
Rotation	Blades rotate around a horizontal axis.	Blades rotate around a vertical axis.
Wind direction adaptability	Requires yaw mechanism to turn towards the wind.	Omni-directional; can capture wind from any direction.
Installation height	Tall towers, often over 100 m for large turbines.	Lower height; suitable for residential or small-scale.
Efficiency	Higher; more efficient at capturing energy from high winds.	Lower; less efficient, but can work with variable winds.
Cost of installation and maintenance	Higher due to size, complexity, and height.	Generally lower, easier access to components.
Typical application	Large-scale wind farms, both onshore and offshore.	Small-scale applications, urban environments.
Footprint	Larger; requires more space due to longer blades.	Smaller; vertical design takes up less horizontal space.
Noise emission	Generally higher noise levels due to blade size and speed.	Typically, quieter, which may be preferable for urban settings.

TABLE 3.4 Comparison of Upwind and Downwind Turbines

Feature	Upwind Turbines	Downwind Turbines
Rotor position	Rotor is positioned upwind of the tower.	Rotor is positioned downwind of the tower.
Wind direction alignment	Requires a yaw mechanism to keep the rotor facing into the wind.	More tolerant to changes in wind direction; yaw mechanism still present.
Mechanical stress	Generally, experiences more uniform mechanical stress.	Experiences a range of mechanical stresses, particularly during gusts.
Tower shadow effect	Less affected due to rotor facing the wind.	More affected; blades pass through the wind shadow of the tower.
Noise and vibration	Potentially more noise and vibration due to proximity to tower.	Less noise and vibration transmitted to tower due to downwind position.
Design complexity	Often simpler with fewer components for the yaw system.	May require more complex design to accommodate for variable wind loads.
Turbine stability	Generally, more stable as the rotor acts as a weather vane.	Can be less stable; requires careful design to avoid dynamic instabilities.
Self-orientation	Relies on active systems to maintain optimal orientation to the wind.	Can passively align with the wind due to the aerodynamic forces on the rotor.
Implementation	More common in commercial wind turbine installations.	Less common; used in specific situations where wind conditions favour it.
Blade flexibility	Blades are typically stiffer to avoid striking the tower when deflected.	Blades can be more flexible, which can be a benefit in high-wind conditions.

In addition, wind turbines can be classified as upwind or downwind turbines (see Table 3.4) based on the direction in which they face relative to the wind (see Figure 3.2). Upwind turbines face into the wind with the help of yaw control, while downwind turbines are positioned with the wind to their backs, allowing the rotor to follow the wind direction passively. This classification is significant for design considerations, especially in terms of the structural demands placed on the turbine by the wind's force.

Downwind turbines can be advantageous in situations where wind directions change frequently or rapidly, as they can passively align with the wind without the use of active yaw systems, potentially reducing maintenance. However, the mechanical stress from the variable wind load can lead to increased wear and tear over time.

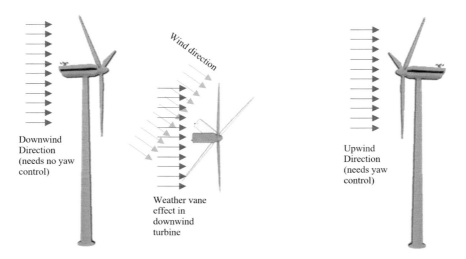

FIGURE 3.3 Horizontal axis downwind turbines and weather vane effect (left); horizontal axis upwind turbine (right).

A weather vane on a downwind wind turbine plays a crucial role in its operation. Downwind turbines are designed so that the wind hits the tower before the rotor blades. Unlike upwind turbines, which use a yaw mechanism—that will be explained later—to keep the rotor facing into the wind, downwind turbines can passively align themselves with the wind direction due to the aerodynamic forces acting on the rotor and tower (as illustrated in Figure 3.3. When the wind direction changes, the tail of the turbine (much like a weather vane) helps pivot the turbine so that the rotor is downwind of the tower, reducing the need for active yaw systems to adjust the turbine's position. This can lead to a simpler and potentially less expensive design, as the turbine naturally seeks the correct orientation with respect to the wind.

While downwind turbines have some advantages, the benefits of upwind turbines generally outweigh these in most commercial wind farm applications for several reasons:

- Higher aerodynamic efficiency due to the absence of the aerodynamic turbulence and wake caused by the wind passing over the tower first.
- Downwind turbines can experience additional mechanical stress as the blades pass through the wind shadow of the tower.

- Upwind turbines have a simpler design because they do not need to be as flexible or as durable against the turbulent wind that can affect downwind turbines.

- Although it may be complex, an active yaw system is well-developed and mature enough to efficiently turn the upwind turbine to face the wind.

- Downwind turbines can generate more noise and vibration due to the interaction of the blades with the turbulent wind after it passes the tower.

- Upwind designs are typically better suited for locations with high wind shear and turbulence since the rotor is the first part of the turbine to encounter the wind.

3.2.1 Shadow Effect in Turbine Types

The shadow effect in wind turbines refers to the impact of the tower on the airflow that reaches the rotor blades.

In upwind turbines, the rotor blades face the wind before the tower, so there is minimal disruption to the airflow from the tower's shadow. However, the tower can still cast a wind shadow that slightly reduces the wind speed and increases turbulence at the blade when it passes in front of the tower, potentially causing dynamic loading hence a torque ripple. This effect is generally brief and can be managed with robust blade design and control systems.

Downwind turbines (see Figure 3.3) experience the shadow effect more directly since the wind passes the tower before reaching the rotor blades. As the blades pass through the tower's wind shadow, they encounter a drop in wind speed and increased turbulence, resulting in a cyclic variation of the forces on the blades. This can cause fluctuating stresses on the turbine structure, known as dynamic loading, which may lead to fatigue over time.

VAWTs experience the shadow effect differently from their horizontal-axis counterparts due to their design and the way they interact with the wind.

In VAWTs, the blades rotate around a vertical axis, and unlike horizontal-axis turbines, they are omnidirectional, meaning they do not need to be oriented towards the wind to be effective. This design characteristic eliminates the need for a yaw system and the associated shadow effect caused by a tower blocking the wind.

However, VAWTs do experience a type of shadow effect as the blades rotate. When a blade moves against the wind direction, it encounters a relative wind speed that is less than the actual wind speed because the motion of the blade subtracts from the wind speed. Conversely, when the blade moves with the wind direction, the relative wind speed is higher. This variation in relative wind speed during each rotation can cause cyclic variations in aerodynamic forces, similar to the dynamic loading in downwind turbines, but for different reasons.

Despite this, the effect is inherently part of the VAWT's operation and design, and these turbines are typically designed to withstand the associated stresses. The impact of this shadow effect is also a consideration in the design and placement of VAWTs, particularly in urban environments where wind conditions can be highly variable. While VAWTs do offer certain design and situational advantages, they typically fall short of the efficiency levels of HAWTs due to inherent aerodynamic and operational factors.

3.3 MAJOR COMPONENTS OF MODERN WIND TURBINES

Modern wind turbines are composed of several major components, each playing a critical role in harnessing wind power. The efficiency and cost associated with each of these components are critical considerations, as they significantly impact the overall performance, energy yield and economic viability of the wind turbine. Table 3.5 offers an exploration of these major components of modern wind turbines. This table not only provides brief descriptions of each component but also presents information on their efficiency and cost ratios. Such insights are necessary, as they not only highlight potential bottlenecks but also indicate areas within the system that can be optimized to enhance performance and efficiency.

It should be noted that the efficiency ranges provided in the table are general estimates and may vary depending on the specific models of turbines and the advancements in technology. In addition, the cost ratios presented are average values and are subject to variation due to factors such as turbine size, design, geographical location and prevailing market conditions.

In addition to the primary components listed in the table, there are several other critical elements that contribute to the overall cost. The electrical infrastructure, including internal and external cabling and transformers, accounts for about 5%–10% of the total cost. The balance of the system, which includes supporting infrastructure such as access

TABLE 3.5 The Components of the Modern Wind Turbines

Component	Description	Efficiency Range	Cost Ratio
Rotor blades	Designed to capture wind energy efficiently with advanced aerodynamic design and materials.	35%–45%	20%–30%
Nacelle	Houses the critical component(s) for energy conversion: generator, control systems and gearbox (if needed).		
Generator	Recently brushless outer-rotor permanent magnet types (which replace asynchronous and synchronous types).	90%–95%	5%–10%
Gearbox	Transfers and converts rotational speed and torque from the rotor to the generator.	94%–98%	10%–15%
Tower	Supports the nacelle and rotor; height influences access to higher wind speeds.	Increases turbine efficiency by 1%–2%	20%–30%
Control systems and sensors	Optimize turbine's response to wind conditions; includes anemometers, wind vanes and temperature.	Improves overall efficiency by 1%–3%	1%–2%
Power electronics	Condition generator output for maximum power transfer.	95%–98%	5%–10%
Foundation	Provides stability and durability. Its type varies based on location (onshore/offshore).	Not applicable	Onshore: 5%–10%, Offshore: ~30%

roads and additional foundations, can represent 15%–30% of the total cost. Installation and logistics, crucial for transporting and assembling the turbine, typically amount to 5%–10%. Operations and maintenance costs are ongoing and are generally estimated at 1%–3% annually of the initial capital cost over the turbine's lifespan. Project development costs, encompassing planning and permitting, can vary but often fall in the range of 5%–10%. Finally, grid connection and integration expenses, essential for linking the turbine to the electrical grid, can constitute an additional 2%–5%. Note that, these components form a significant portion of the wind turbine's overall investment, highlighting the multifaceted nature of wind energy projects where the total cost is distributed across various segments beyond just the core turbine components.

In the subsequent subsections, a brief explanation of each of these critical components will be given. This explanation aims to provide a comprehensive viewpoint, emphasizing their significant roles in determining the cost of energy and enhancing power efficiency. The energy yield is the prime consideration for any wind turbine, underscoring the importance of attaining the highest possible system efficiency in the entire electromechanical energy conversion process. To illustrate this point more clearly, Figure 3.4 has been included, showcasing the power efficiency of the major energy-converting devices within wind turbines.

It is important to recognize that while this figure illustrates the efficiencies of individual and typical components within a wind turbine, a wind farm, whether onshore or offshore, comprises multiple turbines of varying sizes. These turbines are distributed across extensive areas and are arranged in different topologies. As such, the design and optimization of

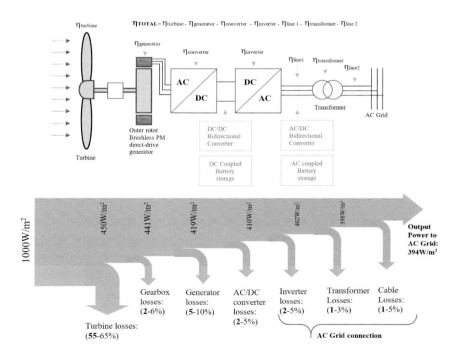

FIGURE 3.4 The major components of modern wind turbine system (top) and the associated power loss range in them (bottom).

cabling and connections to a substation are of critical importance. These aspects are critical not only for achieving the highest efficiency (hence the highest possible power output) from the wind farm but also for ensuring the system's reliability and cost-effectiveness.

In addition, it should be noted that especially in offshore wind farms, the use of High Voltage Direct Current (HVDC) connections is increasingly becoming standard practice. This is complemented by the adoption of variable speed brushless PM generators without a gearbox, which are paired with back-to-back three-phase inverter configurations, which will be discussed later.

3.3.1 Rotor Blades

To capture wind energy efficiently and convert it into mechanical energy, several specific characteristics and design principles are employed to optimize wind turbine blade performance. These are:

- Aerodynamic shape with a curved surface on one side and a flatter surface on the other to create lift as the wind passes over the blade, causing it to turn.
- Blades are typically wider at the base and taper towards the tip, with a twist along their length. The twist helps to ensure that each section of the blade is oriented properly to the wind, maximizing lift and minimizing drag across the entire length of the blade.
- Modern blades are made from composite materials, such as fibreglass or carbon fibre, which offer a good balance between strength, durability and weight.
- The length of the blade hence the swept area is directly related to the amount of energy it can capture. Therefore, longer blades need to be designed and built appropriately since they are subject to higher structural stresses.
- Large turbine blades are designed to rotate on their long axis (pitch) to control the angle of attack, which is utilized to maximize the power output under varying wind speeds and loading conditions as well as to protect the turbine from damage in high winds.

Figure 3.5 illustrates the difficulties in transporting (via ship and truck) large turbine blades and the unique design characteristics that aim to

166 ■ Reinventing the Power Grid

FIGURE 3.5 The typical size of HAWTs together with the transportation difficulties (left), and the winglet design (right).

increase the lift-to-drag ratio. Note that similar to the albatross, winglets which have wingtips can reduce drag as shown in the figure. Wind turbine blades incorporate winglets at the tips that also reduce vortex formation and improve aerodynamic performance. In addition, turbine blade designs that include flexibility can respond dynamically to gusts and turbulence, reducing stress and increasing efficiency. Furthermore, wind turbine blades with a higher aspect ratio (meaning they are long and narrow) can also be more efficient, particularly in locations with steady wind speeds. However, translating these natural adaptations to mechanical design can be complex, and engineers must balance these innovations with factors such as manufacturing costs, material limitations and the specific conditions at the turbine's location.

Table 3.6 provides a high-level view of the factors related to the blades of large and very large wind turbines. Advances in materials and manufacturing processes continue to evolve, aimed at enhancing the performance and environmental footprint of wind turbine blades.

TABLE 3.6 The Aspects of Wind Turbine Blades Used in Large and Very Large Turbines

Aspect	Description
Type	Blades for large and very large turbines are typically of the horizontal-axis type, designed for high efficiency and to withstand various environmental conditions.
Material	Predominantly made from composite materials such as fibreglass-reinforced polyester or epoxy, and carbon-fibre-reinforced epoxy for added strength and stiffness with reduced weight.
Construction method	Blades are often constructed using resin infusion or pre-preg techniques, where the fibres are laid into a mould and then infused with resin or pre-impregnated with resin, respectively and then cured in place.
Assembly issues	Challenges include transportation due to length, the need for precision in alignment and attachment to the hub, handling and lifting during installation, and ensuring structural integrity is maintained.
Lifetime	Designed for a service life of 20–25 years, but this can vary based on maintenance, usage, and environmental factors.
Recycling	Recycling is complex due to the composite materials used. Techniques include mechanical grinding, pyrolysis, and chemical processing, but research is ongoing to improve the sustainability of blade end-of-life options.

3.3.2 Tower and Foundation

Figure 3.3 in the referenced material provides a visual illustration of some of these elements. This helps in understanding the practical application and structural design of wind turbines. Table 3.7 offers a concise overview of various aspects of wind turbine design, including tower types, materials used, foundation types for both onshore and offshore installations and additional key components like internal lift systems and power electronics.

Figure 3.6 illustrates the main components of a wind turbine and their integration into the overall structure. The foundation and tower with an access door in Figure 3.6 (left) capture the base of the wind turbine where the tower is anchored to the foundation, and the access door is visible, which is used for maintenance and servicing of the turbine. The middle image shows the tower extending upwards, leading to the nacelle, which houses the key components of the wind turbine such as the generator. The right image provides a close-up view of the top part of the turbine. It highlights the nacelle; the hub, which is connected to the blades and has pitch angle control to adjust the blades according to wind conditions; the yaw angle control section, which allows the turbine to rotate and face the wind and a wind speed sensor, which is essential for turbine operation as it measures wind speed to maximize the turbine's power output.

TABLE 3.7 A Structured Overview of the Different Aspects of Wind Turbine Design and Construction

Component	Description
Tower types	• Tubular Steel Towers: Common, durable, sectional assembly. • Concrete Towers: For taller turbines, cost-effective, on-site construction. • Lattice Towers: Lighter, less expensive and less common. • Hybrid Towers: Combines steel and concrete, greater heights.
Tower materials	• Steel: Strong, flexible, durable. • Concrete: Cost-effective, heavy, on-site construction. • Composite Materials: Lightweight, strong, used in parts like the nacelle.
Onshore foundations	• Shallow Foundations: Mat or raft, for strong soil. • Deep Foundations: Piling or drilled shafts, for less stable soil.
Offshore foundations	• Monopile Foundations: Single pile, for shallower waters. • Gravity-Based Foundations: Concrete bases, for shallow water. • Jacket Foundations: Lattice structure, for deeper water. • Floating Foundations: Anchored, for very deep water.
Additional components Inside tower	• Lift and Ladder: Internal ladders, and sometimes lifts for maintenance access. • Cabling: Internal for power and control, external for offshore grid connection. • Power Electronics: Inverters, converters, controllers in the nacelle.

FIGURE 3.6 The foundation meets the tower with access door (left), the tower joints the nacelle (middle) and the closed up the nacelle and hub highlighting yaw angle and pitch angle and the wind speed sensor (right).

Wind Energy Systems ■ 169

FIGURE 3.7 Different elements of the wind turbine's lifecycle, from transportation, substation connection, switchboard inside the tower, and an internal lift and the ladders for maintenance and operations.

Figure 3.7 captures a different element of the wind turbine's lifecycle, from transportation and infrastructure to internal mechanisms for maintenance and operations. Figure 3.7 (top left) shows the transport of a large wind turbine blade on a specialized trailer with a warning sign, indicating the logistical challenges due to the blade's size. Note that a wind turbine rotor blade adapter can be used to transport long blades in narrow and complicated terrains. Suitable adapters are also used for lifting and attaching blades to the turbine hub.

The image on the top right captures wind turbines at a wind farm, with a focus on the electrical substation which is used for stepping up the voltage of the electricity generated by the turbines for transmission to the grid. The electrical switchgear shown in the bottom left contains the switches and circuit breakers necessary for controlling and protecting the electrical circuits connected to the wind turbine. The last two images are an elevator or lift system and a view of the vertical ladder. Both systems are used to transport maintenance personnel and equipment to and from the nacelle at the top of the tower. Note that technicians access various parts of the turbine via the ladder, especially in the case where a lift may not be present or operational.

3.3.3 Nacelle and Hub

The nacelle of a wind turbine is the large structure placed at the top of the tower that houses all of the generating components of the turbine (see Figure 3.6-right), including the gearbox (if any), the generator, the controller and the brake. Some of its functions include housing the generator and protecting components. The nacelle also supports the rotor and, through the pitch system, can adjust the angle of the blades to control the rotational speed and optimize power generation.

Moreover, the hub is the central component to which the turbine blades are attached. It is connected to the rotor shaft which extends into the nacelle. Its function is to transfer the rotational force from the blades to the rotor shaft and subsequently to the generator. In all modern turbines, the hub includes a pitch control mechanism that can change the angle of the blades relative to the wind to control the speed of the turbine hence to operate at the highest efficiency as well as to prevent damage in high winds. These critical components of the wind turbine directly influence the efficiency and functionality of the turbine.

Furthermore, the nacelle contains major subcomponents. For example, the yaw system allows the nacelle (and thus the rotor) to rotate on top of the tower so that the blades can be optimally positioned relative to the wind. This system includes yaw motors and bearings.

The cooling system maintains the optimal temperature for all machinery within the nacelle, especially critical for the generator and electrical components. Control systems include sensors, controllers and processors that monitor and adjust various turbine functions, such as blade pitch and yaw orientation. The brake system is used to stop the rotor in emergencies or for maintenance. Typically, this is a mechanical disc brake system.

In modern wind turbines, the placement of power electronics, including inverters and converters, can vary depending on the design of the turbine and the preferences of the manufacturer, whether they are inside the nacelle, on the ground level inside the tower or outside the tower.

When power electronics are located inside the nacelle it minimizes the length of high-current cables between the generator and the power electronics. This can reduce power losses that occur over cable lengths. Having the electronics close to the generator can also improve the responsiveness and efficiency of the power control systems. However, such arrangements impact their reliability and lifespan due to vibration and temperature variations.

When power electronics are located at ground level inside the tower, it facilitates easier maintenance and repair, as accessing the nacelle can be more difficult and time-consuming. Such a setup can also result in better cooling conditions for the power electronics, potentially improving their performance and durability. The downside is that longer cables are needed to connect the generator to the power electronics, which may lead to slightly higher transmission losses.

In some cases, especially for larger wind farms, power electronics might be placed in a separate building or container on the ground. This provides the easiest access for maintenance and may offer the best environment for controlling the operating temperature of the equipment. This setup also allows for a central location where power from multiple turbines can be collected and converted for grid compatibility.

The decision on where to place the power electronics can depend on factors such as the size of the wind turbine, the climate of the location (which affects cooling requirements) and the cost associated with longer cabling versus the benefits of easier maintenance. As technology progresses, the trend is to improve the reliability and efficiency of power electronics to allow for more flexibility in their placement.

The other subcomponents in the nacelle include an anemometer and a wind vane to measure wind speed and direction, providing data to optimize turbine performance. The pitch system includes pitch bearings, motors and control mechanisms. Rotor bearings facilitate the smooth rotation of the hub and blades. These are critical for the effective operation and longevity of the turbine. In addition, lightning protection is often integrated into the blades and hub, this system protects the turbine from lightning strikes.

3.3.4 Generator Types and Choice

The prime criteria for selecting wind turbine generators are based on several essential characteristics including torque/power density, efficiency, reliability, cost and controllability.

Note that the generator's weight and volume are the most critical due to its placement atop the tower while minimizing physical constraints. In addition, the maximum possible energy has to be harnessed from the wind, hence the generator system must operate with high efficiency. Moreover, due to the operating conditions of wind turbines in remote or offshore locations, a high degree of reliability is crucial as maintenance and repair are difficult and expensive. Furthermore, the cost-effectiveness of the generator is a key consideration. Finally, the ability to control the generator system is important due to the intermittent and unpredictable nature of wind for optimal energy capture and the production of high-quality electrical power to comply with grid regulations.

A comparative table of the major electric generators used in wind turbines is given in Table 3.8. Wound Rotor Synchronous Generators (WRSGs) are less common in wind turbines compared to other types of generators like Asynchronous (Induction) Generators, Doubly-Fed Induction Generators (DFIGs) and brushless Permanent Magnet Synchronous Generators (PMSGs). The usage of WRSGs is often seen in applications that require specific optimizations, such as achieving high efficiency over a wide range of operating conditions or where grid independence is necessary.

TABLE 3.8 A Comparative Table of the Electric Generators Used in Wind Turbines

Generator Type	Grid Connection Required	Maintenance	Complexity	Cost	Efficiency at Low Speeds	Remarks
Asynchronous (induction) generators	Yes	Low	Low	Lower	Poor	Older/smaller turbines
Doubly-fed induction generators (DFIG)	Yes	Moderate	High	Moderate	Good	Used to be the common choice. Partial-rated rotor power converters and in need of synchronization, need gearboxes.
Wound rotor synchronous generators (WRSG)	Optional	Moderate	High	Moderate	Good	Not common, specific optimization needs,
Permanent magnet synchronous generators (PMSG)	**No**	**Low**	**High**	**High**	**Best**	**Outer rotor design, full-rated power electronics, dominant in offshore, direct drive (no gearbox)**

Brushless permanent magnet generators (BPMGs) have become increasingly common in large wind turbines for several reasons: higher efficiency, particularly at partial loads; higher power density (volumetric and gravimetric) than their equivalents for the same power rating; with no brushes or slip rings, which improves reliability and reduces maintenance requirements and better control, hence capturing the maximum energy from the wind across a range of speeds.

Therefore, Permanent Magnet Synchronous Generators (PMSGs) are now becoming the norm in large wind turbines due to their high efficiency across a range of wind speeds, low maintenance needs and also the ability to operate without a grid connection for excitation. Their design is well-suited for the larger turbines being developed, especially for offshore applications where maintenance is more difficult and costlier. Despite the higher initial cost and complexity, the long-term benefits of reliability and efficiency make PMSGs a favourable choice for modern wind energy projects.

Regarding the rotor design, both inner rotor and outer rotor PMSG designs have their applications in wind turbines. The choice between an inner rotor and an outer rotor design is often a trade-off between complexity, cost, efficiency and maintenance requirements.

In the design of BPMGs for wind turbines, the configuration of the rotor (whether it is an outer rotor or an inner rotor) and the number of poles is chosen based on several factors including the desired speed, torque characteristics and the physical size constraints of the generator.

Inner rotor designs, with the rotor inside the stator, tend to have fewer poles and are typically used with a gearbox. Gearboxes are used to increase the rotational speed from the slow-moving turbine blades to the higher speeds required for efficient electricity generation in traditional generators. Fewer poles also mean the generator itself operates at a higher speed to produce the same frequency of electricity, which is suitable for connection to a standard electrical grid. Inner rotor generators with fewer poles can be smaller and lighter than high-pole-count generators, making them somehow easier to support at the top of a wind turbine tower. However, they require a gearbox to step up the speed at the cost of added complexity and maintenance while increasing the weight and space requirement. Such a fixed-speed operation is simpler but less efficient, as it does not allow the turbine to adapt to changing wind speeds.

While direct-drive systems with outer rotor generators are less common than inner rotor designs due to their higher manufacturing costs and complexity, the technology has advanced, and several manufacturers have developed and installed wind turbines that use this concept, and the trend is increasing towards this direction.

Outer rotor designs are often used in direct-drive wind turbine generators. These generators have a higher number of poles due to the larger diameter of the rotor structure. By using a high-pole-count outer rotor design, the generator can operate at the same rotational speed as the turbine blades, eliminating the need for a gearbox, which can reduce maintenance while increasing efficiency. Variable-speed wind turbines can take advantage of the fluctuating wind speeds. In addition, a higher number of poles can produce higher torque at lower speeds.

Note that outer rotor generators can also be made close to the turbine blade hub and integrated into the hub in wind turbines. This integration offers several advantages for the design and function of wind turbines including better space utilization, weight reduction, simplified design, improved dynamics (due to reduced moment of inertia around the yaw axis, making it easier to turn the nacelle to align with the wind direction), better aerodynamic efficiency and lower installation cost.

It can be concluded that variable speed operation allows the turbine to capture more energy from the wind by adjusting the rotor speed, but it requires more complex power electronics and control systems, which will be explained later in this chapter.

3.4 PHYSICS OF WIND ENERGY/POWER IN THE WIND
3.4.1 Power in the Wind

Figure 3.8 illustrates the principles of obtaining power from moving air, in the context of wind energy conversion in wind turbines. In the figure, the cylinder represents a pack of air with a base area A (which is equal to the swept area by wind turbine blades). The air moves with a velocity v along the x-axis, which is the direction of the wind. Note that since the mass m can be expressed as the product of air density ρ (in kg/m^3), cross-sectional area A (in m^2), and the length of the cylinder × (in meter), the kinetic energy (in Joules) of the moving air can be given by

$$E = \frac{1}{2}mv^2 = \frac{1}{2}(\rho Ax)v^2 \tag{3.1}$$

The power (in watts) generated from this moving air is the rate of change of kinetic energy with respect to time. Therefore, by differentiating the kinetic

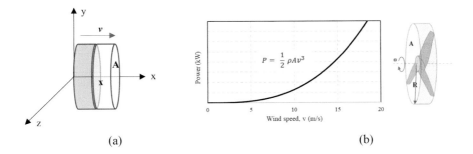

FIGURE 3.8 For the calculation of power in wind: (a) a packet of air moving through an area A and at a speed of v travelling x distance; (b) power in the wind as a function of wind speed that will be converted by a turbine rotating at an angular speed of w.

energy equation with respect to time, and given that the length of the cylinder x changes with time as the air moves, the power equation becomes

$$P = \frac{dE}{dt} = \frac{1}{2}\left(\rho A \frac{dx}{dt}\right)v^2 = \frac{1}{2}\rho A v^3 \qquad (3.2)$$

Note that with dx/dt being the velocity of the air, the power equation has been simplified.

The final result indicates that wind power is directly proportional to the cube of air velocity, and also to the air density and the swept area of the turbine blades. The cubic dependence means that even modest increases in wind speed can significantly enhance power output, an essential principle in wind energy harvesting. However, it is important to understand that not all the kinetic energy can be converted into electrical power due to losses in the electromechanical conversion process, as it was indicated in Figure 3.3 and the subsequent discussions on turbine blade efficiency.

Note that in this theoretical power value in the wind, the swept area A increases with the square of the rotor diameter (since $A = \pi D^2/4$). This means that doubling the diameter D increases the power available by a factor of four.

3.4.2 Temperature and Altitude Correction for Air Density

As it can be observed, the "heavier" the air, the more energy is received by the turbine. For example, the density of the air is 1.225 kg/m³ at normal atmospheric pressure and at 15°C and it is 1.293 kg/m³ at 0°C. However, it is critical to indicate that the density of the air also varies with temperature and altitude.

At higher altitudes, air density decreases because the atmospheric pressure is lower. This means there are fewer air molecules in a given volume, reducing the mass flowing through the wind turbine's swept area, and thus reducing the available wind power. In addition, warmer air is less dense than cooler air because it expands as temperature increases, again meaning fewer air molecules in a given volume. Therefore, the air density value should be corrected to consider these conditions.

The corrections can be done using the ideal gas law for temperature, and the barometric formula to estimate the change in air pressure and hence air density with altitude. In addition, there are empirical values and models that can be used to correct air density for altitude and temperature variations when calculating wind power. The International Standard Atmosphere (ISA) provides standard conditions and formulas for pressure and temperature at different altitudes.

In practical applications, wind turbine performance calculations often use standard atmospheric conditions at sea level for initial estimates and then apply correction factors for local temperature, pressure and humidity to obtain accurate values of air density for the specific location and conditions of the wind turbine site. These corrections are often enough for practical engineering calculations. Table 3.9 provides a set of data of typical air density correction values for various altitudes and temperatures based on ISA (International Standard Atmosphere) conditions. Therefore, the air density correction factor can be included as

$$r = 1.225 K_{Temp} K_{Alti} \quad (3.3)$$

TABLE 3.9 Typical Values of Air Density Corrections Using Density and Pressure Ratios

Temperature (°C)	Air Density (kg/m³)	Density Ratio ($K_{Temp} = \rho/\rho_0$)	Altitude (m)	Pressure (Atmosphere)	Pressure Ratio ($K_{Alti} = P/P_0$)
−10	1.341	1.095	0	0.101	1.000
0	1.275	1.041	500	0.954	0.942
10	1.247	1.018	1,000	0.898	0.887
15	1.225	1.000	1,500	0.845	0.834
20	1.204	0.983	2,000	0.795	0.785
30	1.164	0.950	2,500	0.746	0.737
40	1.127	0.920	3,000	0.701	0.692
			3,500	0.657	0.649
			4,000	0.616	0.608
			4,500	0.576	0.569
			5,000	0.538	0.532

Note that these are approximate values and should be adjusted for local atmospheric conditions. However, for an accurate calculation, it is recommended to measure the local atmospheric conditions directly or to refer to detailed atmospheric models that account for local variances.

3.4.3 Impact of Tower Height and Terrain on Wind Energy Production

It is shown above that the power available from the wind is heavily dependent on wind speed, which is known to increase with height above ground level. Therefore, wind turbines are typically mounted on tall towers to capture more energy from the faster-moving air at higher altitudes. However, there is a balance between capturing more wind and confronting the structural and economic challenges of building taller towers.

In wind energy studies, the reference height and the wind speed at that height are essential for extrapolating the wind profile to the heights at which wind turbines operate. Since power in the wind varies as the cube of wind speed, the power law equation obtained previously can be rewritten to indicate the relative power of the wind at height H versus the power at the reference height of H_0 as

$$\frac{P}{P_0} = \frac{\frac{1}{2}\rho A v^3}{\frac{1}{2}\rho A v_0^3} = \left(\frac{v}{v_0}\right)^3 \qquad (3.4)$$

Table 3.10 outlines the basic differences between the two methods that can be used to find the wind speed ratio hence the change in power due to the height variation. The choice of which model to use would depend on the context of the assessment, the level of detail required and the available data regarding the site's atmospheric conditions and surface roughness.

The values for the Hellmann exponent α (also known as wind shear exponent) in the equation for the power law profile and roughness length (l) for the logarithmic wind profile are often determined empirically through site measurements. For engineering and design purposes, site-specific data should always be used when available to ensure accuracy.

The reference height H_0 and wind speed at that height v_0 are not fixed and must be determined for each site based on local wind data. Therefore, accurate measurement of wind speed at the reference height is crucial for the proper design and placement of wind turbines to ensure they capture the maximum amount of energy.

TABLE 3.10 Comparison of Two Different Methods to Estimate the Impact of Tower Height

Feature	Power Law Profile	Logarithmic Wind Profile
Equation	$\left(\dfrac{v}{v_0}\right)=\left(\dfrac{H}{H_0}\right)^{\alpha}$	$\left(\dfrac{v}{v_0}\right)=\dfrac{\ln(H/l)}{\ln(H_0/l)}$
Variables	v: Wind speed at height H v_0: Wind speed at reference height H_0 α: Hellmann exponent	H: Height for wind speed estimation H_0: Reference height l: Surface roughness length
Surface Roughness Consideration	Not explicitly included; α is adjusted to account for it.	Explicitly included through the roughness length l.
Typical Hellmann exponent (wind shear exponent) α	Open calm sea: 0.10–0.12 Open land: 0.14–0.16 High crops, hedges: 0.22–0.24 Wooded countryside and trees: 0.25 Urban areas: 0.28–0.30 Large city with tall buildings: 0.4	N/A
Typical roughness coefficient values, l	N/A	Open calm sea: 0.0002–0.001 m Snow surface: 0.003 m High crops, hedges: 0.01–0.03 m Wooded countryside and trees: 0.2 m Urban areas with trees: 0.25 m Large city with tall buildings: 2–3 m
Accuracy	Less accurate, especially close to the ground and over complex terrain.	More accurate near the surface and over terrain with significant roughness.
Ease of Use	Simpler to apply with fewer parameters; used for quick estimates.	Requires knowledge of the surface roughness length, which may not always be available.
Preferred Application	Preliminary studies, quick assessments, areas with limited data.	Detailed wind resource assessments, wind farm design, research purposes.
Assumptions	Assumes a fixed power law exponent over the height.	Assumes a logarithmic increase in wind speed with height, which changes based on surface characteristics.
Atmospheric Stability	Typically, does not account for atmospheric stability variations.	Assumes neutral atmospheric stability.

The commonly used reference height (H_0) in wind studies is 10 m above the ground level. This standard is based on the average height of meteorological measurements for wind speed. The actual wind speed (v_0) at the reference height can vary greatly depending on the location, time of day, season and prevailing weather patterns. For initial assessments, average wind speed data is often used. This data is typically obtained from local meteorological stations or databases. For example, in a moderate wind regime, the wind speed at 10 m might average around 5–7 m/s. However, in coastal or offshore areas known for higher wind speeds, this average could be higher, say 7–10 m/s or more.

Both models given in Table 3.10 are fundamental to the field of boundary-layer meteorology and are commonly applied in wind engineering, environmental studies and for the design and analysis of wind farms. They are simply used to estimate wind speeds at different heights to optimize the height and placement of wind turbines. The specific equation chosen for use in a project can depend on the level of detail required, the terrain complexity and the availability of local wind data. For instance, in complex terrains or for detailed site assessments, the logarithmic wind profile may be preferred due to its more accurate representation of wind flow characteristics near the ground. For preliminary assessments or in the absence of detailed roughness data, the power law might be more commonly used due to its simplicity.

In the USA, the power law exponent and roughness length values are integrated into wind resource assessment tools and standards provided by the American Wind Energy Association (AWEA) and the National Renewable Energy Laboratory (NREL). Similarly, in Europe, these models conform to guidelines provided by organizations such as the European Wind Energy Association (EWEA) and are used in compliance with the standards set by the International Electrotechnical Commission (IEC), particularly IEC 61400, which is a part of the standard concerning wind turbines.

Figure 3.9 shows a graphical representation of wind speed profiles at different heights above the ground for various types of terrain. Each curve corresponds to a specific terrain category and shows the percentage of the geostrophic wind speed that is reached at different altitudes. The geostrophic wind speed is given as 25 m/s at 60° latitude, which serves as a reference wind speed at the top of the boundary layer in the atmosphere.

From left to right in the figure, the terrains are categorized as open sea, open terrain, forest/suburbs and city centre, each with a corresponding

180 ■ Reinventing the Power Grid

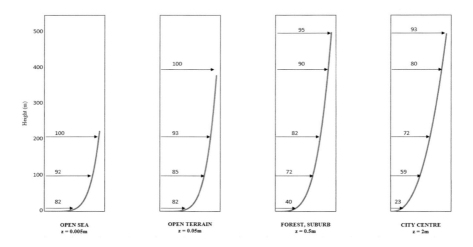

FIGURE 3.9 Graphical representation of wind speed profiles at different heights for various terrains.

roughness length value that indicates the typical surface roughness for that category. The roughness length values are the smallest for the open sea and the largest for the city centre, reflecting the increase in surface obstacles and friction that the wind encounters. The general conclusion is that the higher the altitude, the less influence the surface roughness has, and thus the higher the wind speed percentage indicated by the length of the arrows.

Table 3.11 is also given to classify various wind profiles based on different environmental and geographical conditions, which are essential considerations for the placement and operation of wind turbines. It should be noted that these are general patterns and the actual conditions can vary widely based on local topography and other factors.

3.5 CHARACTERISTIC FEATURES OF WIND TURBINES
3.5.1 Axial Momentum Theory and Maximum Efficiency of the Turbine (Betz Limit)

It can be observed that as the wind approaches a wind turbine, it slows down and spreads out. This change is necessary because the turbine captures a portion of the wind's energy. If a turbine were to take all of the wind's energy, the air would stop moving after passing through the turbine. However, if that occurred, no more wind could flow through, making the turbine inoperative. Therefore, the air must continue moving after

TABLE 3.11 Classification of Various Wind Profiles Based on Different Environmental and Geographical Conditions

Feature	Open Field	Forest/ Suburbs	City Centre	Hill	Sea Breeze Circulation
Wind profile	Steady, increases with height	Disrupted by trees, buildings	Highly disrupted, turbulent	Speed-up over crest, turbulent wake	Consistent pattern, driven by thermal differences between land and water
Surface roughness	Very low (~0.005 m)	Moderate (~0.5 m)	High (~2 m)	Variable, depends on vegetation	Low over water, variable over land
Speed change with height	Gradual	Rapid initial increase, then stable	Rapid initial increase, then stable	Sharp increase over crest	Consistent sea breeze during daytime
Impact of terrain	Minimal	Moderate to high	High	Significant	Low over water, higher near coast
Preferred turbines locations	Higher altitude preferred	Clearings or edges preferred	Generally unsuitable	Crest and upwind side	Coastal areas for consistent breeze
Typical issues	Low turbulence	Increased turbulence and wake effect	Very turbulent, complex wind flows	Eddy formation, dynamic loads on turbines	Site-specific, depends on land-water temperature gradient

it passes the turbine to maintain the flow of wind. In addition, the wind speed cannot remain unchanged because the turbine utilizes some of the wind's energy to spin its blades. As a result, the wind exiting the turbine is slower than the wind that enters.

The literature on fluid dynamics offers in-depth explanations of axial momentum theory. However, a simplified theoretical foundation will be outlined here to understand how to achieve maximum power extraction from turbine blades and their response to changing wind speeds, which is essential for determining control parameters.

The axial momentum theory is rooted in fluid mechanics, as it applies to both air and water flows due to their fluid characteristics. The theory is based on the concept that the wind (or any fluid flow) can be treated as a series of stream tubes where the flow is steady and incompressible. This theory applies the following principles from fluid mechanics:

- Conservation of mass (Continuity Equation): This principle indicates that the mass flow rate through a stream tube is constant. For a wind turbine, this means that the amount of air entering the front of the turbine is equal to the amount of air leaving behind the turbine, assuming the air is incompressible.
- Energy conservation (Bernoulli's Equation): Bernoulli's principle describes the conservation of energy in a flowing fluid and is used to relate the velocities upstream and downstream of the turbine to the pressure changes across the turbine.
- Conservation of momentum (Newton's Second Law): This is applied to the control volume of air passing through the turbine's rotor plane to relate the change in momentum of the wind to the forces acting on the turbine blades.

The slowing of the wind speed at the rotor plane leads to an increase in pressure just upstream of the rotor and a corresponding decrease in pressure immediately downstream of the rotor. The pressure difference is what actually drives the flow of wind through the rotor disc.

The theory also introduces the concept of induced velocities, which are the velocities that are induced in the wind due to the presence of the turbine. These are used to determine the effective wind speed at the rotor and to calculate the aerodynamic forces on the blades.

The axial momentum theory is often associated with the work of German physicist Albert Betz, who in 1919 derived what is now known as Betz's Law. Betz's Law states that the maximum theoretical efficiency of a wind turbine is 59.3%, meaning that no turbine can capture more than this percentage of the kinetic energy in the wind. In the paragraphs below, this maximum limit will be formulated and its variation with the rate of changes in wind speeds will be studied.

Figure 3.10 illustrates the concept of the axial momentum theory as applied to a horizontal-axis wind turbine. The wind turbine plane is illustrated in the middle with multiple blades (ideally with infinite numbers) where the turbine intersects the wind flow. The figure shows stream tubes (areas of airflow) before and after they intersect with the rotor plane

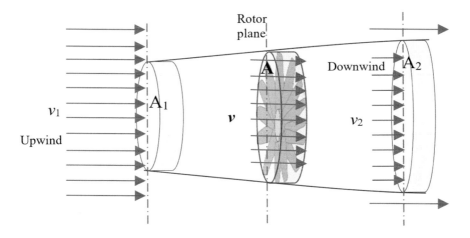

FIGURE 3.10 The axial momentum theory as applied to a horizontal-axis wind turbine.

(turbine-blade plane), where the wind's velocity and pressure change due to the presence of the turbine.

In addition, v_1 represents the wind velocity far upstream of the turbine, before it has been affected by the rotor, and v_2 represents the wind velocity far downstream of the turbine after it has been affected by the rotor. Furthermore, v indicates the velocity of the wind at the rotor plane (turbine-blade plane), which is lower than v_1 because some of the wind's energy has been extracted by the turbine.

In addition, the air pressure far upstream of the turbine (atmospheric pressure), the air pressure far downstream of the turbine (which is considered to return back to atmospheric pressure after some distance) and the air pressure at the rotor disc (which would typically be less than the upstream pressure due to the energy extraction by the turbine) needs to be considered. Furthermore, A_1 is the cross-sectional area of the stream tube far upstream of the rotor, where the flow has not yet interacted with the turbine, A is the cross-sectional area of the stream tube at the rotor plane and A_2 is the cross-sectional area of the stream tube far downstream, which theoretically should be equal to A_1 assuming incompressible flow.

Therefore, the theory assumes that the flow is steady and incompressible and that the rotor plane induces a pressure change that decelerates the wind as it approaches the rotor, resulting in a drop in velocity at the rotor plane, v. As the wind leaves the rotor, it is assumed to have a lower kinetic energy, indicated by the reduced velocity v_2.

Due to the conservation of mass, the mass of air passing through any two cross-sections of the stream tube in a given time is the same (since air is incompressible in this model). Therefore

$$r_A A_1 v_1 = r_A A v = r_A A_2 v_2 \qquad (3.5)$$

Where ρ_A is the air density. Hence the following can be considered

$$A_1 v_1 = A_2 v_2 \qquad (3.6)$$

The principle of mass conservation means that the mass flow rate, denoted as *dm/dt*, remains unchanged across every section of the stream tube. Using the mass flow rate, the "thrust force" F_T by the fluid can be obtained at the rotor plane as follows.

$$F_T = \frac{dm}{dt}(v_1 - v_2) = \rho_A A v (v_1 - v_2) \qquad (3.7)$$

Either "equating powers" or "equating thrusts", the velocity at the rotor (which will be needed to determine the power extracted by the turbine blades) can be determined.

3.5.1.1 Equating Powers

Due to the conservation of energy, the power extracted by the turbine P is the difference between the kinetic energy of the incoming wind and the outgoing wind. Therefore, the power on the turbine can be given by

$$P = \frac{1}{2}\rho_A (v_1^3 A_1 - v_2^3 A_2) = \frac{1}{2}\frac{dm}{dt}(v_1^2 - v_2^2) \qquad (3.8)$$

Since the extracted power is the work rate of the thrust force

$$P = dW/dt = F_T v = \frac{dm}{dt}(v_1 - v_2) v \qquad (3.9)$$

Hence the wind speed at the rotor plane can be calculated as

$$v = \frac{v_1 + v_2}{2} \qquad (3.10)$$

3.5.1.2 Equating Thrust

The thrust force can be represented as a function of the pressure drop Δp across the rotor plane.

$$F_T = A\Delta p \tag{3.11}$$

Hence applying the specific pressure drop, the thrust equation evolves into:

$$F_T = \frac{1}{2} \rho_A A \left(v_1^2 - v_2^2\right) \tag{3.12}$$

By equating the thrust expressions derived from the pressure drop and the change in momentum, a similar conclusion can be obtained, which indicates that the wind speed v on the rotor plane is the average of v_1 and v_2. This would also mean a smooth and gradual deceleration of wind without any energy lost to turbulence or other inefficiencies.

At this stage the axial induction factor "a" can be introduced to quantify the reduction in wind speed through the rotor plane relative to the free-stream wind speed. This factor is a measure of how much the wind slows down as it approaches the turbine blades, which is crucial for calculating the power that can be extracted from the wind.

$$a = \frac{v_1 - v}{v_1} \tag{3.13}$$

Therefore, $v = v_1(1-a)$ and $v_2 = v_1(1-2a)$ can be given. Using these definitions, the power in watts and thrust force in Newtons can be expressed as

$$P = \frac{1}{2} \rho_A A v_1^3 \left[4a(1-a)^2\right] \tag{3.14}$$

$$F_T = \frac{1}{2} \rho_A A v_1^3 \left[4a(1-a)\right] \tag{3.15}$$

From these two equations, the dimensionless power coefficient and thrust force coefficient can be obtained as

$$C_P = \frac{P}{P_1} = 4a(1-a)^2 \tag{3.16}$$

$$C_{F_T} = \frac{F_T}{F_{T1}} = 4a(1-a) \tag{3.17}$$

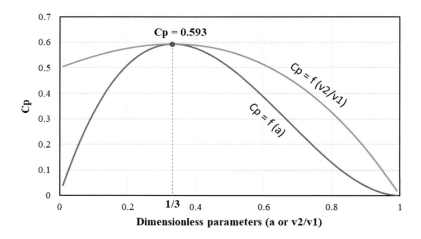

FIGURE 3.11 The power coefficient C_p with respect to two dimensionless parameters: the induction factor a $((v_1-v)/v_1)$ and the velocity ratio v_2/v_1.

The power coefficient can be given as a function of the wind speed ratio between the downstream and upstream wind speeds as

$$C_p = \frac{P}{P_1} = \frac{1}{2}\left[1-\left(\frac{v_2}{v_1}\right)^2\right]\left[1+\frac{v_2}{v_1}\right] \qquad (3.18)$$

Note that the value of "a" can vary between 0 and 1. Therefore, variations of C_p can be plotted as a function of "a" and the speed ratios v_2/v_1 as illustrated in Figure 3.11.

Figure 3.11 is fundamental in wind turbine design, showing the limitations of power extraction efficiency and the importance of optimizing both the induction factor (a) and the velocity ratio (v_2/v_1) to approach the Betz limit. It illustrates that C_p peaks at a certain point for both parameters, which is the optimal point for energy extraction from the wind. This peak value is the Betz limit, which is the theoretical maximum efficiency for a wind turbine, indicating that no turbine can capture more than 59.3% of the kinetic energy in the wind. According to the graph, the maximum value of C_p is 0.593, and it is achieved when both the induction factor and the velocity ratio at 1/3, are in line with the relationship given earlier ($v_2 = v_1(1-2a)$)

Note that the Betz limit can also be calculated using the power coefficient obtained above. For example, if we define the speed ratio as $a = v_2/v_1$,

$$C_p = \frac{1}{2}\left[1-a^2\right]\left[1+a\right] \tag{3.19}$$

This C_p expression can be used to find its local maximum which also defines the Betz limit. This can be obtained by taking the derivative of C_p with respect to the axial induction factor defined above and then setting it equal to zero.

$$\frac{dC_p}{da} = \frac{1}{2}(1-3a)(1+a) = 0 \tag{3.20}$$

There are two potential solutions of this standard quadratic equation ($ax^2 + bx + c = 0$), but since the downstream velocity v_2 must be positive and less than v_1, the positive root has to be considered, which is

$$v_2 = \frac{v_1}{3} \tag{3.21}$$

This solution indicates that, for maximum power extraction, the downstream wind speed v_2 should be one-third of the upstream wind speed v_1.

Using this result, "v" can also be calculated as

$$v = \frac{2v_1}{3} \tag{3.22}$$

If these solutions are substituted into the turbine power, P equation, the maximum efficiency of the turbine can be obtained given as

$$\eta_{turbine} = C_P = \frac{P}{P_1} = \frac{16}{27} \approx 0.593 = 59.3\% \tag{3.23}$$

If no blades exist, the kinetic energy of the air passing through area A during the time dt is ½ $\rho_A v_1^3 dt$. The Betz limit states that no more than 59.3% of this kinetic energy in the wind can be captured by a wind turbine. Note that this is a theoretical limit and assumes that the turbine has an infinite number of blades and there are no energy losses due to drag and mechanical inefficiencies that occur in real-world turbines. In practical terms, modern three-bladed wind turbines usually have a maximum power coefficient of 0.49 (see Figure 3.11) and often operate with power coefficients under 0.4 when the wind speed is above their specified rating.

It can be emphasized here that in a broader context, the axial momentum theory and its derived models, such as the Blade Element Momentum (BEM) theory, are used to design not only wind turbines but also marine propellers and other types of turbines, such as hydrokinetic turbines that extract energy from water currents as in tidal energy generation. The common factor across these applications is the behaviour of the fluid—air or water—as it interacts with a turbine in a similar manner, governed by the above-described laws of fluid mechanics.

In Table 3.12, a comparative table of the dynamics between wind speed and air pressure near wind turbines, water flow speed and pressure in pipes and for water pipes when energy is extracted (such as through a hydroelectric turbine or a tidal turbine) is summarized. Therefore, this table provides how the principles of fluid dynamics are similarly applied

TABLE 3.12 The Dynamics between Speed and Pressure in Air and Water

Aspect	Wind Turbines (Energy Extraction)	Water Pipes (Energy Extraction)	Water Pipes (Water Flowing Normally)
Fluid Speed	Decreases as wind approaches and passes through the turbine due to energy extraction.	Decreases at the site of energy extraction, such as a turbine in a dam, due to conversion of kinetic energy to mechanical energy.	Increases in narrow sections due to conservation of flow rate, decreases in wider sections.
Pressure	Increases in front of turbine blades due to deceleration of wind and decreases behind the blades due to acceleration.	Decreases at the turbine due to energy conversion, creating a pressure differential across the turbine.	Decreases when flow speed increases (narrower sections) and increases when flow speed decreases (wider sections).
Energy Conversion	Converts kinetic energy of wind into mechanical (and then electrical) energy, resulting in a slowdown of wind speed.	Converts kinetic energy of water into mechanical (and then electrical) energy, resulting in a slowdown of flow speed.	Energy conserved within the flow; kinetic and potential energy interconvert due to pipe geometry or elevation changes.

(*Continued*)

TABLE 3.12 (*Continued*)   The Dynamics between Speed and Pressure in Air and Water

Aspect	Wind Turbines (Energy Extraction)	Water Pipes (Energy Extraction)	Water Pipes (Water Flowing Normally)
Resulting Effect on Fluid	Wind speed slows down at the turbine's rotor plane, creating a pressure differential that is harnessed for energy.	Water flow speed slows down at the turbine, creating a pressure differential that is harnessed for energy.	Water flow speed changes due to pipe diameter, leading to variations in pressure that drive the flow.
Pressure Differential	Created to harness energy from the wind; higher pressure in front of the blades and lower pressure behind them.	Created to harness energy; higher pressure in front of the hydro turbine and lower pressure behind it.	Drives the water flow; managed by pipe geometry, valves, and pumps.
Bernoulli's Principle	Applied in the sense that an increase in speed results in a decrease in pressure (and vice versa) around the turbine.	Applied at the turbine; the decrease in speed due to energy extraction results in an increase in pressure ahead of the turbine.	Applied as water moves through different pipe sections, with pressure changes corresponding to speed changes.
Practical Application	Pressure differences are critical for the wind turbine's ability to extract energy from the wind.	Pressure differences are used to generate electricity, with controlled flow rates to optimize energy production.	Pressure differences are used to ensure water flows where needed and at the desired rate.

in hydroelectric energy extraction, mirroring the process of wind energy conversion in wind turbines.

3.5.2 Tip Speed Ratio (TSR)

The shaft and rotor speed is an integral part of the overall design and operational control, specifically concerning the mechanical and electrical aspects of the turbine system. It is related to the generator's rotational speed requirement hence the gearbox design if used. Therefore, the torque produced by the wind turbine is also related to the shaft speed, which together is used for monitoring turbine performance and for the control systems that manage blade pitch and other operational parameters.

The tip speed of a wind turbine blade refers to the speed of the outermost part of the blade, the tip, as it moves through the air. It is the fastest point along the length of the blade, R, because it travels the largest circumference in the same period of time as any other part of the blade. The tip speed is a critical factor in the design and operation of wind turbines as it affects the efficiency, noise levels and structural integrity of the turbine due to the lift and drag on the blades.

Therefore, Tip Speed Ratio (TSR) in wind turbines is defined as the ratio of the speed of the tips of the wind turbine blades to the speed of the wind.

$$\text{TSR} = \lambda = \frac{\text{Blade tip speed (m/s)}}{\text{Wind speed (m/s)}} = \frac{\omega R}{v} \qquad (3.24)$$

A well-optimized TSR can maximize the conversion of wind energy into mechanical energy. Since the power coefficient C_p is a measure of how efficiently a wind turbine converts the kinetic energy in the wind into electrical power, TSR is also related to the ratios used in the formulation of the power coefficient, C_p.

Note that TSR and the factors $(v_2-v)/v_2$ and v_2/v_1 represent similar concepts related to the operation of wind turbines. Since these factors are dimensionless, without physical units, they are applicable and useful for comparison purposes, hence contributing to a comprehensive understanding of the wind turbine's interaction with the wind and its performance. They are related to how effectively a wind turbine interacts with the wind and converts kinetic energy into mechanical energy:

- A high TSR generally indicates a blade design that is optimized for aerodynamic efficiency at a certain wind speed. It is most relevant when the wind speeds are consistent with the design point of the turbine.

- These ratios $(v_2-v)/v_2$ and v_2/v_1 are more focused on the wind speed changes due to energy extraction by the turbine. Both describe the reduction in wind speed across the rotor disc and the increase in rotational speed behind the rotor. They can be used to understand the wake effects and the performance of the turbine in terms of how much kinetic energy is converted as the wind passes through the rotor.

These parameters contribute to a good understanding of the wind turbine's interaction with the wind and its performance. Therefore, the value of C_p depends on both the wind speed variations before and after the turbine blades and the TSR, since these factors influence the distribution of wind speeds and the angle of attack experienced by the blades, hence the lift and drag forces that determine the turbine's power output. In conclusion, the TSR is one of the key non-dimensional parameters that characterize the operation of a wind turbine as it is directly related to the efficiency and design of the turbine, encapsulated in the C_p value.

However, it is important to note that the tip speed of the blades is not directly proportional to the downstream wind speed. Instead, the relationship between the two is influenced by the turbine's operation and design.

The tip speed is the speed at which the tip of a wind turbine blade moves through the air and is a function of the blade length and the rotational speed of the turbine. The downstream wind speed, on the other hand, is the speed of the wind after it has passed through the turbine rotor. This speed is affected by the energy extraction process as the wind transfers some of its kinetic energy to the turbine blades.

The wind speed changes represent the fraction by which the wind speed is reduced as it passes through the rotor disc. The greater the energy extraction (which means the more efficient the turbine), the greater the reduction in wind speed downstream. However, if a turbine extracts too much energy (which means if the blades are rotating too fast), it can cause turbulent wake effects and reduce the efficiency of energy capture, known as the Betz limit.

Therefore, while the tip speed and the downstream wind speed are related through the energy extraction process, they are not simply proportional to each other; the relationship is more complex and governed by the principles of aerodynamics and the conservation of momentum.

3.5.3 Turbine/Blade Arrangements, Solidity

Solidity in the context of turbine blades refers to the ratio of the total blade area to the area traced by the rotating blades (the swept area, A). The solidity, a key parameter in turbine design, can be modified by altering the number of blades or the blade chord (the width of the blade at a given distance from the centre of rotation). Figure 3.12 illustrates the specific impacts of changing solidity on turbine performance. It can be concluded that:

- Low solidity in turbine blades means consistent power output across various speeds but with a lower maximum output due to increased drag.

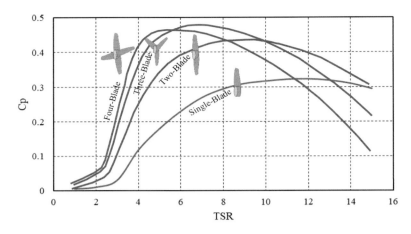

FIGURE 3.12 C_p characteristics as a function of TSRs for different blade numbers (~solidity).

- High solidity leads to a sensitive performance curve with a high peak, but too much solidity can reduce maximum power output because of stalling.
- As used in all modern turbines, three blades are best for balance in power efficiency and the curve width that impacts the power generation in wider wind speed ranges.
- Two blades could be considered due to a wider peak which may capture more power, especially at higher wind speed regions, but it results in a lower maximum C_P.
- A large number of blades with small solidity can be made, but the result will be costly and less durable.
- High-solidity turbines are good for specific uses like water pumps or battery charging, where starting torque and low-speed power are key.

Therefore, power performance in wind turbines is typically obtained through the non-dimensional C_P–λ curve. Figure 3.12 illustrates a curve for a typical three-blade turbine, where the maximum C_p value is 0.47 at a TSR of 7. Although this value falls short of the Betz limit due to factors like drag, tip losses and stall effects at lower TSRs, the Betz limit cannot be reached because of imperfections in blade design.

The graph illustrates that as the number of blades increases, the peak C_p generally decreases, and the efficiency range widens. However, the three-blade turbine indicates a balance between high efficiency and a broad operating range, which is why it is widely used in modern wind turbines. The peak efficiency points move to the left (to lower TSR values) as the number of blades increases, showing that turbines with more blades are optimized for lower rotational speeds relative to the wind speed.

Various wind turbine designs have been developed to meet specific operational requirements, environmental conditions and performance targets. However, the choice of turbine and blade arrangement is influenced by various factors such as cost, efficiency, intended use and environmental considerations. The three-blade HAWT design offers a good balance of efficiency, reliability and cost-effectiveness, which is why it is the most prevalent in large-scale wind power generation. However, alternative designs like the Savonius and Darrieus have their own niches, particularly in urban and low-wind-speed environments.

3.5.4 Yaw Angle

In the real world, wind direction also varies, and turbine rotors cannot instantaneously adapt to these changes, typically trailing the actual wind direction by a small margin. This misalignment reduces rotor efficiency and increases the stress on the turbine.

Note that while the wind power varies as a function of wind speed, some areas experience wind primarily from one direction, while others have multiple prevailing wind directions or none at all. Therefore, to consider this variability effect, the turbine's yaw system aligns the rotor with the prime wind direction to improve efficiency. Areas with a dominant wind direction typically require less yaw system operation. The distribution of wind direction does not, however, indicate how frequently wind direction shifts occur, which would necessitate yaw adjustments.

Wind speed also impacts wind direction stability. At lower speeds, wind direction is more likely to change significantly, whereas at higher speeds, the direction tends to be more stable. Consequently, the likelihood of the turbine yawing varies with the different wind speeds.

Yaw misalignment, the angle difference between the turbine's nacelle and the actual wind direction, typically forms a normal distribution around the average during power generation, with a standard deviation generally under $10°$.

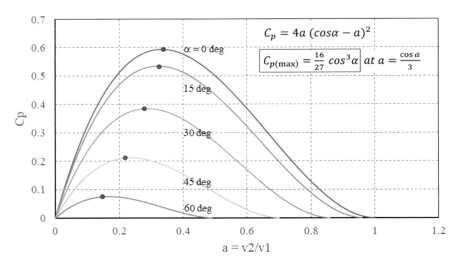

FIGURE 3.13 Coefficient of performance C_p and variations with yaw angle (yaw misalignment).

This yaw misalignment diminishes rotor efficiency. The reduction in efficiency C_p as yaw misalignment increases is shown in Figure 3.13. Theoretical equations, which include the yaw misalignment angle α, can predict this power coefficient for a turbine rotor experiencing steady yaw.

The yaw angle affects the turbine's ability to extract energy from the wind, and an optimal yaw angle is typically aligned with the wind direction. In the figure, the different curves represent the C_p values at yaw angles of 0° (aligned with the wind), and others (significantly misaligned with the wind). Therefore, in practical terms, this figure demonstrates the importance of maintaining proper alignment of the wind turbine with the wind direction. Active yaw control systems in wind turbines adjust the nacelle's orientation to minimize the yaw angle and maximize power output.

In practice, generating such plots would require either a detailed aerodynamic model that can predict performance at various yaw angles or empirical data obtained from a turbine operating under controlled yaw conditions. These plots are valuable for understanding how yaw misalignment affects the performance of a wind turbine and for designing control systems that can mitigate these effects.

3.5.5 Pitch Angle, Power versus the Rotor Speed and the Cp versus TSR Characteristics

To obtain the power versus the rotor speed characteristics for wind turbines, the C_p characteristics should be obtained as a function of λ (TSR) and the blade pitch angle, β. The C_p value can be determined through experimental methods or computed numerically using various aerodynamic models such as Blade Element Momentum (BEM), Computational Fluid Dynamics (CFD) or General Dynamic Wake (GDW) models. However, these methods can be complex and computationally intensive. Therefore, numerical approximations based on empirical relations can be used as they offer a simplified method to estimate the C_p that can be represented by a general form as $C_p(\lambda, \beta) = f(\lambda, \beta)$. This form can provide the power coefficient based on inputs of the TSR and blade pitch angle.

Table 3.13 lists various coefficients from different sources, which are used in the parametric power coefficient models for wind turbines. Note that these constants strictly depend on the studied machine. The general

TABLE 3.13 The Parametric Power Coefficients for Wind Turbines from Different Sources

Coefficient	Method 1 (Slootweg, 2001)	Method 2 (Heier, 2014)	Method 3 (Thongam, 2009)	Method 4 (De Kooning, 2013)	Method 5 (Ochieng, 2014)	Method 6 (Dai et al., 2016)
c_1	0.73	0.5	0.5176	0.77	0.5	0.22
c_2	151	116	116	151	116	120
c_3	0.58	0.4	0.4	0	0	0.4
c_4	0	0	0	0	0.4	0
c_5	0.002	0	0	0	0	0
c_6	13.2	5	5	13.65	5	5
c_7	18.4	21	21	18.4	21	12.5
c_8	0	0	0.006795	0	0	0
c_9	−0.02	0.089	0.089	0	0.08	0.08
c_{10}	0.003	0.035	0.035	0	0.035	0.035
x	2.14	0	0	0	0	0

General Formulas

$$C_p(\lambda, \beta) = c_1 \left(\frac{c_2}{\lambda_i} - c_3 \beta - c_4 \lambda_i \beta - c_5 \beta^3 - c_6 \right) e^{-c_7/\lambda_i} + c_8 \lambda$$

$$\lambda_i^{-1} = (\lambda + c_9 \beta)^{-1} - c_{10}(\beta^3 + 1)^{-1} \qquad i: \text{incremental values of TSR}$$

Formulas used to generate Figure 3.14b.

$$C_p(\lambda, \beta) = 0.5176 \left(\frac{116}{\lambda_i} - 0.4\beta - 5 \right) e^{-21/\lambda_i}$$

$$\lambda_i = \left[\frac{1}{\lambda + 0.08\beta} - \frac{0.035}{\beta^3 + 1} \right]^{-1}$$

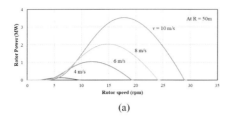 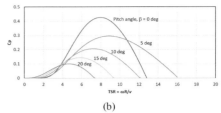

FIGURE 3.14 The power versus rotor speed characteristics for various wind speeds and at a zero-pitch angle (a), and C_p versus TSR (λ) characteristics for different pitch angles at a given wind speed (b).

$C_p(\lambda, \beta)$ function and an approximation used to generate Figure 3.14b are also given in the table. λ_i is an intermediate variable used in the calculation.

To plot power versus rotor speed characteristics (Figure 3.14a):

- Select an appropriate empirical model for C_p from the table which fits the turbine's design and operating conditions.
- Calculate the TSR λ for various rotor speeds. This requires knowledge of the wind speed and the blade length (radius R).
- Calculate C_p for each λ and β combination.
- Compute the power extracted by the turbine for a given air density, the swept area of the rotor, the wind speed and C_p values.
- Plot the calculated power P against the corresponding rotor speeds to visualize the power versus rotor speed characteristics for the wind turbine without the need for extensive experimental or high-fidelity computational analyses.

Similarly, C_p versus TSR (λ) characteristics (Figure 3.14b) can be obtained for different pitch angles under a given wind speed as described above for the power characteristics.

3.6 THE EQUATION OF MOTION, BLADE AND GENERATOR TESTING AND MODELLING

3.6.1 The Equation of Motion

As indicated in Figure 3.15, the basic turbine setup includes a mechanically coupled turbine/generator system that can be described by the mechanical equation of motion. This equation also forms the principle control concept

Wind Energy Systems ■ 197

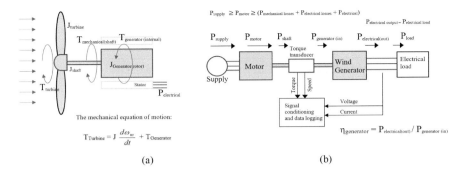

FIGURE 3.15 The background of the equation of motion and the principle structure of the mechanical network in the direct drive wind turbine systems (a), and the components of a dynamometer setup for testing a generator for model parameters (b).

by utilizing the characteristic behaviours of the components of the system which will be discussed in the next section.

In the figure, $T_{turbine}$ and $T_{Mechanical(shaft)}$ is the instantaneous value of the torque developed by the turbine and the torque on the shaft in Nm respectively, and J is the total polar moment of inertia of the system referred to the generator shaft in kg.m^2, ω_m is the angular velocity of the motor shaft in rad/s and $T_{Generator}$ is the instantaneous value of braking torque on the generator shaft in Nm that opposes to the mechanical (shaft, rotor) torque. The internal torque generated by the generator due to electromagnetic forces often referred to as electromagnetic torque, generates internal power in the generator.

Note that in the equation of motion, it is assumed that the system is with constant inertia, hence $dJ/dt=0$. Note that the generator torque contributing to the internal power also has multiple components: friction, windage and internal useful torque. The equation ignores the value of B, the damping coefficient or the viscous friction coefficient.

The equation of motion balances the mechanical power input and losses with the rate of change of kinetic energy in the system. It shows that torque developed by the turbine is counter-balanced by the generator (internal) torque and a dynamic torque $J(d\omega_m/dt)$ that is present only during transient operations and has a negative sign during deceleration (which assists $T_{Generator}$ and maintains motion). The equation is used to model the dynamic behaviour of the generator's rotor and to design control systems that can manage its speed and output.

Acceleration or deceleration of the turbine drive depends upon whether $T_{Turbine}$ is greater or less than $T_{Generator}$. If $T_{Turbine} > T_{Generator}$ the speed increases with time, which may be due to increasing wind speed or reduced electrical load or reduced generation by the control system. If $T_{Turbine} = T_{Generator}$, this may indicate a steady-state operation (when angular acceleration is zero, $d\omega/dt = 0$) at a constant wind speed and at the highest efficiency point of the turbine (at the Betz limit). If $T_{Turbine} < T_{Generator}$ this can indicate that the speed decelerates which occurs when the wind speed drops as the generator is delivering power to the electrical load or power grid.

3.6.2 Generator Testing and Modelling

The mechanical energy from the wind turbine's rotor is converted into electrical energy by the generator. An electrical equivalent circuit of a BPMG, whether it is an outer rotor or inner rotor type, used in wind turbines is illustrated in Figure 3.16a. Note that since such a generator operates on the principle of electromagnetic induction, with the rotor containing permanent magnets providing a constant magnetic field, there is no need for field excitation as in traditional synchronous generators. This simplifies the equivalent circuit.

The equivalent circuit includes the stator windings, which are represented as inductances L in series with resistances R. The inductance represents the magnetic field's effect on the stator coil, and the resistance represents the electrical losses in the stator coil. The relative motion between the permanent magnets of the rotor and the stator windings induces a voltage in the stator coils. This induced voltage is called back Electromotive Force (EMF) and is represented in the equivalent circuit

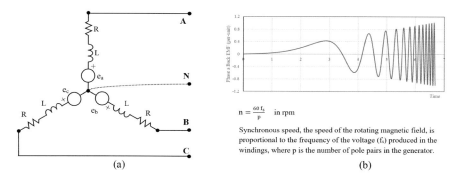

FIGURE 3.16 The equivalent circuit of a three-phase BPMG (a) and the induced voltage (back EMF) of one of the phases as the turbine starts from standstill (b).

by a voltage source *e*. Note that *e* should be in phase with the current for an ideal generator. For a three-phase generator, there will be three sets of windings, each represented by its own equivalent series *R* and *L*, and the back EMF sources will be phase-shifted by 120° relative to each other. The key difference between an outer rotor and an inner rotor type in terms of their equivalent circuits is the value of the inductance and resistance, which can differ due to the different magnetic path lengths and winding structures. The basic topology of the equivalent circuit, however, remains similar for both rotor types.

Figure 3.16b shows the induced voltage over time in one phase of a wind turbine generator as it starts from a standstill and the rotor begins to turn and accelerates until it reaches its operational steady speed. The exact shape and progression of such a voltage would depend on the characteristics of the wind turbine and generator (defined by the equation of motion), as well as the wind conditions at the time. Note that at the very beginning, the plot is flat, indicating that there is no induced voltage. This corresponds to the wind turbine being at a standstill with no movement in the rotor and, consequently, no electromagnetic induction occurring in the stator coils, which defines the cut-in speed.

The cut-in speed of a wind turbine is the minimum wind speed at which the turbine will start generating power. It is influenced by various factors, including the cogging torque of the generator, the inertia of the system and losses in the mechanical system (defined by the equation of motion). Cogging torque is the torque due to the interaction between the permanent magnets of the rotor and the stator slots of a generator. It can create resistance to the initial rotation of the generator. High cogging torque can increase the cut-in speed because more wind force is required to overcome this resistance and start the rotor turning. The moment of inertia of the rotor system, including the blades, the shaft, the coupling and the generator rotor, determines how much torque is needed to accelerate the system from standstill to operating speed. A higher inertia means that more energy is required to start the motion, which could lead to a higher cut-in speed. The equation of motion for a wind turbine generator, incorporating input torque, moment of inertia, damping coefficient and internal torque, provides a dynamic model of the system that includes these factors. The cut-in speed is determined by the point at which the input torque from the wind exceeds the sum of the resistive torques, including cogging torque and mechanical losses, and provides enough acceleration (considering the system's inertia) to start generating electricity.

As illustrated in Figure 3.16b, the voltage amplitude increases as the turbine hence the generator starts rotation. Voltage fluctuations indicate the alternating character of the induced voltage, a result of interactions between the rotor's magnetic fields and the stator coils. As defined by Faraday's Law, the magnet flux and induced voltage are 90° out of phase. Towards the graph's right end, the frequency markedly rises, implying the turbine's progression towards its designated operational speed. Due to the direct mechanical link, the correlation between wind speed and the generator's back EMF is direct. Therefore, any increase or decrease in wind speed causes a corresponding acceleration or deceleration of the rotor, altering the magnetic flux's rate of change through the stator coils, which in turn increases or reduces the magnitude of back EMF while the profile remains the same, sinewave.

3.6.2.1 Dynamometer Setup

Using a dynamometer setup given in Figure 3.15b, the relationship between power output and rotor speed, as well as efficiency versus various operational parameters of the generator, can be determined through the application of appropriate electrical loads. It is important to note that the dynamometer test arrangement can also serve to emulate the wind turbine's control systems. This emulation is achievable by integrating a variable-speed drive system between the power supply and the drive motor, which allows for the simulation of wind speed fluctuations. These variations in simulated wind speed are then mirrored by corresponding adjustments in the electrical load on the generator side, enabling a comprehensive analysis of the generator's response to changing wind conditions.

The figure includes a power supply and motor setup that emulates the mechanical power input similar to what a wind turbine would generate. It shows that the mechanical power input by the motor exceeds the combined mechanical losses, electrical losses and the actual power delivered to the generator. This highlights the selection of power ratings of the dynamometer components. In the setup, an inline torque transducer measures the torque applied from the motor to the generator, which is pivotal since it directly correlates to the mechanical power delivered to the generator shaft, a key determinant of the generator's efficiency.

Torque transducers also measure rotor speed, which is used to define the power transmitted through the shaft. An electrical load simulates the load on the generator, representing either the electrical grid or other loads the wind generator might typically serve, such as a converter charging a battery storage system. Furthermore, the electrical power generated is defined by its voltage and current which are conditioned and logged for the accurate documentation of the generator's performance across various operational scenarios.

Furthermore, a blade testing platform can be utilized in a wind tunnel to simulate and assess the control systems of the wind turbine. This setup provides a means to test and optimize the aerodynamic performance of the blades under controlled conditions, which is vital for the overall efficiency and effectiveness of the wind energy conversion system.

3.6.2.2 Back EMF Constant and Modelling

To measure the back EMF constant k_e of a BPMG, a test is conducted where the generator is driven by a motor at a set constant speed. During this test, the voltage generated across the terminals of each phase is measured. This voltage also referred to as the no-load voltage, can be expressed either as a root mean square (RMS) value or as a peak value of the sinusoidal voltage waveform.

The generator's shaft speed can be monitored via the torque transducer. It can also be determined accurately by measuring the frequency of the no-load voltage for a known number of pole pairs in the generator. This no-load phase voltage measurement, taken between one of the phase terminals (for example, terminal A) and the neutral point (N), is conducted at varying speeds. The collected data is plotted to confirm the back EMF's linearity with respect to speed and to accurately calculate k_e for RMS or peak values.

For instance, in Figure 3.17, two sets of back EMF test results are given for generators equipped with two different rotor types—those with sintered magnets and those with bonded magnets. It should be noted that the

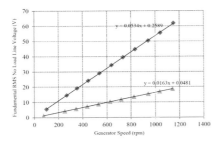

FIGURE 3.17 Two sample back EMF measurements in two small BPMGs with sintered (top line) and bonded (bottom line) PM rotors.

voltages shown in the sample graph represent the fundamental RMS values of the line-to-line voltages, measured because the neutral point N was not accessible. Therefore, the phase no-load voltage in a star-connected winding configuration should be derived as indicated alongside the figure using the term $\sqrt{3}$.

Comparing the slopes of the lines in the graph, it can be defined which generator more effectively converts mechanical speed into electrical voltage. A steeper slope signifies a higher back EMF constant, indicating greater voltage output at a given speed, which generally correlates to improved performance. The relatively straight lines suggest a linear relationship between speed and voltage, which is expected for such generators until saturation effects or other nonlinear factors become prominent. Furthermore, while the y-intercepts of the lines are not precisely zero, their low values suggest minimal measurement errors or design peculiarities. It is also important to mention that the back EMF constants are sometimes represented in units of V/rad/s, which will require a conversion from rpm to rad/s.

As indicated in the sample figure, the type of magnet used in the generator influences the back EMF magnitude. Generators with sintered magnets, which usually possess a higher magnetic flux density, tend to exhibit a larger back EMF constant in comparison to those with bonded magnets. Moreover, the actual back EMF waveform might not be a perfect sinusoid due to harmonics, cogging and other nonlinear effects. An oscilloscope can be employed during testing to capture the waveform, and the Fourier analysis may be applied to model the waveform with high fidelity accounting for any harmonics. This information is essential for engineering the generator's drive electronics and for accurately predicting its operational performance.

3.6.2.3 Measuring Winding Resistance and Inductance

The final two parameters in the electrical model of the generator are the resistance and inductance of the three-phase windings. To consolidate the identification methods for these parameters, Table 3.14 has been provided. For measurement precision, calibration of equipment and an appropriate range of measurements are critical. In addition, appropriate safety measures must be established to mitigate the risk of overheating the windings during the testing process.

TABLE 3.14 The Methods to Determine Winding Resistance and Inductance in the BPMGs

Parameter	Equations	Remarks
Resistance (R)	$R_{DC} = V/I$ $R_{std} = R_{meas}(1+\alpha(T_{std}-T_{meas}))$ $R_{AC} = R_{DC} \cdot K_s$ V: Applied DC voltage I: Measured current R_{meas}: Measured resistance at the actual winding temperature. R_{std}: Resistance corrected to the standard temperature (20°C). α: Temperature coefficient of the wire material. T_{std}: Standard reference temperature for resistance measurements. T_{meas}: Measured winding temperature during resistance testing.	• While the rotor of the generator is locked, DC voltage is applied across a winding and current is measured. • This can be done in star configuration from line to neutral, and in delta configuration from line to line. • High-current injection up to rated current should be considered for accuracy but temperature should be monitored to prevent overheating. • The applied voltage should be adjusted carefully to prevent excessive current flow that is possible due to the absence of back EMF voltage. • Average resistances over all phases should be calculated and lead resistances should be subtracted if significant, especially in low winding resistance machines. • The resistance should also be calculated for temperature effects. • The AC resistance of a winding, often referred to as the effective or "real" resistance when AC is applied, is typically higher than the DC resistance primarily due to frequency-dependent skin effect. For BPMG operating at high frequencies, the increase in AC resistance can be significant, leading to increased I^2R losses and reduced efficiency. • In practice, the AC resistance can also be measured directly at the operating frequency using an LCR meter or an impedance analyser.
Inductance (L)	$L = \tau R$ $\tau = \dfrac{t}{\ln\left(\dfrac{I_{final}}{I_{initial}-I_{initial}}\right)}$ L: Inductance of the winding.	• Inductance of a phase winding can be measured by applying step voltage and observing current transient on an oscilloscope. Then time constant is measured by using transient current response. • Inductance should be measured at a low current level to avoid core saturation and should reflect average inductance.

(*Continued*)

TABLE 3.14 (*Continued*)   The Methods to Determine Winding Resistance and Inductance in the BPMGs

Parameter	Equations	Remarks
	τ: Time constant determined from the transient current response after a step voltage is applied. t: Time at which the current is measured. $I_{initial}$: Initial current just after the step voltage is applied. I_{final}: Steady-state current.	• If the inductance is position-dependent, this should be included in the model. Therefore, inductance should be measured at multiple rotor positions. • Note that the variable inductance model is crucial for control systems in variable speed applications.

3.6.2.4 Electrical Loading

The primary objective in a wind turbine system is to extract the maximum "active power" from the generator, analogous to Maximum Power Point Tracking (MPPT) in solar PV cells, while accommodating changes in wind speed and rotor speed. This is illustrated in the equivalent circuit of the generator in Figure 3.18a, where maximum power transfer can be achieved when the phase current is in phase with the voltage, resulting in the highest power factor (ideally 1.0). Due to the inductive nature of the generator windings, achieving the condition requires the addition of a capacitive load impedance, which compensates for the winding's inductance at various phase current values. The appropriate capacitive impedance can be determined through pre-calculations based on the known

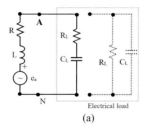
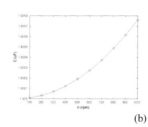
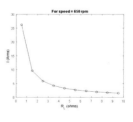

FIGURE 3.18   Per phase equivalent circuit of a BPMG with the electrical load options for maximum "active power" generation (a), and a typical set of analytical results using a parallel load capacitance with a 10 Ω load resistance (for the phase resistance of 78.4 mΩ and the inductance of 762 mH) and the RMS stator current versus load resistance at a constant rotor speed.

winding parameters of resistance R and inductance L as illustrated in Figure 3.18b for a small wind turbine generator.

Furthermore, it is crucial to note that the essence of the complex control system in a wind turbine is to adjust the equivalent electrical load side impedance to optimize active power extraction under varying wind conditions. This is accomplished by manipulating the turbine's yaw and pitch as well as the load current (hence the load). Section 3.8 will discuss these control strategies in detail.

3.6.3 The Blade Testing

Figure 3.19 illustrates a test setup that is used for evaluating the performance of the small-scale wind turbine blades, specifically for determining the power coefficient C_p characteristics.

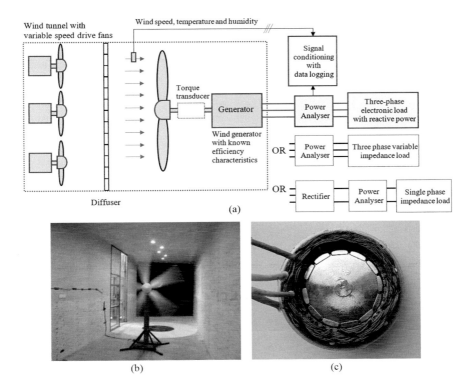

FIGURE 3.19 The principal diagram of the test setup used for determining the power coefficient C_p characteristics of the turbine blades (a) and the photos showing the turbine operating in a wind tunnel (b) and the small-scale brushless PM generator (c).

The test is done in a wind tunnel where variable-speed drive fans generate a controlled airflow that simulates the wind. In the tunnel, the diffuser spreads the airflow evenly across the cross-section of the tunnel as it approaches the turbine blades under test, ensuring uniform wind conditions.

The wind turbine blades are attached to the generator to load the blades using the electrical loads. If the efficiency characteristics of the generator are known C_p of the blades can be calculated. If the generator's efficiency characteristic is unknown, an inline torque transducer needs to be included between the turbine and the generator to calculate the rotor power.

Accurate wind speed measurements are typically obtained using anemometers or non-invasive hot-wire sensors or using air pressure tubes (Pitot tubes). Note that hot-wire anemometers are sensitive and can measure very low speeds, which makes them suitable for turbulence measurements and low-speed wind tunnel testing.

Signals from the sensors are conditioned for accurate and reliable data logging. The power analyser measures the electrical power output from the generator including the frequency of the terminal voltage that can be used to calculate the rotor speed. Then, data obtained from the power analyser and environmental sensors are logged and are used to calculate the power coefficient C_p of the turbine blades by comparing the mechanical power input (as derived from wind speed and blade characteristics) with the electrical power output of the generator.

It is known that the C_p is a critical parameter for wind turbine blade design and performance evaluation, which represents the efficiency across different wind speeds. With the data collected from wind tunnel testing, an equivalent model of the blade can be obtained which can predict how the full-scale blade would perform in real-world conditions. However, such models are often refined with the empirical data to ensure they accurately reflect physical performance. If the scale model in the wind tunnel includes a pitching mechanism, the testing can be extended to determine how the blade's C_p characteristic changes with pitch angle.

However, testing large turbine blades in a wind tunnel typically involves scaled-down models due to the large size of the actual blades used in commercial wind turbines. The process is carefully designed to simulate real-world conditions as closely as possible within the confines of a controlled environment. Table 3.15 provides a list of the processes and their details under specific categories that should be followed for performing the large blade testing.

TABLE 3.15 The Basic Process for Testing Large-Scale Turbine Blades

Category	Details
Model creation	Accurate scale model based on full-size blade is made.
	Materials mimic actual blade properties.
Wind tunnel setup	A wind tunnel that can accommodate the size of the model and generate the appropriate wind speeds is chosen.
	The model is equipped with sensors such as pressure taps, strain gauges, and possibly flow visualization systems like tufts or smoke generators.
	Capable of simulating real-world wind speeds and profiles, including capable of reproducing the turbulence and shear profiles found in the actual operational environment of the turbine.
Mounting system	Model mounted on force balance system to measure forces and moments in multiple axes.
	Allows for pitch and yaw angle changes.
Data acquisition	All sensors are calibrated to ensure accuracy.
	Baseline tests are conducted in the empty tunnel without the blade to account for any background noise or bias in the measurements.
Testing procedures	Testing at multiple wind speeds and orientations.
	Ensuring consistent results through multiple trials.
	Utilizing particle image velocimetry or smoke for flow pattern analysis.
Data analysis	Data analysed to determine performance metrics such as lift, drag and moment coefficients.
	Results are compared to computational models or field data to validate and refine the aerodynamic models of the blade.
Iterative process	Iteratively improving blade design based on test outcomes.
Safety and protocols	Strict adherence to safety guidelines.
	Managing test data with discretion.
Challenges	Ensuring scale model accuracy.
	Reynolds number should match that of full-size blade operation.
Final validation	Correlating tunnel data with actual field performance.

3.7 WIND CHARACTERISTICS, RESOURCES AND ANALYSIS OF WIND REGIMES

Figure 3.20 is a graphical representation of a typical wind turbine's power curve, which relates wind speed to the power output of the blades when driving a generator.

On the far left, the curve represents the theoretical power that can be extracted from the wind for a given swept area of the blade. The next curve represents the power output after accounting for the theoretical Betz limit. Since the practical blades do not capture the theoretical limit and due to the design and material limitations and the aerodynamic losses (such as

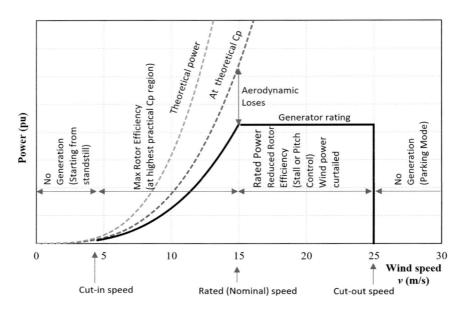

FIGURE 3.20 The graphical representation of a typical wind turbine's power curves from the theoretical power to the generator ratings.

drag on the blades and non-optimal air flow paths), the power curve is further reduced as illustrated on the third curve. The third curve does not start from the origin indicating the cut-in wind speed where the start of the power generation begins.

A horizontal line extends from the peak of the actual power curve to the right, which represents the rated power output of the generator selected. The turbine control systems will adjust operational parameters, such as blade pitch, to ensure that the power output does not exceed this level, even if wind speeds continue to increase.

The point where the actual power curve meets the generator rating line corresponds to the rated wind speed. At this speed, the generator produces its rated power, and it will maintain this power level across a range of wind speeds by managing the aerodynamic performance of the blades. Far to the right, beyond the rated speed range, there is a point where the solid curve drops to zero, which is the cut-out speed, where the turbine is designed to shut down to prevent damage from high wind conditions.

This final solid curve provides valuable insights into the performance of a wind turbine across different wind speeds, laying the groundwork for broader discussions on wind characteristics, resources and the analysis of wind regimes. Understanding the variability and consistency of wind speeds is crucial for optimizing the design and placement of wind turbines to maximize energy capture. By analysing the quantity and quality of wind at a potential site and correlating them with the turbine's power curve, developers can make informed decisions about turbine selection, site layout as well as on energy yield forecasts. An analysis of patterns in wind behaviour primarily includes both speed and direction over time, which will be discussed after a brief background study on measurement aspects.

3.7.1 Wind Quality and Measurements

Table 3.16 provides an overview of the various factors that determine and influence wind quality and levels across different regions and altitudes. In addition, local wind map data is integrated into this table which provides insights into each of the factors which can help to make informed decisions about infrastructure and other human activities and wind farm developments.

TABLE 3.16 An Overview of the Factors That Impact Wind Quality and Levels

Factors	Descriptions	Impact on Wind Quality/Level	Detailed Local Wind Map Data
Geographical distribution	Refers to patterns influenced by latitude, such as trade winds, westerlies, polar easterlies.	Coastal regions or flat terrains may have more consistent and stronger winds.	Provides specific wind patterns for a region, helping in identifying best locations for wind farming.
Topographical influence	The effect of landforms, such as mountains and valleys, on wind patterns.	Mountains can funnel wind, increasing its speed, while large obstacles might block or redirect prevailing winds.	Can show how specific hills, valleys or urban structures influence wind flow, crucial for local wind energy projects.
Climatic factors	Temperature differences leading to pressure variations, influenced by land-sea contrasts and seasons.	Temperature and pressure differences drive wind patterns. Seasonal changes, like monsoons, can shift wind directions.	Offers insights into local temperature-driven winds, such as land-sea breezes in coastal areas.

(Continued)

TABLE 3.16 (*Continued*)   An Overview of the Factors That Impact Wind Quality and Levels

Factors	Descriptions	Impact on Wind Quality/Level	Detailed Local Wind Map Data
Temporal scales	The time frame of wind data, ranging from real-time to historical averages.	Real-time and historical data aids in understanding long-term patterns.	Provides hourly or daily data, crucial for activities like aviation or local event planning.
Altitude levels	Wind patterns can vary significantly with altitude.	Surface winds might differ from those at higher altitudes.	Can provide data on wind shear or variations in altitude, crucial for aviation and tall structures.
Infrastructure and human activity	The presence of urban and suburban infrastructures can influence local wind patterns.	Urban areas can create microclimates, altering local wind patterns due to the heat island effect and physical barriers.	Essential for urban planning, showing how buildings or parks influence local wind patterns.
Global wind patterns	Large-scale wind patterns, influenced by Earth's rotation and atmospheric circulation.	These dictate the primary wind directions and strengths across large regions, influencing weather and climate systems.	Provides context, showing how global patterns influence or are modified by local factors.

Note that global wind map data is also related to the variation of prevailing winds across our planet, revealing the influence of mountains and plateaus on continental breezes, and affecting the flow of seasonal gusts. These patterns, traced over years, offer an insight into the shifting rhythms of the world's wind quality and levels. In addition, Table 3.17 provides a structured overview of the process for measuring wind resources, detailing each step from site assessment to compliance with international standards. Note that the final analysis of wind data will be covered in detail in the consecutive subsection.

Note that LiDAR technology is increasingly used in wind resource assessment due to its accuracy, flexibility and ability to provide detailed data on wind patterns. It uses laser light and the Doppler shift principle to measure wind speed at various heights. It can provide a 3D profile of the wind, including wind speeds, turbulence and wind shear at different

TABLE 3.17 A Structured Overview of the Process for Measuring Wind Resources

Process	Remarks
Site assessment	• Historical wind data from weather stations or databases.
Measurement setup	• Strategic points for data collection based on topography and obstacles. • Met masts with anemometers, wind vanes and possibly sonic anemometers, LiDAR (Light Detection and Ranging) or SoDAR (Sonic Detection and Ranging). • Temperature, pressure and humidity sensors.
Data collection	• High-frequency recording with data loggers, averaged over standard periods (e.g., 10 minutes). • Minimum of 1 year to capture seasonal variations.
Maintenance and quality	• Regular checks and maintenance of all measurement equipment. • Data validation processes to ensure accuracy.
Analysis and compliance	• Wind data analysis and visualization. • Consideration of local wildlife protection and safety regulations during installation and maintenance.

altitudes, and the data can inform the optimal placement of turbines to maximize energy production and minimize wake effects between turbines. Post-installation, LiDAR can also be used to monitor and verify the performance of wind turbines. In addition, the technology is particularly useful for offshore wind farms, where traditional meteorological masts are more challenging and costlier to install, which is done by the floating LiDAR systems.

Accurate wind resource assessment is critical in forecasting performance and determining the return on investment for wind farms. To achieve this, the setup for wind resource measurement utilizes a range of specialized tools and methodologies designed for accurate data collection. Following the detailed guidelines during the measurement setup enables developers to collect essential data, which is required for evaluating the feasibility and potential success of wind energy projects. Table 3.18 presents an overview of the tools, techniques, and standards employed in wind resource measurement, emphasizing the essential components and approaches required for an accurate evaluation of wind energy potential.

It can be concluded that while these guidelines and methods are structured to optimize accuracy and reliability, historical instances reveal shortcomings in their application. These issues include complexities arising from the site's terrain, variability in wind patterns over time, constraints due to technology, impacts of climate change, limitations imposed

TABLE 3.18 An Overview of Tools, Techniques and Standards in Wind Resource Measurement

Tools and Techniques	Remarks
Tools and hardware for measurement	• Anemometers are used for measuring wind speed; cup anemometers for general use, sonic anemometers for precision in complex terrains. • Wind vanes are used to measure wind direction. • Data loggers are needed to store data from anemometers and wind vanes. • Meteorological masts (Met Masts) tall towers are used for mounting anemometers and wind vanes. • LiDAR sensing technology. • SoDAR utilizes sound waves to measure wind speed and direction at various heights. • Measure temperature, humidity, atmospheric pressure • Ensure precise location alignment, via GPS units. • Transmit data to a central database
Height of measurements and selection of locations	• Anemometers and wind vanes at multiple heights on met masts, commonly at 10m, 30m, 50m, and above • Measurements typically at or above hub height, often around 80-100 meters for large turbines • Based on wind farm layout, land topography and avoidance of obstruction • Placement away from trees, buildings to prevent turbulence or shadowing effects
Measurement duration and frequency	• Continuous measurements for at least 1 year for annual variability • High-frequency recording, averaged over standard intervals (such as 10 minutes)
Installation and compliance with standards	• Performed by specialized teams • Equipment carefully calibrated and securely installed • Following international standards such as IEC 61400-12-1 for wind velocity measurement and data analysis

by financial and policy considerations, errors in data processing and analysis and oversights regarding wildlife and environmental impacts.

3.7.2 Analysis of Wind Resources

Wind turbine energy production can be estimated using two of the fundamental methodologies: statistical techniques and wind regime analysis.

Statistical techniques rely on large historical datasets of wind speed and direction and often also consider temperature, air density and turbine-specific performance data. They can include simple regression models, time series analysis, probabilistic models and machine learning algorithms, which aim to identify patterns, trends and correlations in historical data to forecast future energy generation. They also assess the variability and uncertainty in wind energy production.

Wind regime analysis involves a more fundamental examination of the local atmospheric dynamics that drive wind patterns. It includes the study of wind speed distributions, directional distributions and the occurrence of extreme wind events. These techniques often involve the use of wind resource assessment tools and models that can simulate the physical processes governing the wind patterns in a given area. The results are used to define the optimal siting of wind turbines, the selection of turbine models and the design of wind farms.

The principles of analysis of wind resources are illustrated in Figure 3.21. When the wind speed data is known, the percentage occurrence of different wind speed classes at the site can be generated as in the histogram figure at the top. Each bar represents a range of wind speeds (in m/s) and the percentage of time that wind speeds within that range occur.

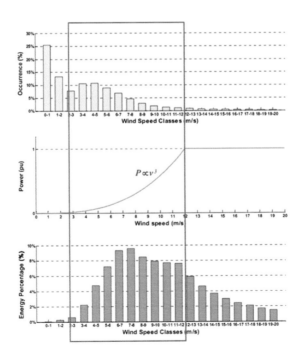

FIGURE 3.21 The principles of the analysis of wind resources.

A power curve of a wind turbine curve illustrates how the power output of a wind turbine increases with those wind speeds. As it is known the power output follows a cubic relationship and is low at low wind speeds, begins to increase significantly at moderate speeds and eventually levels off at higher wind speeds. The dashed vertical line indicates that the power output does not increase due to the generator's power limit.

The bottom histogram graph shows the distribution of the contribution of each wind speed class to the total potential energy generation. As expected, even though lower wind speeds occur more frequently, they contribute less to the overall potential energy generation. The most significant contribution to energy generation comes from wind speeds in the mid-range, where the power output of the turbine increases rapidly with wind speed. The graph demonstrates that the greatest energy production potential is not necessarily from the most frequently occurring wind speeds but rather from the moderate to high wind speeds that are capable of generating more power.

Figure 3.22a presents a sample of wind data measured at an offshore site in Australia, recorded at a height of 10 m above the sea surface.

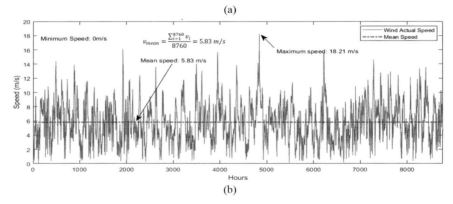

FIGURE 3.22 A sample measured wind data format and associated units averaged over 1 hour (a), and the time series of the measured wind speed and the critical values (b).

The original dataset includes a full year of wind data (8,760 hours), including both wind direction and wind speed measurements. Utilizing this data, a time series of wind speed, a wind rose and a wind speed histogram can be generated as given in the subsequent figures. Figure 3.22b illustrates the time series of wind speed, highlighting the average (mean) wind speed along with the recorded minimum and maximum wind speeds.

3.7.2.1 Wind Roses

A wind rose is an important tool in wind analysis and system design, providing key insights into wind direction, speed and frequency. It is instrumental in the strategic design and placement of wind turbines, including determining the optimal range for yaw angles. There are several tools available that can automate the process of creating a wind rose from raw wind data. These range from specialized meteorological software to general data analysis tools like Matlab, Excel, Python or R.

To construct a wind rose, wind directions are categorized into segments, typically based on 8 or 16 cardinal points such as North (N), Northeast (NE), East (E) and Southeast (SE). Then, wind speeds are grouped into defined ranges appropriate for the analysis. For each directional segment, the frequency of occurrences for each wind speed range is determined by calculating the instances within each category. Figure 3.23a shows the wind rose corresponding to the wind data used in the preceding time series analysis.

Note in the figure that the compass points (N, E, S and W) with the segments show the direction from which the wind is blowing. As it is shown in the legends, the wind rose uses colour to differentiate wind speed ranges,

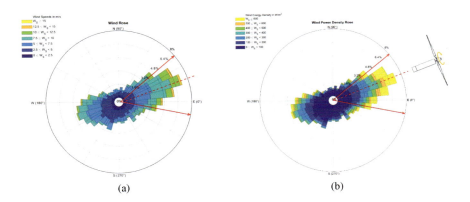

FIGURE 3.23 The velocity wind rose (a) and the power density wind rose with the turbine position (b) of the given wind regime at the given location.

measured in meters per second (m/s). The length of each segment indicates the frequency of winds coming from that direction within the specific speed range. The longer the segment, the more often winds of that speed range occur from that direction. The result shows a higher frequency of winds coming from the eastern direction (towards the west), as indicated by the longer segments in that direction. The highest wind speeds (red segments) are most prominent in the easterly direction. The percentages along the edges represent the proportion of time that the wind blows from a particular direction at any speed.

In Figure 3.23b, a wind power density rose is given, which is similar to a wind rose and is a measure of the energy available in the wind at a specific site and is proportional to the cube of the wind speed.

Synthesizing the data from both wind roses, it can be observed that the eastward winds are both the most common and the most powerful at this location. This is evidenced by the extended segments and the specific colour coding for the easterly direction seen in both charts. The initial wind rose illustrates the prevalence of various wind speeds, while the subsequent wind power density rose sheds light on the potential for energy generation from these winds, which is a key factor for wind energy development. The data clearly indicates that the eastern direction not only predominates but also carries the highest wind power density, establishing it as the optimal direction for siting wind turbines. The significant wind power densities highlight the necessity of an eastward orientation in turbine design and placement to optimize energy yield, including the adjustment of the yaw angle to harness the most dominant and frequent wind currents. From the analysis, it is determined that setting the yaw angle at 20° eastward is optimal to maximize the annual wind power capture.

3.7.2.2 Probability Density Functions (PDFs)

The Rayleigh and Weibull PDFs are two statistical tools used in various fields, including wind energy analysis, to model wind speed distributions. The Rayleigh distribution is a special case of the Weibull distribution, used primarily when modelling wind speed data in regions where the variance is approximately equal to twice the square of the mean wind speed. Table 3.19 provides a comparative overview of the two distributions.

As it is highlighted in the table above, the Weibull distribution is widely regarded as the standard for wind energy applications. The wind speed histogram and Weibull probability density function of the wind

TABLE 3.19　A Comparative Overview of Rayleigh and Weibull PDFs

Distribution	Remarks
Rayleigh distribution	• A special case of the Weibull distribution suitable for modelling wind speed when the variance is approximately twice the square of the mean. • The Rayleigh PDF for wind speed is given by $$f(v) = \frac{v}{\sigma^2} \exp\left(-\frac{v^2}{2\sigma^2}\right)$$ • where σ is the scale parameter related to the mean wind speed. • Best used in areas with minimal obstructions to wind flow and where the variance is about twice the square of the mean wind speed. • Applications include preliminary analysis of wind speed distribution and energy potential estimation in simple terrain. • It is a simpler model with only one parameter, hence easier to fit with less computational resources. However, it is less flexible and may not accurately represent wind speeds in areas with complex terrains or significant obstructions.
Weibull distribution	• It is a versatile distribution that provides a better fit for wind speed data across a wide range of conditions. • The Weibull PDF for wind speed is given by $f(v) = \frac{k}{c}\left(\frac{v}{c}\right)^{k-1} e^{-\left(\frac{v}{c}\right)^k}$ • where k is the shape parameter, indicating the distribution's skewness and c is the scale parameter that is related to the mean wind speed • It is best used in various terrains and wind conditions, widely accepted for wind speed modelling. • Applications include detailed wind resource assessment, turbine selection and placement, energy production estimation as well as risk analysis. • It is flexible, can model different types of wind regimes and provides a more accurate representation of actual wind speed distributions. However, it is slightly more complex due to the two parameters, requiring more data for accurate parameter estimation.

data mentioned above are shown in Figure 3.24, where the shape factor k and scale factor c of Weibull PDF are identified as 2.3 and 8.93 respectively.

In the figure, the x-axis represents wind speed in m/s and the y-axis represents the probability of occurrence as a percentage. The bars represent the actual measured wind speed data as a histogram mentioned previously, which indicates the percentage of time that the wind speed fell within a particular speed interval over the measurement period. The tallest bar indicates the wind speed range that occurred most frequently.

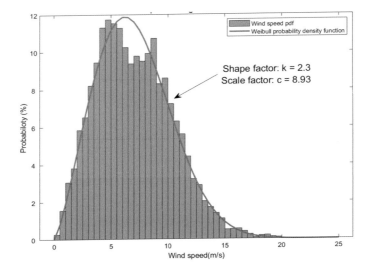

FIGURE 3.24 The wind speed histogram and Weibull probability density function of the wind data studied.

The line represents the Weibull PDF fit to the wind speed data, which is a model that estimates the probability distribution of wind speeds based on the shape and scale factors. Note also that the shape factor k determines the shape of the distribution, with higher values indicating a more peaked distribution, and the scale factor c is related to the wind speed at which the maximum occurrence is observed. The Weibull curve appears to fit the actual data quite well, particularly around the peak, and the most frequent wind speeds are around the scale factor of 8.93 m/s, and the distribution of wind speeds is asymmetrical, with a longer tail extending towards higher wind speeds.

Computational Fluid Dynamics (CFD) is also used to model wind flow over complex terrains and around obstacles to understand the impact on wind speed and turbulence. In addition, various commercial software tools are available that can help in the analysis of wind regimes.

It can be concluded that by combining these analysis techniques, developers can optimize the placement of wind turbines to maximize energy production. Additionally, understanding wind resources and regimes is key for integrating wind energy into the power grid, including energy storage solutions and demand management strategies to handle the variability in wind power generation.

3.8 WIND POWER TO ELECTRICITY: CONTROL ISSUES WITH SENSORS

Wind turbines feature a broad array of designs characterized by various attributes: they can be small or large in size; have a vertical or horizontal axis structure; face upstream (towards the wind) or downstream (away from the wind) and include different numbers of blades, commonly one to three. They are also designed for onshore or offshore applications. In addition, the operational control of large turbines is managed through one of four main strategies: fixed speed with fixed or variable pitch, or variable speed with fixed or variable pitch. Furthermore, large turbines incorporate yaw angle control.

Figure 3.25 illustrates the spectrum of wind turbines, to summarize the possible diverse control schemes. However, this section will specifically address the control of large modern wind turbines powered by BPMGs, which typically feature a three-bladed, horizontal-axis, upstream design with variable-speed and variable-pitch capabilities, as highlighted in the figure.

3.8.1 Blade Design and Regulation

The intricacies of blade design and the interplay of lift and drag forces (see Figure 3.26), as affected by external wind conditions and the rotational speed of the blades, that is highly complex and beyond the coverage of this

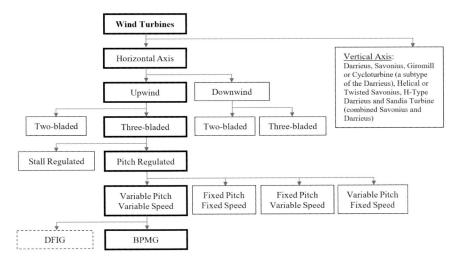

FIGURE 3.25 The spectrum of wind turbines illustrating the possible diverse control schemes.

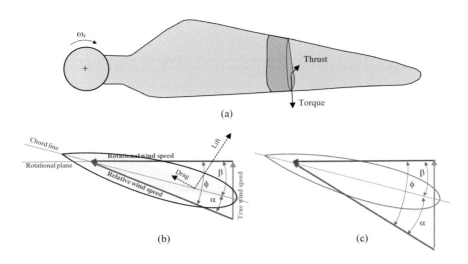

FIGURE 3.26 The blade element (a) with wind speeds and forces, the diagram of wind speeds (b) and the diagram under higher wind speed, close to stall-state (c). b: Local pitch angle, a: Angle of attack, f: Inflow angle. a=f−b, if b=0, a=f.

book. However, the figure illustrates the fundamental relationship between the actual wind speed, the wind generated by blade rotation and the resulting relative wind speed, which together determine the angle of attack.

The importance of the lift-to-drag ratio (L/D ratio) in wind turbine blade design is highly critical. The blades should generate the maximum possible lift while incurring the minimum drag. Although there is no universally "ideal" L/D ratio, it can vary from around 40:1 to 80:1.

This ratio directly affects the turbine's efficiency (hence energy output), its performance in low wind, the mechanical load on the turbine structure and components and its operational speed range.

Note that the L/D ratio is intrinsically linked to the airfoil design which is the cross-sectional shape of the blade designed to produce lift when air flows around it, which is optimized during the airfoil design process involving CFD simulations, wind tunnel testing as well as field testing. Table 3.20 provides the key aspects of airfoil design that affect the lift-to-drag ratio in wind turbine blades.

It can be concluded that wind turbine blades are primarily designed based on either lift or drag principles. Lift-based blades are aerodynamically crafted, resembling the airfoil of an aeroplane wing, to generate a differential pressure that results in the lift, propelling the blades to rotate rapidly and efficiently, which is also accommodated in modern large-scale

TABLE 3.20 The Fundamental Elements of Airfoil Design That Influence Lift and Drag Forces

Airfoil Parameter	Impact on Lift	Impact on Drag	Remarks on Wind Turbine Blade Design
Shape and profile	Determines pressure distribution	Influences boundary layer flow	Optimized for efficient airflow around the blade to maximize lift and minimize drag.
Angle of attack	Increases lift up to a certain point	Minimal impact until near stall	Critical for maximizing energy capture; excessive angles can lead to stall, dramatically increasing drag.
Camber	Increases lift at specific angles	May increase drag	More camber generally means more lift but can also lead to higher drag, so it must be balanced.
Aspect ratio	Less direct impact	Higher ratio reduces drag	Long, slender blades (high aspect ratio) are usually more efficient, with a higher L/D ratio.
Surface roughness	Can reduce lift if excessive	Increases drag	Smooth surfaces are crucial for maintaining laminar flow and reducing turbulence.
Thickness	Affects structural integrity	Thicker profiles increase drag	A balance between structural strength and aerodynamic efficiency is sought; too thick can be detrimental to performance.
Leading/ trailing edge	Affects lift at low angles	Affects airflow detachment	Well-designed edges help minimize turbulence and vortex formation at the trailing edge, reducing drag.

wind turbines. Conversely, drag-based blades rely on the force of the wind pushing against their surface to cause rotation. These blades are simpler and more robust, but less efficient and are typically used in smaller applications where cost and ease of maintenance take precedence over maximum energy output.

It is worth mentioning that stall and pitch regulations are two different methods used in wind turbines to control the power output and rotational speed of the rotor in high wind conditions.

Stall regulation uses the inherent aerodynamic properties of the blades to self-regulate the power output of the turbine without the need for active control systems. It relies on the principle that beyond a certain angle of attack, the blade will stall, resulting in a decrease in lift and an increase in drag, thereby limiting the power extracted from high wind speeds.

As it is illustrated in Figure 3.26, as the wind turbine blade rotates, it encounters airflow from two directions: the incoming true wind and the wind flow due to the blade's rotation. The vector sum of these two airflows gives the apparent wind direction relative to the blade (Figure 3.26b). As the wind speed increases (Figure 3.26c), the rotational speed of the blade increases, which changes the apparent wind direction and thus the angle of attack—that is the angle between the chord line of the blade (an imaginary straight line between the leading and trailing edges of the blade) and the direction of the apparent wind. Stall-regulated turbines are less efficient at lower wind speeds and do not capture as much energy as pitch-regulated turbines over a wide range of wind conditions.

At a certain high angle of attack, the airflow over the blade surface can no longer follow the contour of the blade due to the increased turbulence. This results in a condition known as "stall". As the actual wind speed increases, the rotational component of the airflow also increases. The vector sum of these two components of airflow alters the angle of attack further, leading to a deeper stall if the wind speed continues to increase. Once the blade stalls, the lift coefficient decreases sharply, and drag increases, causing the efficiency of the blade to drop. This natural reduction in lift helps to limit the rotational speed of the turbine and consequently, its power output.

The turbulence around the blade can lead to increased vibration, which can become a source of mechanical stress on the turbine structure, potentially leading to maintenance issues or reduced lifespan. Therefore, such turbines have a simpler design with fewer moving parts, which can potentially lead to lower manufacturing and maintenance costs. They are more common in small wind turbine applications.

In a lift-based wind turbine, the rotor blades are designed with airfoil profiles to interact with the incoming wind, generating both lift and drag forces. These forces are significantly influenced by the blade's angle of attack. As the angle of attack increases, there is a corresponding increase in both lift and drag. Therefore, it is desirable to identify an optimal angle of attack that maximizes lift while minimizing drag, ensuring efficient turbine performance.

Furthermore, there are three methods used in wind turbines to control the rotor speed and power output: active stall regulation, passive stall regulation and pitch control, particularly during high wind speeds. Each method has distinct characteristics and operational approaches (see Table 3.21).

Figure 3.27 visually compares how each regulation strategy manages power output across varying wind speeds, with pitch regulation offering the most consistent power output over the widest range of conditions. Active stall allows for continued operation at higher speeds by inducing

TABLE 3.21   Comparison of Three Major Regulation Methods Used in Wind Turbines

	Passive Stall Regulation	Pitch Regulation	Active Stall Regulation
Description	Relies on the fixed aerodynamic design of the blades.	Blades are actively rotated along their longitudinal axis.	A less common approach where blades are actively turned in the opposite direction of pitch control.
Objective	To naturally reduce lift and power output when wind speeds exceed a certain threshold.	To vary the angle of attack, particularly in high wind speeds to reduce lift and power output.	To deliberately induce stall on the blades to reduce power output.
Control systems	No moving parts or active control systems; the stall is a result of blade design.	Requires sophisticated mechanisms and control systems for blade pitch adjustments.	Utilizes sensors and control systems to actively monitor wind conditions and adjust blade angles.
Complexity	Simpler and more cost-effective than active control systems but offers less flexibility.	Provides precise control over turbine performance, with higher complexity in control systems.	More complex than passive stall control, with a need for reliable control systems and actuators.
Flexibility and efficiency	Less adaptable to changing wind conditions but effective in specific scenarios.	Allows for optimal efficiency across a broad range of wind speeds with greater operational control.	Offers more control than passive systems but less common due to complexity.
Typical application	Commonly used in smaller turbines where the cost and complexity of active systems are not justified.	Widely used in modern large-scale turbines, especially where efficiency and adaptability are key.	Suitable for scenarios where active control can be justified despite the complexity.

stall, while passive stall has a more pronounced drop in power output once the wind exceeds the design speed for natural stall conditions.

Pitch regulation is an active control method as it involves mechanically adjusting the angle (pitch) of the blades to control the lift and the rotational speed in large wind turbines. As studied previously, the blades can be pitched to an optimal angle for energy capture at different wind speeds, or feathered (increasing the angle of attack where the blade no longer produces lift effectively) to reduce the energy capture in high winds. As expected, the mechanism for pitching the blades adds complexity, more

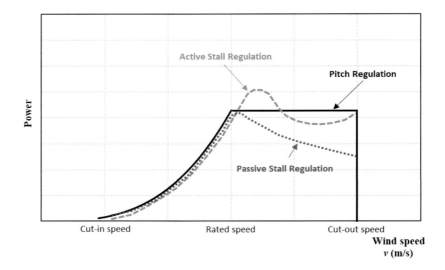

FIGURE 3.27 Principal power curves under active-stall, passive-stall and pitch control.

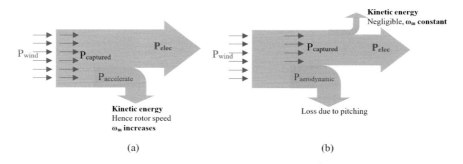

FIGURE 3.28 Power flow diagram in wind turbine under two different pitch control settings: (a) constant pitch operation and (b) pitch to feather operation.

moving parts and requires a control system, which increases initial costs and maintenance. However, pitch-regulated turbines offer greater control over the turbine's performance, allowing high efficiency and for optimization of power output across varying wind conditions. The pitch system is also used to actively feather the blades at dangerously high wind speeds to protect the turbine from damage.

Figure 3.28 demonstrates the role of pitch control in regulating the energy conversion process in wind turbines, optimizing the power output and protecting the turbine from excessive speeds during high wind conditions. For example, if the pitch angle of the blades is held constant (Figure 3.28a), the wind power is partly transformed into captured power which is the useful power harnessed by the rotor. Some of the captured power accelerates the rotor hence increasing the kinetic energy of the system. The remaining part of the captured power is converted into electrical power by the generator. The goal in this state is to capture the maximum possible power from the wind without over-speeding the rotor.

If the pitch angle of the blades is adjusted to the 'feather' position, which is an angle that reduces the lift produced by the blades, this reduces the power extracted from the wind (see Figure 3.28b). In this mode, the kinetic energy in the system is negligible, and the rotor speed is kept constant. However, since the blades are not at an optimal angle due to pitching, there are significant aerodynamic power losses that reduce the electrical output. Note that feathering is a proactive blade control technique to manage turbine speeds and loads, while stall is a condition that can result from excessive angles of attack and is either used as a passive control mechanism or avoided as part of a turbine's control strategy.

3.8.2 Modelling of Wind Turbines

There are a number of key disciplines that are associated to wind turbine systems. These include aerodynamics, structural dynamics, control engineering, electrical engineering, mechanical engineering, material science, meteorology, environmental science, hydrodynamics and economics. Although this highlights the complexity of the wind turbine system, it is the interplay between these fields that allows for comprehensive modelling.

Figure 3.29 illustrates the major disciplines involved in the modelling specifically for an offshore wind turbine system. Aerodynamics involves the interaction of wind with the turbine blades and the aerodynamic forces and moments generated as a result. Rotational dynamics refers to the dynamics of moving parts in the turbine, such as the rotor and the drivetrain. The electrodynamics aspect covers the generation of electricity from the rotational motion and its conversion for grid compatibility. Structural dynamics involve analysing the responses of the turbine structure to dynamic loads. For offshore turbines, hydrodynamics includes the interaction with water currents and waves. Soil dynamics consider the interaction between the foundation of the turbine and the soil. Note that since the generator is a direct-drive BPMG, the rotational dynamics directly connect the rotor to the generator without the need for a gearbox.

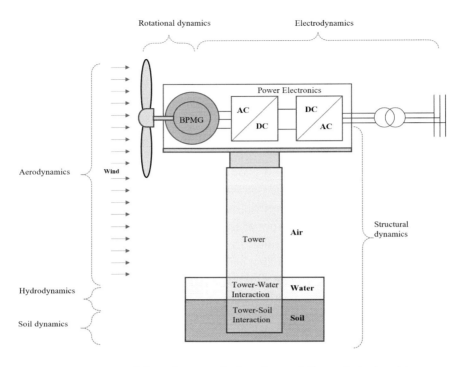

FIGURE 3.29 General disciplines applied to the modelling of wind turbines.

Modelling a wind turbine is an important part of the design, analysis, optimization and operation of wind energy systems. The reasons for modelling include design optimization (to maximize efficiency; power output and reliability while minimizing costs); to assess the suitability of a site; to understand the effects of aerodynamics, gravitational and operational loads on turbine structures; to develop and test advanced control systems; to estimate the energy production of a turbine over time to assess economic viability; to predict the stresses and strains on turbine components to prevent failures; to test new concepts in aerodynamics, materials and turbine configurations without the need for expensive prototypes and to be able to simulate grid integration as well as to study its impact on grid stability. Finally, the models can be used for training purposes.

Figure 3.30a shows a system decomposition of a wind turbine, with the major components and their interactions. The aerodynamics section represents the interaction of the wind (v_w) with the turbine blades. The aerodynamic thrust forces (F_t) and the aerodynamic torque (T_a) are generated as the wind moves past the blades. The pitch angle (β) of the blades

Wind Energy Systems ■ 227

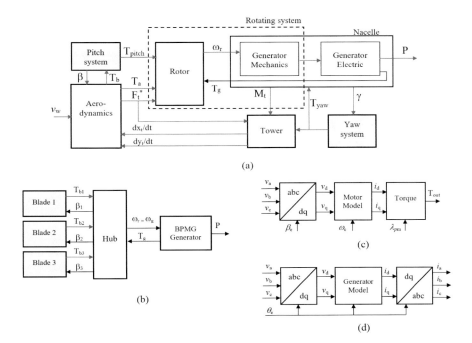

FIGURE 3.30 The components of wind turbines for modelling (a), the sub model of the rotating subsystem in a three-blade direct-drive with a BPMG (b), and a brushless PM motor model (c) and BPMG (d) in the dq-reference frames (see Section 4.6.1 for reference frames).

is controlled by the pitch subsystem. T_{pitch} represents the pitch control torque applied to adjust the blade angle for optimal performance or to protect the turbine from excessive wind speeds. The rotor is turned by the aerodynamic forces acting on the blades. The rotational speed of the rotor is shown by ω_r. The mechanical energy from the rotor is transferred through the drivetrain, for the generator. Note that when considering a BPMG, the drivetrain complexity is reduced, eliminating the gearbox and simplifying the transmission of mechanical energy to electrical energy conversion. The tower supports the nacelle and rotor and transmits mechanical forces M_t from the nacelle to the foundation. The yaw subsystem controls the orientation of the nacelle γ with respect to the wind direction to maximize energy capture. T_{yaw} is the torque applied by the yaw system to rotate the nacelle.

Tower modelling in wind turbines is a complex task that requires detailed analysis in mechanical engineering and structural dynamics.

Even though tower flexibility is moderate compared to blades, it significantly influences the overall structure. Therefore, modelling complexity varies and advanced models can also incorporate blade–tower interactions and various external forces, highlighting the nuanced dynamics involved, which are highly critical in offshore wind farms.

The rotating subsystem can be modelled as a two-mass system (rotor, generator) or a single-mass system (rotor+generator). The components of the rotating subsystem in a three-blade turbine direct drive with a BPMG is illustrated in Figure 3.30b.

While physical models of actuators and the electrical subsystem of the generator are not commonly included in wind turbine modelling, they are integral to control system design due to the electrical power output as considered in this chapter. Considering the modern technology in large turbines, the basic models for electrical drives used in yaw and pitch systems are similar as shown in Figure 3.30c, which consists of a brushless PM motor drive and a gearbox connected to the blade over a pinion, a gear rim and a bearing. In the figure, β_e is the electrical rotation angle proportional to the number of pole pairs of the motor, the mechanical pitch angle and the gearbox and gear rim ratios. In addition, the rate of change in pitch is related to the electrical speed of ω_e. Note that the model shown in the figure is given in a rotating dq-reference frame of the brushless PM motor, where v_a, v_b and v_c are the three-phase input voltages and λ_m is the flux linkage between the rotor and the stator.

In such subsystems, advanced control and fault-tolerant systems are commonly adapted, which accommodate geared brushless PM motor drives. Such motor drives enhance the accuracy, reliability and efficiency of these subsystems. In addition, as previously explained, the BPMG's electrical model is critical for accurately estimating the generator's electromagnetic torque using voltage and current quantities and managing changes in load, which is particularly important in regulating the rotational speed during periods of overrated wind speed.

It should be noted that the yaw system includes multiple motor drives synchronized by the same control signal, a planetary yaw gear, a gear rim and a bearing, and they are designed to generate a large torque to turn the nacelle while considering the aerodynamics of the wind rotor. In addition, yaw brakes are included in the yaw rim in order to lock the position of the rotor when it is perpendicular to the wind direction or during maintenance.

The electrical model of the BPMG, as shown in Figure 3.30d includes the dynamics of voltage and current in both the stator (abc) and the rotor (dq) reference frames, linked by the electrical equations governing the generator's behaviour, which is critical for the design of control systems that can efficiently harness wind energy.

3.8.3 Control of Large Wind Turbines

The aims of wind turbine control can be divided across four domains as given in Table 3.22 to ensure efficient and sustainable operation. Economic objectives seek to maximize financial returns, protective goals prioritize the turbine's longevity and safety, technical goals focus on power quality in general and operational goals address wind intermittency and energy optimization. This approach also highlights the complexity of the control system that is essential for modern wind energy generation.

Note that as the size of the wind turbines gets bigger, it requires higher towers and larger blades. This leads to lighter and flexible structures hence lower natural frequencies. Therefore, the control system, in addition to the energy transformation, should provide damping to avoid resonance and hence vibrations.

In addition, unlike stochastic or random loads, which are unpredictable and can vary due to erratic wind gusts, turbulence or ocean wave loading in offshore turbines, there are significant predictable-deterministic loads which are generally consistent, well-defined and are associated with the turbine's design and the environment in which it operates. Such loads occur due to the constant force exerted by the wind and turbine

TABLE 3.22 The Goals of the Large-Scale Wind Turbine Control

Economic	Protective	Technical	Operational
• Enhancement of profitability • Optimization of energy extraction	• Minimization of structural loads • Mitigation of vibrational forces • Preservation of structural integrity • Implementation of fault-tolerant control mechanisms	• Improvement of power quality including harmonics reduction • Voltage and frequency stability • Implementation of grid code compliant technologies	• Response to intermittency • Integration of energy storage and predictive systems • Efficiency in energy conversion • Adaptive/Flexible operational strategies

components under steady winds, the constant downward force exerted by gravity on the turbine components, centrifugal forces acting on the blades and the rotor, forces that occur during routine operations (such as yawing and pitching of the blades), predictable aerodynamic forces that result from the blade passing in front of the tower (known as the tower shadow effect), forces resulted from static and dynamic imbalance and pitch variations between blades. The types of vibration sources and deterministic loads in large turbines are summarized in Figure 3.31, which are considered in the control of the wind turbine systems.

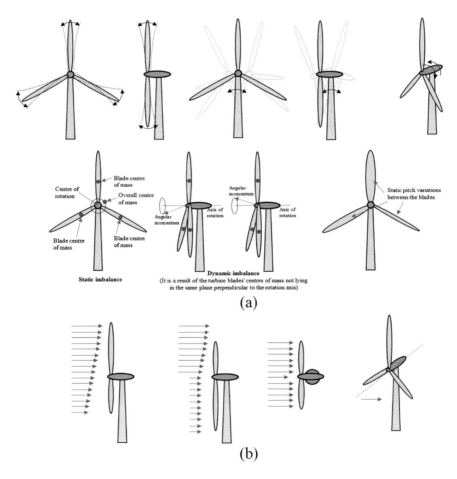

FIGURE 3.31 The types of vibration sources and deterministic loads in large turbines: (a) blade, tower and torsional drivetrain vibrations; (b) deterministic loads: wind speed related and yaw misalignment.

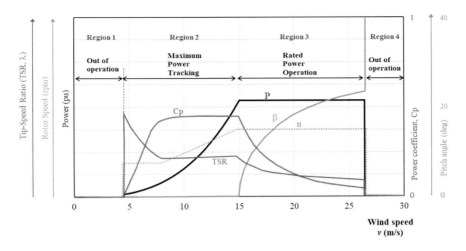

FIGURE 3.32   Operation regions in control as a function of wind speed.

Understanding how control parameters such as the power coefficient Cp, pitch angle β and TSR interact is critical for the development of primary control systems in a wind turbine's power generation process. The control systems should ensure that the generator remains within its mechanical constraints and optimizes power generation. The operational dynamics of variable speed and variable pitch wind turbines are defined by specific strategies across four defined turbine-specific wind speed regions, as illustrated in Figure 3.32. In the initial control action, the control strategy aligns the yaw angle to capture the maximum available wind. Then subsequent primary adjustments include:

- Refining blade pitch to enhance *Cp* across varying wind speeds,
- Keeping the TSR in a range that maximizes efficiency,
- Regulating power output to prevent exceeding the turbine's rated capacity during high wind conditions.

In Region 1 (Out of Operation) of the figure, wind speeds are too low to generate power, hence the turbine is not operational.

In Region 2 (Maximum Power Tracking), the wind speed is between the cut-in and rated speeds. The turbine generates power, but not at its rated capacity. The focus in this region is on maximizing power output through a constant TSR, and control is mainly achieved by adjusting the generator torque. The control system optimizes the TSR to maintain the *Cp* at its peak, which involves adjusting the rotor speed and the blade pitch as wind speed increases.

In Region 3 (Rated Power Operation), wind speeds are high enough that the turbine can produce its rated power. The region is also called the full load region, the wind speed exceeds the rated value, and the turbine generates power at or near its maximum capacity. The goal is to maintain a constant generator speed across a range of wind speeds by altering the blade pitch, thereby adjusting the aerodynamic torque. The operational approach may involve maintaining a constant power output or generator torque.

In Region 4 (Out of Operation), wind speeds are too high for safe operation, hence the turbine is shut down to avoid damage, often using pitch control to feather the blades.

However, it should be noted here that there are also two major transition regions between the main operational regions of a wind turbine that serve as critical phases where the control strategy adapts to changing wind conditions. These transitions are necessary for a smooth changeover in control strategy, to avoid resonance as wind speeds approach certain thresholds, to prevent overshoot to avoid instability and for load management to reduce wear and extend lifespan. The associated processes during these transitions also involve blade pitching, modification of the electromagnetic torque produced by the generator to control the rotational speed of the rotor generator and continuously aligning the turbine with the wind direction.

The transition regions are between Regions 1 and 2 and between Regions 2 and 3.

The first transition begins when $v_{wind} = v_{cut\text{-}in}$ but it is a design decision when it finishes. Contrarily, the second transition region occurs when the wind speed approaches the rated level (but the start point has to be established during the control system design), and the turbine nears its full power potential. To manage noise and mechanical stress, the tip speed is capped by design, impacting the generator speed. This phase delays the turbine's entry into full production and involves strategies to manage a spike in thrust force, which could increase tower loads. One such strategy, known as peak shaving, involves pre-emptively pitching the blades to temper the force exerted on the tower. It can be summarized that at the end of the first transition region, the torque control starts and at the end of the second transaction (beginning of Region 2) the pitch control begins.

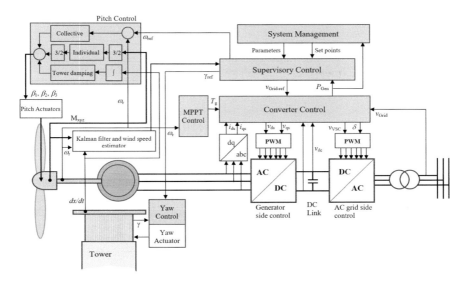

FIGURE 3.33 The components of control system diagram in large wind turbines.

The control system diagram given in Figure 3.33 shows the complexity and interconnectivity of the subsystems necessary for the efficient and reliable operation of a modern wind turbine. These subsystems balance the mechanical and electrical aspects of the turbine, integrating estimation and control techniques to optimize performance across all wind speed regions. Table 3.23 summarizes the prime functions of each block.

Although power electronics in wind turbines play a crucial role in managing and converting the electrical energy generated, the supervisory control requires various parameters to be measured to form reliable feedback. A summary of critical sensors used in the wind turbine control is given in Table 3.24. The converter-associated sensors will be discussed in the next chapter.

Figure 3.34 shows the relationship between a turbine's rotor speed and both output power in watts and torque in Nm for a wind turbine at various wind speeds, from 5 to 12 m/s. Such curves are fundamental in designing wind turbine components as well as in the real-time control systems that govern their operation.

Figure 3.34a the "Maximum power locus" line connects the peak of each power curve, indicating the turbine speed at which the turbine generates maximum power for a given wind speed. The cubical relationship between wind speed and power is demonstrated by the steep increase in

TABLE 3.23 The Prime Functions of the Control Sub-Blocks Are Shown in Figure 3.30

Subsystem	Remarks
Pitch control	• In traditional wind turbines, the pitch of all blades is adjusted simultaneously and identically, which is known as collective pitch control. Individual pitch control (IPC) is employed for several key reasons: to minimize uneven loads (such as due to wind shear and tower shadow) on the turbine blades, to minimize the noise generated by the turbine, to improve the stability of the turbine, especially during high winds or turbulent conditions and to mitigate mechanical stress hence increase turbine lifespan. • It is active in wind speeds between cut-in and rated speeds, in Region 2 (partial load), which involves optimizing the rotor speed to maximize power. It is used to adjust the angle of attack on the blades, typically operates to maintain an optimal tip-speed ratio TSR, optimizing the power coefficient C_p. • In Region 3 (full load), engages when wind speeds are at or above the rated speed. The pitch control actively regulates the blade angle to maintain rated power output and prevent the turbine from generating excessive power that could cause mechanical stress or exceed the electrical system's capacity.
Kalman filter and wind speed estimator	• The Kalman filter processes sensor data, such as acceleration and rotor speed, to estimate wind speed, and it is a critical parameter for adjusting control strategies during transitions between regions. • The speed estimation methods like Kalman Filtering is critical for accurately assessing the wind profile that the turbine experiences, and they are preferred due to the complex and rapidly changing wind patterns, to eliminate physical sensors to mitigate their limitations and improve durability and to seamlessly integrate other measurements (acceleration, rotor speed, pitch angle). • The estimator informs the pitch and yaw systems about the wind speed and direction, which is essential for optimizing the blade position across all regions.
Yaw control	• Although not specific to any region, yaw control keeps the rotor facing into the wind to maximize power. • In Region 4 (shutdown), it is also used to move the blades out of the wind to minimize loads when the turbine is taken out of operation due to excessive wind speeds.
Supervisory control	• It acts as the central management system, coordinating between subsystems based on real-time data and predefined setpoints and ensures that transitions between operational regions are smooth and that the turbine operates within safe limits.

(*Continued*)

TABLE 3.23 (*Continued*)   The Prime Functions of the Control Sub-Blocks Are Shown in Figure 3.30

Subsystem	Remarks
Generator side and converter control	• Generator side power electronics manages the electrical characteristics on the generator's side, such as voltage and frequency, which is highly critical in Region 3, where maintaining the rated output is important despite varying wind speeds. • Generator side converters consist of two inverters connected back-to-back with a common DC link. The AC power from the PM generator is converted to DC power using a rectifier that forms the DC link and DC power that is then converted back to AC power with the desired frequency and voltage to be synchronized with the electrical grid. • The converter control, along with MPPT, ensures that the generator operates at its maximum power point in Region 2 and manages the electrical output to maintain grid compliance. • In direct battery charging, the inverter stage is eliminated, which improves the conversion efficiency.

TABLE 3.24   A Summary of Sensors Used in the Wind Turbine Control Systems

Sensor Type	Description	Components/Functions	Location
Wind speed sensors (anemometers)	Measure wind speed for operational suitability and performance optimization.	Essential for determining operational thresholds and efficiency. They provide real-time wind measurements that are used by the wind turbine control.	Mounted on nacelle or top of the turbine.
Wind direction sensors (wind vanes)	Measure wind direction to assist yaw control.	Integral for aligning the turbine with prevailing wind directions.	Often co-located with wind speed sensors.
Rotor speed sensors	Monitor rotor speed, essential for generator and blade pitch control.	Key in controlling the power output and operational safety of the turbine.	–
Vibration sensors	Detect and monitor vibrations to prevent mechanical failures and plan maintenance.	Placed on critical mechanical components to pre-emptively identify issues.	Gearbox, generator.

(*Continued*)

TABLE 3.24 (*Continued*) A Summary of Sensors Used in the Wind Turbine Control Systems

Sensor Type	Description	Components/Functions	Location
Temperature sensors	Monitor temperatures to prevent overheating and optimize cooling.	Important for the longevity of mechanical components and overall safety.	Nacelle, bearings, gearbox, generator.
Strain gauges	Measure strain on structural components to avoid overloads and failures.	Crucial for the structural integrity and resilience of the turbine.	Blades, tower.
Tilt and acceleration sensors	Monitor stability and alignment of the turbine structure.	Essential for ensuring the structural stability and operational safety of the turbine.	Nacelle, tower.

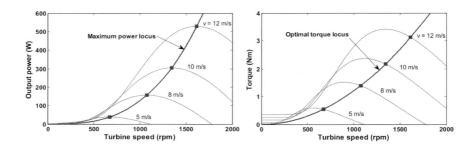

FIGURE 3.34 The relationship between a turbine's rotor speed and both output power (a) and torque (b) under varying wind speeds.

peak power with higher wind speeds. Figure 3.34b displays the torque produced by the turbine at various wind speeds, which can be generated using the relationship of Power = Torque × Angular Rotor Speed. The "Optimal torque locus" line connects points representing the optimal turbine speed for maximum torque production at different wind speeds. It should be emphasized here that for a given wind speed, the torque corresponding to max power is less than max torque, or it can be noted that the maximum torque does not occur at the maximum power point.

These graphs are the keys in wind turbine control systems to operate the turbine at optimal efficiency. The MPPT algorithm would use these curves to adjust the turbine's operational point to where the maximum power is produced. The torque curves inform the control system on how to adjust the generator's resistance to maintain the optimal rotor speed, especially in variable wind conditions.

In addition, the peak points on both graphs are critical for control. The peak power points determine the optimal operating speed of the turbine to maximize power. Peak torque points are essential for the generator and, if exists, gearbox design, ensuring that these components can handle the highest loads expected during operation. Note that the stall torque $\ll$ the rated torque.

The peak power points are used to select the generator's rating, which must be rated to handle the maximum power output of the turbine without exceeding its thermal and mechanical limits. As the blade diameter increases, the rotor sweeps a larger area, capturing more wind energy. This increases the rate of change in the peak power curve, allowing for greater power production at lower wind speeds, enhancing the turbine's performance, especially in regions with lower average wind speeds.

In the wind energy industry, manufacturers typically offer a range of turbines with different rotor diameters and generator ratings, rather than an infinite variety. This approach allows customers to choose the best option for their specific wind speed distributions. However, this selection involves trade-offs in terms of wind speed and power output. If the rotor diameter is increased while maintaining the same generator, the power curve shifts left, enabling the turbine to reach its rated power at lower wind speeds. Conversely, keeping the rotor size constant but upgrading to a larger generator extends the power curve upwards, allowing the turbine to achieve a higher-rated power. It should be noted here that the generator rating directly defines the ratings of the power electronics.

3.9 WIND FARM LAYOUTS, CONNECTION AND SUBSTATIONS

Wind turbine placement significantly impacts energy production. Onshore turbines are often scattered strategically on hills and ridges to catch prevailing winds, while offshore farms can have a more uniform layout due to steadier winds over water. Clustering turbines in wind farms reduces development costs by sharing infrastructure and simplifies maintenance. However, spacing them too close together can hinder power generation as upwind turbines disrupt airflow for downwind ones. As illustrated in Figure 3.35, ideally, turbines should be at least 10 rotor diameters apart. Onshore wind farms typically use an irregular row-and-column structure

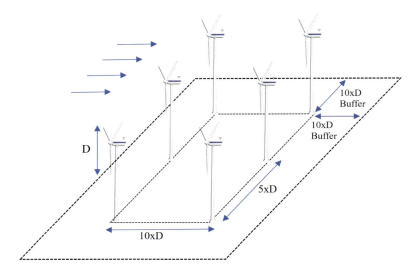

FIGURE 3.35 Optimized wind farm layout with directional wind flow.

to adapt to the terrain, while offshore farms on flatter surfaces can benefit from a more regimented 5D × 10D layout (where D is the rotor diameter). This optimized spacing allows for efficient wind capture while minimizing negative interactions between turbines.

Note that for both onshore and offshore wind farms, the layout must consider several factors. These include wind direction since turbines placed facing into the prevailing wind maximize energy capture. In addition, the arrangement must account for the wake effect, where upstream turbines can decrease wind speed and increase turbulence for downstream turbines, thus reducing their efficiency. Moreover, for maintenance and operation, onshore wind farms need to consider road access, while offshore farms must plan for boat or helicopter access, and the layout should minimize the impact on the local environment and wildlife.

In both cases, the layout of the turbines and the design of the electrical collection system must be optimized to reduce cable lengths and power losses, which can be significant, especially for large offshore wind farms. The integration into the grid also requires careful planning to ensure that the variable nature of wind energy does not adversely affect grid stability and reliability.

Offshore wind farms offer the added advantage of being closer to coastal load centres, reducing transmission needs. Smaller offshore systems may use multiple undersea AC cables, while larger ones further out likely require high-voltage DC transmission for efficient power delivery. As wind power grows, energy storage using batteries and potentially hydrogen in the future is becoming increasingly important to address the intermittent nature of wind generation.

3.10 CAPACITY FACTOR (CF)

The capacity factor is a critical metric that reflects the efficiency of a wind farm over a specific period, typically a year. It is essentially the ratio of the actual electrical energy a wind farm produces to the theoretical maximum it could generate if it operated at full capacity during that same timeframe. The CF is calculated using the following equation.

$$CF = \frac{\text{Actual Energy Produced (kWh)}}{\left(\text{Rated Power (kW)} \times \text{Time Period (hours)}\right)}$$

Note that actual energy produced is typically obtained from the wind farm's monitoring system. Rated power is the maximum power output capacity of a single wind turbine within the wind farm, multiplied by the total number of turbines. For example, if a wind farm has 50 turbines with a rated power of 2 MW each, the total rated power would be 100 MW. The time period is the total number of hours in the chosen period. For annual calculations, this would be 8,760 hours.

There are several factors that influence a wind farm's capacity factor: wind resource availability, turbine technology and design, operational maintenance and site-specific conditions (including local topography and surrounding vegetation). Therefore, examining successful wind farms can also provide valuable insights for strategic planning and operational practices that can impact CFs. In addition, by analysing wind data, selecting efficient turbines and implementing effective maintenance routines, wind farm operators can maximize energy production and achieve high CFs. Note that using direct drive PMSGs improves power output and minimizes maintenance.

Wind farm location also plays a role in CF. Offshore wind farms, situated in open seas, often experience stronger and more consistent winds, leading to potentially higher CFs. However, constructing and maintaining offshore wind farms present greater logistical challenges. Onshore wind farms, while potentially facing more variable wind conditions, benefit from easier access for maintenance and grid connection. The choice between onshore and offshore locations involves a careful evaluation of wind resource availability, technical feasibility and economic considerations.

Therefore, it can be concluded that understanding CF helps wind farm developers and operators to make informed decisions. By carefully selecting locations with strong wind resources, investing in advanced turbine technology, implementing efficient maintenance practices and considering the trade-offs between onshore and offshore options, wind energy production can be optimized.

3.11 WIND TURBINE SYSTEM STANDARDS AND WIND FARM INCIDENT CATEGORIES

As discussed previously, wind turbine systems are highly complex interdisciplinary marvels of modern engineering. This roadmap can be classified under two Table 3.25 summarizes this roadmap throughout the entire

TABLE 3.25 Wind Turbine Standards: A Roadmap

Stage	Category	Key Focus Areas
From blueprint to blade	Design standards	• Withstand specific environmental conditions. • Structural integrity (towers, blades, nacelles). • Material selection (strength, durability, weather resistance). • Safety features (lightning protection, emergency shutdown).
	Manufacturing standards	• Consistent quality throughout production. • Quality management systems (material and component specifications). • Welding and fabrication procedures (robust construction). • Quality control measures (defect identification and correction).
Testing and certification	Established Testing procedures	• Safe and efficient turbine performance. • Structural integrity testing (extreme wind loads). • Performance testing (power generation and efficiency). • Noise emission testing (compliance with regulations). • Grid connection testing (safe and reliable integration).
Connecting to the grid	Grid interconnection standards	• Effective communication with the grid. • Meeting power quality requirements.
	Power system stability standards	• Maintaining grid stability. • Addressing power generation fluctuations (wind variability).

process, from design and manufacturing to testing and grid connection, the wind industry ensures the safety, reliability and efficiency of wind turbine systems.

Wind farms, with their towering turbines and expansive arrays, lead renewable energy technology. However, as in solar PV systems, they are not immune to incidents and operational issues. Therefore, understanding the potential incidents is critical for ensuring safety, reliability and efficiency within the industry. Incident categories range from equipment failures and maintenance issues to environmental impacts and grid integration challenges. By establishing a clear description, wind farm operators can systematically address potential risks, implement preventive measures, and maintain high standards of operation to minimize downtime and ensure the safety of both personnel and the environment. This categorization also facilitates standardized reporting and analysis, paving the way for continuous improvement and innovation in wind farm management.

Table 3.26 highlights the different types of failures that can occur in real-world applications of wind turbine systems.

TABLE 3.26 Potential Failure Categories in Wind Turbine Systems

Category	Description	Potential Impacts
Equipment Failure		
Turbine component failures	Malfunctions within the turbine itself (gearbox, bearings and generator).	Reduced power output, downtime, potential safety hazards.
Electrical system failures	Issues with electrical components like transformers or cabling.	Reduced power output, downtime, potential fire hazards.
Control system malfunctions	Errors in the software or hardware that controls turbine operation.	Reduced power output, downtime, potential safety hazards.
Structural Issues		
Tower integrity problems	Cracks, corrosion or other issues affecting the tower's stability.	Catastrophic failure, downtime, safety hazards.
Foundation weaknesses	Problems with the foundation that supports the turbine.	Reduced stability, potential tipping, downtime.
Blade structural failures	Cracks, delamination or other issues affecting blade integrity.	Reduced power output, blade detachment (safety hazards), downtime.

(*Continued*)

TABLE 3.26 (*Continued*) Potential Failure Categories in Wind Turbine Systems

Category	Description	Potential Impacts
Environmental and Weather-Related		
Lightning strikes	Direct strikes or induced currents can damage electrical components.	Reduced power output, downtime, potential fire hazards.
Ice accumulation	Ice build-up on blades can affect their performance and stability.	Reduced power output, potential blade throws (safety hazards), downtime.
Extreme wind conditions	Winds exceeding design limits can damage turbine components.	Reduced power output, downtime, structural failures (safety hazard).
Safety and Health Incidents		
Workplace accidents	Slips, falls or other incidents during construction, maintenance or operation.	Injuries, fatalities.
Fires and explosions	Electrical faults, lubrication issues or other factors can lead to fires.	Turbine damage, downtime, potential safety hazards.
Electrical hazards	Exposure to live wires or malfunctioning electrical components.	Injuries, fatalities.
Operational Errors		
Human error	Mistakes during operation or maintenance.	Reduced power output, downtime, potential safety hazards.
Incorrect installation	Improper assembly of turbine components.	Reduced performance, downtime, potential safety hazards.
Faulty maintenance procedures	Improper maintenance practices can lead to component failures.	Reduced performance, downtime, potential safety hazards.
Grid Integration Issues		
Power quality problems	Fluctuations in voltage or frequency can affect grid stability.	Reduced power output for wind turbine, potential grid instability.
Grid connection failures	Loss of connection between the wind turbine and the power grid.	Reduced power output, lost revenue.
Voltage fluctuations	Rapid changes in voltage can damage turbine components.	Reduced power output, downtime, potential component failures.

(*Continued*)

TABLE 3.26 (*Continued*)   Potential Failure Categories in Wind Turbine Systems

Category	Description	Potential Impacts
External Factors		
Vandalism	Deliberate damage to the wind turbine.	Damage, downtime, safety hazards.
Wildlife	Bird strikes can damage blades.	Blade damage, wildlife fatalities.
Land use conflicts	Issues with local communities or regulations.	Project delays, reduced power output.
Cybersecurity and Data Management		
Cyber-attacks	Malicious attempts to disrupt turbine operation or steal data.	Reduced power output, downtime, potential safety hazards.
Data breaches	Unauthorized access to sensitive wind turbine data.	Privacy concerns, potential operational disruptions.
Communication network failures	Loss of communication between the turbine and control centre.	Reduced visibility into turbine operation, potential safety hazards.

BIBLIOGRAPHY

E. Branlard, *Wind Turbine Aerodynamics and Vorticity-Based Methods: Fundamentals and Recent Applications*. Springer, New York, 2017, https://doi.org/10.1007/978-3-319-55164-7.

A. Gambier, *Control of Large Wind Energy Systems: Theory and Methods for the User*. Springer Nature, Switzerland,AG, 2022, https://doi.org/10.1007/978-3-030-84895-8.

Q. Gao, A. Bechlenberg, B. Jayawardhana, N. Ertugrul, A. I. Vakis and B. Ding, Techno-economic assessment of offshore wind and hybrid wind-wave farms with energy storage systems. *Renewable and Sustainable Energy Reviews*, 192, p. 114263, 2024

Q. Gao, B. Ding, and N. Ertugrul, "Techno-Economic Assessment of Developing Combined Offshore Wind and Wave Energy in Australia," *International Conference on Offshore Mechanics and Arctic Engineering 86922*, Melbourne, Australia, June 11–16, 2023.

Q. Gao, B. Ding, N. Ertugrul and Y. Li, Impacts of mechanical energy storage on power generation in wave energy converters for future integration with offshore wind turbine. *Ocean Engineering*, 261, pp. 112136, 2022.

Q. Gao, N. Ertugrul and B. Ding, "Analysis of Wave Energy Converters and Impacts of Mechanical Energy Storage on Power Characteristic and System Efficiency," *2021 31st Australasian Universities Power Engineering Conference (AUPEC)*, Perth, Australia, 2021, pp. 1–6, https://doi.org/10.1109/AUPEC52110.2021.9597825.

Q. Gao, N. Ertugrul and B. Ding, "Method and Analysis of Short-term Intermittency in Hybrid Wind and Wave Power Unit," *2022 32nd Australasian Universities Power Engineering Conference (AUPEC)*, Adelaide, Australia, 2022, pp. 1–6, https://doi.org/10.1109/AUPEC58309.2022.10215630.

Q. Gao, N. Ertugrul, B. Ding and M. Negnevitsky, "Offshore Wind, Wave and Integrated Energy Conversion Systems: A Review and Future," *2020 Australasian Universities Power Engineering Conference (AUPEC)*, Hobart, Australia, 2020, pp. 1–6.

Q. Gao, N. Ertugrul, B. Ding, M. Negnevitsky and W. L. Soong, Analysis of wave energy conversion for optimal generator sizing and hybrid system integration. *IEEE Transactions on Sustainable Energy*, 15(1), pp. 609–620, Jan. 2024, https://doi.org/10.1109/TSTE.2023.3318010.

Q. Gao, N. Ertugrul, B Ding, M. Negnevitsky and W.L. Soong, Analysis of wave energy conversion for optimal generator sizing and hybrid system integration. *IEEE Transactions on Sustainable Energy*, 15(1), pp. 609–620, 2023.

Q. Gao, J.A. Hayward, N. Sergiienko, S.S. Khan, M. Hemer, N. Ertugrul and B. Ding, Detailed mapping of technical capacities and economics potential of offshore wind energy: A case study in South-eastern Australia, *Renewable and Sustainable Energy Reviews*, 189(113872), p. 202.

Q. Gao, S.S. Khan, N. Sergiienko, N. Ertugrul, M. Hemer and M. Negnevitsky, Assessment of wind and wave power characteristic and potential for hybrid exploration in Australia, *Renewable and Sustainable Energy Reviews* 168, p. 112747, 2023.

Q. Gao, R. Yuan, N. Ertugrul, B. Ding, J.A. and Hayward, Y. Li, Analysis of energy variability and costs for offshore wind and hybrid power unit with equivalent energy storage system. *Applied Energy*, 342, p. 121192, 2023.

E. Hau, *Wind Turbines – Fundamentals, Technologies, Application, Economics*, 2nd edn. Springer-Verlag, Berlin, 2006.

Y. L. Lim, W. L. Soong, N. Ertugrul and S. Kahourzade, "Embedded Stator End-Windings in Soft Magnetic Composite and Laminated Surface PM Machines," *2018 IEEE Energy Conversion Congress and Exposition (ECCE)*, Portland, OR, 2018, pp. 5387–5394, https://doi.org/10.1109/ECCE.2018.8558168.

E. Muljadi and C. P. Butterfield, Pitch-controlled variable-speed wind turbine generation. *IEEE transactions on Industry Applications*, 37(1), pp. 240–246, 2001.

M. Pathmanathan, Modelling and Control of Phase Advance Modulation for Small-Scale Wind Turbines, PhD Thesis, School of Electrical and Electronic Engineering, Faculty of Engineering, Computer and Mathematical Sciences, The University of Adelaide, April 2012.

M. Pathmanathan, W. L. Soong and N. Ertugrul, "Maximum Torque Per Ampere Control of Phase Advance Modulation of a SPM Wind Generator," *2011 IEEE Energy Conversion Congress and Exposition*, Phoenix, AZ, 2011, pp. 1676–1683, https://doi.org/10.1109/ECCE.2011.6063984.

A. Pemberton, T.D. Daly and N. Ertugrul, On-shore wind farm cable network optimisation utilising a multiobjective genetic algorithm. *Wind Engineering*, 37(6), pp. 659–673, 2013. https://doi.org/10.1260/0309-524X.37.6.659.

B. Petersen et al., Evaluate the Effect of Turbine Period of Vibration Requirements on Structural Design Parameters: Technical Report of Findings, Applied Physical Sciences, Report No: M10PC00066-8, September 1, 2010.

Y-M. Saint-Drenan et al, A parametric model for wind turbine power curves incorporating environmental conditions, *Renewable Energy*, 157, pp.754–768, 2020. https://doi.org/10.1016/j.renene.2020.04.123.

C. Tang, Analysis and Modelling of the Effects of Inertia and Parameter Errors on Wind Turbine Output Power, Master's Thesis, School of Electrical and Electronic Engineering, Faculty of Engineering, Computer and Mathematical Sciences, The University of Adelaide, Australia, 2009.

CHAPTER 4

Power Electronics and Control of Power System Components

4.1 POWER ELECTRONICS (PE) AND WIDE BANDGAP DEVICES (WBG)

Power electronics (PE) is the key to the evolution and transformation of power grids, a role magnified by the growing integration of distributed energy resources (DERs), evolving electrical loads, microgrids, electrical vehicles (EVs) and the broader trend of electrification. Converters in PE, known for their fundamental role, efficiently modify a given form of supply to produce an output current and voltage waveforms determined to the specific needs of various applications, whether stationary or mobile. This enables a wide spectrum of applications to utilize similar circuit topologies, performing tasks like renewable energy conditioning, grid integration, battery charging/discharging and management, generator control and motor control across diverse forms of e-mobility, including sea, air, land transport and space.

Furthermore, as power systems evolve to become more hybrid, with AC and DC systems co-existing alongside a diverse array of DERs, they inevitably grow in complexity. One often overlooked that the advantage of PE lies in its contribution to fault clearance and cybersecurity within the power systems. PE offer exceptional benefits in terms of voltage and angle stability, enhancing overall safety, reducing fire hazards and minimizing damage to electrical equipment. This aspect of PE plays a critical role in maintaining the integrity and reliability of modern, complex power systems, which are growing fast. In addition, PE-based power conditioning offers seamless intelligent control opportunities that are likely to shape the Internet of Things (IoT) and Internet of Everything (IoE) concepts.

The current trends demonstrate that the transformation in power systems is directly connected with the integration of renewable energy sources and a shift towards more sustainable energy practices. Therefore, after highlighting the growing trend in WBG device utilization in PE, this chapter will analyse a selected set of PE circuits to highlight the operating principles of common circuits applicable to a broad array of applications within this power system transformation.

4.1.1 Wide Bandgap Devices Status and Application Benefits

It is important to highlight the dynamic nature of technological advancements in PE switches as well. WBG devices, with their distinctive features, are already altering the landscape of PE circuits. In the coming decades, as illustrated in Figure 4.1, the utilization of WBG devices is anticipated to reshape current practices in PE adding a few alternative circuit topologies viable in a range of applications. However, it should be anticipated that the future power system is expected to be primarily dominated by inverters/converters, comprising DC networks, grids and appliances, while the renewable energy sector is already DC.

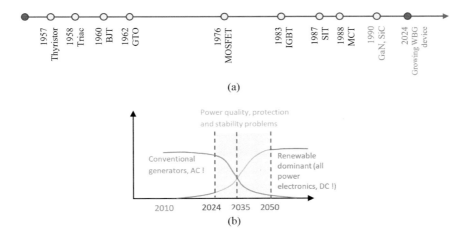

FIGURE 4.1 The historical time line in power electronics switches (a), and a rough time line of an intuitive prediction of the grid transformation from AC to DC grid (b).

Electronvolt (eV) is a key metric in defining the characteristics of semiconductors and their efficiency as switches. While Si has been widely used, it struggles with the high voltages and power levels required for various applications such as EVs, fast chargers, renewable energy systems as well as specialized military and mining uses. The eV levels of common semiconductor materials are shown in Table 4.1. As the WBG and Ultra-WBG materials became commercially viable they are gaining increased utilization in numerous applications.

The main lines of benefits of WBG devices are shown in Figure 4.2 together with the impacts of the device properties on the final PE system. Note that since reduced cooling requirement would lead to higher temperature swings

TABLE 4.1   ElectronVolt-Levels of Some Typical Solids

Solids	eV-Level	Remarks
Germanium (Ge)	0.67 eV	Pure semiconductor
Silicon (Si)	1.1 eV	Pure semiconductor
Gallium-Arsenide (GaAr)	1.42 eV	Compound semiconductor
Silicon-Carbide (SiC)	3.3 eV	Compound (dopped) semiconductor, WBG
Gallium-Nitride (GaN)	3.4 eV	Compound semiconductor, WBG
Gallium-Oxide (GaO)	5 eV	Compound semiconductor, Ultra-WBG
Diamond (C)	5.5 eV	Pure semiconductor, Ultra-WBG
Aluminium-Nitride (AlN)	6.2 eV	Compound semiconductor, Ultra-WBG
Glass	>4.4 eV	Insulator ! (very high resistivity and very low conductivity)

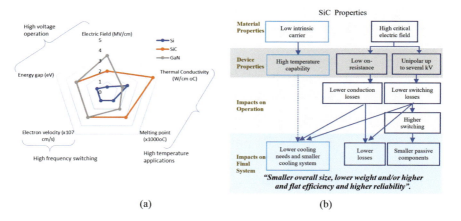

FIGURE 4.2   Comparison and status of specific properties of Si (conventional device) and SiC and GaN WBG devices (a), and the impacts of the SiC device properties on the final power electronics system (b).

for the same loss level, the link between high-temperature capability and a smaller cooler is a dotted line, and if the correct trade-offs are implemented, it is possible to obtain an end product which is smaller in overall size and has lower weight and/or higher efficiency and higher reliability.

Table 4.2 summarizes how WBG device-based PE offers benefits across various sectors, contributing to more efficient, reliable and safer systems in renewable energy, grid transformation, EVs and battery storage and charging technologies.

TABLE 4.2 Typical Benefits of WBG Devices When Used in PE across Four Major Sectors

Benefit	Renewable Energy	Electric Vehicles	Battery Storage and Charging	Grid Transformation
Higher efficiency	Maximizes conversion efficiency	Increases vehicle range and performance.	Improves round-trip efficiency, reducing losses.	Improving power efficiency
Improved thermal performance	Reduces cooling system space and cost.	Minimizes thermal management requirements.	Stable thermal performance during charging/discharging cycles.	Operation under high thermal stress, enhancing grid reliability.
Greater power density (volumetric and gravimetric)	More compact and lightweight systems.	Reduces size and weight of the drive and on-board chargers	Compact battery systems.	Higher power and smaller grid devices
Faster switching speeds	Reduces size and cost of passive components.	Enhances powertrain responsiveness.	Quick and precise charging control.	Improves response times to grid fluctuations and disturbances.
Lower on-state resistance	Lowers conduction losses.	Decreases power losses in electronic components.	Minimizes power losses during battery charging and discharging.	Reduces power losses in grid components.
Enhanced safety and improved reliability	Reduces overheating and longer lifetimes in harsh conditions.	Better thermal and electrical stability, improved PE reliability and lifespan.	Increases safety in operation and longer lifespan and reliability.	Improves safety under high loads and enhanced component durability and reduced maintenance.

It should also be noted that while higher switching frequencies are often associated with increased electromagnetic interferences (EMI), WBG devices can mitigate this by enabling more efficient filtering and shielding, primarily due to their ability to work with smaller, more effective passive components. Their faster switching actions are inherently cleaner, producing less EMI than the slower transitions of traditional devices. Moreover, WBG devices are known for their "flat efficiency", maintaining high performance across diverse temperatures and loads, making them ideal for applications with fluctuating power demands, such as in renewable energy systems, battery storage solutions and EVs.

Furthermore, in DC circuit breakers, WBG technology allows for operation under higher voltages and currents with faster response times and better efficiency. Their excellent thermal properties mean that cooling systems and heat sinks can be downsized, leading to more compact circuit protection solutions essential in high-power DC systems, including solid-state transformers for substations. WBG devices are also driving the evolution of solid-state transformers in intelligent substations, facilitating the integration of renewables and enhancing power flow management. For High Voltage Direct Current (HVDC) systems, their ability to handle high voltages and switch efficiently allows for more compact converter designs. This not only reduces the physical footprint of these systems but also diminishes power losses, making even short-distance power transmission more feasible and improving the integration of renewable energy. In addition to these applications, WBG PE is anticipated to play a critical role in the production of green hydrogen, which is relevant in electrolysis processes, which demand a high power and well-regulated DC source.

4.2 CLASSIFICATIONS AND SELECTION OF CONVERTERS AND SWITCH CONTROL PRINCIPLES

As stated above, PE converters are essential components in various applications, from renewable energy systems to grid interfaces and storage solutions. A broad classification of converter topologies in the context of power grid applications can be done based on:

- Direction of power flow (unidirectional or bidirectional converters).
- Type of conversion (AC-DC, DC-AC, DC-DC or AC-AC).
- Number of phases (single-phase or three-phase converters).

- Voltage source and current source converters.
- Application-specific topologies (such as multi-level or modular multi-level converters).

These classifications are not mutually exclusive, and a particular system may involve multiple types of converters. For instance, as it is summarized later in Section 4.5, a grid-tied solar PV system uses a DC-DC converter for maximum power point tracking (MPPT), followed by a DC-AC inverter to supply power to the AC grid. Similarly, a wind turbine system uses two back-to-back inverter topologies for rectification and AC grid integration, and a wind battery storage system uses a bidirectional DC-AC converter to charge the batteries with excess wind energy and then feed energy back to the grid when wind generation is low. Solid-state transformers, which are also emerging in smart grid applications, may use advanced multi-level converter topologies to interface with high-voltage grids efficiently.

As for the well-established PE circuit topologies—like a switch with a reverse-parallel diode, two switches in series, H-bridge and three-phase bridge topologies —these are widely used due to their proven reliability, ease of control and broad base of design experience. These topologies have been refined over the years to deliver good performance across a range of applications, making them a safe choice for many designers. However, as technology advances and new materials and components become available, shifts may be seen in the preferences for certain topologies based on the evolving trade-offs between various factors. However, it can be summarized that the selection of a PE circuit for a given application is driven by several key factors that are listed in Table 4.3.

In addition, the progression towards PE is not only physically more integrated but also smarter, safer and more communicative, combining the advanced requirements demanded by many applications. The evolutionary trends and features of PE building blocks are also summarized in Table 4.4 which highlights the necessity of such changes which are indirectly linked to the WBG utilizations in PE.

In the forthcoming sections, an analysis and discussion of the operational aspects of a chosen array of PE circuits are made. These circuits have been identified as fundamental components in the power conversion processes utilized across a spectrum of renewable energy and power system applications, ranging from solar PV systems to battery storage technologies.

TABLE 4.3 The Factors That Influence the Selection of PE Building Blocks

Factors	Remarks
Cost	Including purchase, installation, maintenance and operation over the life.
Availability	The ease of acquiring components and the impact on production and maintenance.
Maturity	The proven reliability and predictability based on the technology's track record.
Modularity	The ability to modify the system to adapt to changing needs.
Efficiency	The efficiency, which affects operating costs and heat generation.
Performance	The necessity for specific power quality, control accuracy and dynamic response.
Complexity	The intricacy of the circuit design, which can impact reliability and maintenance.
Size and weight	The physical constraints of the application area, crucial in space-limited scenarios.
Thermal management	The capability to dissipate heat effectively, essential in high-power applications.
Regulatory standards	The need to comply with industry-specific electrical and safety standards.
Existing system integration	The compatibility with existing components and systems, particularly in upgrades or expansions.
Innovation	The potential to provide unique features or competitive advantages.

TABLE 4.4 Summary of the Evolutionary Trends and Features of PE Building Blocks

Evolutionary Trend	Features	Impacts
Integration	Reduction of leakage inductance and parasitic components.	Enhances efficiency and performance by minimizing energy losses and electromagnetic interference.
	High bandwidth.	Allows for rapid response to load or input changes, crucial for precise power control.
Measurement accuracy	Accurate current and voltage sensing.	Ensures effective system control and enables complex but effective control algorithms.
Protection and diagnostics	Integrated protection features.	Prevents damage and enables predictive maintenance, reducing downtime and repair costs.
	Diagnostic sensors.	Monitors module health and performance for proactive system management.
Safety and noise immunity	Signal isolation.	Protects control circuits from high-power noise and disturbances, ensuring reliable operation.

(Continued)

TABLE 4.4 (*Continued*) Summary of the Evolutionary Trends and Features of PE Building Blocks

Evolutionary Trend	Features	Impacts
Signal interface, control and intelligence	Auxiliary power supplies. Microprocessor-based control.	Minimizes drive circuits and associated switching and improves power density. Facilitates advanced decision-making, real-time adjustments, and supports digital control techniques.
System communication	System-level integration and communication.	Enables coordinated operation with other system components and external control systems for optimal performance.

4.2.1 DC-DC Converters

As given in Chapter 2, MPPT is required in solar PV systems that establish pulse width modulation (PWM) control schemes, and modern DC-DC converters often have integrated MPPT capabilities, performing both the function of power point tracking and voltage regulation. In addition, as emphasized above, selecting an appropriate DC-DC converter topology for a given application depends upon a number of factors and specific system requirements, including voltage levels, the necessity for isolation, efficiency, system complexity and cost, voltage polarity, step-up voltage gains, scalability and modularity and a high-frequency operation which is also related to volumetric power density. Figure 4.3 shows generalized

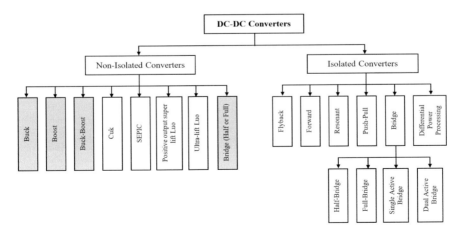

FIGURE 4.3 Classification of DC-DC converters for solar PV and battery applications.

classifications of DC-DC converters that can be used in a range of applications from basic power supplies to solar PV, where the converters discussed are highlighted in shaded boxes.

Although it is not covered in this section, isolation between the input and output in DC-DC converters is often a critical requirement across numerous applications. For instance, in off-line power supply systems including high-voltage solar PV panel arrangements, regulatory standards typically mandate isolation for safety reasons. However, due to the excessive size of the conventional transformers (50–60 Hz), a high-frequency transformer is needed which enables the transformer to operate at the converter's switching frequency (up to MHz). Such transformers use ferrite or powdered iron materials in the core to minimize hysteresis and eddy current losses to improve efficiency. However, they require complex design considerations, such as skin effect, proximity effect and parasitic capacitance, which become significant at higher frequencies. In addition, when high step-up or step-down voltage ratios are needed, a high-frequency transformer within the converter design allows for an optimal balance of the voltage and current stresses on the switching devices, which enhances efficiency and leads to cost savings. By incorporating additional secondary windings on the transformer, different voltages can be obtained without significant additional costs.

Table 4.5 also provides a brief description of DC-DC converters together with their practical efficiency ranges that can be used to justify the selection criteria. It should be noted that solar PV inverters contain several of these converters within a single housing, and high-efficiency converters often use WBG semiconductors like SiC and GaN to achieve efficiencies at the higher end of the spectrum.

4.2.2 Typical Common Converters: AC-DC (Rectifiers) and DC-AC (Inverters)

A general classification of "converters" used in the renewable energy sector can be given in Figure 4.4. The choice of converter topology in power grid applications depends on the specific requirements of the application, such as power level, quality, efficiency and control complexity. The ongoing advancement in PE continues to evolve these classifications with the development of new topologies and control strategies. However, to explain the basic concepts in solar PV, wind and battery storage power applications, a few common converter topologies (see Figure 4.5) that illustrate the principles of power conversion and management in these systems will

TABLE 4.5   Description and Efficiency Ranges of the DC-DC Converters

Name	Description and Application	Efficiency
Non-Isolated Converters		
Buck (step-down)	For higher DC input voltage into a lower DC output. Suited for solar systems with variable panel voltage, ensuring a constant output to downstream devices.	90%–98%
Boost (step-up)	For lower DC input voltage to a higher DC output. If voltage falls below the required input level for downstream devices or inverters.	90%–97%
Buck-boost	Combination of buck and boost converters. For fluctuating irradiance.	88%–97%
Ćuk	As the buck-boost but with an inverted output. To reduce output voltage ripple.	90%–97%
Positive-output Super-lift Luo	In applications if much higher voltage level is needed.	85%–95%
Ultra-lift Luo	Offers even higher voltage multiplication.	85%–95%
SEPIC (single-ended primary inductor converter)	Can step-up and step-down without inverting voltage.	88%–97%
Isolated Converters		
Flyback	Produce multiple output voltages (both polarities) from a single input.	80%–93%
Forward	More efficient than flyback for high power levels Preferred in larger solar installations.	85%–95%
Push-pull	Ideal for high power cases, offers a non-inverted output, suitable for both isolated and non-isolated configurations. Commonly used in high-power and high-efficiency solar PV systems.	90%–98%
Full-bridge and half-bridge	Employed for high-power commercial or utility-scale solar PV applications, for high efficiency when used WBG devices and component durability.	90%–99%
Resonant converters: zero-voltage switching ZVS and zero-current switching ZCS	Designed to reduce switching losses, vital for high-frequency setups when accommodating WBG devices.	90%–99%
differential power processing (DPP)	Effective on module or submodule level and in mismatch management, optimizing individual module performance.	85%–96%

be discussed briefly in the subsequent subsection. Note that these basic topologies can also be considered as the building blocks for more complex systems. For example, renewable energy systems use variations or combinations of these topologies, such as the buck-boost converter, which

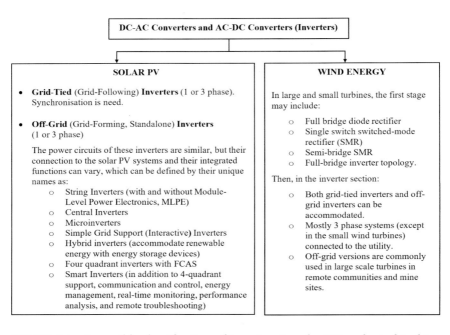

FIGURE 4.4 A possible classification of inverters in solar PV and wind turbine with BPMGs.

combines the functions of both stepping up and stepping down voltages or the multi-level inverter for handling higher power levels with lower harmonic distortion.

Figure 4.5 presents a range of PE converter topologies that are fundamental to the infrastructure of renewable energy systems and battery storage applications. The figure includes DC-DC converters (Figure 4.5a), such as the Buck Converter that steps down voltage for MPPT in solar PV systems, the Boost Converter that steps up voltage to interface low-voltage renewable sources with higher voltage systems or for battery storage, and a non-isolated flexible Up-Down Converter that offers voltage adjustment in both directions.

Within the wind turbine array (Figure 4.5b), various rectifiers are shown for enhanced efficiency and control over power flow. The inverter section in Figure 4.5c features a Single-Phase Voltage Source Inverter for residential solar applications converting DC to AC, and a Three-Phase Voltage-Source Inverter for larger, industrial-scale applications requiring three-phase power grid compatibility. Additionally, the Bidirectional Isolated Battery Storage Converter (Figure 4.5d) is essential for managing

Power Electronics and Control of Power System Components ■ 257

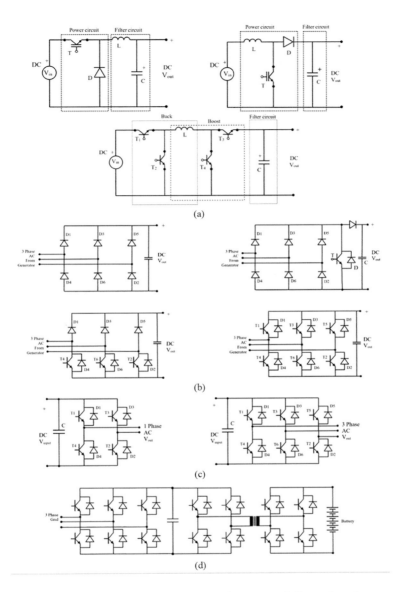

FIGURE 4.5 (a) Three DC-DC converter topologies: Buck (Step-down) converter, Boost (Step-up) converter, non-isolated Synchronous Buck-Boost converter; (b) Rectifiers (AC-DC) for BPMGs in wind turbines: a full bridge diode rectifier, single switch switched-mode rectifier (SMR), semi-bridge SMR, full-bridge inverter topology (also known as three-phase PWM rectifiers/inverters or PWM rectifier); (c) Single-phase voltage source inverter (DC-DC), three-phase voltage-source inverter; (d) Bidirectional isolated battery storage converter system connected to a three phase AC grid.

battery charge and discharge cycles in a grid-connected system, facilitating safe and efficient energy exchange.

The simplest topology that pairs with a three-phase brushless permanent magnet generator (BPMG) in wind turbines is the uncontrolled rectifier, noted for its cost-effectiveness due to the absence of controllable switches, sensors and control circuitry. However, its drawback is the inability to produce output power at low generator speeds because it cannot control the phase shift between the generator's phase voltage and current as will be shown later. The switched mode rectifier (SMR), by integrating a boost switch, allows for output voltage adjustment through duty cycle modulation, enabling power production when the generator's back emf voltage is below the DC link voltage. This increased control results in higher complexity and costs.

4.2.3 Types of Switch Control and Modulation Techniques

The operation of the converters as shown in Figure 4.5 relies on the accurate control of switching transistors, which can be broadly categorized into two types as described and illustrated in Table 4.6.

Both control methods adjust the operation of the switching devices to manage the output characteristics of the converter, whether for voltage, current or power regulation. The selection between time-ratio control and limit control depends on the specific requirements of the application, considering factors such as efficiency, power quality and system compatibility.

Voltage Source Converters (VSC) are commonly used and preferred in renewable energy integration into the AC grid, due to control flexibility (as a result of rapid changes in voltage and current), ease of integration with energy storage systems (including batteries and supercapacitors), high efficiency in low to medium power applications, less stress on PE components, simpler topology and easier implementation of modulation techniques such as PWM, more scalable and modular when expansion or adaptation is needed, advances in power semiconductor devices, better protection against overcurrent conditions and their compatibility with variable renewable energy sources.

Modulation techniques using the switch control mentioned previously allow for the rapid and accurate control of power semiconductor devices in PE circuits at high efficiency, which regulates the power flow and conversion by manipulating voltage and current waveforms to achieve desired properties such as frequency, amplitude and phase.

Figure 4.6 summarizes commonly used VSC modulation techniques in PE.

TABLE 4.6 Control Types for the Accurate Control of Switching Devices

Description and Illustration

Pulse Width Modulation (PWM) (Constant Frequency/Period):
Allows for precise control over the output voltage and current, and is widely used due to its ability to provide a stable frequency for filter design and system integration.

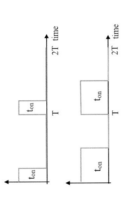

Time ratio control Frequency Modulation (FM), (Variable Frequency/Period Control):
It changes the frequency or the time period while keeping the ON time or the OFF time constant. It has few disadvantages:

- Variable frequency complicates the design of output filters, which also creates the possibility of electromagnetic interference (EMI) with other electronic systems.
- A potentially large OFF time could result in discontinuous load current, not desirable.

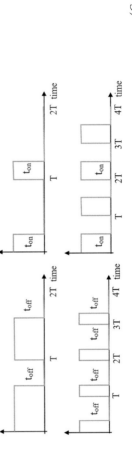

(Continued)

260 ■ Reinventing the Power Grid

TABLE 4.6 (*Continued*) Control Types for the Accurate Control of Switching Devices

Description and Illustration		

Current or voltage limit control	• This control is utilized to regulate the current or the voltage within specified upper and lower limits:
	    • When the current or voltage reaches an upper threshold, the switch is turned OFF, allowing it to decrease, often through a freewheeling path available.
	    • As the current or voltage falls to the lower threshold, the switch is turned back ON.
	    • It can be implemented with either a constant frequency or a constant ON time.
	    • It is particularly suitable for loads with energy storage components like inductors and capacitors and is often used to control the load current or voltage.
	    • Minimizing the difference between the upper and lower limits can reduce the ripple but this may increase the switching frequency, leading to higher switching losses.
	    • Suitable for inductive loads and continuous conduction mode operations, and also to prevent overvoltage conditions, and suitable in applications that demand stable voltage.
		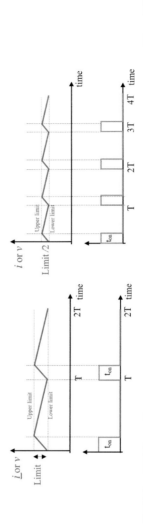

Power Electronics and Control of Power System Components ■ 261

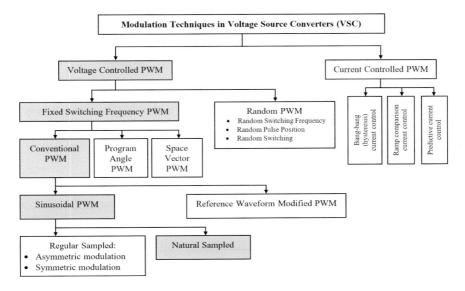

FIGURE 4.6 Common modulation techniques in voltage source converters.

The selection of switch control methods and modulation techniques in PE is a critical factor in determining the efficiency of converters. These aspects impact not only the quality of the power being converted but also the overall system performance, including factors like power losses, heat generation and the ability to handle transient conditions.

The control method and modulation technique must be carefully selected to align with the controlled parameters of the converter, which include the voltage, current, frequency and power factor. This maximizes efficiency, ensures reliability and service life.

The timing of the switch has a substantial impact on the converter's efficiency. Therefore, the control method should minimize overlap, reduce switching losses and minimize harmonics to minimize losses.

Modulation techniques listed in Figure 4.6 can be used to regulate the output voltage and current of converters. The precision of these techniques significantly affects the harmonics introduced into the system, which requires an optimization of the switching frequency that is specific to the application and the hardware's capabilities.

The choice of the most suitable switch control method and modulation technique for a converter is dependent on the particular attributes of the application. These include:

- The type of load, as different loads require distinct current and voltage characteristics.

- The nature of the power source, since intermittent and variable sources like wind and solar require different control strategies compared to stable sources.

- The desired efficiency, especially when higher efficiency is a priority over factors such as cost or size, as seen in applications like aerospace or EVs.

- Thermal management, because efficient switching can reduce thermal stress and the need for cooling, which, in turn, can enhance overall system efficiency.

- Regulatory standards, since grid codes and standards may impose specific control requisites to maintain power quality and safety.

4.3 STEP-DOWN (BUCK) CONVERTER: ANALYSIS AND OPERATION

The step-down converter is integral to solar PV systems, in which the solar PV panel is the power source for the converter. Such converters are crucial for adapting the variable output of solar PV panels to consistently meet the specific voltage requirements of the downstream components including batteries or loads.

The principle operation of the converter is that the switching element(s) turns on and off at higher frequencies to control the power flow from the input to the output while regulating the voltage and current. The duty cycle δ is defined as the ratio of the on-time of the switch to the total switching period T ($T = 1/f$) hence determines the average voltage output relative to the input voltage. Inductors and capacitors in the circuit act as energy storage devices that are used to smooth the output to provide a steady DC voltage. In addition, feedback parameters (such as output voltage) are often used to adjust the duty cycle dynamically to compensate for changes in input voltage or load to maintain a stable output.

As shown in Figure 4.5a, the typical Buck converter circuit includes components such as a transistor T (as the switch), a diode D, an inductor L and a capacitor C. To design the circuit and identify parameters for achieving the desirable characteristics, Figure 4.7 illustrates the steady-state circuit operation (when minimum and maximum values of the currents at

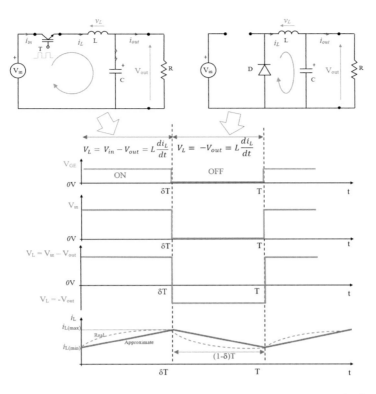

FIGURE 4.7 The circuit states of the Figure 4.5a and associated electrical waveforms.

the end of each switching cycle are equal) for one cycle under the continuous current conduction mode (the current value never reaches to zero). In addition, Table 4.7 summarizes the associated formulas (the output voltage and current, ripple voltage and hence efficiency), in which the critical parameters are also given, including switch duty cycle, inductor value, switching frequency f and load resistance R.

It can be summarized that, in practice, the components of a real converter have a degree of losses, such as the transistor's switching and conduction losses, the diode's conduction losses and the inductor's loss due to its resistance (known as winding loss or copper loss).

The converter can regulate the fluctuating input voltage down to a stable level that is suitable for charging batteries or for further conversion by an inverter while minimizing voltage ripple using an LC filter circuit. Using the converter as an MPPT controller via its duty cycle (by PWM) in response to the changing voltage from the solar panels can optimize power transfer.

TABLE 4.7 Analysis of the States of the Power Electronics Circuit in the Buck Converter Given in Figure 4.5a with Associated Formulas for the Circuit Design and Operation

Transistor ON-State ($0 < t < DT$)	**Transistor OFF State ($DT < t < T$)**
The current through the inductor rises exponentially, and if the switching frequency is high and/or the inductance is large enough, it can be assumed that the current in the inductor increases linearly: $$v_L = v_{in} - v_{out} = L\frac{di_L}{dt}$$ $$\Delta i_L = \frac{v_{in} - v_{out}}{L}\delta T$$	The voltage across the inductor is $-V_{out}$, which makes the diode forward biased and the inductor current decreases linearly: $$v_L = -v_{out} = L\frac{di_L}{dt}$$ $$\Delta i_L = -\frac{V_{out}}{L}(1-\delta)T$$ In continuous current conduction mode, diode conducts current until the transistor is turned on again:

The energy stored in the inductor can be given by $E_L = \frac{1}{2} L\, i_L^2$

The energy stored in each component at the end of the switching cycle T is equal to that at the beginning of the cycle, which means that the overall change in the current (the sum of the changes) is zero:

$$(\Delta i_L)_{ON} + (\Delta i_L)_{OFF} = 0 \quad \frac{V_{in} - V_{out}}{L}\delta T - \frac{V_{out}}{L}(1-\delta)T = 0$$

Hence $V_{out} = \delta V_{in}$ "The duty cycle, δ (ON-ratio) is independent of the output load"
If we assume that all the components are ideal (no losses!), using the power balance:
$V_{in}\, i_{in} = V_{out}\, i_{out}$ Hence $i_{out} = \delta\, i_{out}$
Since the average capacitor current is zero under steady-state operation, the inductor's current becomes the output current

$$i_L = i_{out} = \frac{v_{out}}{R}$$

Using this equation and the variation of the inductor current during ON and OFF times, the maximum and the minimum values of the inductor currents can be obtained as

$$i_{L(max)} = i_L + \frac{|\Delta i_L|}{2} = \frac{V_{out}}{R} + \frac{1}{2}\left(\frac{V_{out}}{L}(1-\delta)T\right) \quad \text{since } |\Delta i_L| = \frac{V_{out}}{L}(1-\delta)T \text{ and } T = 1/f$$

$$i_{L(max)} = V_{out}\left(\frac{1}{R} + \frac{(1-\delta)}{2Lf}\right)$$

Similarly

$$i_{L(min)} = i_L - \frac{|\Delta i_L|}{2} = \frac{V_{out}}{R} - \frac{1}{2}\left(\frac{V_{out}}{L}(1-D)T\right)$$

$$i_{L(min)} = V_{out}\left(\frac{1}{R} - \frac{(1-\delta)}{2Lf}\right)$$

Therefore, the relationship to obtain continuous current conduction can be found using

(Continued)

TABLE 4.7 (Continued)   Analysis of the States of the Power Electronics Circuit in the Buck Converter Given in Figure 4.5a with Associated Formulas for the Circuit Design and Operation

Transistor ON-State ($0 < t < DT$)	Transistor OFF State ($DT < t < T$)

$$i_{L(\min)} \geq 0 \quad V_{out}\left(\frac{1}{R} - \frac{(1-\delta)}{2L_{\min}f}\right) \geq 0 \text{ and}$$

$$L \geq L_{\min} = \frac{(1-\delta)}{2f}R$$

"This equation ensures the continuous current conduction in such converter for better utilization, and indicates the relationships related to the circuit components, $L_{\min}$, f, δ and R. In addition, the current and voltage ratings of the power circuit can be estimated and selected using the ratings of solar PV panel configurations".

In the topology discussed here, replacing the diode with a GaN-type switch also allows synchronous switching, which can switch at higher frequencies with the precise timing of the transistor's drive signals. Such switches are sensitive to parasitic impedances but provide the benefit of reduced losses across a wide range of voltages, leading to a cooler and more efficient design. Such a switch also requires less energy to activate, contributing to a higher efficiency of the converter. In addition, the potential for miniaturization with GaN switches is a desirable characteristic for solar PV applications, allowing for integration directly at the panel level or within the inverter, contributing to a more compact system.

4.4 STEP-UP (BOOST) CONVERTER: ANALYSIS AND OPERATION

The boost converter, a widely utilized switching power supply topology, is designed to increase a DC voltage to a higher value as the irradiance in solar PV panels or arrays drops. Similar to the previous analysis, assuming a continuous current conduction mode, where the current through the inductor never falls to zero, the boost converter operates efficiently across two distinct modes as illustrated in Figure 4.8, in which the ON and OFF states of the switch, illustrate the converter's ability to step-up voltage and maintain a continuous flow of power through the system. The analysis of the circuit is summarized in Table 4.8 as described in the previous section for the step-down converter. Note that the diode can also be replaced by a second switch that can be synchronized to the main switch and operated in a complementary manner.

266 ■ Reinventing the Power Grid

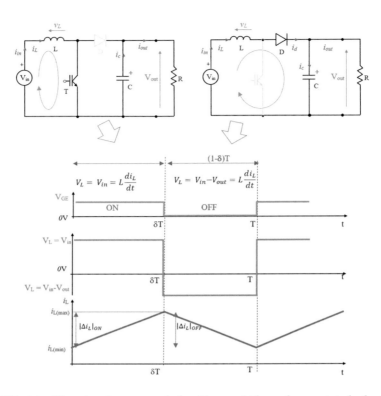

FIGURE 4.8 The circuit states of the Figure 4.5b and associated electrical waveforms.

TABLE 4.8 Analysis of the States of the Boost Converter Given in Figure 4.5a

Transistor ON-State ($0 < t < DT$)	**Transistor OFF State ($DT < t < T$)**
When the transistor is turned on the current through the inductor rises and the energy is stored in the inductor. $$v_L = V_{in} = L\frac{di_L}{dt}$$ Hence, the linear variation in the inductor current becomes: $$\Delta i_L = \frac{V_{in}\delta T}{L}$$ Note that when the transistor is on, the charged capacitor provides the required power to the load by discharging.	When the transistor is turned off, the closed loop current via the diode includes the input DC supply, the inductor and RC load. $$v_L = V_{in} - V_{out} = L\frac{di_L}{dt}$$ If the diode is ideal (zero voltage drop when conducting) and the voltage across the capacitor remains constant, and if it is assumed that the inductor current decreases linearly. $$\Delta i_L = \frac{(V_{in} - V_{out})(1 - \delta)T}{L}$$ Note that the energy stored in the inductor discharges through the diode charges the capacitor and provides the load current simultaneously.

(Continued)

TABLE 4.8 (*Continued*) Analysis of the States of the Boost Converter Given in Figure 4.5a

Transistor ON-State ($0 < t < DT$)	**Transistor OFF State ($DT < t < T$)**

As in the step-down converter, if we assume steady-state operation, the energy stored in each component at the end of the switching cycle T is equal to that at the beginning of the cycle. Therefore, the inductor current has to be the same at the start and end of the switching cycle:

$$(\Delta i_L)_{ON} + (\Delta i_L)_{OFF} = 0$$

$$\frac{V_{in}\, \delta T}{L} + \frac{(V_{in} - V_{out})(1-\delta)T}{L} = 0$$

Hence the relationship between the input and output voltage as a function of δ becomes

$$V_{out} = \frac{V_{in}}{(1-\delta)}$$

Since δ varies between 0 and 1, this equation shows that the output voltage is greater or equal to the input voltage and as δ approaches 1, V_{out} theoretically approaches to infinity. If we assume that all the components are ideal (no losses!), the average value of the input current can be calculated from the output load current by equating the input and the output powers:

$$V_{in}\, i_{in} = V_{out}\, i_{out} \quad i_{in} = \frac{V_{out}}{V_{in}} i_{out}$$

Since the load is resistive and since the inductor current equals to the input current:

$$i_{out} = \frac{v_{out}}{R} \quad i_L = i_{in} \quad \text{Hence}$$

$$i_L = \frac{V_{out}}{V_{in}} i_{out} = \frac{1}{(1-\delta)} \frac{V_{out}}{R} = \frac{1}{(1-\delta)^2} \frac{V_{in}}{R}$$

Using this equation and the variation of the inductor current during ON and OFF times, the maximum and the minimum values of the inductor currents can be obtained as

$$i_{L(max)} = i_L + \frac{\Delta i_L}{2} = \frac{1}{(1-\delta)^2} \frac{V_{in}}{R} + \frac{V_{in}\, T}{2L}$$

$$i_{L(min)} = i_L - \frac{\Delta i_L}{2} = \frac{1}{(1-\delta)^2} \frac{V_{in}}{R} - \frac{V_{in}\, T}{2L}$$

To ensure continuous current conduction $i_{L(min)} \geq 0$. Hence $\frac{1}{(1-\delta)^2} \frac{V_{in}}{R} - \frac{V_{in}\, \delta}{2 L_{min}\, f} \geq 0$

$$L_{min} = \frac{D(1-D)^2 R}{2f}$$

"To ensure continuous current conduction, the inductance value should be higher than L_{min}."

4.5 SYNCHRONOUS BUCK-BOOST CONVERTER: ANALYSIS AND BENEFITS

In the context of the synchronous buck-boost converter given in Figure 4.5a, which employs four transistors in an H-Bridge structure, which is capable of stepping down or stepping up the input voltage to a desired output voltage that can be either higher or lower than the input, depending on the control of the switching elements.

The synchronous switching in the converter improves efficiency by reducing losses that typically occur during the diode's conduction in non-synchronous converters (where T2 and T3 are diodes). While the given converter topology introduces greater complexity in terms of control compared to the previous converters, they enable a significant reduction in the size of the inductor as well, which contributes to a lower overall resistance and thus higher efficiency. The converter is also known for its minimal use of passive components, potentially leading to a smaller design footprint hence very high volumetric power density. In addition, it can offer precise control over the voltage conversion process, and low-loss operation at high switching frequencies when GaN devices are used.

Note that the inductance in the circuit is the key energy storage component. Therefore, Figure 4.9 is given to identify the inductance and switching frequency in relation to an input voltage for the buck-boost converter ensuring a continuous current conduction. As illustrated in the figure, the required inductance varies in the boost converter region (7–24 V) and then linearly increases as the input voltage is higher than the output voltage, in buck converter mode (24–50 V). The inductance reaches its peak value (which is approximately 24 µH at the 48 V mark for the circuit analysed). Similarly, the right-hand y-axis illustrates the required switching frequency, scaled by a factor of $\times 10^5$. Note that starting from the lowest voltage (24 V), the frequency is relatively stable, with minimal fluctuations until around the desirable output level.

Figure 4.10 illustrates the operation of the converter in terms of current flowing in each component in the power circuit, in which the inductor is shared by both the buck and boost modes. For the buck mode operation shown in Figure 4.10a, transistors T1 and T2 are switched complementarily, while T4 is switched off. T1 actively modulates current, indicating its primary role in connecting the input source. T2, serves as a synchronous switch to maintain output voltage, as in the role of the freewheeling diode. The operation of T3 is synchronized with T1 and T2 to provide the current

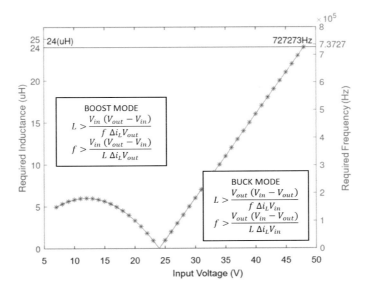

FIGURE 4.9 The relationship between the input voltage (between 7 and 50V) and the inductance or the switching frequency values in a synchronous Buck-Boost converter.

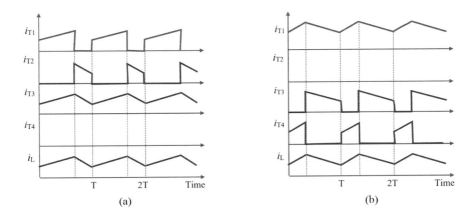

FIGURE 4.10 Current waveforms in the Synchronous Buck-Boost converter: Buck mode (a) and Boost mode of operation (b).

hence the power flow. The switching states and the current waveforms of the boost mode are illustrated in Figure 4.10b, where T4 modulates the current and defines the stored energy in the inductor L for boosting, and T1 is synchronized with the main switches for current flow.

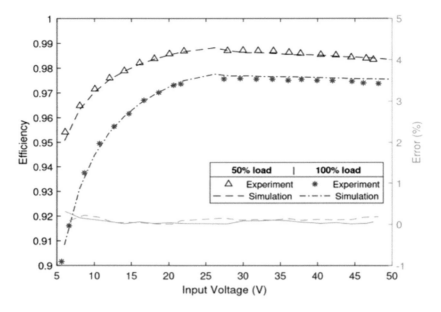

FIGURE 4.11 The Synchronous Buck-Boost converter efficiency when GaN devices are used: Experimental results versus simulations under half and full loads.

Figure 4.11 is also given to show the high efficiency (95%–98.5%) of a synchronous buck-boost converter using GaN devices across a range of input voltages under two load conditions: 50% and 100%. As shown, the high-efficiency performance is consistent in a wide range of input voltages which can represent solar PV inputs. As stated before, the majority of losses are associated with the copper losses of the practical inductance. The results also indicate that when an optimum frequency and inductance value are selected, the efficiency of the converter remains very high across the entire operating voltage range.

Note that in a solar PV system, DC optimizers also perform a conversion task similar to the DC-DC converters, which are usually part of a system that includes an inverter.

Figure 4.12 illustrates the optimal operating points against various current and voltage levels that also occur due to solar PV cell or panel configurations. The figure highlights the current and voltage limits that correlate with the specifications of the PE converter in use. As demonstrated, the broad spectrum of voltage variations in solar PV cells and panel configurations proves that the buck-boost converter is an ideal choice. Therefore,

Power Electronics and Control of Power System Components ■ 271

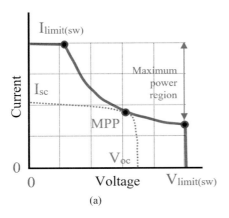 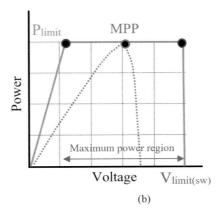

FIGURE 4.12 Optimal operating points in a Buck/Boost converter, which occur in various solar PV cell or panel configurations.

the advantages of employing a buck-boost converter can be listed as optimizing power output, accommodating fluctuations in solar intensity, providing voltage regulation, maintaining high efficiency across diverse conditions and offering design flexibility. This flexibility is specifically important for designers who may need to connect solar panels in configurations that do not align with the voltage demands of the intended load or storage system.

It is important to note that the effectiveness of PV systems is quantified by the performance ratio (PR), which is the ratio of the actual annual energy yield to the hypothetical ideal energy output of a PV array. When multiple PV modules are interconnected to create an array, the aggregate PR may suffer significantly due to mismatches among PV cells. These mismatches arise from a variety of factors, such as partial shading of the cells within a module, the accumulation of dirt, natural ageing, as well as manufacturing discrepancies and defects. In addition, the efficiency of PV cells is influenced by the level of solar irradiance, with the open-circuit voltage experiencing reductions of up to 10% at diminished radiation levels.

Therefore, a high-efficiency buck-boost converter equipped with WBG devices can address these issues, hence enhancing the PR. Such converters can respond to disparities in voltage and current due to the factors mentioned, ensuring that the PV modules operate closer to their maximum power point (MPP) and offer optimized performance across the entire PV array, hence improving overall PR.

4.6 POWER FLOW ANALOGY: VARIABLE SPEED MOTOR DRIVES VERSUS CONVERTERS USED IN WIND TURBINE GENERATORS

Before explaining the converter topologies and their control within PE circuit modules for variable speed BPMG in wind turbines, Figure 4.13 is given to draw a parallel between the well-established variable speed motor drive systems and the converter systems used in BPMG of wind turbines that were illustrated in Figure 4.5b and c.

Note that three-phase PWM inverters are needed in variable-speed three-phase motor drives for generating the three-phase, variable-frequency voltage and current. The commercial motor drive configurations (as shown in Figure 4.13a, Top) use a three-phase diode rectifier combined with a DC-link capacitor to supply the voltage source for the inverter. In such motor drives, due to the unidirectional power flow characteristic of the diode rectifiers, energy generated during regenerative braking of the motor is dissipated through a dynamic braking resistor that also forms a basic step-down DC-DC converter.

To achieve the return of regenerative power to the grid and to improve the power factor of the motor drive, the front-end rectifier can be replaced with a bidirectional switch-based topology, known as a PWM converter/rectifier (see Figure 4.13a, Bottom).

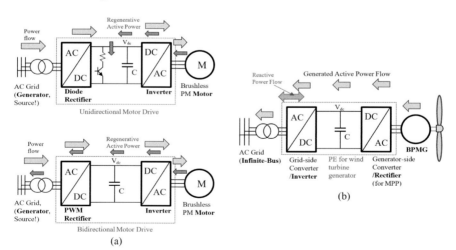

FIGURE 4.13 The parallel between a conventional variable speed motor drive setup (a) and the architecture utilized in BPMG wind turbine systems, back-to-back converter topology (b).

In Figure 4.13b, the motor control system has been repurposed into a generation system for a wind turbine. This system includes a PWM rectifier at the front end, which is capable of controlling both the magnitude and phase of the input current, which allows drawing power from the generator while achieving MPPT, accommodating the variable power outputs arising from changing wind conditions and changing blade efficiencies. Subsequently, the inverter is responsible for converting the DC power back to AC for the purpose of delivering active power to the grid.

Note that Figure 4.13b also introduces the concept of an "infinite bus", an idealized definition assuming that the bus possesses unlimited capacity to absorb or supply power, maintaining constant electrical characteristics (voltage and frequency). The infinite bus can also be considered a perfect voltage source with negligible internal impedance, which simplifies analytical models. However, in practice, at the point of common coupling/connection (PCC), the grid presents a finite impedance. Therefore, the commissioning of a renewable energy source's connection to the AC grid involves fine-tuning control parameters to ensure seamless synchronization, which is primarily based on the grid impedance at PCC.

Furthermore, it is important to emphasize here that, as summarized in Table 4.9, the back-to-back inverter topology given in Figure 4.13b can be

TABLE 4.9 Additional Functions of the Back-to-Back Inverter Topology Used in the BPMG

Purpose of Control	Description
Voltage regulation	By modulating reactive power, it can regulate voltage levels on the grid, which is essential for power system stability under varying load conditions.
Power factor correction	The inverter can improve the power factor towards unity by managing the phase relationship between voltage and current, either by supplying or absorbing reactive power, meeting grid operator standards.
Grid support services	Can provide ancillary services, including frequency regulation, voltage support and black-start capabilities as required by modern grid codes.
Fault ride-through capability	In the event of grid disturbances, the inverter can manage reactive power flow, enabling the wind turbine to maintain grid connection and contribute to grid stability during and post faults.
Harmonic compensation	The inverter can counteract harmonics by injecting phase-shifted currents, to cancel certain harmonic currents in the grid and thus enhancing power quality.
Load balancing	Inverters can distribute reactive power to specific areas as needed, balancing the load across the electrical system.

employed for a variety of functions other than active power generation only, which contributes to grid stability, power quality and regulatory compliance.

Note that the power flow through the converters is also applicable to HVDC transmission systems. The conversion process is similar, with the use of rectifiers and inverters, and the DC link represents the transmission lines.

In addition, the inverter part of the diagram is applicable to solar PV grid connections and grid-connected battery storage systems that are coupled on a DC link. In the case of solar PV systems, the PV cells generate DC that is regulated via DC-DC converters and then converted to AC using an inverter for grid compatibility. Similarly, grid-connected battery storage systems store energy in DC form and use inverters to supply AC power to the grid when needed. The inverter ensures that the power fed into the grid is synchronized with the grid's voltage and frequency.

4.6.1 Utilization of Axes Transformations in Three-Phase Converter Control

Prior to the explanations of axes transformations, it can be indicated that the per-unit (pu) system is also used in the analysis and explanations of PE systems. By normalizing (translating them into a more manageable form) various system quantities (power, voltage, current and impedance), complex circuit analysis can be simplified. Per-unit value(s) is selected in accordance with the system's nominal or rated value(s) and the remaining quantities are defined as fractions of these defined (base) units. This approach not only standardizes the comparative assessment of different systems but also enhances the clarity and convenience of handling large quantities.

Complementing the per-unit system is the concept of "phasors". These "vectors" are also associated with the above-mentioned electrical system quantities, which help in analysing electrical circuits in their sinusoidal steady-state, which is a state signifying the establishment of a uniform pattern of variation in AC circuits. Note that although no ideal sinusoidal waveform is present in PE circuits, the fundamental components of voltage and current waveforms after harmonic analysis can be considered in phasors.

Phasors simply transform these sinusoidal functions into complex numbers, time-invariant constants that simplify the circuit analysis. The magnitude of these phasor quantities represents the root mean square (RMS) or magnitude/peak values of the electrical parameters, while the phasor angle relative to a reference axis denotes the phase angle of the waveform.

The behaviour of three-phase machines is usually described by their voltage and current equations as shown in Figure 3.17 for BPMG.

The coefficients of the differential equations that describe their behaviour are time-varying. The mathematical modelling of such a system is highly complex since the flux linkages, induced voltages and currents change continuously as the electric circuit is in relative motion.

For such a complex electrical machine analysis, mathematical transformations are often used to decouple variables and to solve equations involving time-varying quantities by referring all variables to a common frame of reference.

Two mathematical axis transformations, Clarke and Park transformations, are commonly used in the control of three-phase systems. Both transformations convert three-phase variables (currents and voltages) into a two-dimensional coordinate system to simplify the analysis and control of AC machines (motors and generators). Figure 4.14 provides a visual

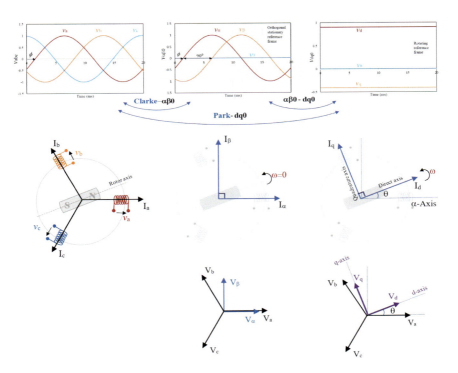

FIGURE 4.14 The visual relationships between Clarke and Park Transforms: when applied to three-phase balanced AC voltage waveforms (top-row); the current phasors (middle-row) from left to right, three-phase reference frame, two-phase stationary reference frame (Clarke) and rotating reference frame (Park); the inverse transformations for voltages (bottom-row).

representation of the Clarke and Park transformations where balanced three-phase voltages result in equal magnitudes for v_α and v_β, and constant values of v_d and v_q, which are all given in per-unit values. Note that current "phasors", space vectors correspond to the original three-phase waveforms, and the transformations are also given in the figure referencing a two-pole BPMG that had the equivalent circuit given previously in Figure 3.16.

It should be noted that in the literature, three different cases are introduced for the definition of the angle θ shown in the phasor of the $dq0$ transformation. The case given in Figure 4.14 indicates that the q-axis is leading the d-axis, and the angle θ is defined between the d-axis with respect to the α-axis. However, two other cases are also given in the literature: when the q-axis is lagging the d-axis, and the angle between the d-axis with respect to the α-axis is used; and when the q-axis is leading the d-axis, and the angle between the q-axis with respect to the α-axis is used. Note also that commonly used "reference-frame speeds" in the Park transformation are $\omega=\omega_r$ (the $dq0$-axes rotate at the rotor speed, mechanical speed) and $\omega=\omega_e$ (the $dq0$-axes rotate at the synchronous speed, electrical speed).

As indicated in Figure 4.14, it is possible to transform between each reference frame: abc-$\alpha\beta0$, $\alpha\beta0$-$dq0$, abc-$dq0$ and reverse. The mathematical formulas associated with Clarke and Park transformations are summarized in Table 4.10.

As illustrated in Figure 4.14, the α-axis represents a direct axis and the β-axis represents a quadrature axis orthogonal to α. The zero component (0) is a common mode that does not interact with the other two and is used to represent the neutral or imbalance in the three-phase system. Therefore, if the three-phase voltages are balanced the zero-sequence component is zero. If the three-phase supply is unbalanced however, this results in unequal magnitudes for v_α and v_β and time-varying v_d and v_q at the second harmonic (at 100 Hz for the waveform studied). The zero-sequence component v_0 at the fundamental frequency and is identical in both transformations.

As illustrated and formulated above, the Clarke transformation converts three-phase quantities into two-phase orthogonal components within a stationary reference frame. This transformation simplifies the control algorithms for electrical machines. As shown in Figure 4.14, the Clarke transformation is often a preliminary step before applying the Park transformation, which further simplifies the control of AC machines by rotating the reference frame to align with the rotor's magnetic field.

TABLE 4.10 The Mathematical Relationships between the Clarke and Park Transformations for the Voltage Functions (the Current Functions Can Also Be Given) and the Instantaneous Active and Reactive Power Calculations

	Amplitude-Invariant	Power-Invariant
Clarke Transf.	$(abc\text{-}\alpha\beta 0)$	$(abc\text{-}\alpha\beta 0)$

$$\begin{bmatrix} V_\alpha \\ V_\beta \\ V_0 \end{bmatrix} = 2/3 \begin{bmatrix} 1 & -\dfrac{1}{2} & -\dfrac{1}{2} \\ 0 & \dfrac{\sqrt{3}}{2} & -\dfrac{\sqrt{3}}{2} \\ \dfrac{1}{2} & \dfrac{1}{2} & \dfrac{1}{2} \end{bmatrix} \begin{bmatrix} V_a \\ V_b \\ V_c \end{bmatrix}$$

$$\begin{bmatrix} V_\alpha \\ V_\beta \\ V_0 \end{bmatrix} = \sqrt{\dfrac{2}{3}} \begin{bmatrix} 1 & -\dfrac{1}{2} & -\dfrac{1}{2} \\ 0 & \dfrac{\sqrt{3}}{2} & -\dfrac{\sqrt{3}}{2} \\ \dfrac{1}{\sqrt{2}} & \dfrac{1}{\sqrt{2}} & \dfrac{1}{\sqrt{2}} \end{bmatrix} \begin{bmatrix} V_a \\ V_b \\ V_c \end{bmatrix}$$

Inverse Clarke $(\alpha\beta 0\text{-}abc)$

$$\begin{bmatrix} V_a \\ V_b \\ V_c \end{bmatrix} = \begin{bmatrix} 1 & 0 & 1 \\ -\dfrac{1}{2} & \dfrac{\sqrt{3}}{2} & 1 \\ -\dfrac{1}{2} & -\dfrac{\sqrt{3}}{2} & 1 \end{bmatrix} \begin{bmatrix} V_\alpha \\ V_\beta \\ V_0 \end{bmatrix}$$

Inverse Clarke $(\alpha\beta 0\text{-}abc)$

$$\begin{bmatrix} V_a \\ V_b \\ V_c \end{bmatrix} = \sqrt{2/3} \begin{bmatrix} 1 & 0 & \dfrac{1}{\sqrt{2}} \\ -\dfrac{1}{2} & \dfrac{\sqrt{3}}{2} & \dfrac{1}{\sqrt{2}} \\ -\dfrac{1}{2} & -\dfrac{\sqrt{3}}{2} & \dfrac{1}{\sqrt{2}} \end{bmatrix} \begin{bmatrix} V_\alpha \\ V_\beta \\ V_0 \end{bmatrix}$$

(Continued)

TABLE 4.10 (Continued) The Mathematical Relationships between the Clarke and Park Transformations for the Voltage Functions (the Current Functions Can Also Be Given) and the Instantaneous Active and Reactive Power Calculations

Park Transf.

Amplitude-Invariant

(abc-dq0)

$$\begin{bmatrix} V_d \\ V_q \\ V_0 \end{bmatrix} = 2/3 \begin{bmatrix} \cos(\theta) & \cos(\theta - 2\pi/3) & \cos(\theta + 2\pi/3) \\ -\sin(\theta) & -\sin(\theta - 2\pi/3) & -\sin(\theta + 2\pi/3) \\ \frac{1}{2} & \frac{1}{2} & \frac{1}{2} \end{bmatrix} \begin{bmatrix} V_a \\ V_b \\ V_c \end{bmatrix}$$

Inverse Park (dq0-abc)

$$\begin{bmatrix} V_a \\ V_b \\ V_c \end{bmatrix} = \begin{bmatrix} \cos(\theta) & -\sin(\theta) & 1 \\ \cos(\theta - 2\pi/3) & -\sin(\theta - 2\pi/3) & 1 \\ \cos(\theta + 2\pi/3) & -\sin(\theta + 2\pi/3) & 1 \end{bmatrix} \begin{bmatrix} V_d \\ V_q \\ V_0 \end{bmatrix}$$

Power-Invariant

(abc-dq0)

$$\begin{bmatrix} V_d \\ V_q \\ V_0 \end{bmatrix} = \sqrt{\frac{2}{3}} \begin{bmatrix} \cos(\theta) & \cos(\theta - 2\pi/3) & \cos(\theta + 2\pi/3) \\ -\sin(\theta) & -\sin(\theta - 2\pi/3) & -\sin(\theta + 2\pi/3) \\ \frac{1}{\sqrt{2}} & \frac{1}{\sqrt{2}} & \frac{1}{\sqrt{2}} \end{bmatrix} \begin{bmatrix} V_a \\ V_b \\ V_c \end{bmatrix}$$

Inverse Park (dq0-abc)

$$\begin{bmatrix} V_a \\ V_b \\ V_c \end{bmatrix} = \begin{bmatrix} \cos(\theta) & -\sin(\theta) & \frac{1}{\sqrt{2}} \\ \cos(\theta - 2\pi/3) & -\sin(\theta - 2\pi/3) & \frac{1}{\sqrt{2}} \\ \cos(\theta + 2\pi/3) & -\sin(\theta + 2\pi/3) & \frac{1}{\sqrt{2}} \end{bmatrix} \begin{bmatrix} V_d \\ V_q \\ V_0 \end{bmatrix}$$

Active (p) and reactive (q) powers

αβ0

$p(t) = 3/2\,(v_\alpha i_\alpha + v_\beta i_\beta + 2 v_0 i_0)$
$q(t) = 3/2\,(v_\beta i_\alpha - v_\alpha i_\beta)$

dq0

$p(t) = 3/2\,(v_d i_d + v_q i_q + 2 v_0 i_0)$
$q(t) = 3/2\,(v_q i_d - v_d i_q)$

αβ0

$p(t) = v_\alpha i_\alpha + v_\beta i_\beta + v_0 i_0$
$q(t) = v_\beta i_\alpha - v_\alpha i_\beta$

dq0

$p(t) = v_d i_d + v_q i_q + v_0 i_0$
$q(t) = v_q i_d - v_d i_q$

The inverse Clarke transformation is used to revert the two-dimensional stationary reference frame back to the original ABC three-phase system. It is important to note that after processing the control algorithms in the two-dimensional frame, the inverse Clarke transformation is applied to generate the appropriate three-phase control signals for the three-phase converters. These control signals are then utilized in the PWM control of power electronic devices, which manage the speed, torque and power.

Both Clarke and inverse Clarke transformations are integral to field-oriented control (FOC) or vector control of AC machines. In this method, the Clarke transformation is used in real-time to apply control algorithms that can rapidly respond to changes. Then the inverse transformation is applied to translate back into three-phase signals to control the machine (motor or generator) by PWM to regulate the power in the electrical machine.

On the other hand, the Park transformation is a mathematical transformation that brings significant benefits to the real-time control of electrical machines. It allows any vector (voltage and current) that rotates on the $\alpha\beta$-plane to be expressed in the dq-reference frame (also known as the synchronous reference frame). Since the dq reference frame rotates at a specific frequency ω and aligns with the angular position $\theta=\omega t$ on the $\alpha\beta$-plane, this characteristic is particularly useful for analysing the behaviour of electrical machines whose states change over time, which is critical for defining air gap flux and torque generation after interacting with the current in the stator windings.

By utilizing the rotating dq-frame, the complex, time-varying nature of the machine's operation can be simplified. This simplification helps to understand the interrelationship between different electrical quantities. As shown in Table 4.10, the transformation facilitates the control of active and reactive power components in three-phase systems.

It should be emphasized here that the transformation ensures the norm of the voltage vector is conserved across different reference frames. This means that the physical quantities represented do not alter in magnitude due to the transformation, ensuring consistency in analysis. Expressing voltages and currents using space vectors in the $dq0$-reference frame allows for the study of instantaneous behaviours in three-phase systems, which is vital for real-time control and dynamic analysis.

Moreover, the orthogonal components in both transformations enable control systems to treat the AC system more like a DC system, where the control of power, torque and flux can be controlled independently. This is particularly useful in systems such as wind turbine generators or electric vehicle drivetrains. This decoupling is powerful in FOC, where it is used to maximize efficiency and responsiveness. Their relationships to torque, flux and power are as follows:

- The flux produced by the stator windings is critical for creating the magnetic field that interacts with the rotor. The d-axis component of the Park transformation (i_d for current or v_d for voltage) is directly related to the control of magnetic flux within the machine.

- The q-axis component (i_q or v_q) is associated with the control of torque. This is because, in many machine types, the torque is produced by the interaction of the stator and rotor magnetic fields, and this interaction is most effective when the stator's magnetic field is orthogonal to the rotor's. By controlling the i_q current, the electromagnetic torque can be directly controlled.

- The real power $p(t)$ in an AC system is a function of both voltage and current and their phase relationship. The apparent power is the product of the d-axis and q-axis currents and the respective d-axis and q-axis voltages. Therefore, the real power can be calculated from the product of the voltage and current in the q-axis, as this axis is responsible for the active power transfer in a synchronous machine, as given in Table 4.10. Reactive power, which is related to the magnetic field associated with the d-axis.

4.7 OPERATION PRINCIPLES OF BPMG, THREE-PHASE CONVERTERS OPTIONS IN WIND TURBINE SYSTEMS AND THEIR CONTROL

Using the equivalent circuit presented earlier in Figure 3.16 for a three-phase BPMG, the voltage equation for a single phase can be expressed using Kirchhoff's voltage law as $E = V + (R + j\omega L) I$, which has the vector representation shown in Figure 4.15. It is important to note that the generator features a surface-mounted permanent magnet rotor; therefore, the stator inductance L is assumed constant and does not vary with the rotor position. By ignoring losses and saturation, the dq-axis current and voltage equations can be given as shown in the figure. In addition, the formulas in the figure outline the total mechanical input power P_m for a three-phase ideal generator ($R=0$), the total torque T_m, and the electrical power produced P_e, without ignoring the winding resistance R. Here, p represents the number of pole pairs of the generator, and Ψ_{pm} is the magnet flux linkage of the generator.

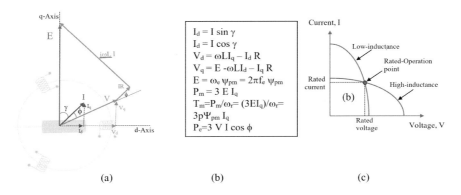

FIGURE 4.15 Single-phase phasor diagram of the surface mount BPMG (a) associated equations as a function of dq-axis quantities (b) and Voltage–Current values for the inductance of the surface-mount PM generator.

Although the formulas presented in the figure are primarily utilized in the control algorithms of converters, they are equally applicable in defining the ratings of PE, including their maximum or rated operating voltage and current values. For example, the characteristic short-circuit current of the generator at high speeds is determined by the formula Ψ_{pm}/L. Therefore, low-inductance generators typically operate at voltage levels near their rated back electromotive force (EMF) value, resulting in current values significantly lower than the characteristic short-circuit current. Conversely, high-inductance generators operate at voltage levels significantly lower than the rated back EMF voltage, and their current levels are close to the characteristic short-circuit current.

Therefore, a high-inductance surface PM generator can be considered as a constant current source, while a low-inductance generator presents the behaviour of a constant voltage source. The typical voltage-current values for high and low-inductance surface PM generators, when connected to a resistive load, are shown in Figure 4.15c. It is important to note that if a generator displays a high-inductance value, the generator's rated current equals the characteristic short-circuit current, which results in a desirable field-weakening performance at short-circuit fault currents.

In summary, field-weakening performances and short-circuit fault currents are useful for the safe, efficient and reliable operation of BPMG in wind turbines. Moreover, they are directly associated with Low Voltage Ride-Through (LVRT) capabilities and the management of fault currents in wind turbine systems. Table 4.11 shows the relationships among these factors, indicated by the application of Park and Clarke transformations.

TABLE 4.11 The Interconnections between Various Aspects of BPMG Operation and Control in Wind Turbine Systems

Aspect	Role of Park and Clarke Transformations in BPMG/Wind Turbines
Field-weakening performance	Decoupling of control variables: Separate control of torque and flux for field weakening. Efficient control strategies: Implementation of field-weakening by adjusting d-axis current. Real-time analysis and adjustment: Dynamic control during field-weakening operations. Relation to LVRT: Adjusting generator output during voltage dips for grid stability.
Short-circuit fault currents	Fault detection: Detecting faults by analysing transformed vectors. Rapid response to faults: Quick identification and response in dq-frame. Analysis of fault dynamics: Insights into fault impact for protective strategy. Relation to LVRT: Managing currents during voltage sags for turbine and grid safety.
Low voltage ride-through (LVRT)	Dynamic response to voltage dips: Maintaining operation during grid voltage reductions. Control of reactive power: Decoupling active and reactive power for effective grid support. Enhanced grid support: Adjusting control strategies for grid stability during voltage dips.
Fault current management	Compliance with grid requirements: Meeting stringent grid code requirements for fault handling. Limitation and control of fault currents: Managing excessive currents during grid faults. System resilience: Ensuring turbine operation safety and effectiveness during adverse grid conditions.

4.7.1 Common Features of Generator-Side Converters/Rectifiers

The most frequently used three-phase AC/DC power electronic converters in wind turbines, regardless of scale, are shown in Figure 4.5b. Table 4.12 highlights the operational principles of these circuit topologies, utilizing their single-phase equivalent circuits and including both low and high-speed phasor diagrams. It is important to note that these diagrams include resistive and capacitive loads to simulate inverter operation and optimize power transfer while omitting the generator winding losses.

Figure 4.16 presents a comparative analysis of three circuit topologies' performance under varying loading conditions across different speeds. It illustrates the voltage, current, power factor (pf) and active power against generator speed characteristics, all expressed in per-unit (pu) for each circuit, which aims at the selection of wind turbines of different sizes and cost

TABLE 4.12 Three Different Types of Rectifier Circuits, Their Equivalent Circuits and Phasors

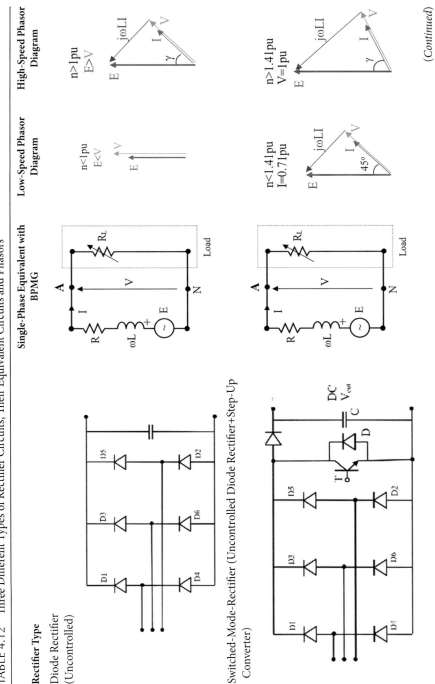

(Continued)

TABLE 4.12 (Continued) Three Different Types of Rectifier Circuits, Their Equivalent Circuits and Phasors

Rectifier Type	Single-Phase Equivalent with BPMG	Low-Speed Phasor Diagram	High-Speed Phasor Diagram
PWM Rectifier (Bidirectional)	[circuit diagram with R, ωL, E, R_L, $-1/j\omega C_L$, Load, and PWM bridge with transistors T1–T6 and diodes D1–D6, DC V_{out}]	$n<0.71\text{pu}$, $I=1\text{pu}$; phasors I, E, $j\omega LI$, V at $45°$	$n>0.71\text{pu}$, I, $V=1\text{pu}$; phasors E, I, $j\omega LI$, V with angles θ, γ

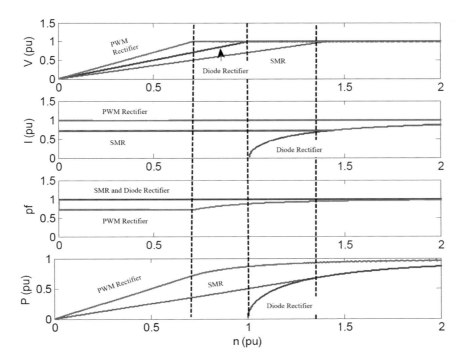

FIGURE 4.16 A comparative analysis of three rectifier topologies' performance under varying loading conditions across different speeds.

considerations. Despite the potential for highly distorted waveforms in PE circuits, this simplified analysis assumed sinusoidal waveforms for ease of comparison.

As illustrated, the generator's phase voltage increases linearly with speed until it reaches the rated voltage, and the PWM rectifier exhibits the fastest increase, followed by the diode rectifier, and the SMR.

The phase current for the diode rectifier presents a unique characteristic at its rated speed (1 pu). In contrast, the SMR starts generating current at a lower speed (0.71 pu), with its magnitude progressively rising with speed, reaching 1 pu. However, the PWM rectifier consistently generates a current of 1 pu magnitude, irrespective of speed.

It is noted that the power factor for both the diode rectifier and the SMR remains at unity for all speeds. However, the PWM rectifier's power factor commences at 0.71 pu at lower speeds and increases to unity as the speed increases.

The generator's output power is normalized to the rated active power. The output power for the diode rectifier is similar to the phase current behaviour, with both voltage and power factor at 1 pu. However, the PWM rectifier, due to its capacity to induce a leading phase shift at lower speeds, is capable of generating significantly more power than the other two circuits. The PWM rectifier produces the maximum output power at lower speeds, up to twice that of the SMR, and as speed escalates, the power converges to 1 pu.

As can be seen, the choice of power electronic converters is influenced by the scale and cost considerations of the system. Large wind turbines use the PWM rectifier topology as the generator side converter due to its ability to handle higher power outputs and provide improved control over the generator's speed and power factor. On the other hand, small-scale wind turbines, where cost is a significant concern, choose either rectifiers or SMR.

4.7.2 Control of Generator-Side PWM Rectifiers

Control methods in power systems and renewable energy applications range from traditional approaches like Proportional-Integral-Derivative (PID) control, Vector Control (also known as Field-Oriented Control or FOC) and Direct Torque Control (DTC), to more complex methods such as Model Predictive Control (MPC), Adaptive Control and Sliding Mode Control (SMC). There are also advanced computational methods like Neural Network Control, Fuzzy Logic Control, Genetic Algorithms and Reinforcement Learning. Furthermore, the use of Expert Systems and Predictive Analytics are also reported.

Choosing control methods for power systems and renewables relies on application demands. Although simplicity makes PID control preferred in many applications, vector control offers precision for controlled rectifiers and inverters. In addition, there are emerging control techniques which offer real-time optimization in complex settings, aiding variable and intermittent renewable integration.

As was studied previously, a direct drive variable speed wind turbine system that contains mechanical components like blades and hub, and electrical components such as the generator and converter, is interconnected through electromagnetic coupling. In such a system, the control system's function is to continuously monitor both mechanical and electrical parameters, apply motion equations and regulate the PWM rectifier. This regulation is important for maximizing power capture from the wind turbine across varying wind conditions, which is achieved by altering the electrical load and modulating the pitch angle of the blades.

Figure 4.17a presents the rotor flux-oriented vector control block diagram for the generator-side converter in a back-to-back converter topology, which is used in wind turbine systems for optimal power extraction. This mature control concept, alongside the PE block and other wind turbine components, also aims to enhance system efficiency and resilience.

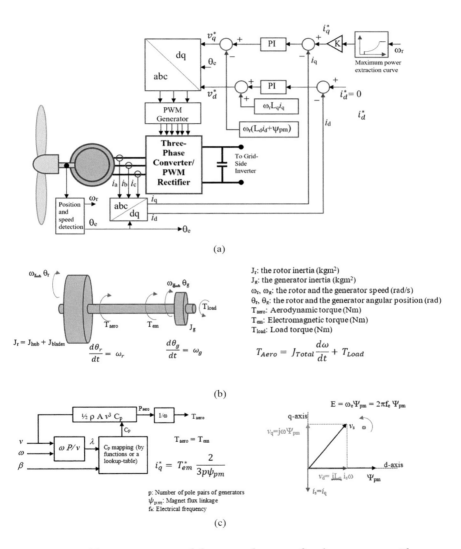

FIGURE 4.17 The components of the control system for the generator-side converter (PWM Rectifier): the vector control block diagram (a), two mass-model of the direct drive turbine system (b), and aerodynamic system model and the phasor diagram for maximum torque per ampere control (c).

Using the electrical equivalent circuit of the BPMG, the generator equations can be obtained in the *dq*-frame as below,

$$v_d = i_d R + \frac{d}{dt} i_d L_d - \omega L_q i_q \qquad (4.1)$$

$$v_q = i_q R + \frac{d}{dt} i_q L_q + \omega L_d i_d + \omega \psi_{pm} \qquad (4.2)$$

where v_d and v_q are the equivalent stator voltages and i_d and i_q are the equivalent stator currents in the *dq* frame, ω is the electrical speed in rad/s, L_d and L_q are the equivalent self-inductances of the stator in the *dq* frame in henrys, Ψ_{pm} is the flux linkage of the permanent magnet volt-seconds. It is assumed that the stator windings are balanced and saturation and parameter change due to temperature and frequency are neglected. The model also assumed that there is no saliency.

If it is assumed that the generator is in the steady state and the winding resistances are ignored, the given equations can be reduced as

$$v_s = v_q + j v_d = -\omega L_q i_q + j \omega \psi_{pm} \qquad (4.3)$$

Since $i_q < 0$ in the generating mode

$$v_s = \omega L_q i_q + j \omega \psi_{pm} \qquad (4.4)$$

This is used in the control system as illustrated in Figure 4.17a. The stator flux components in the *dq* reference frames can also be expressed as

$$\psi_d = L_d i_d + \psi_{pm} \qquad (4.5)$$

$$\psi_q = L_q i_q \qquad (4.6)$$

Since the generator is a round rotor type, the electromagnetic torque of the generator can be given in the *dq* rotor reference frame as

$$T_{em} = \frac{3}{2} p \psi_{pm} i_q \qquad (4.7)$$

In addition, the active and reactive powers of the generator in the same *dq* reference frame become

$$P = \frac{3}{2}\left(v_q i_q + v_d i_d\right) = T_e \omega_m \qquad (4.8)$$

$$Q = \frac{3}{2}\left(v_q i_d - v_d i_q\right) \qquad (4.9)$$

The result shows that the active and reactive powers of the generator can be controlled by the converter using the stator voltages and currents.

The dq transformation reveals that the stator current is split into two parts: one for torque production and the other for magnetic field generation. Vector control employs two current techniques: unity power factor control and maximum torque per ampere (MTPA) control. For generator control, which aims to optimize torque and thus power (by adjusting the speed to ensure it operates at the optimal point on its power curve), MTPA is preferred, achieved by setting the d-axis current to zero ($i^*_{d=0}$) and aligning the q-axis current (i_q) with the stator current.

The phasor diagram given in Figure 4.17c (right) illustrates this method. The figure also shows the stator current and voltage are not phase-aligned, indicating nonzero reactive power due to the generator's magnetizing inductance. Therefore, maximum active power generation also involves some reactive power demand. Therefore, to accommodate the increased total apparent power ($S = P + jQ$), the converter's capacity must be increased accordingly.

The q-axis current reference (i^*_q) is calculated with the formula shown in Figure 4.17c, utilizing the optimal electromagnetic torque, pole pair number and magnet flux linkage for maximum power generation. This torque reference is derived from either a look-up table or a formula as indicated in Figure 4.17c. The figure's two-mass model of the direct drive turbine system given in Figure 4.17b equates the wind turbine's aerodynamic torque T_{aero} with the generator's electromagnetic torque T_{em}.

Despite simplifying assumptions, the aerodynamic model converts wind energy to aerodynamic torque, the generator's input torque. The block diagram in Figure 4.17c demonstrates the T_{aero} calculation, using rotor speed from the generator speed and wind speed to compute the tip speed ratio λ. The pitch angle β and λ determine the C_p value, which is then used to calculate T_{aero}.

The swing equation or the equation of motion describes the dynamic behaviour of the synchronous generator's rotor, linking mechanical input P_{aero} to electrical output P_e or $\underline{P}_{load}$ and rotor acceleration ($d\omega/dt$) shown in Figure 4.17b. The rotor's acceleration or deceleration is proportional to the net power input, dictating that excess mechanical power over electrical conversion causes rotor acceleration and vice versa. Proper control design allows the system to respond to load changes, maintaining stability.

The block diagram in Figure 4.17c also indicates two methods for determining the C_p mapping function: a look-up table or an approximating function, enabling direct correlation with operational parameters like wind speed and blade pitch angle.

Note that pitch control maintains rotor speed within operational limits, preventing turbine and generator damage. When wind speeds exceed the rated speed, the pitch angle is adjusted for protection and rotor speed stabilization.

The pitch actuator control, although not shown in the figure to avoid complexity, involves a PI controller with an anti-windup circuit based on generator speed error, adjusting the pitch angle to modulate aerodynamic torque, ensuring no excess acceleration torque and a steady generator speed. The pitch system controls aerodynamic torque to counteract generator torque and regulate rotor speed, optimizing power extraction at lower wind speeds.

The generator side utilizes a rotor flux-oriented vector control strategy with a dual PI control structure, including an outer speed loop and an inner current loop. As indicated in the control block diagram, after Park Transformation, the direct-axis current is nullified, indicating that the quadrature-axis is current for torque generation. The stator currents are measured against their reference counterparts and discrepancies are addressed by the PI controllers, which then adjust the stator voltages in the *dq* frame, including cross-coupling compensation to ensure separation of the *d* and *q*-axis controls. In addition, any voltage induced by magnetic flux is deducted from the *q*-axis voltage reference. These processed *d* and *q*-axis voltage references are then converted to real-time stator voltages to produce PWM signals for the converter.

It is important to note that the generator's complete decoupling from the grid via the converter allows for the exchange of reactive power solely with the generator side converter, not directly with the grid. This decoupling allows the converter to independently manage active and reactive power, apart from the turbine's rotational inertia. Therefore, the turbine blades' kinetic energy does not contribute to grid inertia during disturbances, differing from traditional synchronous generators that provide direct inertia support.

This separation means that while turbine blade inertia can influence the turbine's local dynamics, such as its response to wind speed variations, it does not offer the same grid inertia response that assists with frequency regulation. Therefore, synthetic or virtual inertia has been developed, employing PE and control algorithms to replicate the effects of traditional inertia to improve grid stability.

Wind turbines are commonly classified by their power ratings, which correlate with rotor diameters, into micro (<1–2 kW), small (<100 kW), medium (<1 MW) and large/utility (>1 MW) categories, although no universal standard for size classification exists. Considering their small size, along with the complexities and expenses and complexities associated with grid interconnection, these turbines are usually operated off-grid, such as in residential uses, battery charging for boats or recreational vehicles and powering remote equipment such as weather stations.

These micro turbines often feature fixed-position blades, a design that reduces mechanical complexity and cost. Power from the rotating nacelle is transferred via a slip ring and brush assembly, allowing for free nacelle rotation and wind tracking without cable twisting. Typically, simple open-loop-control rectifier topologies are implemented within the nacelle.

4.7.3 Control Principles of Grid-Side Inverters

To transfer power from the generator system developed in the previous section to the grid using a voltage source inverter (VSI), two control methods are typically employed: "load angle and magnitude control" and "vector control".

The load angle and magnitude control method adjust the phase angle of the voltage (δ, load angle) and its magnitude at the inverter's terminals relative to the grid. This method is illustrated in Figure 4.18b, which shows an equivalent circuit and phasor diagram ignoring the line resistances. Therefore, the active power in watt and the reactive power in var can be defined as

$$P_{grid} = \frac{V_{inv} V_{grid}}{X_{grid}} \sin\delta \qquad (4.10)$$

$$Q_{grid} = \frac{V_{inv}^2}{X_{grid}} - \frac{V_{inv} V_{grid}}{X_{grid}} \cos\delta \qquad (4.11)$$

Here, V_{inv} is the voltage magnitude at the inverter's output terminals, V_{grid} is the magnitude of grid voltage, X_{grid} is the total reactance between the converter and the grid in ohms. Note that for small values of δ, $\sin\delta \cong \delta$ and $\cos\delta \cong 1$. Therefore, the power equations can be simplified as

$$P_{grid} = \frac{V_{inv} V_{grid}}{X_{grid}} \delta \qquad (4.12)$$

$$Q_{grid} = V_{inv} \left(\frac{V_{inv} - V_{grid}}{X_{grid}} \right) \qquad (4.13)$$

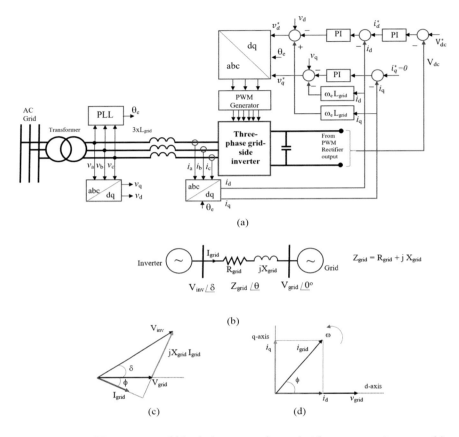

FIGURE 4.18 Vector control block diagram of a grid-side converter/inverter (a); single-line diagram of an inverter connected to the grid (b); the phasor diagram for the load angle and magnitude control ignoring line resistance (c); and the phasor diagram of the vector control principle (d).

These equations indicate that the active and reactive powers can be independently controlled by adjusting the angle δ and the voltage magnitudes of the voltages.

In Figure 4.18c, the phasors of the inverter voltage and grid voltage are represented as vectors on the complex plane, with their angles relative to a horizontal reference axis showing the phase angles ϕ. The load angle δ is the angle difference between V_{inv} and V_{grid}.

As it can be observed, "load angle control" operates on the principle that when V_{inv} leads V_{grid} by δ, the inverter, acting as the generator, delivers active power to the grid ("overexcited" condition in conventional synchronous generator). If V_{inv} lags behind V_{grid}, the inverter draws active power from the grid, indicating an "under excited" condition. With "magnitude control", altering the inverter's voltage amplitude affects reactive power flow. A higher voltage magnitude results in the inverter supplying reactive power, whereas a lower magnitude leads to the absorption of reactive power by the inverter.

However, in real-life applications, the active and reactive power exchange between a wind turbine and the grid can be managed through different methods.

- *For load angle control*: The control system locks the phase angle of the generator output to the grid frequency by using a Phase-Locked Loop (PLL) structure by adjusting the reference signal, then the phase angle hence the load angle can be controlled. Alternatively, the control algorithms within the VSI can adjust the timing of the inverter switching to either advance or delay the phase of the output voltage relative to the grid voltage, hence controlling the load angle δ. Note that in conventional synchronous generators (with the DC exciter) the rotor angle can be adjusted physically but this method is not applicable to brushless PM synchronous generators.

- *For voltage magnitude control*: The VSI can modulate the magnitude of the voltage by varying PWM signals, which vary the average voltage magnitude. In addition, the voltage level on the DC link can be varied by a DC-DC converter before inversion. Furthermore, on-load tap changers transformers can be used to adjust the grid voltage.

Vector control (also known as field-oriented control) is commonly used in a wide range of variable speed motor control applications from home appliances to industrial applications and EVs. In wind turbines, it is also used on the grid-side converter due to its high efficiency and enhanced performance.

The basis of the control algorithm for vector control illustrated in Figure 4.18a is based on the d-axis and the q-axis currents that can be obtained using the Park Transformation as explained and used in the generator-side converter previously.

Using the equivalent circuit of the three-phase inverter and the line parameters, the grid voltages in the ABC reference frame can be written as

$$\begin{bmatrix} v_a \\ v_b \\ v_c \end{bmatrix} = R_{\text{grid}} \begin{bmatrix} i_a \\ i_b \\ i_c \end{bmatrix} + L_{\text{grid}} \frac{d}{dt} \begin{bmatrix} i_a \\ i_b \\ i_c \end{bmatrix} + \begin{bmatrix} v_{a(\text{inv})} \\ v_{b(\text{inv})} \\ v_{c(\text{inv})} \end{bmatrix} \qquad (4.14)$$

and using the Park transforms for the grid voltages (v_a, v_b and v_c), the grid currents (i_a, i_b and i_c) and the inverter voltages ($v_{a(\text{inv})}$, $v_{b(\text{inv})}$ and $v_{c(\text{inv})}$), the grid voltages in dq reference frames can be obtained as

$$v_d = i_d R_{\text{grid}} + L_{\text{grid}} \frac{d}{dt} i_d - \omega_e L_{\text{grid}} i_q + v_{d(\text{inv})} \qquad (4.15)$$

$$v_q = i_q R_{\text{grid}} + L_{\text{grid}} \frac{d}{dt} i_q + \omega_e L_{\text{grid}} i_d + v_{q(\text{inv})} \qquad (4.16)$$

Here ω_e is the synchronous speed $2\pi f_e/60$ in rad/s, where f_e is the grid frequency.

Using the active and reactive power definitions given previously in (Eqs. 4.8 and 4.9)

$$P = \frac{3}{2}\left(v_q i_q + v_d i_d\right) \quad Q = \frac{3}{2}\left(v_q i_d - v_d i_q\right) \qquad (4.17)$$

The electrical position of the supply voltage can be calculated using the Clarke Transformation, the $\alpha\beta$ stationary transformation,

$$\theta_e = \int \omega_e dt = \tan^{-1} \frac{v_\beta}{v_\alpha} \qquad (4.18)$$

By aligning the d-axis of the reference frame with the position of the stator voltage as specified by the above equation, the q-axis voltage component v_q is reduced to zero. This is reflected in the phasor diagram in Figure 4.18d, where the grid voltage vector is aligned along the d-axis, resulting in a q-axis voltage component of zero. With the supply voltage amplitude being held constant, the d-axis voltage component v_d is also maintained at a constant value. Given these conditions, the power equations previously described in the equations 4.8 and 4.9 can be simplified, showing that the active power is directly proportional to the d-axis current i_d, and the reactive power is directly proportional to the q-axis current i_q.

$$P_{grid} = \frac{3}{2} v_d i_d \qquad (4.19)$$

$$Q_{grid} = -\frac{3}{2} v_d i_q \qquad (4.20)$$

In the operation of an inverter that transforms DC to AC, managing power flow is the key. If ignoring harmonics from inverter switching, other inverter-related losses and losses due to line resistance, the power on the DC side is equivalent to the power on the AC side.

$$V_{dc} I_{dc} = 3/2 \left(v_d i_d \right) \qquad (4.21)$$

As mentioned in Section 4.2, the modulation technique is used in inverters to control the output voltage and current of the inverter to ensure it matches the frequency, phase and amplitude of the grid's AC voltage and inject power into the electrical grid efficiently. After synchronization with the grid, the key points after about modulation in grid-connected inverters are:

- To create a waveform using PWM that approximates the desired AC sinusoidal voltage.
- To compare a sinusoidal reference signal (proportional to the desired output voltage) with a triangular carrier wave to determine the switching times of the inverter's power devices.
- Choosing the best technique (such as Space Vector Modulation) to provide better utilization of the DC bus voltage hence producing quality AC output with lower harmonic distortion (which is also to meet the grid code requirement).

Note that the "modulation index", in the context of PE and inverter design, is a measure that indicates the ratio of the magnitude of the reference or modulating signal to the magnitude of the carrier signal in a PWM scheme. In the case of voltage source inverters, the modulation index (the symbol m or M_a is used for the amplitude modulation index) defines "how much of the DC bus voltage is being used" to create the AC output voltage. For example, a modulation index of 1 implies that the peak AC output voltage is equal to the DC bus voltage, while a modulation index of less than 1 means the AC output is lower than the DC bus voltage.

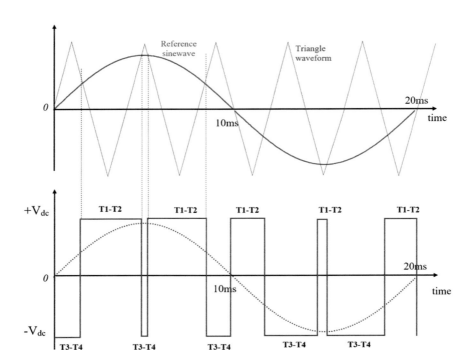

FIGURE 4.19 The principle of the modulation index for a single-phase inverter (at 50Hz). $M_a = v_{ref}/v_{carrier}$, where v_{ref} is the peak value of the modulating (reference) waveform (the sine wave in the context of PWM), and $v_{carrier}$ is the peak value of the carrier waveform (the triangle wave in the context of PWM).

Figure 4.19 shows the principle of the modulation index for a single-phase inverter, as shown in the H-Bridge configuration in Figure 4.5c. The top graph shows a reference sine wave overlaid with a triangle waveform, used in PWM. The points where the sine wave intersects the triangle wave determine the switching points for the inverter transistors.

The bottom graph in the figure shows the output voltage of the inverter. When the sine wave is above the triangle wave, the output voltage is positive $+V_{dc}$, and when the sine wave is below the triangle wave, the output voltage is negative $-V_{dc}$. The width of the voltage pulses in the bottom graph corresponds to the modulation index, representing the ratio of the reference sine wave amplitude to the triangle wave amplitude. The resulting waveform, after filtering, approximates the AC sine wave desired for the load.

Therefore, from the power balance equation and the voltage source circuit in the three-phase inverter, there is a need to find a link between the

DC link voltage and I_{dc} through the d-axis current i_d. In a three-phase inverter, assuming $M_a \leq 1$, the peak grid phase voltage is equal to $v_d + jv_q$ in d_q-reference frame. As v_q is zero since the peak grid phase voltage is in phase with the d-axis as illustrated in Figure 4.18d, the d-axis voltage v_d becomes,

$$v_d = M_a \frac{V_{dc}}{2} \tag{4.22}$$

When this equation is substituted into the power balance equation given earlier I_{dc} can be found as

$$I_{dc} = \frac{3M_a}{4} i_d \tag{4.23}$$

And the DC link voltage across the DC link capacitor is

$$C \frac{dV_{dc}}{dt} = I_{dc} - I_{load} \tag{4.24}$$

As can be seen, the DC link voltage, a critical parameter in the vector control system, is regulated by the d-axis current id.

This regulation is integral to the vector control strategy used in grid-side inverters found in solar PV and battery storage systems as well. The strategy utilizes the d-axis current to adjust the DC link voltage, thereby managing the active power supplied to the grid, as shown in the control loop in Figure 4.18a. In addition, the q-axis current controls the reactive power supplied to the grid, required for grid support functions.

In renewable energy systems, minimizing reactive power is standard practice to achieve a near-unity power factor and maximize active power transfer to the grid. With the resistance R_{grid} and inductance L_{grid} of the filter being small, the control model can be simplified without compromising system stability. However, the nonlinear relationship between the DC link voltage V_{dc} and the d-axis current i_d, along with the variable nature of power generation, means generated power can disrupt the DC link voltage control. High open-loop gain is employed to counteract these disturbances, but it can introduce voltage noise and AC ripple. Therefore, control measures like dynamically adjusting the voltage controller's bandwidth are necessary.

In summary, the vector control method shown in Figure 4.18a is essential for the management of active and reactive power in grid-connected wind turbines, highlighting the stability of the DC voltage. It is designed for a responsive and compliant integration with the grid, aiming for both power factor improvement and reactive power support. This strategy, shown in the figure, uses a PLL to synchronize the system with grid frequency, which is required in the voltage transformations. The PI controllers within the system regulate the d-axis and q-axis currents, and these regulated currents are then used by the PWM generator to control the inverter's switching. By compensating for voltage equation cross-couplings, the vector control ensures precise modulation of the turbine's output, allowing for efficient operation across various grid scenarios and maintaining a balance critical for effective power delivery.

Finally, it should be emphasized here that grid-forming inverters are designed to establish a voltage waveform with a specific frequency and amplitude to either create or emulate a power grid. The output voltage amplitude is controlled by the amplitude modulation index, which adjusts the height of the voltage waveform relative to the DC bus voltage. Frequency control is essential for synchronization with the grid or for setting the frequency of a standalone grid. Therefore, in grid-forming inverters, the output frequency is accurately controlled to match the grid frequency (usually 50 or 60 Hz) or to dictate the frequency in an off-grid or island-mode operation of such inverters.

4.8 COMMON INVERTER CHARACTERISTICS, CLASSIFICATIONS, OTHER CONTROL METHODS AND THE FUTURE

The three-phase inverter topology in PE has become a predominant converter across various applications. Initially developed for motor drives and to utilize the accuracy of vector control concepts, these inverters have significantly broadened their scope in power systems over the past two decades. They now play a major role in HVDC transmission systems, renewable energy sources and battery storage integration. With the power grid evolving towards an inverter-dominated infrastructure, understanding the versatility of these converters becomes essential. Although they primarily utilize similar control schemes with minor modifications, their overall integration requires further system control to adapt to a diverse range of large-scale applications.

To facilitate understanding and adaptation to the evolving landscape of power systems, inverters can be systematically analysed and classified. This approach highlights both their similarities and differences, simplifying the complexities associated with the dynamic nature of modern power

TABLE 4.13 The Choice between a VSI and a CSI and Typical Applications

Inverter Type	Realization Method	Applications	Reasons for Use
Voltage source inverter (VSI)	By a large DC voltage source, such as a capacitor or a battery providing a stable voltage supply. Inverter switches convert DC to AC.	Motor drives, renewable energy systems, UPS, active power filtering.	Control flexibility, power quality management, simplicity in control, high efficiency at variable loads
Current source inverter (CSI)	Realized with a large inductor in series with the DC power supply, ensuring a stable current flow. Inverter switches convert DC to AC waveform.	Induction heating, high-power drive applications, utility-scale renewable energy plants	Robustness against short circuits, performance at high power levels, natural inductive filtering, suitability for constant current applications

systems. An essential step in this process is presented in Table 4.13, which classifies inverters based on the characteristics of their input sources. This classification includes detailed descriptions of realization methods and the reasons behind specific application choices, offering a comprehensive framework to understand the diverse roles and functions of inverters in current power systems. Additionally, it's important to note that such classifications are closely linked to power supply options available to inverters, often associated with various renewable energy and distributed generation sources.

Several key factors can be defined that accelerated the frequent utilization and improvements of the inverters in power systems as follows:

- The gradual phasing out of conventional synchronous generators helped the use of inverters to replicate inertial and damping characteristics.

- As the proportion of inverters in the power mix increases, there is a need for controls that ensure grid stability in the absence of physical inertia.

- The shift towards decentralized grids has led to a growing demand for localized control strategies.

- Technological advancements in PE, control systems and software have made more sophisticated inverter control approaches practical.

- The development of high-power density PE converters has facilitated the creation of highly integrated and hybrid systems.

- Evolving control strategies are required to effectively manage power flow in DC-centric grids, catering to DC loads such as battery storage systems and EVs.

- The increasing reliance on renewable energy sources, which often produce DC power, has prompted the need for new control strategies.

During the adaptation of grid-connected inverters that are primarily voltage source types, there has been a significant shift in the control approaches. This shift supports the transformation of the grid from basic grid-following operations to more complex and active grid-forming roles with multiple inverters connected in parallel at different points of the connections. This evolution is marked by an increasing integration of intelligent features and enhanced safety and security measures which are gradually integrated into the inverter functions. These advancements are driving the technology towards hybrid structures that incorporate autonomous, nano and microgrid functions. To provide a clear insight and highlight the ongoing nature of this change, Table 4.14 is presented. The table outlines the chronological development of inverter technologies, illustrating the continual evolution. Therefore, it can be foreseen that as the grid components and loads become increasingly electrified, a concept elaborated upon in Chapter 1, the PE inverters will evolve greatly.

In summary, while the timeline given in the table broadly reflects the technological evolution, it is important to note that these categories of inverters have seen overlapping and sometimes parallel developments, driven by both technological advancements and evolving grid requirements.

Early concepts of grid-forming inverters (like load-frequency control with battery storage systems (BESS)) have been introduced before more advanced grid-supporting features. However, the practical and widespread implementation of these technologies usually follows a more complex trajectory primarily defined by the regional characteristics of the power grid. In addition, it should be highlighted that inverter technology does not evolve linearly. Therefore, developments in grid-following inverters often occurred alongside advancements in grid-forming types. Furthermore, as it may apply to other technological developments, grid-forming capabilities

TABLE 4.14 The Time Frame in Inverter Technologies

Timeframe	Grid-Following Inverters	Grid-Supporting Inverters	Grid-Forming Inverters
1980s	Basic synchronization for solar applications; narrow voltage tolerance suitable for stable grids.	–	Introduction of load-frequency control with BESS.
1990s	Enhanced performance and synchronization control.	–	Development of droop-controlled parallel converters in standalone AC networks and UPS.
2000s	Integration of MPPT for optimized solar energy conversion; improvements in semiconductors for wider voltage tolerance.	Introduction of basic grid-supporting functionalities including reactive power and voltage regulation.	Introduction of virtual impedance methods and adaptive control strategies: Droop-controls in PV inverter units and in parallel BESS units, with virtual impedance and with adaptive impedance, adaptive droop control, virtual synchronous machine and generator, power synchronization control, virtual resistance for UPS units in 1-ph LV grid and synchronverter (inverter technology designed to mimic the behaviour of a synchronous generator).
2010s	Advancements on broader renewable sources; focus on LVRT capabilities and reactive power support.	Control for power quality; fault ride-through and smart inverter features.	Equivalency of droop control, adaptive inertia control, matching control, virtual oscillator control and synchronverter and virtual impedance.
2020s	Developing grid codes and operation under wide voltage variations.	Integration of DERs; real-time monitoring and control.	The smart inverter advancements with dynamic support ability for the grid.

have been developed out of technological possibility in PE, rather than as a direct grid need. As grid systems have become more complex, with more intermittent renewable sources, the need for advanced grid-forming and grid-supporting inverters has become more pronounced.

Finally, it should be emphasized that a similar PE converter topology is commonly used with some minor alterations in the control schemes that will be explained in the next sections primarily involving software-level alterations.

Table 4.15 shows the distinct control concepts and features of grid-forming, grid-following and grid-supporting inverters in the context of vector control studied earlier. Note that all grid-forming inverters can be considered standalone inverters when they operate in off-grid systems, but not all standalone inverters are grid-forming. Some standalone inverters may require an external source or generator to set the grid frequency and voltage. Furthermore, off-grid inverters cover both grid-forming and other types of inverters that can operate without a grid connection. Off-grid inverters are more generic and can refer to inverters with various functionalities suitable for off-grid applications.

Note that grid-forming inverters can be unstable on weak grids due to PLL oscillations but operate well with stiff grids, ideally suited to an ideal voltage source. They offer good current control and manage faults efficiently but can

TABLE 4.15 Comparative Overview of Inverter Control Strategies in Vector Control Systems

Characteristics and Features	Grid-Forming	Grid-Following	Grid-Supporting
Primary functions	Establishes grid voltage and frequency, provides black start capability and ensures power quality in the absence of a grid voltage reference.	Injects active power into the grid in accordance with grid voltage and frequency, but typically does not provide voltage or frequency regulation services to the grid.	Enhances grid stability by offering voltage support, frequency regulation and reactive power compensation, actively contributing to the grid's operational stability.
Control strategy	Uses vector control to independently set and regulate grid voltage and frequency.	Employs vector control to inject current in phase with the grid voltage. Relies on a PLL for synchronization.	Utilizes vector control to adjust output based on grid demands, including voltage and frequency regulation within limits.
Grid interaction	Acts as a primary voltage and frequency source in island mode or during grid formation.	Synchronizes with the grid without influencing the grid's fundamental characteristics.	Adjusts output dynamically to maintain grid stability and resilience, providing ancillary services.

decrease grid strength and lack black start capability. In contrast, grid-following inverters, unstable on strong and series-compensated grids, are ideal with an ideal current source and improve grid strength. They require current limitation techniques and can perform black starts with energy storage. Grid-supporting inverters are more stable across varying grid conditions, provide voltage and frequency support, and enhance system damping. They feature advanced current control, limited black start capability, and can balance voltage and current needs dynamically for grid stability.

Smart inverters have emerged within the last decade as critical components, adapting to the roles of all inverter types based on the demands of the evolving grid, whether it is at the nano, micro or macro scale. This adaptability describes the integral role smart inverters play in the progressive adaptation of power systems to accommodate changing energy landscapes and demands.

Note that smart inverters are not a fundamentally different type of hardware compared to traditional inverters. They are standard inverters that have been enhanced with software and control capabilities to perform advanced grid services. The "smart" aspect derives from their ability to communicate with the grid and other devices, respond to grid conditions, and actively manage power flow for various applications. Over the years, the evolution of standard inverter functionalities has improved the integration of DERs into the grid, enhanced grid stability and provided a range of critical services.

Smart inverters can operate in grid-following, grid-supporting and grid-forming modes, which enables seamless integration into a variety of DER systems, from microgrids to grid-tied renewable installations. For example:

- In grid-following mode, they synchronize with the grid's frequency and voltage, typically used in basic DER systems.

- In grid-supporting mode, they provide ancillary services such as voltage and frequency regulation.

- In grid-forming mode, they can establish a grid's voltage and frequency, especially useful in islanding scenarios or areas without a stable grid presence.

- They can also regulate numerous functions to support the hybrid grid in conjunction with battery storage systems and DERs.

While the PEs used in these inverters are conventional, it is indeed the software and controls that define an inverter with "smart" capabilities to improve grid integration and support. Their intelligence and versatility primarily come from:

- Advanced control algorithms that allow the inverter to adjust its output in real-time to support grid stability, match supply with demand and provide ancillary services.

- Communication capabilities that enable the inverter to receive and interpret signals from the grid or a central control system, allowing coordinated operation with utility needs and market signals.

- Software-defined settings that can be adjusted remotely or automatically to provide a dynamic response to changing grid conditions.

- Interoperability protocols that adhere to standardized communication protocols, making it possible for smart inverters to integrate with utility distribution management systems and SCADA systems.

The key functions of smart inverters are summarized in Table 4.16.

TABLE 4.16 Common Smart Inverter Functions and Features

Functions	Descriptions
Device settings and limits	Setting power output levels, voltage limits and the rate at which power can ramp up or down, ensuring the inverter safely and efficiently.
Connect/disconnect function	Determines the conditions under which the inverter connects to or disconnects from the grid, critical for safety and regulatory compliance.
Limit DER power output	Caps the maximum power that can be output by DERs, essential for preventing grid overload and managing energy flow.
Energy storage: direct charge/discharge management	Management of charging/discharging, energy use and health detection.
Energy storage: Price-based charge/discharge	Controls storage operations based on energy pricing, which can maximize economic returns or cost savings.
Energy storage: Coordinated charge/discharge management	Coordinates energy storage activities based on grid demand, renewable generation levels and other factors using sophisticated algorithms.
Fixed power factor function	Adjusts and maintains the power factor of the inverter, vital for efficient energy transfer and reducing losses.

(*Continued*)

TABLE 4.16 (*Continued*)   Common Smart Inverter Functions and Features

Functions	Descriptions
Volt-var function	Manages reactive power (var) in relation to voltage levels, helping to stabilize the grid and improve power quality.
Volt-Watt function	Balances the relationship between voltage and power output; important for maintaining grid stability under varying voltage conditions.
Frequency-Watt function	Aiding in frequency regulation.
Watt-Power Factor function	Manages how active power output affects the power factor, ensuring efficient operation and minimal energy losses.
Price or temperature-driven	Optimizing performance and cost-effectiveness based on external factors.
Low/high-voltage ride-through	Enables the inverter to continue operating under abnormal voltage conditions, ensuring reliability and continuity of supply.
Low/high frequency ride-through	Maintains operation during abnormal grid frequencies, contributing to grid resilience.
Dynamic reactive-current support	Provides reactive power dynamically, responding to immediate grid needs for voltage stabilization.
Dynamic real-power support	Smoothens real power output to mitigate fluctuations, crucial for integrating intermittent renewable sources like solar or wind.
Dynamic volt-watt function	An advanced version of the Volt-Watt function, offering more nuanced and responsive control over the interplay between voltage and power output.
Peak power limiting	To avoid overloading the grid during high production periods.
Load and generation following	To adapt its operation to demand/generation patterns, improve stability.
DER settings for multiple grid configurations.	Manages various operational settings for different grid conditions, including scenarios where the DER operates in islanding mode.
Watt-var function	Adjusts reactive power output in response to changes in active power, further aiding in voltage regulation and grid stability.
Features	
Multifunctionality	Grid forming, following and supporting to ensure stability and reliability.
Integrated battery storage	BESS interfacing to provide energy shifting, peak shaving and backup.
DER compatibility	Seamless integration with DERs.
Secure communications	Encryption for data exchange, operational control and monitoring.

(*Continued*)

TABLE 4.16 (*Continued*) Common Smart Inverter Functions and Features

Functions	Descriptions
Cyber security	To protect against unauthorized access and cyber threats.
Autonomous operation	Intelligent control for autonomous operation and optimal performance and self-healing and the creation or support of nano and microgrids.
Advanced grid services	Provides ancillary services such as voltage and frequency regulation, reactive power support and inertia emulation to maintain grid stability.

Finally, it should be indicated that the choice of inverter topology is based on the balance between cost, complexity, efficiency, reliability and compliance with grid codes. In both wind and solar PV systems, almost identical inverter topology is used since the fundamental function of converting DC to grid-compatible AC is common. However, the details of the operation and integration are quite distinct, which are summarized in Table 4.17 for wind and solar PV systems.

4.8.1 Droop Control

Figure 4.20 presents a graphical illustration of a load and a generation characteristic in a grid, given for explanations of the need for droop control. The vertical axis shows the grid frequency ω as a function of the active power. Note that there are two intersecting lines: one showing the frequency-demand (load) characteristic ($\omega = f(P_{demand})$), indicating a linear decrease in frequency with increased load, and the other showing the frequency-supply (generation) characteristic ($\omega = f(P_{Supply})$), which is flatter, demonstrating how supply (generation) frequency is less sensitive to power changes. This may be due to the fact that supply has a large power system inertia and/or is large relative to the load variation.

The nominal operating frequency ($\omega 0$) is where these two lines intersect, signifying a balance between power generation and load, including consumption and network losses. This equilibrium point is crucial for maintaining grid stability, as any demand (load)-induced frequency deviation triggers generator adjustments to restore balance through primary and secondary control mechanisms. The flatter generator characteristic curve ensures that the network remains frequency-stable even as the load varies.

Droop control, initially developed for synchronous generators due to the direct relationship between frequency and active power, has been

TABLE 4.17 Comparison of Inverters in Wind and Solar PV Systems

Aspect	Wind System Inverters	Solar PV System Inverters
Variable speed operation	Must handle a wide range of input frequencies and voltages due to variable speeds.	More constant output characteristic, allowing for simpler input stage design.
Control algorithms and MPPT strategy	Advanced control algorithms for capturing maximum power across varying wind speeds. Perform MPPT for the entire turbine.	Less complex due to steady input, but must compensate for temperature coefficient. Often include integrated MPPT for each string or panel.
Integration with control system	Closely integrated with turbine control for blade pitch and yaw adjustments.	Not as tightly integrated with control systems, simpler operation.
Response to input variation	Require fast response to changing wind conditions to prevent overloading and optimize performance.	Less variation in input, leading to less dynamic response requirements.
Size of passive components	Might need bulkier inductors and capacitors for smoothing due to higher power fluctuations.	Can use smaller passive components due to less severe power fluctuations.
Cooling requirements	Potential for high power throughput and variable operating speeds, influencing cooling requirements.	Cooling requirements influenced by consistent output, potentially less stringent.
Communication protocols	More complex due to integration with turbine control systems.	Incorporate simple interfaces for system monitoring.
Protection circuits	Can differ based on turbine design and operational environment.	More emphasis on anti-islanding protection and suitability for distributed deployment.

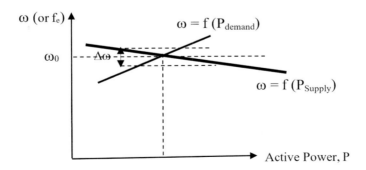

FIGURE 4.20 Demand and supply characteristics of a grid and balancing.

commonly adapted for inverter-based systems as well. This ensures that increases in load lead to proportional reductions in frequency, a critical adaptation for grid stability and efficient operation, particularly in grids increasingly dependent on renewable energy sources and distributed generation.

The primary function of droop controllers is to facilitate stable and proportional load sharing among generators. In scenarios, where multiple generators are connected in parallel, droop control allows them to work together harmoniously, adapting to load changes.

Inverters are a key technology across a broad spectrum of applications, from providing uninterrupted power supplies to integrating renewable energy sources such as photovoltaic arrays, wind turbines and HVDC. The shift towards renewable energy integration in the power grid also requires droop control within inverter-dominant systems.

Integration of droop control into vector control schemes is a standard approach to facilitate power sharing in parallel inverter configurations, necessary for both grid-connected and islanded systems. Technically, droop control is essential for several reasons: it aids in load sharing for system stability and efficiency, contributes to voltage and frequency regulation, manages power quality by controlling voltage and current dynamics hence THD, enhances grid resilience in response to load fluctuations, provides equipment protection by limiting current during fault conditions and ensures compliance with industrial standards through precise tuning of inverter outputs.

A simplified model of a single-line equivalent circuit of an inverter connected to a grid via an output impedance was given in Figure 4.17b. Output filters, integral components in grid-connected inverters, serve to eliminate harmonics and restore the desired voltage waveform between the inverter and the load. In addition, it was demonstrated in Figure 4.17b that the inverter can be modelled as a voltage source with a series output impedance, which is connected to the grid.

Therefore, understanding the interaction between the inverter's output impedance and the chosen filter type is important for designing effective droop control strategies. Every filter type influences the droop controller's behaviour and its ability to manage power sharing in parallel inverter operations.

In practice, inverter output impedance is usually inductive, contributed by the filter inductor and the line impedance. However, in low-voltage scenarios, line impedance can be predominantly resistive. Control strategies are employed to adjust the output impedance as needed, allowing it to be set as resistive, resistive-inductive or other type depending on the application. This versatility in control is vital for power sharing and overall system performance.

Therefore, the following paragraphs will explain two filter options and how to find the corresponding droop controllers that can adapt to the filter's characteristics, contributing to stable and efficient inverter operation.

As given in Figure 4.17b, the grid side inverter can be modelled as a reference voltage source with an impedance $Z_{grid} = R_{grid} \pm jX_{grid}$. Therefore, the active and reactive power that can be dispatched to the grid can be given by

$$P_{grid} = \left(\frac{V_{inv} V_{grid}}{Z_{grid}} \cos\delta - \frac{V_{grid}^2}{Z_{grid}} \right) \cos\theta + \frac{V_{inv} V_{grid}}{Z_{grid}} \sin\delta \sin\theta \quad (4.25)$$

$$Q_{grid} = \left(\frac{V_{inv} V_{grid}}{Z_{grid}} \cos\delta - \frac{V_{grid}^2}{Z_{grid}} \right) \sin\theta - \frac{V_{inv} V_{grid}}{Z_{grid}} \sin\delta \cos\theta \quad (4.26)$$

For an output filter impedance composed solely of inductance ($\theta = 90°$), as previously analysed, active power is proportional to the load angle δ and reactive power to V_{grid}. Hence, the conventional droop controller is defined by the following equations

$$V_{inv} = V_{inv}^* - nQ_{grid} \quad (4.27)$$

$$\omega = \omega^* - mP_{grid} \quad (4.28)$$

Similarly, if the grid impedance consists only of resistance ($\theta = 0°$), the power equations are rearranged, and for small load angles δ, the relationships become $P_{grid} \sim V_{grid}$ and $Q_{grid} \sim -\delta$. The droop controller equations are then

$$V_{inv} = V_{inv}^* - nP_{grid} \quad (4.29)$$

$$\omega = \omega^* + mP_{grid} \quad (4.30)$$

Where V_{inv}^* is the rated RMS voltage of the inverter, ω^* is the rated grid frequency, and ω is the measured frequency. The coefficients n and m are the droop coefficients for voltage and frequency respectively.

These relationships indicate that the droop control approach integrated into the vector control system must be adapted according to the specific grid impedance characteristics.

It should be noted that the grid impedance in most of the cases (around the fundamental frequency) is inductive but can also be in other configurations. The variation in droop control methodologies is illustrated in Figure 4.21a for three types of filters L, R and C. In addition, the table included in Figure 4.21b summarizes all potential grid impedance scenarios that may arise from different converter topologies, filter configurations or inverter control strategies implemented.

As shown in Figure 4.21d, with an inductive output impedance, active power is proportional to the load angle δ, and reactive power is proportional to the inverter voltage V_{inv}. Moreover, it has been demonstrated that in voltage source inverters, active and reactive power can be regulated using droop control. In addition, the power regulation characteristics for this system model, as illustrated in Figure 4.21a, indicate that power management can be achieved by assigning appropriate droop coefficients to the converter.

Droop control can be implemented by integrating the block diagram presented in Figure 4.22, where power calculations are performed using the instant power theory previously developed with dq-axis currents and voltages. In the figure, v_d and v_q represent the grid voltages in the dq reference frame, and i_d and i_q are the inverter output currents in the dq reference frame, which are injected into the grid. Then, using the equations provided in Figure 4.21d, the output frequency and voltage amplitude are generated using the vector control block diagram of the grid-side inverter. The droop coefficients are defined within the limitations of power ratings as illustrated. Note that $\Delta\omega$ and ΔV_{inv} represent the maximum allowable deviations for grid frequency and grid voltage, respectively. Furthermore, $P_{grid(max)}$ and $Q_{grid(max)}$ are the maximum active and reactive power. Low-pass filters are also included, which determine the droop control loop's bandwidth. Then, the output voltage command, which is sent to the voltage control loop, is set as $V_{d(ref)} = V_{inv}$ and $V_{q(ref)} = 0$.

When employing the droop method in power systems, there is an inherent trade-off between achieving good voltage regulation and effective load sharing. Optimal voltage regulation needs a high DC gain, yet this can

Power Electronics and Control of Power System Components ■ 311

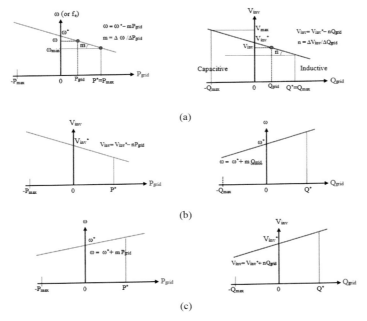

FIGURE 4.21 Primary droop control (power regulation) characteristics with L_{grid} and (a), with R_{grid} (b), with C_{grid} (c), and relationships between the grid impedance and droop controller (d). Where V_{inv}^* is the rated rms voltage of the inverter, ω^* and ω are the rated and measured grid frequency, and n and m are the droop coefficients.

lead to poor load sharing, particularly if the converters in the system are not identical. Consequently, while settings that prioritize voltage regulation can maintain voltage levels effectively, they are sensitive to even minor discrepancies in droop characteristics, potentially leading to significant

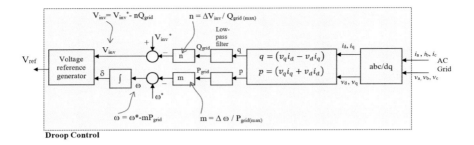

FIGURE 4.22 The droop control block diagram to be integrated into the vector control block diagram of the grid-side inverter.

power imbalances. On the other hand, configurations optimized for load sharing can tolerate variations in converter characteristics but may result in considerable voltage fluctuations. This limitation can be addressed by incorporating a droop characteristic into the DC controller as well. By doing so, the slope of the droop curve becomes proportional to the DC gain. As a result, power can be regulated within a defined range by managing the voltage within specific intervals. It is important to emphasize that this solution requires independent control of both the generator-side and grid-side converters.

In addition, it is important to consider that as the number of inverters connected in parallel with various filter options (see Figure 4.21d) increases within the grid, the complexity of droop control is likely to increase, which heavily depends on the droop control coefficients m and n. Therefore, the droop control parameters in vector-controlled grid-connected inverters are vital for effective load sharing and maintaining stability in a power system, particularly in configurations with multiple inverters or generators.

Vector control is also useful for synchronizing inverters (grid-connected, grid-following, grid-supporting and grid-forming) with the grid in various operational modes.

Synchronization in grid-connected inverters is necessary for seamlessly integrating power into the electrical grid while maintaining its stability and safety. It involves aligning the inverter's output with the grid's voltage, frequency and phase. It is also critical for preventing power quality issues such as voltage fluctuations and ensuring regulatory compliance. In addition, synchronization plays a role in safely managing grid outages through anti-islanding protection and maximizes the efficient utilization of generated power, particularly from renewable sources.

In grid-following inverters, the inverter's output aligns with the grid's phase and frequency using a PLL for accurate detection and vector adjustment. Grid-supporting inverters use vector control to independently manage reactive and active power vectors, aiding in voltage support and frequency regulation by continuously adapting to grid conditions. Similarly, grid-forming inverters employ vector control to establish voltage and frequency references in microgrids or islanded systems, ensuring stable power distribution and load sharing.

4.9 FOUR-QUADRANT CONTROL CONCEPT OF CONVERTERS/INVERTERS

"Quadrant control" in PE systems can be explained through the capabilities of capacitors and inductors, fundamental energy storage components known for their filtering and smoothing roles. They store energy which can be redirected back to the source through appropriate current paths in the PE converter.

Figure 4.23a illustrates a PE converter interfacing between a supply, AC or DC, and a load, which can be active, such as a battery or motor/generator, or passive, like a resistor. The converter simply manages the direction and magnitude of current and voltage outputs, hence power flow, with directionality dependent on the presence of an active load. Flywheel loads and EVs are additional active load examples, but the converter must include PE device arrangements for bidirectional current flow.

The supply type dictates the converter design, adjusting voltage and current to the load's requirements (see Figure 4.23b). Converter topology decides its bidirectional or unidirectional capability, and the load's ability to consume or generate power, such as a battery charging or discharging, or rotating machines acting as motors or generators, linked to the converter enabling reverse power flow.

Figure 4.23c displays the four-quadrant operation of converters in a power-plane crucial for grid-connected inverters and renewable energy and DERs integration. In the first quadrant, converters supply both active and reactive power during peak output or energy storage discharge. The second quadrant involves active power supply while absorbing reactive power, offsetting excess reactive power in the grid. The third quadrant sees the absorption of both active and reactive power during renewable downtimes and energy storage charging, aiding power factor correction. The fourth quadrant is where the system regenerates active power while

314 ■ Reinventing the Power Grid

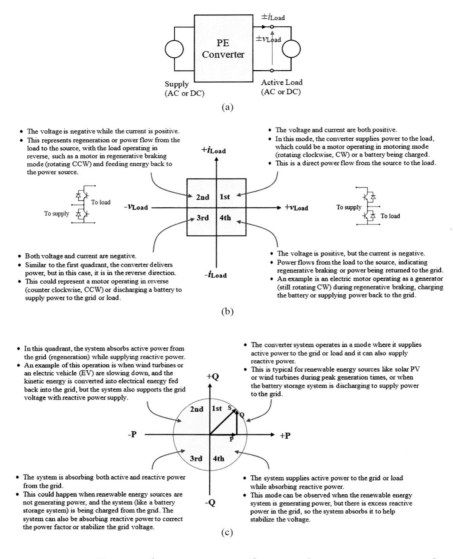

FIGURE 4.23 Four-quadrant operation of power electronics converters for active and reactive power management.

providing reactive power support, as in decelerating wind turbines, feeding energy back to the grid and supporting voltage levels.

For grid-connected battery storage systems, operating across all four quadrants facilitates a dynamic grid response, including during peak and off-peak periods, offering grid services like frequency regulation, voltage support and load levelling. Four-quadrant converters are critical for reactive power compensation, improving power quality, achieving a unit power factor and promoting efficient transmission with minimized losses and enhanced stability.

Table 4.18 shows a wide array of applications that highlight the versatility of power electronics converters.

Despite the diverse nature of these applications, from solar PVs and wind generators to EVs and VSDs, they all share similar power electronics converter topologies. These converters simply regulate voltage and current outputs, ensuring that the specific demands of each application are met with accuracy. Moreover, almost standardized feedback mechanisms are adapted to applications. This unifying approach to power conversion indicates the adaptability and efficiency of power electronics in catering to the energy needs of a broad spectrum of technologies from basic power supply to variable speed motor drives in EVs or industrial applications to grid-side inverters in renewable energy.

As stated previously, the evolution of PEs is being influenced by the emergence of WBG devices, which are enabling the development of high-power density, frequency and efficiency PE converters. As alternative converter topologies become more viable, there is a growing trend towards the adoption of DC grids connected to loads that are primarily DC types or disguised. The inherent advantages of DC power distribution, such as the elimination of reactive power and the associated losses, along with the avoidance of synchronization challenges present in AC systems, make it a compelling choice for the future of energy systems. As can be seen in Table 4.18, the transition to DC grids by removing DC/AC converters opens up new possibilities for a seamless integration of a multitude of DERs and loads. This can also streamline the connection process, as many renewable energy sources, such as PV panels, and energy storage systems, like batteries, inherently generate or store power in DC. This compatibility reduces the need for complex conversions and allows for more straightforward control strategies. Furthermore, DC systems can enhance the overall security of the grid. With fewer conversion stages, there are fewer potential points of failure, which can improve system reliability. Additionally, DC microgrids can operate independently from the main grid, which can be beneficial for critical infrastructure and can provide more control over energy resources at a local level.

The efficiency gains, combined with ease of control, position DC systems as a preferred choice in automation, smart grid technologies and data

TABLE 4.18 Embedded Generation Sources and Converter Topologies Using Power Electronics to Couple with the AC Grid

Application	Embedded Converter Topologies	Remarks
Solar PV		A DC/DC converter is used for MPPT to optimize the power output before it is converted to AC for grid integration.
Wind Generators		They utilize an AC/DC rectifier to optimization of power extraction from wind then a DC/AC inverter to connect to the grid.
Fuel Cells		A DC/DC converter regulate the DC output from the fuel cell as it changes with the load.
Supercapacitors		Bidirectional DC/DC converters are used for charging and discharging supercapacitors, providing a quick response to power demands.
Flywheel systems		Similar to supercapacitors, they use bidirectional converters for storing energy in a rotational mass and releasing it when needed.
Microturbine Systems		Microturbines generate AC which is rectified to DC, and then inverted back to AC for grid compatibility or direct use.
Stationary Battery Storage Systems		These systems use a DC/DC converter to manage battery charging and discharging, ensuring optimal energy storage and retrieval.
EV to Grid		EVs can connect to the grid using a DC/DC converter for charging and discharging the battery, contributing to grid stability and energy management.
Solid State Transformer (SST) for AC Grid		SSTs employ front-end AC/DC conversion, a high-frequency transformer for voltage scaling, and DC/AC conversion to deliver transformed power.
Transformerless Intelligent Power Substation (TIPS) for AC Grid		This topology uses a similar approach to SSTs but is designed for high-voltage power substation applications.
Variable Speed Drives for MV Motors in AC Grid		This is the topology from your previous image, where a front-end rectifier converts AC from the grid to DC, which is then inverted to AC to drive a high-power motor.

centres. They are well-suited to support the emergence of autonomous energy management systems that can optimize power flow in real-time, ensuring stability and resilience.

4.9.1 Converters in Battery Storage, Electric Vehicles and Others

Figure 4.5d was given previously to highlight a grid-connected bidirectional converter setup with isolation through a high-frequency transformer, specifically designed for battery configurations. Commonly employed in grid-tied energy storage systems, this arrangement allows for the charging of batteries during periods of low demand and their discharge during peak demand or power outages. In addition, the setup is instrumental in integrating renewable energy sources. This can be achieved either by coupling to the AC side or to the DC link, where the DC source is created using a capacitor. The switches within the system are managed through PWM signals. This ensures that the output voltage waveforms are of the correct amplitude and phase relation, enabling synchronization with the grid. Furthermore, the system utilizes a three-phase inverter bridge and H-bridge topologies, which facilitate both charging and discharging functions by operating in various quadrants, as previously described.

In larger-scale grid battery storage systems, multiple identical setups are used. This approach not only provides high-power capacity but also offers redundancy, compensating for potential downtimes due to maintenance or hardware failures. In certain applications, the conventional three-phase inverter is substituted with three single-phase inverters. In these scenarios, three identical units are employed, each connecting to one of the phases of the grid. This configuration is particularly effective for handling operations involving unbalanced loads.

The converter given in Figure 4.24a is a multi-level inverter type typically more than the two levels (as in Figure 4.5d), which is also suitable for electrochemical storage like batteries and supercapacitors. Common configurations of multi-level inverters include diode-clamped (neutral-point clamped), capacitor-clamped (flying capacitors) and cascaded H-bridge designs, each offering unique benefits and challenges. The selection criteria for multi-level inverters over traditional two-level inverters involve several considerations, including the quality of the AC output required, as smoother waveforms reduce the need for filtering and improve the efficiency of energy conversion.

Multi-level inverters are also advantageous in high-voltage applications, as they can manage higher voltages without exceeding the voltage ratings of individual PE components. Additionally, lower switching frequencies contribute to reduced switching losses and EMI, making multi-level inverters suitable for applications where these factors are critical. However,

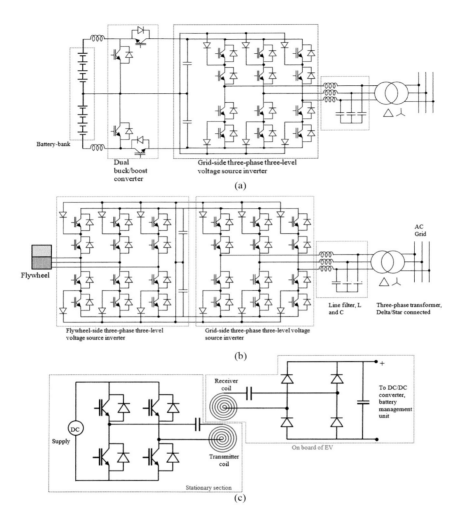

FIGURE 4.24 The power electronics architecture for battery storage, flywheel systems.

these benefits come at the cost of increased complexity, higher component count, and potentially greater expense. Therefore, the decision to use multi-level inverters is typically justified in scenarios where the performance improvements align with the application's demands and outweigh the drawbacks of added complexity and cost.

As illustrated in Figure 4.24, a three-phase inverter connects to the grid via a three-phase transformer and line-side filter, as well as employing a dual buck/boost DC/DC converter for efficient charge/discharge control and to maintain DC bus voltage.

The selection of Δ/Y or Y/Δ configurations in the line-side transformer is typically based on the specific requirements of the grid connection, including voltage levels, power quality standards and the electrical characteristics of both the inverter system and the grid infrastructure. The right choice of transformer winding configuration can significantly impact the efficiency, reliability and longevity of the grid-connected inverter system.

Delta/star winding configurations in multi-level inverter outputs are strategically selected for grid connections to address critical power system requirements. For example, a star (Y) connection is often used to step-down the voltage from the inverter to a level suitable for the grid. The phase voltage in a star connection is $1/\sqrt{3}$ times the line voltage, which can be useful when the inverter's output voltage is higher than the grid voltage. Conversely, a delta (Δ) connection can step-up the inverter output voltage to match the grid voltage levels. Since the phase voltage in a delta connection is equal to the line voltage, it facilitates higher voltage transmission.

In addition, these transformer configurations facilitate harmonic cancellation, enhancing power quality when interfacing with the grid. For example, a delta connection on the inverter side can provide a path for triplen harmonics (multiples of the third harmonic), effectively filtering them out and preventing them from entering the grid. They also contribute to system stability and fault tolerance, with delta windings maintaining operations under unbalanced loads or single-phase faults, and star connections providing neutral points for load balancing. Furthermore, transformer-based isolation between the inverter and the grid increases safety and protects against grid disturbances, ensuring a reliable and efficient integration of the inverter system into the power grid.

Flywheel energy storage systems (Figure 4.24b), utilize an electric machine to generate sinusoidal voltage waveforms. Therefore, the PE circuit comprises two three-level voltage source inverters: one machine-side for active power control and one grid-side for reactive power regulation. Across these applications, the PE converters regulate voltage and current outputs, with the type of energy storage influencing control dynamics but minimally impacting the power circuit design, due to the standardized feedback sensors and control techniques used in motor drives and wind turbine systems.

Figure 4.24c is a simplified representation of a unidirectional wireless charging system for EVs. The system uses resonant inductive coupling to transfer power wirelessly from a transmitting coil to a receiving coil, which is a fundamental concept in wireless power transfer (WPT) systems. WPT for EVs offers the advantages of increased convenience and safety by removing the need for cables, reducing wear and potential electric shock risks. It facilitates easier integration into various environments due to the absence of exposed equipment, and it can provide accessibility benefits for individuals with physical disabilities. Current WPT systems typically achieve efficiency ranges between 70% and 90%, with the exact figures depending on the system design, coil alignment and distance between the transmitter and receiver. Note that the standard single-phase bridge configurations are used both on the stationary side and on-board units. Both the transmitter and receiver sides include capacitors to form resonant circuits. These capacitors, along with the inductance of the coils, are chosen to resonate at the operating frequency of the inverter. Resonance enhances the power transfer efficiency by minimizing reactive power. The induced voltage in the receiver coil is utilized by the DC/DC converter after the diode bridge is used to charge the battery of the vehicle.

Also, note that replacing the receiver side's diode rectifier with an H-Bridge circuit enables the system to achieve bidirectional power flow. This means the EV can not only be charged but also supply power back to the grid or another system. Such functionality is particularly useful in vehicle-to-grid (V2G) applications, where EVs contribute to grid stability by supplying power during peak demand or absorbing excess energy during periods of low demand. Bidirectional capability requires additional control and communication mechanisms to ensure power flows in the intended direction and to maintain power quality and safety of the grid connection. Furthermore, this system can serve as an emergency power supply if associated grid-forming capabilities are incorporated. EV systems will be explored further in the context of microgrids and DERs in Chapter 6.

As key elements of the modernized grid, the PE converter examples used in three major applications are also given in Figure 4.25. These circuits are commonly used in the integration and control of advanced energy systems such as flow batteries, fuel cells and solid-state transformers for grid applications that are already impacting grid modernization and will play a significant role in future grid transformation. In addition, alternative resonant converters are likely to emerge as WBG devices become more widely used.

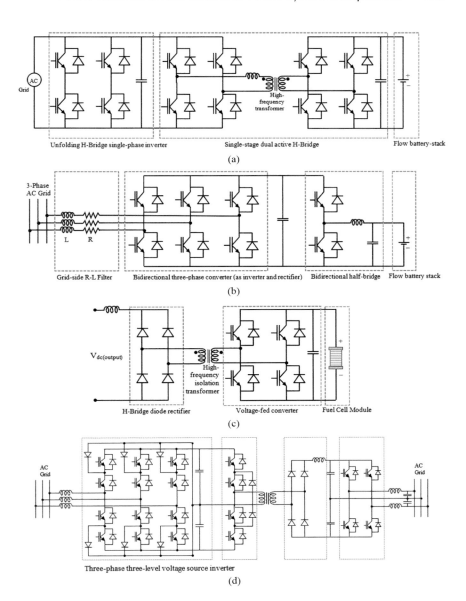

FIGURE 4.25 Power electronics topologies for energy conversion and grid integration: (a) single-phase and dual H-bridge converters for flow batteries; (b) three-phase bidirectional converter for flow batteries; (c) voltage-fed H-bridge converter for fuel cell modules with high-frequency isolation transformer; (d) three-phase three-level voltage source inverter for solid state transformers.

Figures 4.25a and b show two configurations for interfacing a flow battery stack with the AC grid. The first configuration (Figure 4.25a) shows a typical configuration that can enable the flow battery to either supply power to the grid or be charged from it depending on the direction of the current flow, which uses well-known H-Bridge configurations. Both configurations allow for bidirectional power flow, but the high-frequency transformer in the first configuration also provides electrical isolation and voltage scaling. The second circuit (Figure 4.25b) illustrates a three-phase converter capable of operating as both an inverter and a rectifier, and the grid side R-L filter reduces harmonics in the power transferred between the grid and the converter.

Figure 4.25c shows a power electronics circuit used in fuel cell applications. Note that as the fuel cell generates DC power, this voltage-fed H-Bridge converter is used to regulate the DC voltage from the fuel cell. The converter is essential for applications that require a constant voltage despite varying load conditions. As explained before, the transformer provides electrical isolation between the fuel cell and the load or grid if DC is inverted, and it can be used to step-up or step-down the voltage to the required levels, and its high-frequency operation allows for a more compact design. The alternating voltage waveform after the transformer is converted back to DC by the rectifier. The details of the fuel cell will be further explained in Chapter 5.

A variety of solid-state transformer configurations exist, which can be categorized based on their high-frequency isolation transformer arrangements, or in terms of single, modular or multiport setups. A solid-state transformer configuration is given in Figure 4.23d, in which a three-phase three-level voltage source inverter is used to convert DC to AC with higher power quality due to multiple voltage levels.

As can be seen, the architecture of the PE circuits typically demonstrates a subset of the converter topologies that were previously discussed. Moreover, their principal control concepts are fundamentally similar to those used in grid-connected control systems, as used by wind generators explained before. Finally, it is important to emphasize here that the PE circuits presented in Figure 4.7 provide the fundamental building blocks for converters that form nearly all subcomponents of the power system including FACT and STATCOM devices. Power electronics are also used in desalination plants and electrolysis processes as well as in entire electrification processes as discussed in Chapter 1.

4.10 FAULTS, FAILURES AND RELIABILITY

As discussed previously, PE circuits and systems are vital to modern energy systems, enabling electrical energy conversion and control for various applications. They are required for renewable energy, EVs and other grid technologies. Reliability in these systems is highly critical, ensuring stability, safety and uninterrupted power system operation. However, due to their complexity and demanding operational environments, faults and failures are a prime risk.

The costs of such faults in PE are extensive, potentially disrupting power delivery and causing significant infrastructural and economic impacts. Inefficiencies and reduced performance can lead to increased operational costs and emissions. Safety risks also arise, with failures potentially causing fires or electric shocks.

Designers employ various tools and methods to predict and improve PE reliability. Statistical measures offer insights into component and system lifespans, while reliability models simulate failure scenarios and assess system robustness. Stress analysis identifies component operational limits, and derating practices help prolong device life. Redundancy and fault tolerance are design strategies that enhance system resilience, influencing overall cost and return on investment.

For fault prevention and damage minimization, timely detection is essential. Sensor-based monitoring and model-based detection proactively identify issues. Machine learning and predictive maintenance analyse data patterns to forecast and eliminate faults, complemented by remote diagnostics for localized or centralized system management.

Protection strategies incorporate hardware solutions and control algorithms. Overcurrent protection is typically managed by fuses and circuit breakers, while surge protection devices guard against voltage spikes. Effective thermal management is critical for temperature regulation. Moreover, fail-safe and fail-secure modes ensure that systems prioritize safety and prevent further failures during faults.

PE are prone to open-circuit and short-circuit conditions, thermal overloads and component wear. Electromagnetic disturbances can also cause transient or permanent failures. Recognizing these fault types is key to formulating effective protection and mitigation plans.

Faults and failures stem from both inherent design limitations and external operational factors. Design flaws might result in inadequate power handling, and unsuitable component ratings can cause early failures.

Although manufacturing defects have decreased, they remain a concern. External factors like temperature, moisture and mechanical stress, along with operational misuse, contribute to the gradual degradation of PE.

Note that PE design primarily revolves around power density, efficiency, cost and reliability. WBG devices such as SiC transistors have significantly improved power density and efficiency. However, the reliability of these systems is also critical, as failures can lead to increased maintenance costs and impact energy costs, particularly in renewable energy applications.

Figure 4.26 provides a hierarchical breakdown of failure modes in the entire range of power system components, which are becoming highly PE dominant. The figure can provide insights for troubleshooting, maintenance and design improvement purposes and offers a systematic approach to understanding how different parts of the system can fail and how those failures are categorized.

The figure outlines potential sources of failure within a PE system, identifying hardware, software and human error as the primary categories. Hardware failures are further broken down into wear-out and random categories, indicating that some components degrade over time while others may fail unexpectedly. Hardware is then dissected into assembly-level components such as power semiconductors, passive components, battery systems and other components, each of which could potentially contribute to system failure. At the component level, specific items like transistors,

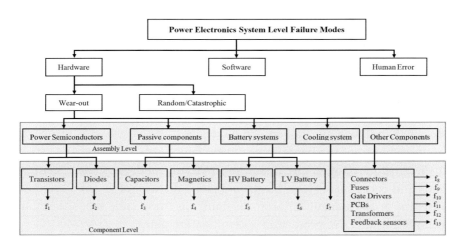

FIGURE 4.26 Power electronics system level failure modes which are used in power system components. f_k stands for failures (k=1–13).

diodes, capacitors, magnetics and various battery and connection components are identified, each with associated failure functions.

By understanding where and how these failures occur, more reliable systems can be designed and selected, which develop targeted mitigation strategies to prevent or address failures in power system components.

The first level of classification distinguishes between three primary sources of failure: hardware, software and human error. This categorization recognizes that failures can originate from physical components, programming and software issues or incorrect human actions.

Within the hardware category, failures are further divided into "wear-out" and "random/catastrophic" types. Wear-out failures occur over time due to the degradation of materials and components, while random or catastrophic failures are unanticipated and can result from various factors, including design flaws, manufacturing defects or extreme operational conditions.

The figure then breaks down the hardware category into five specific assembly groups found within PE systems. Each assembly is then dissected further into its component level. Battery systems and cooling systems are recognized as distinct categories essential for modern grid operation and thermal management.

Each component in the figure is associated with a failure function, denoted by "f" followed by a subscript. In the literature, these functions represent the failure rates or the probability of failure over time for each specific component. However, it should be noted here that the percentages or failure rates for the components are highly specific and often derived from empirical testing and operational history, which are not typically available in the public domain due to their proprietary nature. In addition, the failure rates of electronic components like transistors, capacitors and diodes depend on a variety of factors, including the design of the component, the materials used, the manufacturing process and the conditions under which the component is used. Manufacturers often provide some reliability data in the form of mean time between failures (MTBF) or failure in time (FIT) rates, which can be converted into annual failure rates given the number of components in use and their operating hours. However, as highlighted by the figure, the reliability of a final product depends on numerous parameters, including the operating environment, as power system components operate under highly diverse and varying environmental conditions.

BIBLIOGRAPHY

F.R. Allende, M. A. Perez, J.R. Espinoza, T. Gajowik, S. Stynski and M. Malinowski, Surveying solid-state transformer structures and controls. *IEEE Industrial Electronics Magazine*, 14(1), pp. 56–70, 2020. https://doi.org/10.1109/MIE.2019.2950436.

L. Chen, W. L. Soong, M. Pathmanathan and N. Ertugrul, "Comparison of AC/DC Converters and the Principles of a New Control Strategy in Small-Scale Wind Turbine Systems," *2012 22nd Australasian Universities Power Engineering Conference (AUPEC)*, Bali, Indonesia, 2012, pp. 1–6.

EPRI, *Common Functions for Smart Inverters: 4th Edition*. EPRI, Palo Alto, CA, 2016.

N. Ertugrul, Mine electrification and power electronics: The roles of wide-bandgap devices. *IEEE Electrification Magazine*, 12(1), pp. 6–15, March 2024, https://doi.org/10.1109/MELE.2023.3348254.

N. Ertugrul and D. Abbott, DC is the future. *Proceedings of the IEEE*, 108(5), pp. 615–624, May 2020, https://doi.org/10.1109/JPROC.2020.2982707.

J. Licari, Control of variable-speed wind turbine, PhD Thesis, Institute of Energy, Cardiff University, February, 2013.

M. Parvez, A. T. Pereira, N. Ertugrul, N. H. E. Weste, D. Abbott and S. F. Al-Sarawi, Wide bandgap DC-DC converter topologies for power applications. *Proceedings of the IEEE*, 109(7), pp. 1253–1275, July 2021, https://doi.org/10.1109/JPROC.2021.3072170.

M. Pathmanathan, Modelling and Control of Phase Advance Modulation for Small-Scale Wind Turbines, PhD Thesis, School of Electrical and Electronic Engineering, Faculty of Engineering, Computer and Mathematical Sciences, The University of Adelaide, April 2012.

M. Pathmanathan, W. L. Soong and N. Ertugrul, "V-θ Control of Inverters Used in SPM Wind Turbine Generators," *2013 15th European Conference on Power Electronics and Applications (EPE)*, Lille, France, 2013, pp. 1–10, https://doi.org/10.1109/EPE.2013.6631807.

R. Pena, J.C. Clare and G.M Asher, Doubly fed induction generator using back-to-back PWM converters and its application to variable- speed wind-energy generation. *IEE Proceedings of Electric Power Applications*, 1(143), pp. 3, May 1996.

S. Peyghami, P. Palensky and F. Blaabjerg, An overview on the reliability of modern power electronic based power systems. *IEEE Open Journal of Power Electronics*, 1, pp. 34–50, 2020, https://doi.org/10.1109/OJPEL.2020.2973926.

L. Wang, N. Ertugrul and M. Kolhe, "Evaluation of dead beat current controllers for grid connected converters," *IEEE PES Innovative Smart Grid Technologies*, Tianjin, China, 2012, pp. 1–7, https://doi.org/10.1109/ISGT-Asia.2012.6303109.

P. Wellmann, *Wide Bandgap Semiconductors for Power Electronics: Materials, Devices, Applications*, edited by P. Wellmann et al, Volume 1, WILEY-VCH GmbH, New York, 2022.

D. Wu, F. Tang, J. C. Vasquez and J. M. Guerrero, "Control and Analysis of Droop and Reverse Droop Controllers for Distributed Generations," *Proceedings of the 11th International Multiconference on Systems, Signals & Devices, SDD 2014 IEEE Press,* Castelldefels-Barcelona, Spain, 2014, https://doi.org/10.1109/SSD.2014.6808842.

Y. Zeng, Droop Control of Parallel-Operated Inverters, PhD Thesis, The University of Sheffield, Faculty of Engineering, Department of Automatic Control and Systems Engineering, June 2015.

CHAPTER 5

Batteries and Fuel Cells in Energy Storage

5.1 INTRODUCTION

As renewable energy sources like wind and solar lead the way in distributed energy resources (DERs), efficient methods to store excess energy and mitigate the limitations of their variability and intermittency become critical. Table 5.1 provides a high-level comparison of four distinct energy storage solutions: pumped hydroelectric, battery, thermal (resistance water heating and heat pumps) and their key features. Note that, while the ultimate goal —storing energy —remains the same across these technologies, the chosen storage medium can be diverse. From pumped hydro's use of elevated water reservoirs to electrochemical storage in batteries, the form of storage impacts efficiency and suitability.

This chapter will provide a detailed examination of battery storage, with its impressive efficiency and scalability. However, it is important to acknowledge the significance of other technologies as well. When the dynamics of storing intermittent renewable energy are considered, the advantages unfold in multiple dimensions. For example, pumped hydro and battery storage act as direct electrical options, designed to release energy on demand. However, domestic water heaters on a secondary level, trap thermal energy. Their capacity is contingent on the water tank's volume and its insulation level. Similarly, heat pumps, which effectively move thermal energy for heating or cooling purposes, can be considered a storage solution —especially in well-insulated buildings where heat retention

TABLE 5.1 Comparison of Four Distinct Energy Storage Options That Can Utilize Electricity

	Pumped Hydroelectric Storage	Battery Storage	Resistance Water Heating	Heat Pumps
Efficiency	70%–85% (round-trip eff.)	85%–90% (round-trip eff.)	100%	See the comments below*
Installation Cost	$1,000–$3,000/kW	$500–$1,000/kWh	$300–$2,000/kW	Size and type dependent, typical unit cost $2,000–$8,000
LCOE	$0.05–$0.12/kWh	$0.15–$0.30/kWh	Not typical to calculate	Not typical to calculate
Who May Build	By utilities and large energy companies	Domestic, community level, utilities	Homeowners, business	Homeowners, business
Other general key features	Long life span, grid level	Small to large-scale storage.	Domestic applications, under low-demand tariff	Domestic applications, under low-demand tariff

Notes: The efficiency of a heat pump is typically measured by its coefficient of performance (COP) or, its seasonal energy efficiency ratio (SEER) for cooling. COP is the ratio of the heat output to the electrical energy input, while SEER measures the cooling efficiency over a typical cooling season. For example, if an "air-source" heat pump has a COP of 3.0, and if it is used to produce 10,000 kWh of heat over a heating season, it would consume about 3,333 kWh of electricity. In addition, if a "ground-source" heat pump has a COP of 4.0, and if it is used to produce the same 10,000 kWh of heat, it would consume about 2,500 kWh of electricity.

is maximized. This analogy of heat pumps functioning like batteries with high internal resistance highlights the versatility of the storage concept.

Note that there are numerous types of batteries, some are commercially available and mature and some are emerging, and their primary function in an evolving power system is essentially simple: to store energy and release it when needed.

However, two major parameters, power and energy, define a battery's performance characteristics in any given application, from small electronics to large-scale energy storage systems. The relationship between energy and power, in the context of battery utilization within power system components and applications, is fundamentally tied to the concepts of energy capacity and power capacity.

Energy capacity, measured in Wh, kWh or MWh, represents the total amount of energy that a battery can store and subsequently deliver over time. It determines how long a battery can operate before requiring recharge, hence defining its endurance in a given application. For example, a sustained power supply is required over longer periods in electric vehicles (EVs) for driving range or in energy storage systems for grid stability and load levelling.

Power rating, measured in W, kW or MW, indicates the maximum rate at which a battery can deliver (discharge) or absorb (charge) energy. Power capacity is critical for applications requiring high levels of power for short durations, such as accelerating an EV or providing peak power for frequency regulation in power grids. Therefore, power rating simply indicates how quickly a battery can respond to load demands.

In practical terms, a battery with high energy capacity but low power capacity can supply a low amount of power for a long duration. On the contrary, a battery with high power capacity but lower energy capacity can supply a high amount of power but only for a short time. The ideal balance between these two depends on the specific requirements of the application. For example, for grid storage, a high energy capacity is desirable for energy arbitrage (buying low, selling high), while a high power capacity is needed for frequency regulation.

Therefore, applications must consider both the energy and power capacities to match the battery system's capabilities with the operational demands. For instance, integrating batteries into renewable energy systems involves ensuring sufficient energy capacity to store excess generation and adequate power capacity to meet peak demand swiftly. Similarly, in EVs, the battery must provide enough energy for a desired range and enough power for acceleration and hill climbing.

Let us now look at the battery device more closely from the electrical engineering viewpoint. A battery is well known as an electrochemical device that stores energy in chemical form and converts it to electrical energy through an electrochemical reaction. Batteries are developed as small building blocks, called cells, which are formed in parallel and series connections, as described in solar PV cells, to achieve higher power (or voltage or current) and energy levels. Therefore, this chapter will begin with the basic descriptions of commonly used and emerging battery cells, then expand the explanations from common and critical descriptions to equivalent circuits (to be able to link to the power electronics (PE) converters discussed previously) and to form practical products.

Given their widespread use across a multitude of applications, this chapter will focus solely on rechargeable batteries. These are also referred to as secondary batteries and are designed for repeated charging and discharging. This capability for reversibility allows them to restore their charge by reversing the chemical reaction that depletes the battery during discharge when connected to an electrical load. In addition to their practical utilization, selection criteria and their operation in networks will also be covered, which will form the basis of case studies and DERs in microgrid structures in the next chapter.

5.2 BATTERY CELL STRUCTURES AND BASIC OPERATION

Figure 5.1 illustrates the principles of electrochemical processes in a lithium-ion battery cell, with the components and their respective roles in the battery's operation. The figure also shows the movement of lithium ions between the anode and cathode through an electrolyte medium while electrons flow through the external circuit (electrical load), hence providing electrical power. Such batteries are rechargeable and produce electricity through the movement of lithium ions between the anode and cathode during discharge and the reverse during charging as illustrated in the figure. The collectors ensure the efficient conduction of electrons to and from the anode and cathode materials, which are essential for the overall performance.

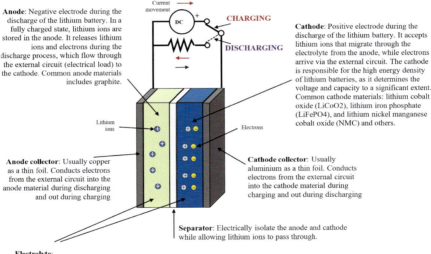

FIGURE 5.1 An operational schematic of a lithium-ion cell during charging and discharging cycles. For example, the basic chemical reactions for a lithium iron phosphate (LiFePO4) battery during discharging can be written as: LiFePO4→ FePO4+Li++e−, and during charging, the reaction is reversed.

The electrolyte within lithium batteries includes two primary components: lithium salts and organic solvents. Lithium salts offer distinct advantages in terms of ionic conductivity, thermal stability and compatibility with other battery materials. In addition, organic solvents dissolve these salts and enable the free movement of lithium ions. The solvents are selected for their low viscosity, which enhances ionic mobility, and high dielectric constant, necessary for effective dissolution of the salts. The selection of a solvent often is based on an optimal balance between electrolyte conductivity, operational temperature range and battery safety.

Graphite is the most preferred anode material due to its capacity for intercalating lithium ions within its carbon layers, hence providing a stable and reversible lithium storage mechanism. In addition, advanced anodes may incorporate silicon-graphite composites to achieve a compromise between capacity and longevity.

Note that separators in batteries are typically constructed from microporous polyethylene or polypropylene, materials that offer chemical resistance and mechanical strength. To improve thermal stability and electrolyte wettability, separators may be coated with ceramic materials or consist of composite layers. In addition, technological advancements have led to the development of nonwoven, nanofiber and specialized separators that integrate safety features, such as thermal shutdown or flame-retardant materials.

Note also that an "ion exchange membrane" is a form of separator designed to selectively allow the passage of certain ions while blocking others based on their charge (cation or anion exchange membranes). They are used in various applications, including electrolysis, water purification and energy generation (like in fuel cells). They contain fixed ionic groups that facilitate the selective transport of ions of opposite charge.

In flow batteries, the terms anolyte and catholyte refer to the electrolyte solutions that flow through the anode and cathode compartments, respectively (see Figure 5.2a). As is illustrated in the figure, an anolyte is an electrolyte that contains the electroactive species (ions) which undergo reduction at the anode during the discharging process. During charging, the anolyte is where oxidation takes place. The anolyte is stored in a separate tank (Negative Tank) and is pumped through the cell where the anode is located. On the contrary, a catholyte is an electrolyte that contains the electroactive species which undergo oxidation at the cathode during the discharging process. During charging, the catholyte is where reduction occurs. Similar to the anolyte, the catholyte is stored in a separate tank (Positive Tank) and is pumped through the cell containing the cathode.

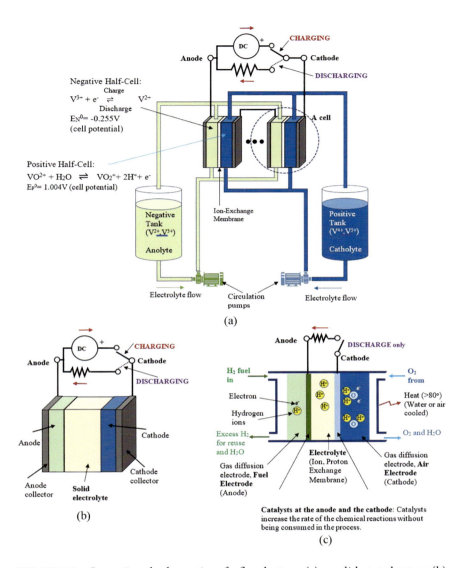

FIGURE 5.2 Operational schematics of a flow battery (a), a solid-state battery (b) and a fuel-cell (c).

Both anolyte and catholyte are subsets of electrolytes. In flow batteries, the anolyte and catholyte are specifically designed to circulate through their respective half-cells and facilitate the respective half-reactions.

Flow batteries are unique in that they separate the energy storage and power generation functions. The amount of energy stored is determined by the volume of electrolytes in the tanks, while the power output is determined by the size of the electrochemical cell that the electrolytes flow through. This design allows for independent scaling of energy and power, which is a significant advantage for certain applications such as grid energy storage.

Similar to lithium-ion batteries in operation, solid-state batteries use a solid electrolyte instead of a liquid one (see Figure 5.2b), which allows for higher energy density and increased safety due to the reduced risk of leaks or fires.

Fuel cells (see Figure 5.2c) generate electricity through a chemical reaction between a fuel, usually hydrogen, and an oxidant, often oxygen, which continuously converts fuel and oxidant into water, producing electricity as long as the reactants are supplied.

Table 5.2 provides a comparison across a range of energy storage and generation technologies, ranging from widely used lithium-ion batteries to emerging solutions like solid-state batteries and ultracapacitors, while highlighting their distinct characteristics, advantages and current stages of technological development.

As can be seen in the table, widely used lithium-ion batteries provide a high energy density and moderate safety profile, making them versatile for various applications. They offer moderate lifespans and charge speeds, continuously improving through ongoing research to enhance efficiency and safety. Moreover, flow batteries stand out with their extended lifespan and scalability, making them ideal for large-scale energy storage such as grid stabilization. Despite lower energy densities and slower response times, however, their suitability for prolonged storage and robust scalability is unmatched. In addition, solid-state batteries promise improved safety and energy density, though still in the research and development phase for most applications. Furthermore, fuel cells offer moderate to high energy density and long lifespans, making them potential cornerstones for future energy solutions. Continuous operation capability, provided with a steady fuel supply like hydrogen, positions fuel cells for transportation and stationary power generation. Fuel cell research focuses on enhancing catalyst efficiency and exploring cost-effective alternatives to precious metals to improve sustainability.

Finally, it can be indicated that ultracapacitors also known as supercapacitors excel in safety, lifespan and charge speed. Unlike conventional capacitors that store energy purely through electrostatic mechanisms, ultracapacitors store energy through electrostatic and electrochemical processes. The combination of these mechanisms allows ultracapacitors to achieve much higher energy densities than conventional capacitors,

Batteries and Fuel Cells in Energy Storage ■ 335

TABLE 5.2 Comparative Analysis of Energy Storage Technologies: Characteristics and Performance Metrics

Features	Lithium-Ion Batteries	Flow Batteries	Solid-State Batteries	Fuel Cells	Ultracapacitors
Electrolyte type	Liquid	Liquid (often aqueous or organic)	Solid (ceramic, glass, or polymer)	Liquid (aqueous) or solid polymer	Organic or aqueous electrolyte
Energy density	High	Low to moderate	Higher potential	Moderate to high	Low to moderate
Safety	Moderate (flammable electrolyte)	High (non-flammable electrolytes, but dependent on chemistry)	Higher (non-flammable electrolyte)	High (non-flammable, but depends on fuel utilization)	High (less chemical reactivity)
Lifespan	Moderate (degrades with cycles)	Very long (can be cycled many times without significant degradation)	Long (less degradation expected)	Long (but depends on catalyst degradation)	Very long (millions of cycles)
Operating temperature	Sensitive to extreme temperatures	Varies (some chemistries are temperature-sensitive)	Stable across a broad range	Wide range, but optimal performance at high temperatures	Wide range
Charge speed	Fast	Slow (limited by pump speed and reaction rates)	Potentially very fast	N/A (continuous operation with fuel supply)	Very fast
Scalability	Limited by size and energy density	Highly scalable (energy capacity is limited by tank size)	Potentially very scalable	Scalable (limited by system design and fuel storage)	Limited by energy density
Stage of development	Mature and widely used	Mature for specific applications (large-scale energy storage)	Emerging and under development	Mature, widely used in specific applications (backup power, transportation)	Mature for short-term power applications

though not as high as batteries. They are characterized by their ability to charge and discharge very rapidly, offering a very large number of cycles without significant degradation, and operate across a wide range of temperatures. These features make them suitable for applications requiring fast power demand, efficient energy capture from regenerative braking systems and providing power stability in various technologies.

5.3 COMMON DEFINITIONS USED IN BATTERIES

Due to the complex electrochemical characteristics of batteries, a number of practical and operational definitions are commonly used for rechargeable batteries, which are utilized to identify their electrical states within the entire life cycle. Table 5.3 provides an overview of the few key parameters to consider when evaluating different battery technologies. Understanding these parameters also helps in making informed decisions about battery selection for specific applications, considering trade-offs among other performance attributes and operational requirements. In addition, due to the comprehensive nature of battery behaviour, other key definitions will also be given in separate groups and tables, including "battery life", "battery efficiency", "state of charge (SOC)" and "battery voltage", which will also be supported by further descriptions and figures.

5.3.1 Battery Voltage

In rechargeable batteries, understanding critical voltage levels is essential for safe and efficient operations. These levels are influenced by various factors, including the anode and cathode materials, electrolyte composition and operational temperature, typically measured at 25°C.

The various voltage definitions presented in Table 5.4 cover the entire spectrum of a battery's operational and performance characteristics, which should be read together with Figure 5.3. These definitions highlight distinct states of battery behaviour under different and practical conditions, such as charging, discharging and resting states and establish safe operating limits.

The variation in voltage values among different battery chemistries and manufacturers' specifications also highlights the importance of these definitions. For example, the open circuit voltage (OCV) provides insight into the battery's SOC when not under load, while the nominal voltage serves as a standard benchmark for the battery's expected operational voltage. The cut-off voltage and the charge cut-off voltage show the lower and upper thresholds of the battery's safe operating range, respectively. Moreover,

Batteries and Fuel Cells in Energy Storage ■ 337

TABLE 5.3 Few Key Battery Parameters, Definitions and Associated Formulas

Parameters	Descriptions	Formulas and Symbols
Capacity	• In Unit Ah, it indicates the amount of electric charge (Q, in Coulombs) a battery can hold and supply over time. It determines how long a battery will last under specific load conditions • For example, a battery with a capacity of 10 Ah can supply 10 A of current for 1 hour or 5 A for 2 hours.	$Q = I\,t$ Capacity = Current (A) × Time (hours)
C-rate	• The C-rate is used to describe the charge and discharge current of a battery relative to its capacity. For example, 1C represents a charging or discharging current equal to the battery's capacity in Ah, leading to a full charge or discharge in one hour. A 2C rate would mean charging or discharging (Discharge-Rate) at twice the capacity (20 A for a 10 Ah battery), completing the process in half an hour. • It defines the speed of charging (slow and fast), which also defines the current ratings of the PE converter in which the battery can safely and efficiently serve the application.	C-rate = Capacity (Ah)/Current (A)
Energy density	• It measures the amount of energy stored per unit volume (Wh/L) or mass (Wh/kg) of the entire battery system, including the cells, packaging and any protective casings. • It defines the physical size and weight of the battery in the limited space available, such as in EVs and mobile devices.	Energy Density, $\rho_E = (C \times \text{Voltage})/\text{Mass or Volume}$
Specific energy	• It solely defines the energy stored per unit mass (Wh/kg) of the battery cells themselves, excluding external components. • It provides a direct comparison of the energy storage capacity of different battery chemistries or cell technologies, highlighting the efficiency of the active materials in storing energy.	Specific Energy, E_s or $SE = (C \times \text{Voltage})/\text{Mass}$
Power density	• Units are power per volume (W/L) or mass (W/kg) of the entire battery system. • It is critical for applications requiring high power output in compact or lightweight formats, as in EVs or in space-restricted DERs.	Power Density, $\rho_P = (\text{Discharge Power})/\text{Mass or Volume}$

(Continued)

TABLE 5.3 (Continued) Few Key Battery Parameters, Definitions and Associated Formulas

Parameters	Descriptions	Formulas and Symbols
Specific power	• Quantified as the power per unit mass (W/kg) of the battery cells. • It is used for evaluating the intrinsic power delivery capabilities of the cell materials or designs themselves. • It is useful for comparing the performance of various battery technologies accurately.	Specific Power, P_s or SP = Power/Mass
State of charge (SOC)	• It is a measure of the current amount of energy in the battery relative to its maximum capacity	$SOC = \dfrac{\text{Current Capacity}}{\text{Maximum Capacity}} \times 100\%$
State of discharge (SOD)	• It is expressed as a percentage, indicating the proportion of the battery's capacity that has been consumed. • SOD is complementary to State of Charge (SOC). For example, if a battery's SOC is 70%, then its SOD would be 30%, indicating that 30% of the battery's capacity has been used.	$SOD = 100\% - SOC$
Depth of discharge (DOD)	• The fraction of the battery capacity that has been discharged relative to the total available capacity, expressed as a percentage, indicating how much of the battery's total energy capacity has been utilized. • A higher DOD means a greater portion of the battery's capacity has been used, affecting longevity and performance over time.	DOD (%) = (Capacity Used/Total Capacity) × 100
State of health (SOH)	• SOH is a dynamic metric that provides a snapshot of the battery's current condition compared to its new condition, expressed as a percentage. • It considers factors like capacity fade and internal resistance increase, offering insights into where the battery is in its lifespan.	$SOH = \dfrac{\text{Current Maximum Capacity}}{\text{Original Maximum Capacity}} \times 100\%$

TABLE 5.4 Summary of Voltage Descriptions and Typical Corresponding Voltage Levels for "Typical" Lithium-Ion Battery Cells

Description	Definition	Typical Voltage Level
Theoretical voltage	Maximum potential energy difference between the anode and cathode of a battery under ideal conditions and is not directly measured in practice.	~4.2 V
Open-circuit voltage (OCV)	Voltage measured with no load, closely approximating the theoretical voltage. It is directly related to the SOC.	3.6–4.2 V
Closed-circuit voltage	Voltage under load, varying from OCV due to internal resistance. It provides a more practical measure of the battery's SOC under operating conditions.	Varies with load.
Nominal voltage	Standard value representing typical operating voltage. It is expected that the battery will operate at this average voltage value and under typical conditions and falls within the "Recommended Region" for the SOC.	3.6–3.7 V
Working voltage	Actual operating voltage under load, lower than OCV. This is the voltage range within which the battery operates effectively and can be associated with the "Operational Region" for the SOC.	3.0–4.2 V
Average voltage	This voltage is the average over the discharge cycle.	~3.7 V
Midpoint voltage	Voltage at the halfway point of the discharge process. It could be considered as the midpoint in the SOC operational range.	3.7–3.85 V
Cut-off voltage	Minimum operational voltage indicating a fully discharged state. This is the minimum voltage at which the battery can operate before it is considered fully depleted, correlating with the "Lower Exception Limit" in terms of SOC.	2.5–3.0 V
Charge cut-off voltage	Maximum safe voltage limit for charging. The voltage at which charging should stop to prevent overcharging, aligning with the "Upper Exception Limit" of the SOC.	~4.2 V
End-of-charge voltage	This is the voltage at which the battery is considered fully charged (charging process should stop), corresponding to the "Maximum Usable State of Charge" in terms of SOC.	~4.2 V

(*Continued*)

TABLE 5.4 (*Continued*)   Summary of Voltage Descriptions and Typical Corresponding Voltage Levels for "Typical" Lithium-Ion Battery Cells

Description	Definition	Typical Voltage Level
Over-discharge voltage	A voltage level below which the battery should not be discharged to avoid damaging the cell, linked to the "Lower Exception Limit" for SOC.	Above 2.5 V
Over-charge voltage	The voltage that should not be exceeded during charging to avoid damaging the battery, associated with the "Upper Exception Limit" for SOC.	Above 4.2 V
charge voltage	This is the voltage level to which a battery is charged under normal conditions and would be within the "Recommended Region" of SOC.	Up to 4.2 V
Peak voltage	The highest voltage a battery reaches during charging, typically just before the charge is complete; this should fall within the "Upper Operational Limit" of the SOC.	Around 4.2 V

understanding these voltage characteristics is important for addressing the "battery balancing" issues inherent in lithium batteries, which will be discussed in the subsequent sections. Therefore, accurate voltage monitoring and management are essential for balanced cell charging and discharging and for optimizing battery performance.

However, it is highly critical to consult the specific battery's datasheet for accurate voltage characteristics, as these values are representative and may vary according to the battery's chemistry, design and manufacturing processes. Therefore, relying on the manufacturer-specified voltage ranges is critical not only for the safety and longevity of the battery but also for ensuring that energy storage solutions follow the associated standards of efficiency, reliability and safety in their respective applications.

5.3.2 State of Charge (SOC)

As briefly described in Table 5.3, the State of Charge (SOC) and Depth of Discharge (DOD) are two critical parameters for understanding a battery's current capacity and the extent of its discharge. In addition, the State of Health (SOH) offers a snapshot of the battery's current condition relative to its original state, indicating its remaining useful life and performance capabilities.

Figure 5.3 illustrates the characteristic values of actual and usable energy in a battery as a function of SOC. Note that while the figure does not show voltage levels directly, it implies them through SOC, which serves as a proxy for voltage. A 100% SOC generally corresponds to the battery's maximum voltage, and a 0% SOC to its minimum before being

Batteries and Fuel Cells in Energy Storage ▪ 341

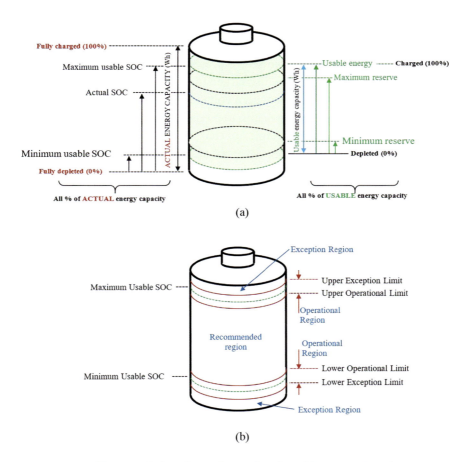

FIGURE 5.3 Characteristic values of actual and usable energy in a battery as a function of SOC (a), and operating regions for energy storage (b).

fully depleted. Since maintaining the battery within the recommended SOC range is vital for its health, it is also tied to sustaining appropriate voltage levels throughout its operation.

In practice, the voltage levels of a battery offer a quick method to assess the SOC. They can also provide insights into the SOH when deviations from expected values occur. Monitoring and managing these voltage levels against the discussed parameters, such as maximum and minimum usable SOC, reserve percentages and the operational and exception regions, can enhance the battery's performance and extend its lifespan.

In summary, in reference to the figure, "Usable Energy Capacity" has been visually illustrated, which differentiates between the actual energy capacity and the usable energy capacity. This indicates that not all stored energy in a battery is recommended for use, as deep discharging can significantly reduce its lifespan. In addition, it is known that as a battery discharges, its voltage decreases. The maximum usable SOC corresponds to the voltage level below which the battery should not be discharged to prevent damage. Similarly, the "minimum usable SOC" aligns with a voltage level above which the battery should be charged to maintain health and ensure a reserve.

There are also "Exception Regions" that indicate voltage levels beyond the normal operational range. Using a battery within these regions can lead to diminished performance or damage. "The Upper and Lower Exception Limits" likely correspond to voltage thresholds that should not be breached to avoid irreversible battery damage. Furthermore "The Upper and Lower Operational Limits" define the SOC range where the battery operates efficiently and safely, which corresponds to a secure range of voltage levels.

Associating actual and usable energy with the voltage levels also highlights the relationship between SOC and voltage characteristics across the battery's discharge and charge cycles, which can be explained using the characteristic values given in the figure.

For example, "Actual Energy Capacity" in Wh represents the total energy the battery can store at full capacity. Therefore, the total energy capacity spans from the over-charge voltage (the highest voltage the battery should attain) down to the cut-off voltage (the lowest voltage before the battery is considered fully depleted). On the other hand, "Usable Energy Capacity" is a portion of the actual energy capacity available for routine use, excluding reserves. Therefore, the usable energy capacity lies between the charge cut-off voltage (where normal charging ceases) and the discharge cut-off voltage (where normal discharging should end). In addition, "Actual State of Charge (% of Actual Capacity)" is the proportion of energy currently stored compared to the battery's full capacity. Therefore, SOC can be obtained from the OCV when the battery is at rest, with a higher OCV suggesting a higher SOC. Finally, the SOC in reference to the usable capacity (in %) is the total energy available to the user, not including energy reserved for special purposes, which is represented by a narrower voltage range within the battery's total voltage range, starting above the cut-off voltage and concluding below the charge cut-off voltage.

Therefore, when establishing operational limits, the "Upper Operational Limit" is set slightly below the charge cut-off voltage to prevent overcharging and maximize battery life, and the "Lower Operational Limit" is set slightly above the cut-off voltage to avoid deep discharging and prolong battery life.

5.3.3 Battery Efficiency

Before discussing battery efficiency, it is beneficial to revisit the fundamental parameters used in electrical circuits, which include current, voltage and power while also remembering that batteries operate on electrochemical principles.

In the context of batteries, "current" refers to the flow of electrical charge into a load during discharging and into the battery during charging. The discharge current varies according to the device's power demand, and the charge rate is also critical since a higher charge current can overheat the battery and reduce its lifespan. Therefore, batteries are often rated with the maximum continuous current they can provide or accept, with distinct rates specified for charging and discharging by using PE converters.

In addition, despite a variety of voltage descriptions provided previously, "voltage" in batteries is the measure of the electrical potential difference between the battery's positive and negative terminals. A battery's nominal voltage, determined by its chemistry, changes during charge and discharge, which is higher when fully charged and gets low as it discharges. Therefore, a battery's operating voltage range must match the voltage requirements of the circuitry it powers, which is also regulated by PE converters.

Furthermore, "power" in the context of batteries is the rate at which energy is supplied during discharge or absorbed during charging, calculated as the product of voltage and current. A battery's power output dictates how much energy it can deliver at a given time. Therefore, "power efficiency" is a relevant concept for batteries, especially when considering the performance of battery systems in power-intensive applications. In the context of batteries, it typically refers to how effectively a battery can deliver or absorb power during discharge or charge cycles, and it considers several factors, including the battery's internal resistance, the efficiency of PE involved in the charge/discharge process, and the battery's ability to maintain performance under high power demands. However, power efficiency is not commonly defined as a standalone term in battery specifications or academic literature as an electrochemical process varies by a number of factors during the charging and discharging process. Instead, its aspects are often covered by the more widely used efficiency descriptions, such as "Coulombic efficiency", "energy efficiency" and "voltage efficiency".

Since "energy" is the total amount of work done or heat transferred, in the context of electricity, it is often measured over a specific time period. Therefore, power is also defined as the rate of energy consumption or generation. While power efficiency focuses on the rate of conversion at any instant, energy efficiency examines the total energy utilized over time. Therefore, "Energy efficiency" is the ratio of the useful output of services from an energy-consuming device to the energy input over a longer period, often considering the energy consumed during the entire life cycle of the process or product.

As stated above, defining efficiency in batteries can be complex due to their electrochemical nature and the influence of external PE system components as described in Chapter 4. In addition, voltage efficiency is also a key metric in battery management since it impacts the battery's energy efficiency, longevity, safety and compatibility with other system components.

Therefore, Table 5.5 is provided to offer a clear and concise overview of efficiency types in batteries, including measures and calculation methods, which is useful for understanding the performance characteristics of batteries.

TABLE 5.5 Overview of "Efficiency" Types in Lithium-Ion Batteries

Efficiency Type	Description	Formula
Coulombic efficiency (CE)	• Reflects charge conservation. • It is the ratio of charge output during discharge to charge input during charging.	$CE = \dfrac{\text{Charge Discharged}}{\text{Charge Charged}} \times 100\%$ $CE = \dfrac{\int I_{dis}(t)dt}{\int I_{char}(t)dt}$
Energy efficiency (EE)	• Measures total energy delivered relative to energy used. • It is the ratio of energy output to energy input.	$EE = \dfrac{\text{Energy Discharged}}{\text{Energy Charged}} \times 100\%$ $EE = \dfrac{\int V_{dis}(t)I_{dis}(t)dt}{\int V_{char}(t)I_{char}(t)dt}$
Voltage efficiency	• Assesses impact of voltage changes on efficiency. • Defined by average voltage during discharge compared to charge.	$VE = \dfrac{\text{Average Discharge Voltage}}{\text{Average Charge Voltage}} \times 100\%$ $VE = \dfrac{\int V_{dis}(t)dt}{\int V_{char}(t)dt}$

(Continued)

TABLE 5.5 (*Continued*)   Overview of "Efficiency" Types in Lithium-Ion Batteries

Efficiency Type	Description	Formula
Round-trip efficiency	• Represents overall efficiency of energy cycle. • It is the product of Coulombic and energy efficiency, representing overall performance.	Round-Trip Efficiency = CE × EE
Charge efficiency	• Evaluates how well the battery accepts charge. • It indicates the effectiveness of the battery to be charged under various conditions.	Varies with SOC and temperature.
Thermal efficiency	• Looks at battery's ability to manage heat. • It defines how temperature affects the efficiency of the battery.	Depends on thermal management system efficiency.
Faradaic efficiency	• Focuses on charge contributing to energy storage. • It is the efficiency of the electrochemical reactions within the battery.	$FE = \dfrac{\text{Useful Charge (Ah)}}{\text{Total Charge (Ah)}} \times 100\%$
Ragone efficiency	• Important for variable discharge rate applications. • It is the efficiency considering power characteristics and energy density.	Specific to discharge rates and energy density, no single formula.
Power efficiency	• Measures power delivered compared to received. • It is the ratio of output power to input power during battery operation.	$PE = \dfrac{\text{Output Power}}{\text{Input Power}} \times 100\%$ $SE = \dfrac{\int_{0}^{t_{\text{dis}}} \left(P_{\text{dis}}(t) - P_{\text{loss}}(t) \right) dt}{\int_{0}^{t_{\text{char}}} \left(P_{\text{char}}(t) + P_{\text{loss}}(t) \right) dt}$
Instantaneous power efficiency	• Offers a snapshot of efficiency at a specific moment. • It is the "Power efficiency" at a specific moment.	$IPE = \dfrac{\text{Instantaneous Output Power}}{\text{Instantaneous Input Power}} \times 100\%$

5.3.4 Battery Life

Battery life can be described using several metrics, each covering different aspects of a battery's performance over time, as listed in Table 5.6. These metrics serve as fundamental indicators of a battery's durability and suitability for specific applications. Whether for daily cyclic use, long-term storage or standby power, they aid in designing systems that meet the operational demands and lifespan expectations of battery-powered applications. It is important to note that the precise formulation of "life metrics" is difficult due to the complex interplay of chemical, electrical and physical factors. Therefore, they are described on a conceptual basis, serving as guidelines based on empirical data and manufacturer's specifications.

It should be emphasized that these key battery life metrics are interconnected parameters that collectively offer a comprehensive understanding of a battery's lifecycle and operational efficiency from various perspectives. "Cycle Life" focuses on the durability of a battery under repeated use, measuring the number of charge-discharge cycles before significant capacity degradation occurs. In contrast, "Calendar Life" accounts for the total time a battery remains functional, considering ageing effects even when not in active use. "Shelf Life" specifically addresses how long a battery can be stored without losing its charge or performance, critical for long-term inventory or emergency preparedness. "Service Life" provides an overarching view, incorporating both cycling endurance and ageing factors to gauge a battery's operational lifespan under specific conditions. "Floating Life" defines a battery's ability to maintain charge in a standby state, essential for backup power applications. The beginning of life (BOL) and the end of life (EOL) mark the initiation and conclusion of a battery's service period, defining its initial performance standards and the endpoint of its usability respectively. Therefore, these parameters form a detailed framework for assessing a battery's performance, guiding maintenance schedules and optimizing its use across various applications.

In addition, note that "Useful Life" is also an important metric, which is commonly used to describe the period during which the battery operates reliably and above a certain performance threshold before it degrades to an unacceptable level. Among the metrics listed in Table 5.6, cycle life, calendar life and service life are most closely associated with it since they directly measure the performance and longevity of a battery under operational conditions. Floating life can also be relevant, depending on the specific application and operational mode of the battery. Useful life is a

TABLE 5.6 Overview of Key "Battery Life" Metrics

Metric	Description
Cycle life	• Cycle Life = Number of complete "charge-discharge cycles" a battery can undergo before its capacity falls usually below 80% of its original capacity. • It is largely determined through empirical testing under controlled conditions, cycling the battery until it reaches a predetermined capacity threshold. • It is about the durability of a battery under repeated charging and discharging. • For applications requiring frequent cycles, as in EVs, selecting a battery with a high cycle life ensures that it can handle the daily wear without premature capacity degradation. • It helps in predicting the replacement cycle and total cost of ownership over lifespan.
Calendar life	• Calendar Life = Time until a battery reaches its specified end-of-life criteria, not necessarily based on cycles but on ageing factors. • It involves long-term ageing tests and extrapolation of degradation trends under specified storage and operational conditions. • It is related to applications where batteries are expected to last for several years, even if not frequently cycled, as in backup power systems or utility-scale energy storage systems where the battery may remain at a full or partial state of charge for extended periods. • By selecting batteries with a long calendar life, designers can ensure that these systems remain reliable over time, minimizing maintenance or replacement needs.
Shelf life	• Shelf Life = Duration a battery can be stored before use while maintaining a specified minimum performance level. • Determined by storage studies at various conditions, monitoring parameters like self-discharge and capacity loss over time. • This is particularly critical for stockpiling of batteries that are used infrequently. • Batteries with a longer shelf life can be stored for extended periods without significant loss of performance, ensuring they are ready for use when needed. • This metric guides the selection of suitable battery types for products that spend a considerable amount of time in storage before being activated.

(Continued)

TABLE 5.6 (*Continued*)   Overview of Key "Battery Life" Metrics

Metric	Description
Service life	• It combines cycle life and calendar life considerations, factoring in operational conditions, maintenance and possibly restoration processes. • It is a holistic assessment based on both cycling endurance tests and calendar ageing predictions, adjusted for real-world usage patterns. • For systems that are subject to both frequent use and long-term deployment, understanding the service life helps in designing a battery solution that will consistently meet performance standards throughout its intended operational period. This ensures that the system remains functional and efficient, without unexpected failures.
Floating life	• Floating Life = Time a battery can remain on float charge at a specific voltage without falling below a certain capacity level. • Evaluated through long-term float charging tests, observing the rate of capacity decline and other failure mechanisms. • It is essential for designing standby or emergency power, as in UPS for data centres, where batteries must be kept in a charged state and ready to provide power instantly. • Selecting batteries with an extended floating life ensures that these systems can reliably provide backup power whenever it's needed, with minimal capacity degradation over time.
The beginning of life (BOL)	• It refers to the initial state of a battery when it is first put into service. • It is characterized by its performance metrics such as capacity, internal resistance and efficiency, as specified by the manufacturer. • It represents the baseline against which a battery's performance and degradation are measured over time.
The end of life (EOL)	• It is the point at which a battery no longer meets the required minimum performance criteria for its intended application. This typically occurs when the battery's capacity falls below a certain threshold (often 80% of its original capacity) or when other performance metrics degrade beyond acceptable limits, indicating that the battery needs to be replaced.

critical metric for planning and economic analysis, as it helps determine when a battery will need to be replaced based on the operational demands and conditions it faces, rather than just an abstract capacity threshold.

The parameters and formulas described above provide a fundamental framework for comparing and selecting rechargeable batteries, facilitating an analysis of trade-offs between energy storage, power delivery and lifespan. When evaluating batteries for specific applications, it is important to consider how these parameters align with the operational demands and performance requirements of the intended use.

5.4 ELECTRICAL EQUIVALENT CIRCUITS OF BATTERY CELLS

The equivalent circuits in battery technology are important since they can be used in various aspects of battery management and development. These circuits, which represent the complex interplay of a battery's internal processes, enable a broad spectrum of applications, from enhancing the precision of performance simulations to the monitoring of a battery's health over its operational lifespan.

In addition, such models facilitate a deeper understanding and prediction of how batteries respond under different conditions, without the need for extensive physical testing. Such capability not only supports the design and development process but also helps the integration of batteries into more complex systems. As summarized in Table 5.7, the diverse applications of equivalent circuits in batteries highlight their critical roles in advancing battery technology.

While comprehensive battery analysis may require specific complex models that accommodate both electrical and electrochemical behaviours, in this section only electrical equivalent circuits will be given in Table 5.8. Such circuits can provide a sufficient level of detail to facilitate

TABLE 5.7 The Applications of Equivalent Circuits in Batteries

Use Case	Description
Modelling and simulation	Electrical equivalent circuits allow for the simulation of battery performance, enabling analysis of behaviour under various conditions without physical testing.
State of health monitoring	They provide insights into the battery's degradation over time, allowing for estimation of its current health compared to its initial state.
State of charge estimation	They are critical for real-time determination of SOC.
Design and development	During the design process, they help optimize battery characteristics for specific applications by predicting performance outcomes from design changes.
Control systems integration	In real-time systems, they are used by battery management systems to manage power, control charging and protect the battery, based on predictive data.
Thermal management	They can include thermal modelling to anticipate how a battery will respond to temperature changes, informing the design of cooling systems.
Lifetime prediction	Continuous monitoring with the equivalent circuit helps in forecasting the end-of-life for batteries based on usage patterns and historical data.

TABLE 5.8 Electrical Equivalent Circuits of Cells Used in Lithium and Flow Battery Applications

Models	Equivalent Circuits	Remarks
Simple model (Ideal battery)		Simple model, an ideal battery. The terminal voltage $v(t)$ is equal the OCV when current flowing or not.
Charge dependent model		When a cell is charged its OCV is higher than when discharged, which can be modelled by a dependent source. Therefore, f(SOC) is 100% when fully charged and 0% when it is fully discharged. Intermediate values of f(SOC) can be associated to the total capacity Q of a cell and Coulomb efficiency.
Internal Resistance Model		An equivalent series resistance is added to consider the voltage drops under load. R_o represents the total battery resistance hence the losses in the battery as a heat, which is also known as the equivalent series resistance and the infinite-frequency impedance. Model is not suitable for accurate applications. $v(t) = OCV(f(SOC)) - i(t)R_o$.
1 RC Model (Thevenin' model)		It is a widely used model. Addition of R_1 and C_1 (which are equivalent polarization resistance and capacitance respectively) increases the model accuracy by reflecting the polarization characteristics of the cell. In the practical battery, R_o ohmic resistance has nonlinear relationship with SOC.
Second-Order (2 RC) Model		Adding more RC networks to the battery model improve the accuracy but an appropriate compromise is needed between accuracy and simplicity of the model especially for real-time applications. R_2C_2 is used to cover the battery concentration polarization effect. A constant-current source parallel to the terminals can also be added to include the battery self-discharge phenomenon.
1 RC model for a flow battery		R_{res} is represents the internal resistance of the battery, which accounts for the losses during the charge and discharge processes. R_f is related to charge transfer reactions. C represents the battery's capacitance, which relates to the ability to store charge electrostatically. R_{rea} is associated with the electrochemical reactions. I_{diff} represents diffusion currents as the movement of ions through the electrolyte.
2 RC model for a flow battery		I_{shunt} represents parasitic or shunt currents within the battery that do not contribute to the charge-discharge processes but instead lead to inefficiencies. R_o represents the ohmic resistance, accounting for the immediate drop in voltage when current starts to flow, similar to R_{res} in the 1RC model. R_1C_1 and R_2C_2 circuit is similar to the second order model representing different storage mechanisms.

the understanding and optimization of battery integration within PE converters, thus enabling the efficient management and operation of power systems.

Note that polarization in a battery cell is characterized by the deviation of the cell's terminal voltage from its open-circuit voltage when current flows through it. One form of polarization is represented by the product of the current, $i(t)$, and the ohmic resistance, R_0, which shows an immediate reaction to alterations in the input current. However, in a real-world scenario, the response to a change in current is not instantaneous but rather dynamic, similar to the behaviour of a resistor-capacitor (RC) circuit.

After the ceasing of the current flow, the cell's voltage does not revert to its open-circuit value instantaneously but instead reduces gradually, taking a significant duration, potentially over an hour, to near the open-circuit voltage. This effect is attributed to the slow diffusion processes within the cell, referred to as diffusion voltage, which changes slowly. To accurately represent this effect, one or more parallel RC sub-circuits are used in modelling as shown in Table 5.8

Note that the second order (2RC) model is commonly used in battery models. A solution of this circuit can be given by

$$V_{cell}(t) = V_{OCV} - I_{load}\left[R_0 + R_1\left(1 - e^{(-t/\tau_{small})}\right) + R_2\left(1 - e^{(-t/\tau_{large})}\right)\right] \quad (5.1)$$

Here, is the cell voltage at the terminals, and $I_{load}R_0$, $I_{load}R_1$ and $I_{load}R_2$ are the voltage drops across the DC resistance representing the immediate voltage drop, the activation and the concentration polarization circuits, respectively. $\tau_{small} = R_1C_1$ and $\tau_{large} = R_2C_2$ are the time constants representing faster dynamics (such as the response to changes in a current draw) and slower dynamics (like changes in diffusion processes within the battery) respectively.

In this lithium-ion battery model, R_0 represents the ohmic resistance, while R_1 and C_1 correspond to the immediate electrochemical kinetics and double-layer capacitance, respectively. R_2 and C_2 are associated with slower electrochemical processes and diffusion effects. The numerical values for these components can vary significantly based on the battery's chemistry, design, size and SOH. Although some of these parameters may be provided by battery manufacturers, detailed characterization through electrochemical impedance spectroscopy and other techniques is often essential for accurate modelling. This is particularly true when developing models for new types of batteries or for batteries under specific operating conditions. However, typical ranges for these parameters as a starting point for common lithium-ion batteries are as follows: R_0 and R_1:

10–100 mΩ; C_1: 10–1,000 µF; R_2: 50–500 mΩ, depending on the depth of discharge, rate of discharge and SOH.

The values of C_2 are more challenging to generalize, as they depend on the specific battery system's slower diffusion processes and can vary widely. It is important to note that the R_2C_2 time constant is expected to be much larger than the R_1C_1 time constant. This difference reflects that R_2C_2 represents slower processes, such as diffusion within the electrode materials, which have a longer response time compared to the faster electrochemical kinetics and double-layer charging effects captured by R_1C_1. Therefore, C_2 should represent a significantly larger time constant in conjunction with R_2 to accurately reflect slower processes. The typical capacitance for C_2, which models these slower diffusion-related phenomena, might not be precisely represented in µF. Instead, the effective capacitance could conceptually be much larger to accurately reflect the slow diffusion time constants. In practice, this is often modelled through other means (such as Warburg impedance) rather than direct capacitance values.

5.5 BATTERY CONNECTIONS, PACKAGING CONFIGURATIONS AND TEMPERATURE CONTROL

As known in power supplies, circuit element configurations and solar PV cells, there are three ways to connect battery cells as well, which are illustrated in Figure 5.4. The purpose of the connections are:

Series connection: Increases the total voltage of the battery pack to the sum of the individual cell voltages while keeping the capacity (in Ah) the same. It is used when the application requires a higher voltage than what one cell can provide.

Parallel connection: Increases the total capacity of the battery pack to the sum of the individual cell capacities while maintaining the voltage at the level of a single cell. This is useful when the application requires a larger capacity and longer runtimes without increasing voltage. Common in portable electronics and backup power systems where a longer duration is needed without a need for higher voltage.

Series-Parallel connection (Combination): Can be used to increase both the voltage and capacity of the battery pack. It is used when the application requires a specific voltage and capacity that cannot be achieved by series or parallel connections alone. This configuration is typical in large-scale energy storage systems and in EVs.

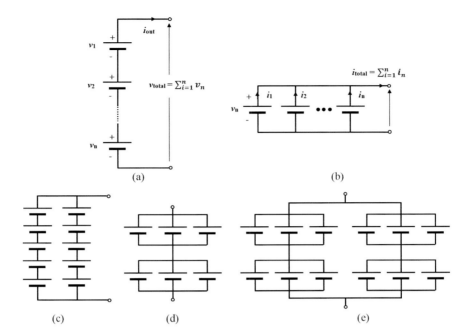

FIGURE 5.4 Three primary ways to connect battery cells: series (a), parallel (b) and combinations (c, d, e).

For example, a commercially available 48 V battery pack, which is commonly used in electric bikes, home energy storage and backup power systems, would generally be assembled from multiple 3.7 V lithium-ion cells connected in series to reach the desired voltage, with 13 cells being a common configuration. If higher capacity is required, groups of these series-connected cells can be connected in parallel. In practice, the energy level is also defined in kWh. Given that the relationship between capacity and energy is defined by Energy = Capacity × Voltage, a battery pack with a capacity of 50 Ah and a voltage of 48 V would have an energy capacity of 2.4 kWh. This calculation assumes that the battery discharges at a constant rate and that the voltage remains stable throughout the discharge period, which represents an ideal scenario.

Although critical voltage levels for cells have been defined previously, it is also important to note the typical voltage levels are defined and given for battery packs as well. For example, the typical voltage levels for a 48 V lithium-ion battery pack are summarized in Table 5.9 (top), and typical cell voltages are also compared across various battery types at the bottom of the table. Therefore, the Battery Management System (BMS) should regulate to prevent the voltage from exceeding safe limits during charge and

TABLE 5.9 Typical Voltage Levels of a 48V Li-Ion Battery Pack (top); Comparative Table of Cell Voltage Definitions for Various Battery Types (bottom)

Voltage Descriptions	Typical Voltage Levels
Theoretical voltage	~54.6 V (assuming 4.2 V per cell)
OCV	48.0–54.6 V (depending on SOC)
Nominal voltage	~48 V (3.7 V per cell × 13 cells)
Working voltage	~39–54.6 V
Average voltage	~48 V
Cut-off voltage	~32.5–39 V (2.5–3 V per cell)
Charge cut-off voltage	~54.6 V (4.2 V per cell × 13 cells)
End-of-charge voltage	~54.6 V
Over-discharge voltage	Above 32.5 V
Over-charge voltage	Above 54.6 V

Battery Type	Nominal Voltage	Cut-off Voltage	Charge Voltage	Peak Voltage	Open Circuit Voltage (OCV)
Lithium-Ion	3.6–3.7 V	2.5–3.0 V	~4.2 V	~4.2 V	Close to nominal
LiFePO$_4$	3.2–3.3 V	~2.5 V	~3.65 V	~3.65 V	Close to nominal
NaS	2.08 V	~1.8 V	2.5 V	~2.5 V	Close to nominal
Vanadium Redox	1.41 V	Varies with chemistry	Varies with chemistry	Varies with chemistry	Varies with state of charge

discharge cycles, to balance the cells effectively, and to protect the battery pack against overcharging and deep discharging.

It should be emphasized here that in practice the configuration of battery cell connections significantly influences the parameters of the individual cells as well, hence affecting the capacity and energy utilization of the entire battery pack. Each connection type also has its implications for the overall BMS design, as balancing the charge and monitoring the health of the cells becomes more complex with increased series and parallel connections. The choice of connection will depend on the specific energy and power requirements of the application, as well as considerations for manufacturability, cost and safety. The balancing and battery management concepts will be discussed later in Section 5.9. Therefore, selecting the appropriate configuration is a critical decision that impacts the performance, safety and cost-efficiency of the energy storage solution and hence requires a detailed study for a given application.

Table 5.10 provides a structured overview of the hierarchy and terminology used in battery packaging technology, indicating the distinctions and connections from individual cells to expendable solutions, accompanied by typical photographs. As the table suggests, the configuration

TABLE 5.10 An Overview of Lithium-Ion Battery Packaging Configurations, Their Functional Descriptions

Configuration	Description	Packaging
Single cell	• The basic electrochemical unit and the primary/simplest building blocks for larger battery configurations/structures. • Used in small electrical/electronic devices. • Typically, voltage range: 3.6–3.7 V, and at a capacity of mAh-Ah.	
Pack/module	• A collection of battery cells assembled together in series and/or parallel configurations. • Used in industrial and domestic-scale applications, may facilitate integrated cooling systems. • Typically, voltage ranges: commonly 48 V to use in racks. Pack capacities range from a few to several tens of kWh.	
Rack/unit/bank	• Used in industrial and domestic-scale applications. • Multiple packs/modules are configured for larger and manageable units, can facilitate integrated cooling systems. • Multiple packs are connected in series to achieve higher voltages (200, 400, 800, 1,000 V) and to obtain multiple MWh capacity.	
Container	• They are the largest but transportable units used in grid-scale and community-scale applications. • A physical structure designed to hold and organize multiple racks/units/banks, facilitating cooling systems, electrical connections, integration with modular PE converters, isolation and integration with larger systems.	

(*Continued*)

TABLE 5.10 (*Continued*)   An Overview of Lithium-Ion Battery Packaging Configurations, Their Functional Descriptions

Configuration	Description	Packaging
	• The configuration is also useful to reduce the cost of safety measures and ensure environmentally reliable systems (against harsh environmental conditions and insulated structures for better air conditioning).	
Multiple containers	• Used in grid-scale applications. • A configuration of multiple containers is connected usually externally together to provide a larger energy storage capacity, achieving desired electrical characteristics for specific applications.	

size of a battery is dictated by various factors, including the intended application, required energy capacity, power demands and physical limitations.

Although the physical sizes of lithium-ion battery cell configurations can vary widely depending on the specific application, industry standards and manufacturer specifications, there are some common sizes and form factors that are widely used in various industries for target applications. These include cylindrical (such as 18,650 and 21,700), prismatic (rectangular or square shape) and pouch (flat shape) cells, primarily for better utilization of available spaces. However, the choice also depends on specific application requirements, cost considerations and manufacturing capabilities.

The critical packaging of the foundational building block (the pack or module) involves several stages and manufacturing steps, which are critical not only for safety but also for the product's lifecycle. These steps include intercell connections, cell encapsulation, case configuration, terminals and contacts and pack formation and housing. In "intercell connections", materials such as nickel, copper or aluminium are preferred for their high electrical conductivity, aiming to minimize resistance, with spot or laser welding techniques being employed to forge strong and reliable joints without harming the cells during the assembly process. "Encapsulation" of cells is essential to provide mechanical protection and chemical resistance with a rigid or semi-rigid casing, which also aids in thermal management. Furthermore, "case configuration" is carefully designed with specific materials like ABS plastic, polycarbonate or aluminium to ensure durability, heat resistance, effective thermal management and accommodation for cell expansion and contraction throughout the charge-discharge cycles. "Terminals and contacts" are constructed from materials that resist corrosion and maintain robust electrical contact over time, such as gold-plated contacts for essential connections. Finally, "pack formation and housing" is designed to ensure uniform temperature distribution, integrating spaces for thermal insulation or cooling channels, selecting materials that enable efficient heat dissipation and contribute to structural integrity, like aluminium for its favourable lightweight and thermal properties, as well as cost-effectiveness.

Note that packs for stationary energy storage applications, such as residential or commercial battery systems, may vary in size based on the desired capacity and power requirements. Common sizes range from a few kWh to multiple MWh. Similarly, some industries, such as data centres and telecommunications, may use standardized rack sizes (such as 19-inch rack arrangement) for mounting and housing battery modules or packs. Modularity is also preferable in these structures for easy scalability, customization and handling (primarily due to heavy weight and safety).

"Rack/Unit/Bank" configurations are engineered to handle the weight of the battery modules while ensuring sufficient airflow or integration with cooling systems (air or liquid) within the racks to maintain optimal operating temperatures. Cooling pathways must be strategically designed for uniform temperature regulation across all modules. The materials selected for rack construction should be robust, supporting the overarching thermal management approach. Additionally, the BMS can be incorporated at the rack level to oversee and regulate performance, temperature and SOH. Appropriate terminals and connectors, capable of handling the electrical load, should be included, with designs that facilitate easy connection and disconnection for maintenance purposes. As a desirable and common feature, these terminals and connectors ought to be safeguarded against environmental elements and be readily accessible.

"Larger battery units" are often constructed in a container format, which utilizes the modularity of the Rack/Unit/Bank configuration, allowing easy expansion according to power requirements. This framework also includes safety provisions such as fire suppression systems and emergency power-off capabilities. Terminals and connectors are designed for high currents with an emphasis on safety and durability, incorporating features such as quick disconnects for emergency scenarios, DC circuit breakers and appropriate voltage, current and temperature sensors (both ambient and at the battery level).

It is important to note that these battery configurations, whether small or large, must follow local and international safety standards and regulations for battery systems. Each stage of the design process should incorporate efficient thermal management to guarantee safety and enhance performance. Moreover, components should be easily accessible for maintenance, monitoring and replacement purposes. Ideally, the design should also be scalable, accommodating future growth or reconfiguration, potentially even incorporating different battery technologies. Container designs should consider environmental factors such as moisture, extreme temperatures, dust, lightning, strong external winds and earthquakes when designing enclosures and protective measures. The final integration step of the battery system involves establishing connections for DC couplings with renewable energy sources and linking to nearby inverters through a DC bus.

Note that, in Table 5.10, Ah is also used for specifying the capacity of individual cells and small-scale batteries. The units kWh and MWh are used for larger scale energy storage systems, focusing on the total energy stored and delivered over time, which is critical for grid integration, billing and system sizing in commercial and utility applications.

Furthermore, battery packaging configurations are also designed to eliminate faulty or failed cells or packs easily and reduce repair/replacement costs (very critical in EVs). The packaging and housing in the configurations include two major auxiliary functions: BMS and supervisory monitoring, detection and control systems. The primary aim of BMS involves charging and discharging sub-units and integrated balancing units (except in the single cell configuration), all based on various PE converter topologies. The monitoring, detection and control systems utilize sensors and PE converters to regulate charging and discharging in the light of SOC and SOH for balancing, safety, protection and temperature control.

Control also plays a critical role in battery energy storage systems (BESS), ensuring safe and efficient operation, optimal performance and integration with external systems. Control hierarchy refers to the organization of control functions within a BESS, from individual cell monitoring and management to system-wide optimization and coordination. At each level of the hierarchy (cell, pack/module, rack/unit/bank, container, multiple containers), control functions may also include monitoring, protection, balancing, state estimation, energy management and communication. Higher-level control systems (supervisory systems) coordinate the operation of lower-level control systems to achieve system-wide objectives such as grid stability, renewable energy integration and cost optimization. Control hierarchy ensures that control functions are distributed and coordinated effectively, providing scalability, flexibility and robustness in BESS, which usually forms a sub-section of a DER, microgrid and larger power system, as it will be covered in Chapter 6.

Note that the modularization approach seen in lithium-ion batteries can also apply to flow batteries, fuel cells and supercapacitors, although at different scales and configurations. For example, in flow battery systems, a single cell consists of two electrolyte tanks and a membrane stack between them, and the voltage and capacity of a single cell depend on the size of the membrane stack and the volume of electrolyte. Similarly, flow battery packs or modules consist of multiple individual cells connected in series or parallel configurations to achieve the desired voltage and capacity. In addition, containerized flow battery systems consist of one or more racks/units/banks housed within standard ISO shipping containers or custom enclosures. Finally, the containers are interconnected to form a single integrated system, enabling seamless operation and management across multiple units.

5.5.1 Battery Temperature Control: Cooling/Heating

Temperature control in batteries is one of the most critical measures, as effective temperature management directly impacts a battery's lifespan, efficiency, optimal operational performance as well as safety. This control can be both passive (such as insulation and heat sinks) and active (including cooling and heating systems) types to maintain the battery within its optimum operating temperature range. Therefore, the thermal design—covering cell spacing, insulation, cooling mechanisms and overall packaging and rack design as previously discussed—plays a significant role.

Table 5.11 provides an overview of internal and external heat sources in batteries, highlighting their causes, effects and implications for thermal management. Furthermore, it highlights potential heating needs, hence the importance of thermal management.

Therefore Figure 5.5 is given to present an analysis of a battery's charge-discharge cycle and the impact of cell packaging on temperature variation and overall battery performance. The upper graph shows the normalized voltage and current throughout a complete battery cycle, which includes charging, a rest period and discharging. During the charge

TABLE 5.11 A Comprehensive Summary of Thermal Control Needs in Batteries

	Description
Internal Heat Sources	
Electrochemical reactions	The fundamental chemical reactions during charging and discharging are exothermic, releasing heat as a by-product.
Internal resistance	Movement of ions within the electrolyte and electrons through the external circuit generates heat due to I^2R losses.
Joule heating	Occurs when current flows through internal resistance, converting electrical energy into heat.
Overcharging	Exceeding the battery's capacity leads to excessive heat generation from the oxidation of electrolytes and other side reactions.
Rapid discharging	Discharging a battery too quickly increases the rate of electrochemical reactions, leading to elevated temperatures.
Cell compression	Mechanical pressure applied to maintain contact between layers contributes to heat generation.
External Heat Sources	
Ambient temperature	High ambient temperatures can increase battery temperatures.
Solar radiation	Direct solar exposure, like in an EV parked outside or BSS exposed to sunlight directly, can heat the battery.
Heat transfer from adjacent components	Heat from neighbouring components in a system can be transferred to the battery, which includes battery packs in a rack arrangement and PE converters.
Poor ventilation	Inadequate ventilation can cause heat accumulation due to insufficient airflow.
Charging systems	Inefficient external charging systems can generate heat during the charging process.
Causes from Design	
Temperature rise from adjacent cells	Heat from one cell can induce additional heat in neighbouring cells. Cells in close proximity can transfer heat to each other (see Figure 5.5)
Inadequate packaging design	Designs not facilitating adequate heat dissipation. Packaging that reduces airflow prevents heat removal.
Orientation of packs in a rack	Improper stacking can block cooling paths. Insufficient space between packs impedes airflow or cooling fluid circulation.
Cold Environment	
Cold temperature	Causes reduced power output and slower charging rates, which is due to the slow electrochemical reactions within the cells.

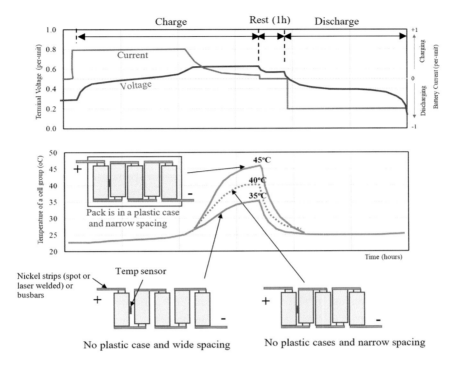

FIGURE 5.5 Temperature characteristics of a battery packed formed by five cells connected in series during a cycle involving charge, rest and discharge.

phase, the current is positive, and the corresponding voltage rises. This is followed by a rest period of 1 hour, where no current flow is present. The discharge phase is characterized by a negative current as the battery delivers power and a corresponding decline in voltage.

In the middle graph, temperature profiles for different battery packaging configurations are illustrated over time. It is clear that a battery pack encased in a plastic housing with narrow spacing between the cells exhibits a rapid and higher temperature increase, reaching up to 40°C and 45°C. However, battery packs without plastic cases and with wider spacing between the cells show a more gradual temperature increase, with broader peaks at lower temperatures, indicating more efficient heat dissipation.

In conclusion, the packaging method and temperature measurement point have a significant impact on the volumetric power density of the pack and the effectiveness of cooling control respectively. Note that the gap between the batteries reduces the power density as well as the accuracy of the battery temperature measurement. On the contrary, if the batteries are packed without the thick plastic housing significant space saving is possible in addition to the effective heat transfer and correct battery temperature measurement, which is typically performed in mobile systems. In addition, prismatic cells swell over their cycle life, specifically at higher temperatures. The designer should consider an additional 10% beyond the nominal thickness to accommodate this swelling. This is very critical when cells are stacked and welded in a battery pack.

Table 5.12 outlines the various cooling and heating methods used in battery thermal management, highlighting their type (passive or active), a brief description and typical application scenarios where each method might be used.

Figure 5.6a shows a graph demonstrating the temperature rise in a battery cell at two different current levels, two thermal models showing temperature distribution under two different loads and a thermal camera image capturing temperature variations across the battery connectors.

The graph indicates a steeper temperature increase at a higher current of 154 A compared to a lower current of 70 A, indicating that higher currents lead to rapid heating within the battery. The 60°C threshold

TABLE 5.12 Overview of Cooling and Heating Strategies for Battery Thermal Management across Various Applications

Thermal Management	Type	Description	Application Scenario
Cooling Methods			
Natural convection	Passive	Utilizes ambient airflow to dissipate heat away from the battery.	Low-power applications
Heat sinks	Passive	Metal fins or plates that increase the surface area for heat dissipation.	Portable electronics, small devices
Phase change materials	Passive	Substances that absorb or release heat during phase transitions to regulate temperature.	Various, including EVs and stationary storage
Thermal insulation	Passive	Materials designed to reduce heat transfer between the battery and its environment.	Cold climates
Air cooling	Active	Forces air over or through the battery pack to remove heat.	EVs, portable devices
Liquid cooling	Active	Circulates coolant fluid through channels within or around the battery to absorb and dissipate heat.	High-power applications, EVs
Thermoelectric cooling	Active	Utilizes the Peltier effect for cooling one side of the battery pack while heating the opposite side.	Specialized applications

(Continued)

TABLE 5.12 (*Continued*) Overview of Cooling and Heating Strategies for Battery Thermal Management across Various Applications

Thermal Management	Type	Description	Application Scenario
Heating Methods			
Insulation	Passive	Minimizes heat loss to the environment, helping the battery maintain its generated heat.	Cold weather operations
Phase change materials	Passive	Release heat as they solidify, providing warmth to the battery.	Cold climates, pre-start heating
Internal heating elements	Active	Embedded elements within the battery pack that generate heat when powered.	Very cold environments, EVs
External heating pads	Active	Applied externally to the battery pack and powered to generate warmth.	Pre-operation in cold conditions
Electrical current	Active	Applies a controlled electrical current to induce resistive heating within the battery.	Quick heating, cold start
Heat pumps	Active	Capable of transferring heat from an external source into the battery system.	Energy-efficient heating in cold environments
Warm coolant circulation	Active	In systems with liquid cooling, the coolant is heated externally and circulated to increase the battery's temperature.	Integrated thermal management systems

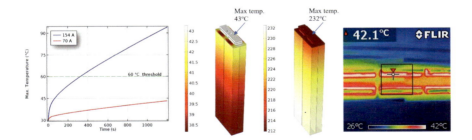

FIGURE 5.6 Thermal management evaluation for battery cells, indicating temperature progression under varied loads, simulated thermal gradients at peak temperatures and real-time thermographic analysis of cell connectors.

line is also given to indicate the critical temperature limit for safe battery operation. The thermal model illustrates how temperature gradients develop within a battery cell, which shows a higher temperature at the top. This is due to heat accumulation away from the thermal path to the cooling system as well as the proximity to the heat source. The camera image reveals hot spots, especially at the points where the connectors meet the cell terminals, indicating regions of higher resistance where most of the heating occurs.

Cell connectors in battery systems are critical to thermal management due to heat conduction, current-carrying capability (linked to the resistance) and mechanical stability. Therefore, these can affect the performance and safety of the battery, as excessive heat can lead to thermal runaway. Note also that anode and cathode can heat differently due to the nature of the reactions and the materials used.

In general, forced air cooling and water cooling are two distinct approaches to managing the thermal performance of lithium battery modules. Forced air cooling involves pushing air through the modules to remove heat, offering a simpler and more cost-effective solution. However, as mentioned in Chapter 1, it is less efficient due to air's lower heat capacity and may struggle with achieving uniform temperature distribution across the battery cells (see Figure 5.7b). This method also faces a higher risk of internal condensation, particularly if the air introduced is humid (such as on rainy days), potentially leading to moisture-related issues within the modules. On the other hand, water cooling circulates water or coolant around the battery cells, significantly enhancing heat dissipation efficiency and ensuring more uniform cooling. Despite its superior performance, water cooling systems are more complex and costlier, primarily associated with leaks, which could lead to safety hazards. However, internal condensation is less of a concern with water cooling, external condensation on the system could still pose risks in high-humidity environments.

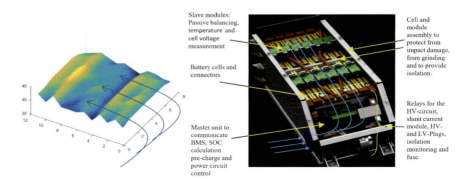

FIGURE 5.7 Airflow simulation in a battery pack (for an EV, 360–750V, 3.3 kW), highlighting the components and airflow dynamics.

Figure 5.7 shows a composite of a simulation result and a graphical representation of a battery module employing forced air cooling. The figure on the left is a 3D plot from computational fluid dynamics (CFD) simulation software, showing the velocity field of air as it moves through a battery module. The colour gradient on the surface represents the velocity magnitude or temperature distribution of the air, with cooler colours indicating lower values and warmer colours indicating higher values. On the right side, the actual battery module assembly with multiple cells shows air being directed through the module, with arrows demonstrating the path of airflow, consistent with the simulation results on the left. This visualization helps to understand how air is intended to traverse the module, helping heat removal from the individual battery cells. The air inlet and outlet are marked, demonstrating that the system is designed to pull in cooler air, force it through the module to absorb heat, and then expel the warmed air.

Therefore, it is important to evaluate the efficacy of the cooling method to identify zones of optimal airflow and detect potential areas of diminished flow that could create hot spots, hence preventing the overheating of battery cells in the module. This approach is equally applicable to entire battery units in larger systems.

To reduce the heating costs of lithium batteries in cold climates, several strategies can be implemented. Insulation, for instance, effectively retains heat, and integrating the battery's heating system with the vehicle's thermal management system allows for repurposing excess heat. Preheating the batteries while charging can optimize the use of available power, and keeping them in temperature-controlled settings diminishes the need for initial heating. Employing phase change materials can offer temperature stabilization with reduced energy consumption, and intelligent battery management systems can customize heating schedules to meet actual requirements, thereby enhancing efficiency. Additionally, the adoption of heat pumps and the selection of battery chemistries and structures that perform better at lower temperatures can decrease the necessity for supplementary heating. Furthermore, aligning battery operation schedules with actual usage patterns can prevent unnecessary heating during periods of inactivity, leading to further energy savings.

Note that different strategies are applicable to mobile systems, such as EVs, and stationary grid-scale battery storage, in cold climates. For example, EVs require lightweight and compact solutions, such as using smart battery management systems. On the other hand, stationary systems often have fewer size and weight restrictions, allowing for more robust insulation and larger installations of phase change materials.

In addition, storing a stationary BES underground can significantly reduce both heating and cooling costs, since the underground environment has a stable temperature throughout the year compared to the surface temperature. This is due to the thermal inertia of the ground, which means it heats up and cools down much more slowly than the air above ground. Furthermore, there are added benefits of underground installations: temperature stability, reduced HVAC, protection from environmental factors (such as wind, rain and snow) and security and safety.

Although the initial costs of underground installation may be higher due to the need for excavation and reinforced structures, the long-term savings on heating and cooling can be substantial.

Thermal management discussions for flow batteries, particularly vanadium redox flow batteries, share similarities with those for lithium-ion batteries, but with distinctions due to their distinct operational principles and components. Since such batteries store energy in liquid electrolytes that circulate through an electrochemical cell during operation, tailored thermal management strategies are necessary.

Therefore, the need for temperature control, employing both passive and active thermal management methods, and the impact of effective thermal management on performance and safety are very similar to those in lithium-ion batteries. However, flow batteries exhibit distinct differences. One notable difference is the thermal dynamics of liquid electrolytes in flow batteries. The circulating electrolytes can, to some extent, act as a self-regulating thermal management system by distributing heat more evenly or removing it from the cell area. However, external cooling or heating may still be essential, especially under high load conditions or in extreme environmental temperatures.

Moreover, integrating cooling systems can be more straightforward in flow batteries, as the flow system can be designed to incorporate a heat exchanger. This setup efficiently manages the electrolyte solution's temperature before it re-enters the cell. Note also that flow batteries are often utilized in large-scale energy storage applications, meaning their thermal management systems can be more intricate, potentially integrating into broader building or facility cooling systems. Additionally, the temperature can influence the chemical stability of the electrolytes in flow batteries. Optimal thermal management ensures the electrolytes remain in a stable condition, preventing degradation that could diminish the battery's effectiveness or lifespan.

It should be noted that advances in battery technology and thermal management systems continue to enhance efficiency, potentially reducing the power required for these operations. Furthermore, accurate information regarding the additional power consumed by thermal control—which reduces system efficiency—may not be readily available, as it depends on specific details about the battery system, thermal management system design, the ambient temperature range, capacity and the intended application.

Although some general insights and typical power ranges in kW are provided in the literature to estimate the power required for cooling and heating, it is more accurate to quantify it in terms of energy in kWh. This is due to the fact that thermal control is influenced by various factors, including the space to be conditioned, the flow and pressure of the cooling or heating fluid, and, most critically, the duration or duty cycle of the operation. Therefore, for an accurate approach, it is necessary to discuss the thermal management energy requirements in the context of the total energy supplied by the battery unit. For instance, in large-scale battery storage systems, which typically operate on a daily cycle, the energy consumed by thermal management can constitute a significant portion of the total energy output. Similarly, in EVs, the driving cycle dictates the thermal management's energy demands, affecting the overall energy efficiency and range of the vehicle. To provide a more comprehensive analysis, the following five points can be considered:

- The size and configuration of the battery system, as larger systems require more energy for thermal management. Consequently, additional space, such as racks placed inside a large room where the room space is thermally managed, should always be minimized or avoided.

- The ambient temperature and operational environment, which can drastically affect heating or cooling needs. Therefore, the colour and paint type of the external enclosure and its location are very critical. For instance, to avoid direct sunlight and create a more stable ambient temperature, racks or containers can be situated at ground level or underground, which also reduces heating needs.

- The specific thermal management technology used, as some systems (like heat pumps under certain conditions) can be more energy-efficient than others (such as resistive heating or basic air cooling). Additionally, the selection of the electric drive system used in fluid (air or liquid) circulation can significantly influence energy loss, as an overrated drive system operates inefficiently even with variable speed drives.

- The efficiency of the battery system itself, since higher efficiency reduces waste heat generation, thereby potentially lowering cooling requirements.

- The operational duty cycle, which includes the intensity and duration of battery use, directly impacts the thermal management system's energy consumption. Note that this will indirectly link to the bullet points given above.

Effective management of battery thermal conditions requires accurate temperature measurements both at the cell level and in the surrounding ambient of the packs, modules or units. This highlights the strategic placement of temperature sensors within a battery pack for accurate monitoring for efficient cooling. The temperature data is essential for informing the thermal management system. Key considerations for sensor placement include detecting hot spots, often near the centre of the battery pack or in areas with less effective cooling, and ensuring uniform temperature monitoring throughout the pack. This uniform monitoring is critical for uniform cooling across both the core and the peripheries. Sensors must be in contact with the cells for accurate readings and should accommodate the cells' expansion and contraction during operation.

In addition, it is important for sensors to identify thermal gradients within the pack, as temperature differences can negatively affect charging and discharging. Placement near temperature-sensitive components, such as separators or electrolytes, is critical to prevent overheating. Distributing a sufficient number of sensors throughout the pack is also needed for a detailed temperature map, specifically in larger packs. To measure cooling effectiveness accurately, sensor placement must also be coordinated with the cooling system design, such as near the inlet and outlet for liquid cooling systems or along the airflow path for air cooling systems. Integration with the BMS is also important since temperature data is required to manage charging rates and initiate safety measures when needed. Furthermore, sensor placement should facilitate easy access for maintenance or replacement without dismantling the entire pack and be positioned away from external heat sources or cooling components to avoid false readings.

Directly measuring the central temperature within lithium and flow batteries presents difficulties due to their construction and operation. For example, for lithium batteries, the sealed environment and the potential risk to cell integrity make internal temperature measurement impractical for regular use. Therefore, manufacturers may embed thermocouples or fibre optics inside the battery to monitor internal temperatures or external sensors are used to estimate internal temperatures using accurate thermal models. However, for flow batteries, the external circulation of the electrolyte allows for easier internal temperature measurement. Sensors placed in the electrolyte stream outside the reaction chambers offer a good approximation of the reaction temperature. External measurements and thermal modelling are also used to estimate internal temperatures.

Common measurement techniques for both battery types include embedding sensors in controlled environments, using infrared thermography for non-contact surface temperature assessment and applying thermal modelling. These approaches consider heat generation rates, thermal conductivities and specific heat capacities to predict internal temperatures from external measurements.

In summary, achieving optimal power density and effective thermal management in battery packs requires identifying both structural and material factors. Reducing gaps between cells, limiting the use of thick plastic housings and arranging cells to facilitate uniform cooling are key. Using high-conductivity thermal interface materials and strategically placing temperature sensors are crucial for accurate monitoring and efficient heat dissipation.

5.6 COMPARISON OF BATTERIES FOR SELECTION

In the contents of energy storage, lithium-based batteries power a wide array of applications. Comparing common batteries enables informed decision-making that aligns with specific application needs, optimizing efficiency and overall performance. Furthermore, such comparisons guide the selection process, ensuring the chosen battery technology not only improves application effectiveness but also addresses environmental and safety concerns.

In the subsequent sections, key parameters will be discussed, including energy density, power density, cycle life, safety, the operational temperature range and cost. As can be seen, comparing batteries to identify the most suitable option for a given application involves considering many aspects that account for the specific characteristics of the electrical load they will support.

Table 5.13 provides a comprehensive comparison of criteria for common lithium-based batteries, detailing the description of each criterion and its relevance to specific applications. It highlights load requirements,

TABLE 5.13 Comparative Analysis of Lithium-Based Battery Criteria and Their Application Relevance

Criteria	Description	Relevance
Load Requirements		
Power demand	To determine the continuous and maximum power.	Essential
Energy consumption	The total energy usage over a specified period.	Essential
Load profile	Variation and patterns of energy usage over time.	High
Usage Conditions		
Temperature	The operational temperature range of the application.	High
Physical space	Size and weight limitations due to the application's design.	High
Maintenance	Maintenance or replacement based on the application's design.	Variable
Battery Characteristics		
Energy density	Energy a battery can store relative to size or weight.	Essential
Power density	Ability to deliver power, especially for high-load applications.	Essential
Efficiency	Percentage of energy input that can be effectively used.	Essential
Safety	Risk assessment for hazards such as fire or chemical leakage.	Essential
Lifespan	Expected operational life in terms of calendar and cycle life.	High
DOD	The degree to which a battery can be used before recharging.	High
Charge/discharge	Speed of charging and power delivery.	High
Total Cost of Ownership		
Initial costs	Cost of battery purchase and installation.	High
Operational costs	Ongoing costs for energy, maintenance and downtime.	High
End-of-life costs	Costs associated with recycling or disposing of the battery.	Variable
Regulatory and Environmental		
Compliance	Adherence to safety, transportation and disposal regulations.	Essential
Environmental	Ecological footprint and recyclability of the battery.	Variable
Manufacturer Specifications		
Performance metrics	Battery performance data provided by the manufacturer.	High
Warranty conditions	The manufacturer's warranty, including duration and coverage.	High

usage conditions, battery characteristics, total cost of ownership, regulatory and environmental considerations and manufacturer specifications.

Note that while all listed parameters in Table 5.13 are necessary, it is the essential ones that directly impact the core functionality and safety of the application. For example, in power grid applications, parameters such as power capability, response time, reliability and safety are considered essential due to their direct impact on grid stability and operational safety. For EVs, however, energy density, cycle life, efficiency and safety are crucial for ensuring sufficient range, longevity and user safety. The key parameters for selecting batteries for these two major applications are outlined in Table 5.14, highlighting their prime importance. This distinction is important as it assists in prioritizing the evaluation of battery technologies based on application-specific requirements, ensuring the chosen solution supports the application's operational needs and safety requirements optimally.

TABLE 5.14   Key Battery Parameters for Power Grid Applications and for EV Performance

Parameters	Reasons	Importance
Power Grid Applications		
Power capability	Ability to deliver immediate high power for frequency regulation and reactive power for voltage support.	Essential
Response time	Rapid response is needed for FCAS to counteract frequency deviations and for mitigating power quality issues.	Essential
Reliability	High reliability ensures continuous operation and less downtime, which is critical for grid stability.	Essential
Safety	The system must be safe to operate, given the high energy and power levels involved in grid applications.	Essential
Energy capacity	Required for periods of intermittency/variability and to provide sustained support during grid disturbances.	High
Charge/discharge rate	Fast rates are necessary for quick services provision, such as voltage support or addressing the intermittency of renewables.	High
Efficiency	High round-trip efficiency ensures that more of the stored energy is available for grid services.	High
Cycle life	A longer cycle life reduces replacement frequency, critical for cost-effectiveness and reliability in grid applications.	High
EV Applications		
Energy density	Determines the potential range of the EV on a single charge.	Essential
Cycle life	EV batteries are cycled daily, so a longer life reduces replacement costs and increases vehicle lifespan.	Essential

(*Continued*)

TABLE 5.14 (*Continued*)   Key Battery Parameters for Power Grid Applications and for EV Performance

Parameters	Reasons	Importance
Efficiency	Impacts the range and performance of the EV, as well as energy consumption costs.	Essential
Safety	Safety is paramount to prevent accidents due to battery malfunctions, which can be severe given the EV's energy content.	Essential
Power density	Affects acceleration and the ability to manage quick response.	High
Temperature tolerance	EV batteries must perform well across a range of temperatures as vehicles are exposed to various environmental conditions.	High
Charge/discharge rate	Affects how quickly the vehicle can be charged and its performance during acceleration and regenerative braking.	High

In summary, essential parameters are non-negotiable characteristics that impact the application's core functionality, safety and reliability directly. In contrast, high-importance parameters, while not as critical as the essential ones, significantly influence the battery's performance, efficiency and overall suitability for the application.

Table 5.15 shows additional practical and operational parameters critical in battery technology evaluation. The table not only explores basic electrical characteristics but also extends to factors critical for a comprehensive assessment, including safety, environmental impact and practical usability. This detailed view ensures that the selected battery technology aligns well with the specific needs and constraints of its intended application, facilitating informed decision-making in battery selection.

Figure 5.8 provides four radar charts to compare multiple quantitative variables: two charts for comparing different battery technologies, and two charts on the materials used for cathodes and anodes in batteries.

As can be seen, in Figure 5.8a, lead acid batteries are recognized for their cost-effectiveness and established technology, but they fall short in energy density and lifetime. Redox flow batteries excel in energy density and cycle life but have lower energy and power densities. Sodium sulphur batteries stand out for their high energy density, making them ideal for large-scale storage, though they operate at high temperatures and present safety risks. Lithium-ion batteries are visible for their high energy density and efficiency, hence a popular choice in various applications.

Similarly, in Figure 5.8b, Lithium Iron Phosphate (LFP) is characterized by its cost-effectiveness, safety and long cycle life, although it has a lower energy density, making it less suitable for energy-intensive

TABLE 5.15 Advanced Battery Performance and Operational Parameters

Parameter	Description	Assessment Method
Fast charge capability	The battery's ability to accept a high charge current for rapid recharging. Essential in EVs.	Maximum safe charge current
Energy output	Total amount of energy that can be delivered over its lifetime, is the key to long-term value and cost-effectiveness.	Total energy delivered over Lifetime
Chemical stability	Indicates the stability of the battery's chemistry over time and under various operational conditions. Impacts safety, performance and lifespan.	Qualitative assessment
Electrochemical window	Voltage range within which the battery is stable and operational. Going outside this range can affect performance or safety.	Operational voltage range
Maximum continuous discharge rate	Highest current at which the battery can be safely discharged, for ensuring battery longevity.	Maximum continuous current
Impedance stability	Changes in the battery's internal resistance over time and under different temperatures and states of charge. Critical for consistent performance.	Variations in Ohms under different conditions
Environmental friendliness	Assessing the materials used in the battery for toxicity and environmental impact. Important for disposal and recycling considerations.	Qualitative assessment
Modularity and scalability	Ability to scale battery capacity and power by combining multiple units or cells. Important for flexibility in energy storage capacity.	Qualitative assessment
Regulatory compliance	Ensuring the battery meets all relevant safety and environmental regulations and standards, for legal acceptance and marketability.	Compliance with relevant standards and regulations
Size and weight	Physical dimensions and weight of the battery. Critical for mobile devices and EVs.	Measured in kg and m^3
Charge retention	The battery's ability to hold a charge when not in use. Higher retention is preferred for intermittent usage.	Percentage of charge retained over a specified period
Charging temperature range	Specifies the temperature range within which the battery can be safely and effectively charged. Different from operating temperature range.	Specified as a range
Expansion and contraction during operation	Physical changes in the battery, especially lithium-ion types, during charging and discharging. Must be considered in design.	Qualitative assessment
Ease of integration	How easily a battery can be integrated into existing systems, considering form factor, electrical characteristics and communication protocols.	Qualitative assessment
Reliability under varying load conditions	Battery performance under different load conditions, with variable power demands.	Qualitative assessment

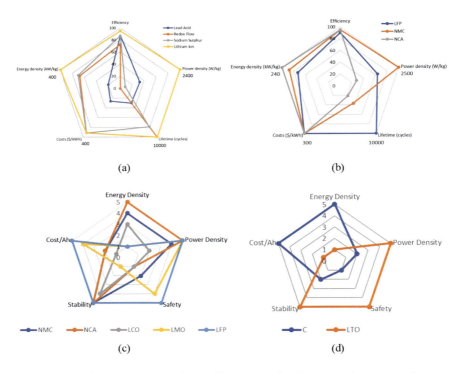

FIGURE 5.8 Comparative analysis of battery technologies and material characteristics: General battery technologies (a and b) across five parameters that are normalized for comparative purposes; and cathode (c) and anode (d) materials evaluated based on five parameters.

applications. Nickel Manganese Cobalt (NMC) displays a balanced profile with good energy and power densities, suitable for electric vehicles, but it incurs higher costs. Nickel Cobalt Aluminium (NCA) stands out with high energy density and efficiency, advantageous for high-performance use, but this comes at a higher cost and potentially shorter lifecycle compared to LFP.

The cathode materials chart in Figure 5.8c indicates the substantial impacts cathode materials have on a battery's energy density, cost and safety. NMC is noted for its high energy density and is desirable in EVs, yet the use of cobalt increases costs. NCA offers similar benefits to NMC but with considerations around aluminium's effect on the battery's performance. Lithium Cobalt Oxide (LCO) provides a high energy density suitable for portable electronics; however, it comes with a high cost and thermal risks. Lithium Manganese Oxide (LMO) is recognized for safety and thermal stability but has a lower energy density. LFP is distinguished by its safety and lifetime but with a lower energy density than cobalt- and nickel-based alternatives.

Finally, the anode materials in Figure 5.8d highlight the role of anodes in determining charge rates, capacity and durability. Graphite anodes are common, offering respectable energy density and stability, but they can be prone to lithium plating at high charging speeds. Lithium Titanate (LTO) anodes are exceptional in cycle life and charging rate, but their lower energy density results in larger and heavier batteries, which may not be optimal for mobile applications.

As explained before, choosing the right battery technology depends on the application's specific parameters, primarily for optimal performance, safety and cost-effectiveness. Therefore, as final remarks, Table 5.16 has also been provided that can be used to identify the suitability of various energy storage technologies for given applications. For example, Li-ion batteries, with their high energy density and fast charging times, are ideal for portable electronics, EVs and renewable energy storage, despite concerns over safety and environmental impact. LiFePO4 batteries on the other hand stand out for their long life and safety, making them suitable for stationary energy storage, low-speed EVs and backup power systems. LTO batteries offer very fast charging and a wide operational temperature range, ideal for high-power applications and grid balancing, although they are preferred in stationary applications due to their lower energy density. However, NaS batteries are best suited for large-scale energy storage, such as grid support, due to their high energy density but require significant control due to their high operating temperature. Flow batteries, as previously stated, are preferred in long-duration storage, like grid energy storage and backup power, due to their scalability and long cycle life. Finally, it can be concluded that, although still under development, solid-state batteries are likely to become the preferred solutions for various mobile applications with their high energy density, enhanced safety and reduced environmental impact.

Note that the conversion from Wh/kg to Wh/m^3 requires the knowledge of the battery's volumetric density in kg/m^3. Since this value can vary depending on the specific cell design and materials, a typical volumetric density is used in the table for each battery type to estimate the energy density in Wh/m^3. The densities in kg/m^3 are Li-ion: 1.6–2.0; LFP: 1.6–2.0; LTO: 2.5–3.0; NaS: 3.0–3.5; Flow Battery: 1.0–1.5; Solid-state Battery: 1.5–2.0 and Supercapacitor: 1.0–1.5.

Note also that while Li-ion batteries currently dominate the market, emerging technologies like Na-ion, Lithium Sulphur and Metal-Air hold promise for

TABLE 5.16 Comparative Overview of Common Battery Technologies

Parameter	Li-Ion	LFP	LTO	NaS	Flow Battery	Solid State Battery	Sup. Cap.
Nominal cell voltage (V)	3.6–3.8	3.2–3.3	2.3–2.5	~2.0	1.0–2.0	Higher than Li-Ion	~2.7
Energy density (Wh/kg)	150–250	100–160	40–80	100–160	20–50	200–350	5–15
Energy density (Wh/m^3)	250–400	160–260	60–120	150–240	100–250	300–500	5–15
Power density (W/kg)	250–500	250–500	3,000–5,000	100–200	20–100	Possibly 250–500	10,000–50,000
Self-discharge rate per day (%)	0.03%–0.1%	0.07%–0.17%	0.03%–0.07%	0.03%–0.07%	<0.003%	Lower than Li-Ion	0.17%–1%
Cycle life	500–2,000	2,000–4,000	10,000–30,000	200–4,000	10,000–20,000	1,000–7,000	500K–1,000K
Op. Temp. range (°C)	−20 to 60	−20 to 60	−40 to 70	250–350	−20 to 40	Wider than Li-Ion	−40 to 65
charge time	Typical 0.5–4 hours	As in Li-Ion	As in Li-Ion	1–8 hours	Few Hours to 1 day	Faster than Li-Ion	sec to min
Safety	Moderate	Good	Very High	Moderate	High	High	Very High
Environmental impact	Moderate	Good	Moderate	Moderate	Good	Good	Moderate

the future, offering the potential for higher energy density, improved safety and reduced environmental impact. As research and development progress, these new battery technologies have the potential for a wide range of applications too from portable electronics and EVs to grid-scale energy storage.

Finally, Table 5.17 provides a comprehensive set of criteria for battery selection and sizing. As stated previously, energy and power requirements

TABLE 5.17 Key Criteria for Battery Selection and Sizing Grouped by Importance

Criteria	Justification
Energy requirements	Essential for determining how long the system can operate on a single charge; must meet or exceed system energy demands.
Power requirements	The battery should handle these without exceeding its maximum discharge rate. Critical for ensuring the battery can handle peak and continuous loads; affects performance and lifespan.
Voltage requirements	Must match system requirements to ensure compatibility and prevent damage, which defines PE converter inputs hence outputs.
Safety	A critical criterion due to risks of fire, chemical leaks, etc.; selected battery must have appropriate safety features for the application's conditions.
Regulatory and compliance issues	Compliance with relevant regulatory standards for the region of operation, and necessary to meet legal standards for transportation, usage and disposal.
DOD	Since it impacts battery life and performance; the battery must be capable of handling expected discharge levels without excessive degradation.
Cycling capability	Important for applications with frequent charge/discharge cycles; affects replacement frequency and costs.
Life span	Aligns with application longevity requirements; affects total cost of ownership and system reliability.
Temperature sensitivity	Affects performance and degradation; the chosen battery must operate effectively within the application's temperature range.
Charge rate	Important for operational flexibility; affects the time required for charging between uses.
Weight and size constraints	Critical for portability and applications with limited space; affects design and usability. Energy and specific energy densities are key.
Maintenance	Affects long-term cost and labour; preference for batteries with minimal maintenance requirements.
Environmental impact	Reflects the sustainability of the battery; lower impact is preferable for reducing the ecological footprint.
Scalability	Ensures the battery system can adapt to changing energy or power needs over time; important for future-proofing the application.

dictate the battery's ability to meet operational demands, while voltage requirements ensure compatibility with the system it powers. Safety is the key to handling potential risks. Regulatory compliance is also necessary, and the battery's DOD must align with its life and performance expectations.

Cycling capability and lifespan are critical for assessing long-term value and operational efficiency. Batteries must also be sensitive to temperature variations to function effectively across different climates. The charge rate affects operational flexibility, and weight and size constraints are critical for portability. Maintenance needs impact long-term serviceability and environmental impact considerations align with sustainability goals. Lastly, scalability ensures the battery system can adapt to evolving energy needs, future-proofing the investment.

5.7 CHARACTERISTIC CURVES OF BATTERIES, DISCHARGING AND CHARGING

Figure 5.9 illustrates the interrelationships between various parameters of a battery cell, which indicates that changes in one can significantly affect the others. As can be seen, discharge current and temperature are two primary external factors that influence the internal electrochemical behaviour of a battery cell, which in turn determines its performance characteristics.

Since understanding the intricate behaviours of lithium-based batteries is important in real-world applications, the characteristic curves play a role. For example, SOC defines the operating voltage range and in real-time SOC estimation, providing a reference for energy management and operational efficiency. The capacity versus discharge rate curve, or the Ragone plot, provides further details about the impact of load conditions on battery capacity, which is critical for applications where discharge rates vary significantly. Moreover, charge and discharge curves provide a window into the battery's time response to charging and load patterns, helping design issues for charging infrastructure and the load.

Furthermore, the cycle life curve can indicate the expected lifespan of a battery under cyclical use, which is valuable for sustainability and cost-effectiveness. Temperature dependence and internal resistance versus SOC curves offer a closer look at how environmental conditions and charge levels influence battery efficiency and power capability, which also impacts safety. Similarly, the efficiency versus SOC curve indicates energy losses in an operational cycle, which helps to predict energy output and to reduce loss. Finally, Ragone plots highlight the balance between energy

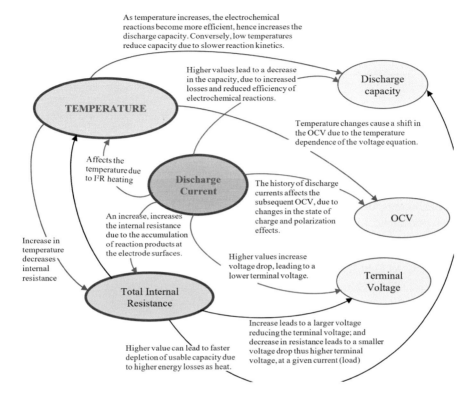

FIGURE 5.9 The interrelationships between various parameters on the characteristics of a battery cell.

storage and delivery, which is needed for the selection of energy storage technologies for power and energy-intensive applications.

In summary, the characteristic curves provide a snapshot of battery performance, to support the development of robust and reliable BMS that can optimize for temperature, discharge rates and load management and to monitor and predict battery performance and end-of-life. In the following figures and paragraphs, a selected set of battery characteristics will be included and explained while linking the real-life operation via PE control.

Figure 5.10 illustrates the relationship between OCV versus SOC and OCV versus DOD for different battery chemistries at 25°C. The OCV curve, a fundamental characteristic of a battery, indicates the voltage output when no load is applied and varies directly with the battery's SOC. As the SOC increases, the OCV generally rises until it reaches a plateau indicative of full charge. It is important to note that the DOD is inversely related to SOC, which can be defined as DOD = 1 − SOC when expressed

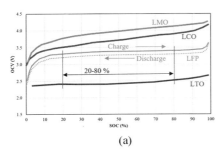 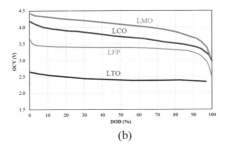

FIGURE 5.10 Typical OCV versus SOC (a) and OCV versus DOD (b) characteristics for battery cells with different chemistries at 25°C; From top down: lithium manganese oxide LMO (LiMn$_2$O$_4$), lithium cobalt oxide, LCO (LiCoO$_2$), lithium iron phosphate, LFP (LiFePO$_4$) and lithium titanate oxide, LTO battery.

as a fraction. In addition, DOD can also be given in Ah, given that DOD = $Q(1 - SOC)$. Therefore, the OCV curves are also given as a function of DOD, as shown in Figure 5.10b.

Note that the 20%–80% SOC range for lithium-ion batteries is established to enhance their lifespan and efficiency, based on empirical and theoretical evidence. This optimal range helps prevent degradation by minimizing risks like lithium plating, electrode material stress, solid electrolyte interface growth and thermal stress, which can occur when batteries are charged or discharged beyond these limits. As it can be observed, the OCV-SOC relationship is notably flat in this middle range—allowing for a predictable performance but varies rapidly beyond these bounds, which increases the risk of overcharging. Overcharging a lithium-ion battery can lead to excessive voltage that the battery cannot safely accommodate, potentially causing overheating, capacity loss or even catastrophic failure due to thermal runaway. Similarly, the steep fall below the 20% level represents another critical point in the discharge cycle that can result in a significant drop in the battery's voltage hence the risk of deep discharge, which can cause irreversible damage to the battery's electrodes, leading to a permanent loss in capacity and, in some cases, failure to recharge. In addition, hysteresis in the charging and discharging characteristics of an LFP battery in the figure refers to a lag in the voltage response when the current direction changes.

Table 5.18 outlines both the discharging and charging methods utilized in battery systems. There are three fundamental methods for discharging batteries until they reach a low SOC, each dictated by the specific energy needs and operational requirements of the loads. It should be noted that

TABLE 5.18 Commonly Used Discharging and Charging Methods for Batteries

Name	Description
Discharging Methods	
Constant power discharge	• Power output remains constant; current adjusts as voltage decreases, $p(t) = v(t) \, i(t)$. • Needed to maintain power output, if it is critical for performance. • It is typically used in grid energy storage systems, frequency regulation, to deliver or absorb power at a constant rate in renewable energy integration, variability and intermittency, demand response, EVs, data centres and in UPS. • PE converters allow for accurate control of the discharge rate, enabling batteries to maintain a constant power output even as their voltage changes with the state of charge (see Figure 5.9).
Constant current discharge	• Occurs at a constant current regardless of voltage changes, and preferred for consistent performance over time. • Used in applications requiring stable operation over time as required in most electronic devices, LED lights, medical equipment such as infusion pumps and in EVs where it facilitates predictable usage patterns. • PE converters regulate current, ensuring constant current even as the battery's internal resistance changes with temperature and ageing (see Figure 5.9).
Constant resistance discharge	• Current decreases as battery voltage decreases, $v(t) = R \, i(t)$ • Simple, passive discharge. • As required in heating elements.
Charging Methods	
Constant current charging	• Initial stage of the charging process for lithium-ion batteries within safe limits. • Standard for most rechargeable lithium-ion battery applications.
Constant voltage charging	• Final stage of the charging process after reaching a voltage threshold. • Avoids overcharging and prolongs battery life. • Universally applied across lithium-ion battery applications.
	• Although distinct names may be given in some cases, the above two charging methods are often employed in tandem as part of a charging strategy in practice. For example, "multi-stage or smart charging" involves pre-charging deeply discharged batteries and maintaining a full charge to compensate for self-discharge, alongside cell balancing for uniform charging, accommodated in BMS. Similarly, "trickle or maintenance charging" uses a low current post full charge to maintain optimal SOC during inactivity as well as in standby or backup systems like emergency lighting and UPS systems. Finally, "pulse charging" applies current pulses with rest periods to enhance battery longevity and mitigate heat build-up if thermal management is vital.

both processes require the use of suitable PE converters to modulate the electrical quantities (voltage and current). This modulation requires feedback and suitable control algorithms, as explained in the previous chapter.

The trends in battery storage applications in the content of a power system indicate a preference for discharge methods that can adapt to changing conditions and demands and the need for efficiency, which PE facilitate. Therefore, constant power and constant current discharge profiles are particularly significant, which offer the flexibility and control required for dynamic renewable energy-dominant systems.

Constant power and constant current discharge profiles can be analysed by three different approaches as summarized and justified in Figure 5.11. This is due to the fact that, under controlled conditions enabled by PE, these approaches provide insights into battery behaviour and are useful for designing battery systems that meet specific application needs. After defining the approach, the most practical discharge profiles can also be

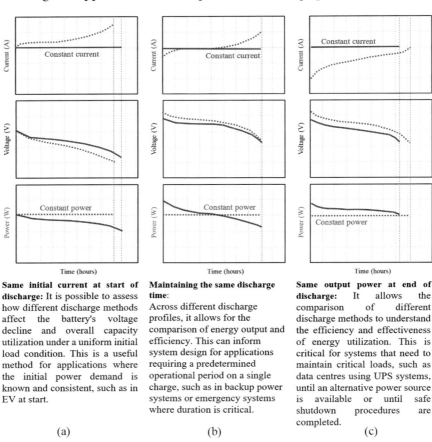

FIGURE 5.11 Three different approaches that best suits to the practical discharge profiles in applications utilising PE converters. Note that $p(t) = v(t)\, i(t)$.

identified by considering the requirements of modern power systems and the capabilities provided by PE converters.

In a real battery, for every SOC, there is a range of possible stable OCV values, which is known as hysteresis voltage. This is due to the fact that diffusion voltages change with time but hysteresis voltages change when SOC changes as illustrated in Figure 5.10a. Therefore, more accurate cell models can contain more than a single parallel RC circuit and a hysteresis model in series with the resistance R_0.

Figure 5.12a represents a voltage-time discharge curve of a battery after a pulse test, which captures the dynamic voltage variation, and offers a detailed assessment of the unique electrochemical properties inherent to each battery chemistry. The process involves subjecting the battery to alternating periods of discharge and rest to identify its performance under dynamic load conditions. During these tests, voltage, current and temperature data is collected and analysed. Such analysis provides insights into capacity retention, voltage sag and variations in internal resistance, along with other equivalent circuit parameters. In addition, it informs on voltage response and energy efficiency, providing a window into potential real-world applications. Note that, pulse (switching) parameters in the test can also be adjusted to simulate specific operational patterns. Such switching characteristics occur repetitively forming a decaying voltage variation while controlling current and power.

The figure simply shows the solution of an electrical equivalent circuit model after a pulse test. Immediately after discharge begins, there is a rapid drop in voltage (ΔV_0), which is attributed to the internal ohmic resistance (R_0) of the battery. This initial voltage drop represents the immediate response of the internal resistance to the discharge current (Δi). Following the initial drop, there is a more gradual voltage decline over time due to the $R_1 C_1$ circuit, which models the polarization effects or the slower electrochemical processes in the battery. The time constant for this part of the circuit is approximately $5R_1 C_1$, which indicates how long it takes for the voltage to stabilize after the initial drop. Finally, the curve flattens out, indicating that the voltage is stabilizing as the battery reaches a new steady state. The final voltage change (ΔV_∞) reflects the total voltage drop from the OCV due to both the internal resistance R_0 and the polarization resistance R_1 as the discharge progresses.

Figure 5.12b presents three distinct voltage-time plots that show the discharge characteristics of three different lithium-based batteries under the same switched pulse width and load conditions.

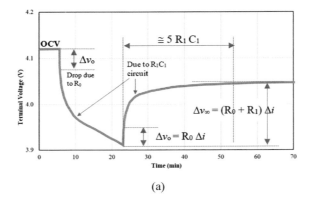

(a)

Battery 1 (Lithium Cobalt Oxide, LiCoO2)
- A rapid response to the load. The subsequent recovery and stabilization suggest the battery's capacity to regain and maintain a reduced voltage under continued load.
- High energy density and potential thermal build-up and capacity fade with high load.

Battery 2 (Lithium Iron Phosphate, LiFePO4)
- A slight dip and recovery in voltage, maintaining a relatively stable profile.
- Lower energy density compared to Battery 1, higher power density, good thermal stability and a more resilient battery that better maintains its voltage under load (at high discharge rates)

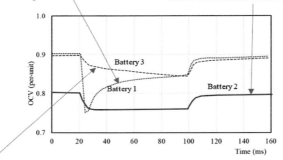

Battery 3 (Lithium Nickel Manganese Cobalt Oxide, NMC)
- The smallest voltage drops and recovery, which is less severe than Battery 1 but slightly more than Battery 2.
- Nearly flat post-recovery line, indicates a small and stable voltage throughout the discharge event.
- Balanced energy and power density and better temperature management compared to Battery 1.

(b)

FIGURE 5.12 Characteristics of discharge voltage-time plots: (a) polarization in a battery is visible during a discharge and rest pulse. Δv_o voltage drop across R_0 with a current pulse of Δi. Note that Δv_∞ is the steady-state voltage change (hence R_1 can be identified), and using the time $5t = 5R_1C_1$ (since voltage reaches to steady-state) C_1 can be identified; (b) discharge voltage-time plots of three different battery cells to a 100ms pulse test. 1 per-unit = 4V.

Note that such a test forms the fundamentals of discharge control as used to identify the optimal PWM control signal for PE converters. As discussed in Chapter 4, the PWM signal is required for the regulation of battery current and voltage throughout the charging and discharging cycles, which, in turn, influences the design of battery packs as well as the PE system. The prime aim is to achieve optimal energy and power densities required in particular applications. For example, EVs require high torque and, therefore, high current at start and during acceleration. Similarly, the need for rapid frequency control in grid-supporting inverters also requires high current (discharge rate).

Figure 5.13 displays the typical behaviour of a lithium-ion battery over numerous charge-discharge cycles, plotting both capacity and impedance against the cycle number. The dashed line indicating capacity shows a gradual decline as the number of cycles increases, which is a common phenomenon due to battery ageing. The dotted line represents impedance, which conversely increases with the number of cycles. This increase in impedance is typically due to factors such as the build-up of the solid electrolyte interphase (SEI) layer, electrode degradation and other irreversible chemical changes within the battery. The overall trend shows that as a battery ages through repeated cycling, its ability to store energy diminishes while the resistance to energy flow increases, both of which are undesirable but inevitable characteristics of battery wear.

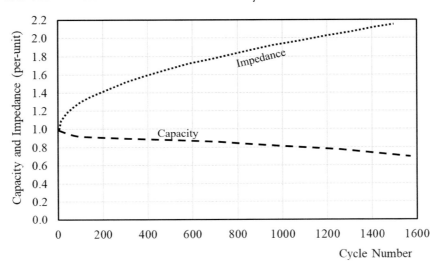

FIGURE 5.13 A typical capacity and impedance variations in lithium-ion batteries with cycle: $Z_{initial}$= 48mW, 100% initial capacity.

Figure 5.14a is given to show a series of discharge curves for a lithium-ion battery at a constant ambient temperature but at various C-rates, which are rates of discharge relative to the battery's capacity. Each curve shows the relationship between the cell voltage and the discharge capacity in mAh. At lower C-rates, the battery maintains a higher voltage for a longer period and provides a greater total discharge capacity. As the C-rate increases, the voltage drops more steeply, and the usable capacity decreases. This is because higher discharge rates lead to greater losses inside the battery due to internal resistance, heat generation and other factors that reduce efficiency. The curves also demonstrate that at very high discharge rates, such as 15C and 20C, the voltage drops rapidly and the total energy output of the battery is significantly reduced. This information is crucial for designing battery systems for various applications, ensuring that the batteries can deliver the required power without exceeding their safe operational limits.

Figure 5.14b illustrates the relationship between discharge rates and battery performance in cold climates, with slower rates yielding more efficient use of the battery's capacity. At a C/10 rate, the discharge is slow, taking 10 hours, and the battery maintains a higher voltage for a longer period, utilizing its maximum capacity. Increasing the discharge rate to C/2 and then to the nominal rate of C results in a faster voltage drop and reduced capacity, showing decreased efficiency. At 2C, the battery discharges in 30 minutes, with a steeper voltage fall-off and further capacity reduction. The fastest rate, 5C, completes discharge in just 12 minutes, resulting in a very steep voltage drop and significantly lower capacity, indicating a substantial loss in efficiency due to the rapid discharge. This reduces runtimes in cold conditions and is why special considerations, like preheating or other methods, might be necessary for batteries operating in such environments.

Figure 5.14c shows the discharge curves at various ambient temperatures, all measured with a constant discharge current of 360 mA. The discharge capacity in mAh is plotted against the cell voltage. At the highest temperature (45°C), the battery maintains the highest voltage for a longer period during discharge, which suggests lower internal resistance at elevated temperatures. As the temperature decreases to 23°C, 0°C, −10°C, and then to −20°C, the curves show a progressively steeper voltage drop-off for the same discharge capacity. This indicates that the battery is less efficient at lower temperatures, likely due to increased internal resistance and reduced electrochemical activity. Moreover, the battery's usable capacity decreases with temperature; it delivers more capacity at 45°C and 23°C compared to the colder temperatures of 0°C, −10°C and −20°C.

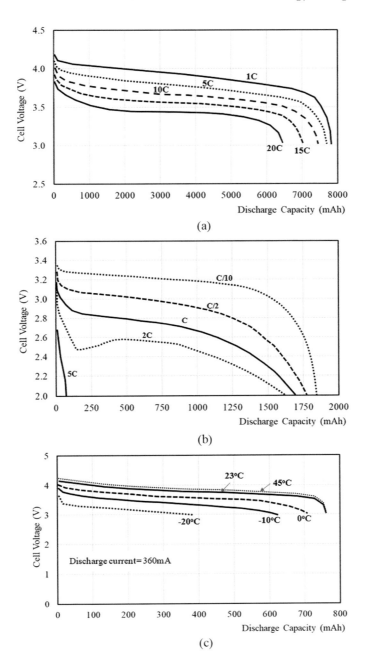

FIGURE 5.14 Cell voltage versus discharge capacity characteristics: (a) at various constant current discharge rates; (b) at various discharge rates of an LFP battery at a cold climate (−20°C); (c) at various ambient temperatures of a polymer lithium-ion cell.

At the lowest temperature (−20°C), the battery shows a significantly reduced discharge capacity and a steeper voltage decline as mentioned previously.

Figure 5.15 shows the discharge capacity retention of a lithium-ion battery over time when stored at a high temperature of 45°C. Two curves are shown, representing batteries stored at 100% SOC and 0% SOC. The solid line at the top indicates that when the battery is fully charged, it maintains a higher percentage of its capacity over the span of a year. However, there is a slight but steady decline, suggesting some loss of capacity due to self-discharge. The dotted line at the top suggests that when the battery is stored at 0% SOC, it retains close to its full capacity for the duration of the storage period.

In addition, Figure 5.16a illustrates how the cell voltage varies over time during various discharge rates. Note that the ideal discharge curve of a battery is also illustrated, with no internal resistance. As expected, at higher current discharge voltage drops more quickly and reaches the cut-off voltage sooner. The solid lines indicate that when the battery is discharging at a higher current and reaches the cut-off voltage, the current is reduced (which could be done by a DC/DC converter in practice). After this reduction in current, the voltage recovers slightly due to reduced internal losses and the discharge curve then follows a path similar to that of a lower current discharge, allowing a little more energy to be extracted from the battery. This behaviour is typical of lithium-ion batteries. The practice

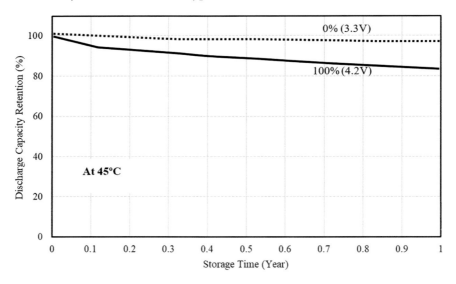

FIGURE 5.15 The impacts of storing a lithium-ion battery on discharge capacity when fully charged (100% at 4.2V) and when fully discharged (0% at 3.3V), (shelf-life).

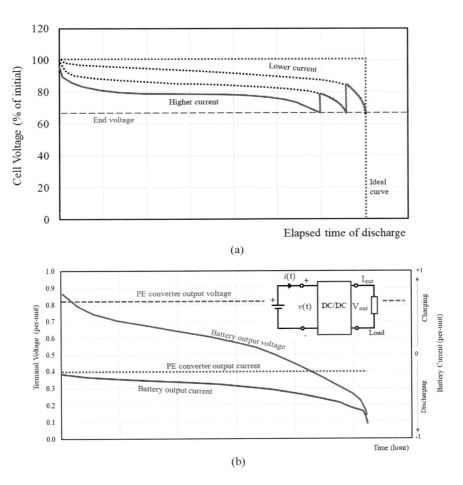

FIGURE 5.16 The control of discharge current: a) Variation of cell voltage as discharging at different rates down to a fixed cut-off voltage; b) Voltage and current profiles of battery and DC/DC converter. Note that $p_{input(t)} = p_{output(t)} + p_{losses(t)}$.

of reducing the discharge current once the cut-off voltage is reached helps to maximize the energy extracted from the battery by taking advantage of the higher efficiency at lower currents.

Figure 5.16b is provided to demonstrate the principles of how the PE converter regulates the output voltage and current to match the load requirements during the battery discharge process. This concept setup is typical in applications where a stable output voltage is necessary despite the variability of the battery's voltage during discharge.

Furthermore, Ragone plots as given in Figure 5.17, are used in evaluating and comparing different energy storage devices by illustrating the

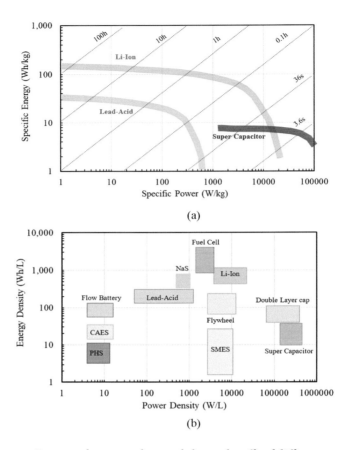

FIGURE 5.17 Ragone plots to understand the trade-offs of different energy storage technologies for various applications, from high-power, short-duration tasks to low-power, long-duration storage.

trade-off between energy density (specific energy, in Wh/kg) and power density (specific power, in W/kg). These plots effectively demonstrate how the discharge load affects the amount of energy delivered by an energy storage system. It maps out the energy-power relationship, highlighting the inherent compromise between energy and power within a given energy storage system. This becomes particularly useful in quantifying performances when the system operates off-design, such as in a constant-power regime, which is common in electric grid applications.

In Figure 5.17a, the Ragone plot illustrates energy versus power envelops for two rechargeable batteries and a supercapacitor on logarithmic scales. As can be seen, Li-ion batteries have a high specific energy but lower specific power compared to capacitors. The band curve represents the range of specific energies and powers for each technology, with the discharge time, from 100 hours to 3.6 seconds, indicated along the bands. The figure highlights that capacitors can deliver power very quickly (high specific power) but do not store as much energy (low specific energy), while Li-ion batteries are more balanced in their energy and power capabilities.

Figure 5.17b presents a wider range of energy storage technologies. Each technology is represented by an arbitrary shape with its area covering the range of specific energy and power densities that the technology can achieve, both on logarithmic scales. For example, Flywheels, superconducting magnetic energy storage (SMES) and capacitors generally offer high specific powers, making them suitable for applications requiring quick bursts of energy. In contrast, pumped hydro storage (PHS) and compressed air energy storage (CAES) systems offer lower specific powers but high specific energies, which make them more suitable for long-duration energy storage.

Finally, Figure 5.18 shows the OCV and the energy efficiency of a battery in relation to its SOC at two different times: at day 0 (initial) and after 1 year. As shown, the energy efficiency decreases over time due to battery ageing and degradation. The OCV during charging and discharging is also given, which do not overlap due to the hysteresis as mentioned previously.

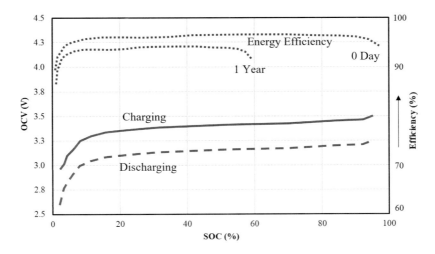

FIGURE 5.18 A typical OCV and efficiency versus SOC characteristics of a new LFP battery.

To obtain the efficiency characteristics, a series of measurements need to be conducted. Initially, the battery is charged and discharged at a controlled rate while monitoring the voltage and current to establish the OCV–SOC relationship. This process is done for a fully rested battery to ensure that the OCV is not influenced by recent charging or discharging. For efficiency, the energy put into the battery during charging is compared to the energy retrieved during discharging, typically over several cycles to account for variability, and repeated after a certain period, to observe changes due to ageing. The measurements are often performed under controlled temperature and load conditions to ensure consistency. The following process summarizes the principles of this process using the figure and the equations given below.

Once the OCV is defined as a function of SOC, the energy can be calculated by

$$E_{net} = \int_{SOC_{initial}}^{SOC_{final}} I \cdot OCV \, dSOC \tag{5.2}$$

Note that charge and discharge energy can be calculated using OCV_{charge} and $OCV_{disharge}$ functions. Since charge and discharge energy efficiencies can be defined with reference to the net energy defined above, the final round-trip (full-cycle: charge and discharge) energy efficiency can be calculated as

$$E_{charge} = \int_{SOC_{initial}}^{SOC_{final}} I \cdot OCV_{charge} \, dSOC \tag{5.3}$$

$$Charge = \frac{E_{net}}{E_{charge}} \tag{5.4}$$

$$E_{discharge} = \int_{SOC_{initial}}^{SOC_{final}} I \cdot OCV_{discharge} \, dSOC \tag{5.5}$$

$$Discharge = \frac{E_{discharge}}{E_{net}} \tag{5.6}$$

$$= charge \cdot discharge = \frac{E_{discharge}}{E_{charge}} \tag{5.7}$$

5.8 BATTERY BALANCING AND BATTERY MANAGEMENT

Before discussing unbalancing in lithium batteries, it is worth revisiting the terminology of "capacity" and "charge" again as they are interrelated concepts defining the amount of energy a battery can store and deliver.

Capacity is measured in Ah and represents the maximum electrical energy a battery can store. It indicates the total duration for which the battery can deliver a specific current until discharged. This is analogous to the water-holding capacity of a container. On the contrary, charge refers to the amount of electrical energy stored in a battery at a specific point in time. Typically measured in Coulombs (C) but expressed in Ah for practical applications. This is analogous to the amount of water in the container. In conclusion, charge is dynamic and changes with battery use, while capacity is fixed.

Therefore, "battery balancing helps maintain similar charge levels across individual cells within a pack", maximizing overall capacity and lifetime.

Unlike other types, lithium-ion batteries do not naturally "self-balance" due to their high efficiency. This imbalance, if left unaddressed, can significantly reduce the battery's overall capacity, power and usable energy. Understanding cell capacity differences, SOC variations and self-discharge rates is critical for designing effective cell balancing systems that maintain optimal battery performance and safety in its lifespan.

Figure 5.19a shows four scenarios assuming a battery pack with three different-sized cells. The cell with the lowest charge absorption limits the pack's ability to accept charge and be discharged completely. These unbalanced scenarios highlight the importance of battery balancing. Balancing techniques ensure that all cells within a pack are charged and discharged to similar levels, preventing individual cells from becoming overcharged or over-discharged. This extends the lifespan and improves the overall performance and safety of the battery pack. However, note that while Figure 5.19a uses different-sized cells to illustrate capacity variations, real-world battery packs typically use identical cells connected in series or parallel to achieve the desired voltage and capacity.

For identical cells, the primary balancing issues are charge-holding limits and discharging limits. As highlighted above, even slight variations in individual cell capacities can limit the overall usable capacity of the pack. The pack can only be discharged until the cell with the lowest capacity reaches its limit, even if other cells still hold charge. Similarly, differences in self-discharge rates can lead to unequal depletion among cells, limiting

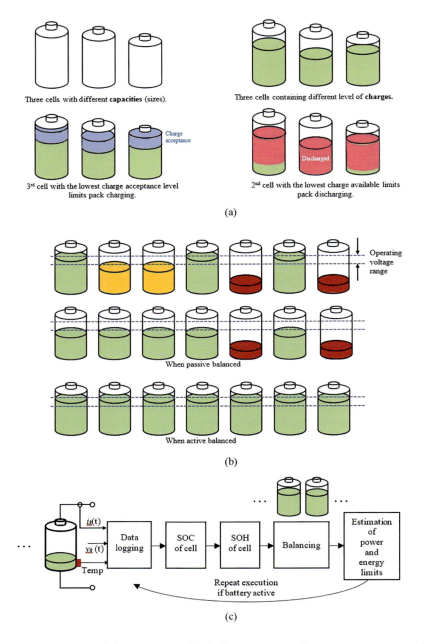

FIGURE 5.19 Visual descriptions of unbalance states in batteries: Capacity and charge variations in three different size battery cells (a); pre- and post-balancing of identical battery cells: without cell balancing (top-row), with passive cell balancing (middle-row) and with active cell-balancing (bottom-row) in (b); and the flowchart of a BMS operations (c).

the usable dischargeable energy if any cell reaches its minimum safe voltage before others.

Therefore, effective cell balancing techniques are critical to mitigate these issues, ensuring all cells within the pack are charged and discharged to similar levels, maximizing overall capacity.

Rechargeable batteries have several previously defined voltage levels crucial for BMSs to regulate charging and discharging processes, maximizing battery life, performance and safety. Manufacturers specify these voltage thresholds based on the battery's chemistry and design to guide users and the BMS in proper battery handling. Li-ion BMSs are designed to closely monitor and control these voltages during operation to ensure the battery operates within safe limits.

Figure 5.19b visualizes the concept of battery cell balancing. It compares a non-balanced identical-capacity array of battery cells with varying charge levels (top row) to an actively balanced array (bottom row) where all cells are uniformly charged within the optimal operating voltage range.

Figure 5.19c **presents a schematic representation of the main control flow for a BMS**. Note that initially, the loop measures current, voltage and temperature, which are critical parameters for monitoring the health and performance of a battery pack. Subsequently, the BMS estimates the SOC of individual cells, defining the remaining charge. The system then assesses the SOH of the cells, which defines the battery's condition, efficiency and projected lifespan. To ensure uniform SOC and voltage levels across the battery pack, the BMS balances the cells. The control flow concludes with the calculation of energy and power limits to define maximum safe charge and discharge rates, preventing damage to the battery cells. This control block operates at each measurement interval when the battery pack is active.

Note that voltage, a more readily accessible balancing metric, often serves as an indirect SOC indicator. However, voltage alone may not accurately reflect SOC due to factors like temperature effects, cell impedance and the nonlinear voltage-SOC relationship throughout the charge and discharge cycle. Figure 5.19c illustrates that balancing is essential to equalize the charge among multiple cells within a pack, compensating for manufacturing variances in capacity and resistance, ageing, temperature or usage history. These variances could lead to overcharging or undercharging of individual cells.

In practice, balancing usually aligns cell voltages at the end of the charge cycle, which equates to SOC equalization. While voltage uniformity is an immediate balancing goal, active balancing methods that redistribute energy from cells with higher SOC to those with lower SOC directly align the charge in each cell, targeting SOC equilibrium.

SOH estimation is also vital for predictive maintenance, timely failure identification and assessing risks in ageing batteries. This information informs lifecycle management, warranty considerations and ascertains residual battery value for repurposing or recycling which will be discussed later. In addition, accurate SOH estimation in battery packs informs the development of balancing strategies by identifying cells with reduced capacity or increased internal resistance, which impact reliability and efficiency.

The type of battery cell connection significantly impacts the choice of balancing techniques. For example, in a series connection, balancing requirements are very high due to the risk of overcharging or over-discharging individual cells even with minor imbalances, leading to premature failure. However, series-connected cells require a complex and expensive BMS. In addition, balancing requirements are low in parallel connections since cell voltages tend to naturally equalize. However, capacity differences can still cause imbalances. Furthermore, in a series-parallel connection, balancing requirements are high, requiring the most complex BMS as both voltage and capacity variations need to be addressed. This configuration is used in high-voltage and high-capacity applications such as EVs and grid-scale storage.

Figure 5.20a describes various SOH estimation methods for batteries. As illustrated, they can be grouped as experimental and model-based methods. In the direct measurement method, internal resistance or capacity through full charge-discharge cycles are calculated via direct physical measurements as they are directly related to SOH. In the indirect measurement method, SOH is estimated indirectly using voltage response or temperature behaviour during specific test conditions.

Furthermore, in model-based estimation methods, adaptive algorithms use models that adapt over time to the battery's usage and ageing data. Examples include Kalman filters and particle filters, which adjust their estimates as new data becomes available. In addition, the data-driven approach relies on historical data and machine learning algorithms to estimate SOH. Algorithms can include neural networks, support vector machines or regression analysis, which identify patterns from past battery performance data to predict current SOH.

As given in Figure 5.20b, at the cell level, passive and active balancing methods are utilized to manage voltage differentials among cells. Passive balancing dissipates excess energy through resistors or bypass diodes, while active balancing employs electronic circuits for energy transfer.

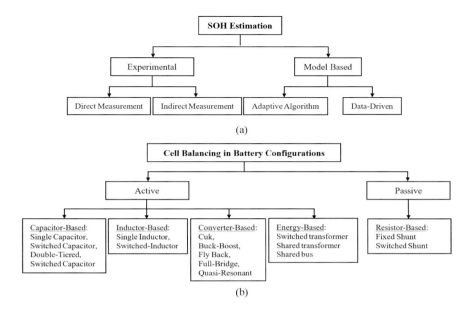

FIGURE 5.20 The common SOH estimation (a) and balancing (b) methods used in battery configurations.

Moving to pack balancing, voltage monitoring and SOC balancing techniques are critical. Voltage monitoring ensures individual cell voltages are within safe limits, while SOC balancing equalizes charge states across cells. At the rack level, master-slave and communication-based balancing strategies come into play. Master-slave balancing involves a central controller adjusting charging and discharging rates, while communication-based balancing uses protocols like CAN bus for coordinated energy distribution. Although it is not the aim to cover each individual balancing method, Figure 5.18b is given to provide an insight into how SOH detection contributes to efficient and reliable battery operations across different scales of energy storage systems using PE converters and electrical circuits.

There are essentially four primary active balancing methods: transferring energy between adjacent cells (utilized in small batteries), moving energy from the cells with the highest SOC to the rest of the battery (which is the simplest method and offers the highest efficiency), channelling energy from the battery to the least-charged cells (preferred when using a charger with an equal number of outputs as there are cells) and facilitating bidirectional energy flow from either cell to battery or battery to cell (an effective solution for redistribution).

Figure 5.21 shows the principal circuit diagram of various cell balancing techniques. **For example, the switched shunt resistor method in Figure 5.21a involves connecting a resistor across the terminals of each battery cell. When activated the associated solid-state switch is activated** based on the voltage level, the excess charge is dissipated as heat, effectively lowering the charge of that cell to match others. It is simple and cost-effective but not energy-efficient as the excess energy is wasted as heat.

The switched inductor method shown in Figure 5.21b uses an inductor to transfer energy from higher-charged cells to lower-charged ones or

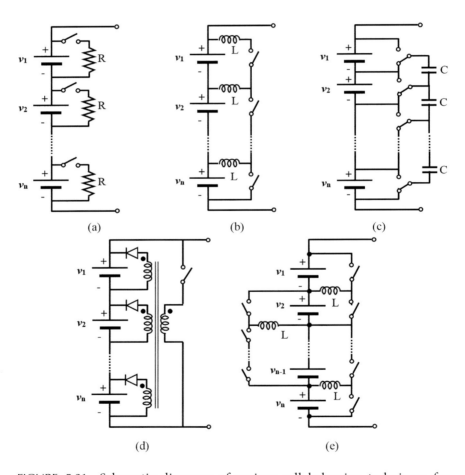

FIGURE 5.21 Schematic diagrams of various cell balancing techniques for Lithium-ion batteries: (a) switched shunt resistor (dissipative, passive), (b) switched inductor, (c) switched capacitor, (d) transformer-based, (e) buck-boost converter.

to a common load. This is more energy-efficient than resistive methods but is typically larger and more complex. Similarly, the switched capacitor method in Figure 5.21c transfers charge from higher to lower-charged cells using a capacitor via the switches. This method can also be less efficient than the resistive method, with moderate complexity.

In addition, the transformer-based method shown in Figure 5.21d utilizes a transformer to couple cells, enabling energy transfer between them. This method can be highly efficient and allows for isolated balancing, which is desirable in high-voltage battery configurations. The transformer comprises a primary winding connected to the entire battery pack and multiple secondary windings, each corresponding to an individual cell. Each secondary winding is connected in series with a diode to its corresponding battery cell. **Finally, the buck-boost converter in** Figure 5.21e can either step-down or step-up the voltage to transfer energy. It is a dynamic method that can be very efficient and can handle a wide range of voltage differences.

However, the need for balancing in flow batteries is not the same as in lithium batteries for several reasons. Firstly, since the electrolyte is typically the same throughout the system, the issue of cell-to-cell variance as in lithium batteries is less prevalent. Therefore, flow batteries can be balanced by ensuring the electrolyte's SOC is uniform. This is managed through the flow and mixing of the electrolyte rather than balancing individual cells. In addition, in flow batteries, the energy capacity (determined by the volume of electrolyte) is separate from the power capacity (determined by the size of the cell stack). This decoupling allows for more straightforward management of the system without the need for cell-level balancing.

The radar chart in Figure 5.22 is also given to summarize some of the cell balancing technologies shown in Figure 5.21b with the parameters that can be utilized in practical applications. Although the higher number in the chart can be considered vital as a single parameter, it is important to note the trade-offs. For example, a system with excellent manufacturability and low cost might have lower efficiency, which could be acceptable for stationary battery applications but less so for EVs. Similarly, a high degree of control might come at the expense of a larger size or higher cost.

In addition, it is important to prioritize these parameters for specific applications. For example, in utility-scale BESS applications, control and reliability may be the highest priorities due to safety and performance requirements, whereas in EVs, efficiency, cost and size may be the most critical factors, as available in Cuk or buck-boost topology.

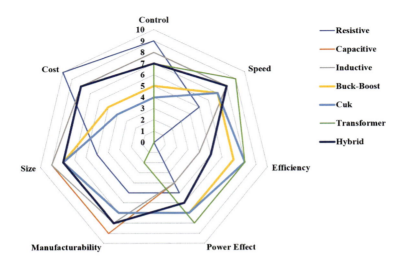

FIGURE 5.22 Comparative analysis of different power electronic converter topologies based on seven key criteria that are required in applications.

Although the most commonly used balancing method in practice is often the resistive or passive balancing method, due to control and cost parameters, in general, the choice of balancing method ultimately depends on the specific requirements of the application, including factors such as cost, complexity, efficiency and the operational demands of the battery system.

Battery configurations that utilize multiple cells, such as packs/modules, racks/units/banks, containers and multiple containers, require comprehensive balancing strategies to maintain efficiency and longevity. Imbalances within cells or modules can arise due to uneven ageing or operational stresses, necessitating cell-level balancing using passive or active methods. At the pack or module level, external chargers or integrated circuits can mitigate voltage differences. For larger units like racks, banks or containers, a BMS plays a crucial role in monitoring and rectifying imbalances, often requiring advanced communication features to handle complex arrays and thermal management systems to address temperature-induced variances. When multiple containers are involved, a centralized BMS ensures uniformity across the entire system, supported by a robust communication infrastructure that allows for synchronized control and balancing actions across different battery configurations.

Finally, it should be reported that while using cells from the same batch can be beneficial for minimizing initial imbalances in a lithium battery configuration, it is not always a strict requirement. Modern balancing systems effectively manage moderate imbalances even between cells from different batches.

5.9 SENSORS USED IN BATTERY SYSTEMS

Sensors play a crucial role in battery storage systems, ranging from individual cells to entire containerized units. Their primary functions include ensuring safety, optimizing performance, maintaining reliability, complying with regulations and enhancing cost-efficiency. These sensors monitor potential threats such as overheating, fires and electrical faults, facilitating early detection and mitigation of risks. They also track environmental conditions like temperature and humidity to maintain optimal operation and prevent degradation. Moreover, sensors provide real-time data for essential functions such as balancing cell voltages, assessing battery health and determining the charge state, hence ensuring operation within safe parameters.

The selection and integration of sensors are influenced by various factors including the specific application, system size, battery technology and environmental conditions. For instance, harsh environments necessitate robust sensors for accurate monitoring, whereas applications requiring precise control and rapid response protection benefit from high-bandwidth current measurement. Integrating sensors with the BMS helps with early fault detection, preventive measures and predicting maintenance needs based on sensor data analysis.

Figure 5.23 lists sensors grouped by their application areas, and an overview of the sensor types used in BSS, along with their characteristic features, is summarized in Table 5.19.

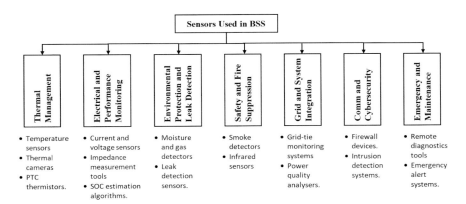

FIGURE 5.23 The sensors grouped by their application areas used in battery storage systems.

TABLE 5.19 The Overview of Sensor Types Used in BSS

Sensor Type	Characteristic Features
Voltage sensors	• Measure cell/module/unit voltage. • Essential for cell balancing, detecting overvoltage or undervoltage conditions. • High bandwidth and electrical isolation required. • Precision amplification, filtering, isolated power supplies are necessary. • Usually measured by isolation amplifiers for electrical isolation.
Current sensors	• Measure charge and discharge currents. • Required for managing charging rates and overcurrent situations for protection. • Commonly closed-loop Hall-Effect types with high bandwidth and amplification. • Shunt resistor types offer much higher bandwidth measurement but without isolation.
Temperature sensors	• Monitor the temperature of cells, modules, units or battery environments, and positioned on the cell surface, embedded within modules or in the battery enclosure. • Used for thermal management, overheating and for protection. • Thermocouples, RTDs and thermistors used. • Linearization, cold-junction compensation for thermocouples.
Humidity sensors	• Monitor moisture levels inside the battery enclosure or in the BMS housing to avoid condensation and moisture-related degradation in harsh environmental conditions. • Capacitive and resistive types are used.
Gas sensors	• Measure concentrations of hazardous gases (hydrogen, electrolyte vapours, CO_2). • Used for safety by early warnings of thermal runaway or electrolyte leakage. • Electrochemical, semiconductor or IR sensor types. • Require stable power supply, temperature compensation required. • Positioned within the battery enclosure or in venting paths.
Smoke detectors	• Identify signs of fire or smouldering within battery storage systems. • Critical for early fire detection and mitigation. • Photoelectric and ionization types are used. • Modulation/demodulation circuits and ion current measurement necessary. • Installed in the battery room or enclosure, near the battery pack.
Pressure sensors	• Monitor pressure within cells or battery enclosures. • Piezoelectric, capacitive and strain gauge types. • Bridge circuit, temperature compensation and amplification required.

(Continued)

TABLE 5.19 (*Continued*)   The Overview of Sensor Types Used in BSS

Sensor Type	Characteristic Features
Insulation monitoring sensors	• Detect insulation breakdowns or leaks to prevent short circuits or ground faults. • Essential for maintaining electrical safety and integrity, especially in large-scale or high-voltage systems. • Based on dielectric or resistance measurement. • High-voltage isolation, leakage current detection needed. • Span across the insulation of high-voltage components. • Measure insulation resistance and breakdown.
Lightning sensors	• Needed in outdoor or exposed installations. • Known as "lightning detection Sensors" or "storm detectors." • Can be simple rod-like structures to complex electronic devices. • Utilize electromagnetic pulse detection or RF detection for sensing lightning/storm. • They activate safety protocols, disconnect systems and trigger alarms.
Grounding sensors	• Also required in outdoor or exposed installations. • Referred to as "ground integrity monitors" or "earth fault detectors". • Electronic devices, often integrated with the grounding system. • Measure resistance or impedance of grounding to ensure safety compliance using methods like the fall-of-potential. • Ensure proper functioning of the grounding system for safety and equipment protection.

Note that infrared sensors are also utilized for detecting heat and hotspots, which require signal conditioning for temperature calculation and come in various formats, from single-point sensors to imaging arrays.

Incorporating the SOA concept in battery operations involves monitoring and controlling three key factors: current, voltage and temperature, essential for SOC estimation algorithms. In addition, sensors for grid and system integration monitor and analyse power quality and grid interactions, measuring voltage and current to assess harmonics and other power quality indicators, available in both fixed installation and handheld units.

Furthermore, in communication and cybersecurity, firewall devices and intrusion detection systems safeguard network integrity, offered as software or hardware solutions, and remote diagnostics tools and emergency alert systems, integrated into BMS or as standalone systems, focus on connectivity and user interface for effective system monitoring and response.

404 ■ Reinventing the Power Grid

Figure 5.24 shows a selection of principal sensors with images and fundamental signal conditioning devices commonly employed in battery storage systems (BSSs). It is important to note that resistance temperature detector (RTD) sensors are available in various formats to suit the mediums in which they are utilized, requiring a signal conditioning circuit that accommodates a bridge and an amplifier. Hall-effect sensors and shunt resistors both can measure current within a conductor, yet isolated measurement is possible via Hall-effect sensors only, which operate on the principle of magnetic fields generated by electrical currents. Signal conditioning circuits employing amplifiers are integral to both sensor types,

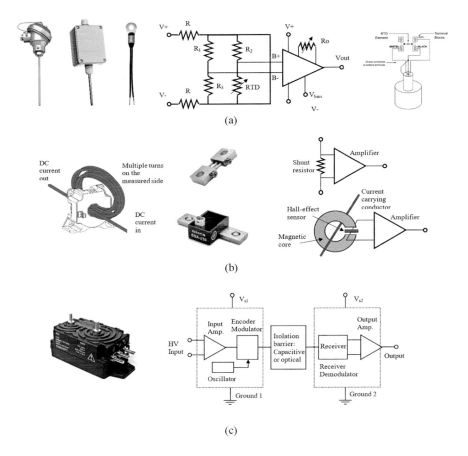

FIGURE 5.24   Some of the core sensors used in BSS: (a) RTD temperature sensors for ambient and direct contact types; and principle signal conditioning circuit with wiring details, (b) Isolated Hall-effect and shunt resistor type current sensors, (c) Isolation amplifier-based voltage sensing.

ensuring signal levels meet the required values. Furthermore, isolation amplifiers are implemented for high-frequency and high-voltage sensing. However, they represent one of the more complex and expensive sensor types, which has an isolation barrier, as well as encoder modulators and receiver demodulators, each requiring independent floating voltage supplies for isolation and safety.

It should be emphasized here that the integration of diverse sensor types along with the requisite hardware and software tools significantly contributes to the increased cost of BSS. Moreover, the complexity of integrating these sensors into a unified BMS requires advanced software capable of real-time data analysis and decision-making, further increasing costs. In addition, the need for secure communication and robust cybersecurity measures introduces additional layers of expense due to specialized hardware and software solutions required to safeguard against unauthorized access and ensure system reliability.

It can be summarized that cost distribution within a BSS project varies, with battery costs typically constituting 40%–60% of the total (can reduce to 30%–50% when connected to the AC grid), reflecting the significant impact of battery technology and capacity on overall expenses. Engineering and procurement can account for 10%–20%, influenced by system complexity and customization levels. Electronics and the BMS, critical for system integration and control, represent another 10%–15% of the cost. Installation and grid connection costs, including, infrastructure modifications and compliance measures, can vary but generally fall between 10% and 20%. Other long-term expenses, such as maintenance and software updates, though not upfront, are critical for sustained efficiency and operation. This distribution highlights the technological demands and complexity of modern BSSs, indicating the balance between the cost and the benefits of enhanced energy storage solutions as a critical industry question.

5.10 FUEL CELLS AND SUPERCAPACITORS: DESCRIPTIONS, MODELLING AND OPERATION

Significant growth in the fuel cell and hydrogen industry can be been seen as a credible alternative within the last decade, driven by its potential to power grids, transportation and portable applications. This makes fuel cells, combined with hydrogen fuel, effective energy storage devices. Since the hydrogen produced can be stored with minimal energy loss for extended periods, it also has a great potential for on-demand energy utilization. Advancements in fuel cell technology focus on high efficiency and low-maintenance operations, which are crucial for their wider adoption.

Hydrogen, the primary fuel for fuel cells, is traditionally produced through hydrocarbon reforming, especially from natural gas. However, electrolysis, which separates hydrogen from water using renewable energy, emerges as a cleaner alternative. This technique enables fuel cell systems to operate as truly low- or zero-emission sources of electricity when paired with renewable energy sources. A detailed breakdown of hydrogen production methods is provided in Table 5.20.

Figure 5.25 shows two hydrogen production routes and their efficiencies: the first from renewable or nuclear energy through electrolysis, with a high generation efficiency of 70%–80%, leading to high fuel cell system

TABLE 5.20 Hydrogen Production Methods

Names	Remarks
Green hydrogen	• Produced by electrolysis of water using renewable energy sources like wind, solar or hydro. • Currently expensive.
Blue (low-carbon) hydrogen	• From natural gas through steam methane reforming (SMR), combined with carbon capture and storage (CCS) to capture the carbon emissions.
Grey hydrogen	• Produced by steam reforming natural gas without carbon capture and storage, leading to CO_2 emissions.
Brown hydrogen	• Produced by converting lignite (brown coal) into a gaseous fuel (called syngas).
Black hydrogen	• Produced by converting black coal into a gaseous fuel' • The most environmentally damaging type. • Considered to produce liquefied hydrogen.
Pink hydrogen (purple hydrogen or red hydrogen)	• Generated through electrolysis powered by nuclear energy. • Very high temperatures from nuclear reactors are also used to produce hydrogen, which is done by producing steam first for more efficient electrolysis or fossil gas-based steam methane reforming.
Turquoise hydrogen	• Not yet commercially proven, but can be made using methane cracking to produce hydrogen and solid carbon (graphite). • May be considered as a low-emission hydrogen. The thermal process can be powered with renewable energy, and the carbon can be permanently stored or used.
Yellow hydrogen	• Produced through electrolysis using solar energy.
White hydrogen (natural Hydrogen)	• There is no scientific evidence for the existence of naturally occurring, geological hydrogen accessible through fracking. • No strategies to exploit this hydrogen at present, but its value chain is similar to natural gas production.

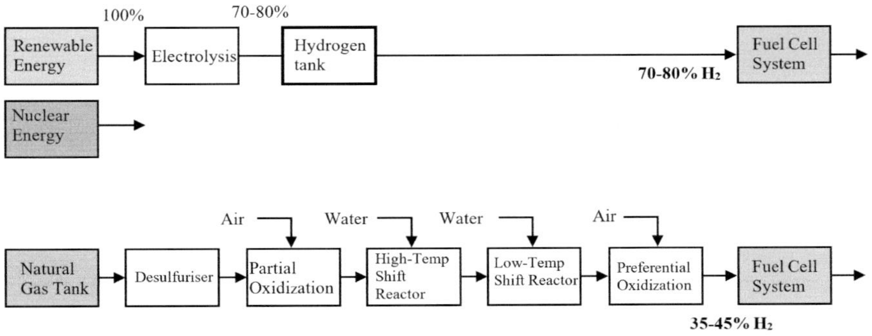

FIGURE 5.25 Hydrogen generation efficiency and processes using renewable energy or nuclear energy (top) and natural gas (bottom).

efficiencies; the second from natural gas, involving desulfurization and reforming processes that result in a lower fuel cell system efficiency of 35%–45%. Although hydrogen storage is nearly lossless, the overall system efficiency is significantly influenced by the method of hydrogen generation, with renewable or nuclear energy offering a more efficient pathway compared to natural gas. However, note that output power is defined by the efficiency of the type of fuel cell used, which results in significantly low system efficiency, hence highlighting the need for efficient fuel cell research.

Although hydrogen is an ideal fuel for fuel cells, several factors limit the technology's widespread adoption: the use of high-cost catalysts like platinum, operating environments, low efficiency and high overall costs. Progress in materials science, membrane technology and catalyst development is expected to make fuel cells economically viable. Table 5.21 highlights the diversity of existing fuel cell technologies, aiming to demonstrate their suitability for various applications. It also identifies areas requiring further development for wider adoption within the next few decades.

In fuel cell systems, efficient power management is achieved using DC/DC converter topologies as mentioned in Chapter 4, which address similar issues as in batteries, to adapt to the low voltage and high current characteristics under load. Commonly used converters, including the buck-boost converter and interleaved boost converters, condition the output voltage and current in fuel cells, which influence the overall system efficiency. Therefore, if the efficiency of a DC/DC converter is assumed at an average of 95% for simplicity, the electrical efficiency given in Table 5.21 will be reduced further.

TABLE 5.21 The Fuel Cells, Efficiencies, Applications with Operating Principles and Limitations

Fuel Cell Type	Efficiency (with CHP)	Applications	Operating Principles and Limitations
Polymer electrolyte membrane (PEM) fuel cells	40%–60%	Portable power, transportation, stationary power generation.	• Low operating temperature, well-developed technology, • A solid polymer electrolyte membrane is used to conduct protons from the anode to the cathode, • 0.6 V per cell, moderate current and high-power density, • Requires pure hydrogen, sensitive to impurities. • Expensive platinum catalyst needed.
Solid oxide fuel cells (SOFC)	60% (Up to 85%)	Large-scale stationary power generation.	• High efficiency, fuel flexibility. • Solid oxide or ceramic electrolyte conducts oxygen ions from the cathode to the anode and fuel flexibility. • 0.7 V per cell, high current density. • High operating temperatures (800°C–1,000°C) can cause slow start-up and material degradation.
Molten carbonate fuel cells (MCFC)	50% (Up to 85%)	Utility-scale power plants, industrial applications.	• Employs a molten carbonate electrolyte, allowing the movement of carbonate ions from the cathode to the anode. • Operates at high temperatures (about 650°C), leading to corrosion and a shorter lifespan.
Alkaline fuel cells (AFC)	60%	Spacecraft, some military applications,	• Uses an aqueous alkaline electrolyte like potassium hydroxide, with hydroxide ions moving from the cathode to the anode. • 0.7 V per cell, moderate current density • Low cost, mature technology, tolerant of impurities. • Limited fuel flexibility, corrosion, low power density, • Sensitive to CO_2 and requires pure hydrogen and oxygen.
Phosphoric acid fuel cells (PAFC)	40% (Up to 85%)	Commercial buildings, utility power plants.	• Phosphoric acid is used as the electrolyte, conducting protons from the anode to the cathode. • Lower efficiency for the cell alone and higher operating temperatures than PEM.

(Continued)

TABLE 5.21 (*Continued*) The Fuel Cells, Efficiencies, Applications with Operating Principles and Limitations

Fuel Cell Type	Efficiency (with CHP)	Applications	Operating Principles and Limitations
Direct methanol fuel cells (DMFC)	20%–30%	Portable power applications where refuelling is difficult.	• Directly utilizes methanol as fuel at the anode, where it is oxidized, with protons conducted through a polymer membrane to the cathode. • 0.3 V per cell, low current density. • Compact, portable, low emissions, fast start-up. • Lower efficiency and power/energy density.

Note that in combined heat and power (CHP) applications of fuel cell systems, the total system efficiency is higher due to the utilization of waste heat for space and water heating purposes in addition to electrical power generation. However, it is important to note that the efficiency of CHP should be considered separately since it combines both electrical and thermal outputs. Note also that CHP systems introduce added complexity and initial costs due to the addition of advanced thermal management and control systems, alongside the standard electrical components. Moreover, they require higher capital, installation and maintenance costs. However, their improved efficiency may offer substantial long-term energy savings and environmental benefits, which are particularly valuable in industrial processes and commercial buildings.

However, prior to discussing the electrical equivalent circuit, it will be valuable to highlight the similarities between flow batteries described previously and fuel cells, which is due to their reliance on fluid dynamics and electrochemistry and is given in Table 5.22. Firstly, it is accurate to say that both fuel cells and flow batteries involve the movement and management of fluids:

- In fuel cells, the "fluids" typically refer to the gaseous fuels (such as hydrogen) and oxidants (such as oxygen from air). In some types of fuel cells (such as PEMFCs and DMFCs), liquid water is also used for maintaining membrane hydration or managing product water. Moreover, liquid cooling systems might be used to dissipate the heat generated during operation.

- Flow batteries inherently rely on the flow of liquid electrolytes. Since the electrolytes are pumped through the system, circulation between tanks and through the cell where the electrochemical reactions occur.

TABLE 5.22 Comparison of Flow Batteries and Fuel Cells across Various Parameters

Features	Flow Batteries	Fuel Cells
Flow management	Require pumps to circulate electrolytes.	• Requires compressor to deliver hydrogen gas as in PEMFCs. • Require pumps to circulate liquid fuel as in DMFCs.
Dynamic response	Fluid flow management introduces delays in response to load changes.	Similar delays due to fluid flow management to supply reactants.
Electrochemical reaction	Depend on electrochemical reactions that have inherent response times.	Also rely on electrochemical reactions with similar limitations.
Control systems	Control strategies for flow rates are crucial for efficiency and lifetime.	Similar control strategies can be applied due to operational similarities.
Efficiency and performance	Efficiency influenced by the balance between reaction rate and reactant supply.	Efficiency similarly influenced, with performance affected by flow of reactants.
Energy capacity	Increased by larger electrolyte storage tanks.	Increased by more fuel storage or larger reservoirs.
Power capacity	Enhanced by larger electrode surface areas and higher electrolyte flow rates.	Boosted by larger active areas of the fuel cell stack and optimized fuel flow.
Cell configurations	Series configurations for higher voltage; parallel for higher current.	Similar series and parallel configurations used to adjust voltage and current output.
Modularity and scalability	Benefit from modularity; additional stacks enhance capacity.	Inherently scalable; modularity allows for adapting to energy demands.
Redundancy and load sharing	Multiple configurations improve reliability and can extend system life.	Similar benefits from redundancy and load sharing across multiple cells.
Current density	Linked to electrode surface area, catalyst loading and electrolyte conductivity.	Also related to active area, catalyst usage and fuel conductivity.
Modelling	Flow rates, pressures and concentrations are modelled to predict behaviour.	Share similar modelling parameters for predicting system behaviour.
Electrical equivalent circuit	• Includes voltage sources (to represent the electrochemical potential), resistors (to model ohmic losses in electrodes and electrolyte/membrane) and capacitors (to account for the capacitive behaviour of the double layer at the electrode/electrolyte interface).	• Similar electrical circuit is used. • But circuit might emphasize the dynamics of gas diffusion and reaction kinetics, represented by specific resistive and capacitive elements related to the fuel and oxidant supply and consumption rates.

(Continued)

TABLE 5.22 (Continued) Comparison of Flow Batteries and Fuel Cells across Various Parameters

Features	Flow Batteries	Fuel Cells
	• But the ion transport between two electrolyte solutions and through the membrane, as well as the redox reaction kinetics at the electrodes, can be included in the model by resistive and capacitive elements based on the chemistry and physics of the flow battery.	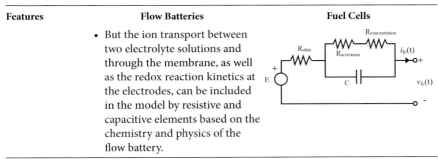

Note that while the primary functions and chemistries of fuel cells and flow batteries differ, they also share underlying electrochemical principles, including the formation and impact of the charge double layer at the electrode-electrolyte interface. This similarity also highlights the fundamental electrochemical nature of these technologies and their behaviour under dynamic operating conditions.

5.10.1 Equivalent Circuit, Electrical Characteristics and Definitions

Fuel cells, as electrochemical devices, have their performance closely linked to electrical equivalent circuits and characteristic curves when interfaced with varying electrical loads and subjected to diverse environmental conditions. Similar to batteries, these aspects present specific limitations that are critical to understanding and optimizing fuel cell operation. The following paragraphs will explain these characteristics and explore how they impact overall performance.

A simplified equivalent circuit diagram of a fuel cell is given in Table 5.22. In the circuit, E represents the open-circuit potential or electromotive force, EMF of the cell. R_{ohm} indicates losses due to electron and ion flow within the cell components. While the activation resistance, $R_{activation}$ models the energy barriers at the electrode surfaces during the electrochemical reactions. The capacitor represents the capacitance of the electric double layer at the electrode–electrolyte interface, needed for understanding the cell's transient response to changes in load. $i_{fc}(t)$ and $v_{fc}(t)$ represent the time-dependent behaviour of the output current and voltage as the cell operates under various conditions.

In reference to the circuit, the dynamic fuel cell voltage behaviour can be given by

$$C\frac{dv_c}{dt} + \frac{v_c - v_o}{v_{act} + v_{conc}} = i_{fc} \tag{5.8}$$

$$v_{fc} = E - v_c - i_{fc}R_{ohm} = E - v_{act} - v_{conc} - i_{fc}R_{ohm} \tag{5.9}$$

where, v_o is the voltage drop at zero current density, v_c is the voltage across the capacitor, v_{fc} is the voltage of a single cell and v_{act} and v_{conc} are voltage across the activation and the concentration resistors respectively.

Figure 5.26 demonstrates how current density and pressure impact the performance of a fuel cell. The solid black line in Figure 5.26a represents a typical fuel cell's characteristic polarization curve, the actual voltage output of the fuel cell at different current densities. As current density increases, the voltage output decreases due to various losses. Note that the cell voltage in the figure is the actual cell voltage, and the current density can be defined as cell current i_{fc} that is equal to the ratio between stack current in A and cell active area in cm², assuming that the stack is formed by connecting the cells in series. In practice, the design of multi-cell systems must consider the complexities of fluid and gas distribution, thermal management, electrical connections and control systems to ensure that all cells operate optimally. Both flow batteries and fuel cells are inherently scalable, which is one of their advantages over conventional batteries. This scalability makes them suitable for a wide range of applications, from small portable devices to large stationary energy storage systems.

The dashed lines in Figure 5.26a represent different types of losses within the fuel cell: the ohmic loss that is related to the internal resistance of the fuel cell and leads to a linear decrease in voltage, the activation loss that is significant at lower current densities due to the energy required to initiate the electrochemical reactions and the concentration loss that gets

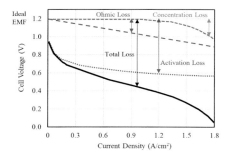
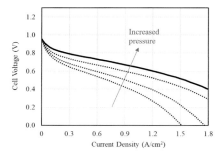

FIGURE 5.26 Fuel cell performance curves by analysing voltage output current density: (a) Losses for efficiency and (b) Impact of operating pressures.

higher at higher current densities when mass transport limitations lead to a shortage of reactants at the reaction sites. The ideal EMF is shown at 1.2 V level, representing the maximum potential difference the cell can achieve without any losses.

Figure 5.26b shows the impact of increased pressure on the voltage output of the fuel cell. As pressure increases, the voltage output at a given current density improves, which is due to enhanced mass transport of reactants at higher pressures reducing concentration losses.

Efficiency in a fuel cell is a key performance indicator during steady-state operation, which is the ratio of the electrical energy generated (considering all other auxiliary system losses: such as pumps, cooling and electronics), to the chemical energy input (such as energy hydrogen consumed in PEMFC) for electrical efficiency.

Note that efficiency is also given in terms of power descriptions. The choice of using power as W or energy as Wh depends on whether one is interested in instantaneous efficiency or efficiency over a period. The key conversion parameter in this choice is 1 kWs = 1 kJ.

$$\text{Fuel cell} = \frac{P_{elec}\,(\text{in kW})}{P_{input}\,(\text{in kW})} = \frac{\left(V_{output} I_{output}\right)}{\left(\dfrac{n_{fuel} \times \text{HHV or LHV}}{3{,}600}\right)} \tag{5.10}$$

Here n_{fuel} is the molar flow rate of the fuel consumed per second (in kmol/s) and Higher Heating Value (HHV) or Lower Heating Value (LHV) is in joules per mole (J/mol or (kJ/kmol). Note that the molar flow rate of fuel can be measured directly using flow meters that are calibrated for the specific type of fuel being used (such as hydrogen in many fuel cells). In addition, HHV and LHV values are obtained through calorimetric measurements or can be found in chemical data tables and literature. It should be emphasized that this approach considers the heating value of the fuel used. This approach uses the molar flow rate of the fuel instead of the mass flow rate, which allows for direct comparison between different fuel types with varying molecular weights. Moreover, HHV and LHV are used depending on whether the latent heat of vaporization is considered in the energy input calculation.

Note that when describing the "Energy efficiency" in fuel cells, all forms of energy output can be considered, not just electricity, including thermal energy recovered, which can be used for heating or other purposes, which improves the overall efficiency.

$$\text{fuel cell}(\text{energy}) = \frac{E_{\text{output}} + Q_{\text{output}}}{n_{\text{fuel}} \times \text{HHV or LHV}} \qquad (5.11)$$

Here E_{output} is electrical energy output in kJ and Q_{output} is thermal energy output in kJ.

The above descriptions can be given as a function of mass flow rate (m_{fuel}) instead of the molar flow rate of the fuel. The correct units should be considered: mass flow rate of fuel consumed in kg/s, and HHV in kJ/kg. In addition, the molar flow rate should be converted to the mass flow rate using the fuel's molecular weight: $m_{\text{fuel}} = n_{\text{fuel}} \times$ (Molecular weight of fuel). Note that although the concepts of voltage and Faradaic efficiency apply to both fuel cells and batteries, the specifics of how they are measured and what they indicate about the device's performance can differ due to the operational differences between these two types of electrochemical cells, hence not commonly used.

Finally, note also that PE converters serve as a bridge between the fuel cell and the load. The use of two-stage converters for grid-connected fuel cell devices requires complex control strategies to manage ripple disturbances and ensure seamless integration with the power grid.

Finally, note also PE converters are needed to match the DC power output from fuel cells to the load. Converters adjust the voltage and current to optimal levels, protecting the fuel cells from fluctuations and enabling efficient operation. They also integrate energy storage by matching the energy levels between the fuel cells and batteries or supercapacitors. Conventional DC-DC converters discussed in Chapter 4 are used to step-up or step-down voltage levels as required.

Figure 5.27 presents the transient response of a proton exchange membrane fuel cell (PEMFC) to a sudden load change, indicative of a scenario where a PE converter plays a role between the fuel cell and

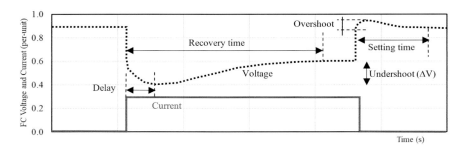

FIGURE 5.27 Transient Response of a PEMFC to load variability: Voltage and current dynamics with PE converter regulation.

the load. Note that, initially, there is a delay in voltage before the fuel cell reacts to the altered demand, illustrating the system's response time. Subsequently, the fuel cell voltage dips, undershooting before it momentarily overshoots the steady-state value, signifying the system's attempt to adjust to the new load. The recovery time is observed as the period for the voltage to approach the steady-state value, while the setting time indicates how long it takes for the voltage to stabilize within a specific range after any oscillations.

A constant current step change directly affects the voltage behaviour. This graph is critical for understanding how PE converters must be designed to handle such dynamic conditions. The converters are responsible for managing parameters such as the rate of change in current (di/dt), voltage stability and ensuring the power supplied to the load is smooth and continuous despite the rapid changes in demand. They achieve this by controlling the output voltage and current, acting as a buffer to absorb the transient fluctuations. This also helps to protect the fuel cell from potentially damaging states.

The operation of a supercapacitor, as the final energy storage device to be discussed, can be as complex as the batteries and fuel cells. However, it can be represented by using simple electrical components that capture its essential electrical characteristics for the purposes of analysis and design. The equivalent circuit of a supercapacitor is commonly composed of two principal elements: capacitance and resistance. Capacitance is indicative of the supercapacitor's capacity to store electrical energy as an electric field, with its value dictating the charge it can retain at a specified voltage. Resistance, on the other hand, is the supercapacitor's intrinsic resistance to current flow, with lower resistance enabling quicker charge and discharge cycles.

However, this model may not be sufficiently accurate for every application, in applications that involve high frequency or high power, which need dynamic components in the model. These include a leakage resistance that accounts for self-discharge, a Warburg impedance that describes the frequency-dependent behaviour of the electrolyte affecting the charging and discharging characteristics at higher frequencies, and an inductance that quantifies the magnetic field produced by the current flow within the supercapacitor, significant in high-power applications.

Figure 5.28 demonstrates the voltage and current characteristics during the charge and discharge of a supercapacitor. On charge, the voltage increases linearly and the current drops by default when the capacitor is full without the need for a full-charge detection circuit. This is true with a constant

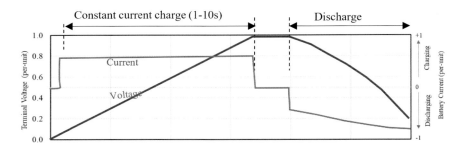

FIGURE 5.28 Constant current charge and discharge characteristics of a super capacitor.

current supply and a voltage limit that is suitable for the capacitor-rated voltage. The voltage drops on discharge and if there is a DC/DC converter, it can keep the power constant hence increasing current with dropping voltage.

In summary, supercapacitor connection arrangements (series and parallel) and balancing considerations are similar to those of batteries. While supercapacitors have a long lifespan, rapid charging and safe operation, ideal for applications demanding frequent high-power discharge, their limited energy storage, uneven discharge profile and higher cost compared to batteries currently limit their widespread use in specific applications.

5.11 PROTECTION, SAFETY AND BATTERY TESTING METHODS

As battery technologies evolve and are utilized in a wider range of diverse and demanding power and energy applications, the protection and safety of these batteries become a serious concern. This defines the entire life cycle of a battery, from post-manufacturing to end-of-life and beyond, including potential second and third-life applications and final integration into the circular economy.

Therefore, there is a need to discuss batteries and associated products in the context of the regulations and standards to ensure that rechargeable batteries are not only effective but also comply with the highest standards of safety and environmental responsibility.

Some of the recognized standards include the IEC (International Electrotechnical Commission) which develops international standards for all electrical, electronic, and related technologies; UL (Underwriters Laboratories) which sets industry-wide standards for new technologies; SAE (Society of Automotive Engineers) which specializes in setting standards for the automotive industry, including standards for EV batteries; ISO (International Organization for Standardization) that creates and publishes standards for various industries, including battery manufacturing and NEMA (National Electrical Manufacturers Association) that provides a forum for the development of technical standards in the electrical manufacturing industry.

Unique testing bodies like TÜV, BSI and ASTM also have a role in certifying and supporting battery technology, to ensure that products are suitable for market use and comply with international and local safety standards. These certifications are often mandatory for batteries, noting that these certifications also help to build consumer trust and facilitate international trade. Furthermore, the standards and testing bodies are crucial for guiding manufacturers in the development of batteries that are not only high-performing but also safe for both users and the environment.

Table 5.23 provides a comprehensive classification of various battery tests categorized by their primary purpose, which also offers insights into the testing procedures employed. The table highlights the critical role of

TABLE 5.23 Comprehensive Battery Test Classification for Protection and Safety

Category	Name of Tests		Justification
Safety and abuse testing	• Overcharge • Short circuit • Crush • Impact • Shock • Over-discharge • Thermal cycling • Vibration • Low-pressure (altitude) • Projectile/external fire	• Drop • Penetration • Separator shutdown integrity • Internal short circuit (propagation) • Thermal runaway	To ensure reliable and safe operation of batteries undergo a series of rigorous tests that simulate various environmental and operational stresses. By identifying potential failure modes and designing batteries with robust safety features, these tests ultimately protect users and prevent accidents.
Performance testing	• Continuous low-rate charge • Moulded casing heating • Temperature cycling • Open circuit voltage • Insulation resistance (Rx) • Reverse charge		These tests are the key to verifying that a battery will function as expected throughout its intended use. The results confirm that the battery can hold and deliver power effectively, operate within designated temperature ranges and respond correctly to charging and discharging scenarios, which are also needed for consumer satisfaction and the longevity of the battery's application.

(Continued)

TABLE 5.23 (*Continued*)   Comprehensive Battery Test Classification for Protection and Safety

Category	Name of Tests	Justification
Electrical and electrochemical testing	• Internal resistance measurement • Electrochemical impedance spectroscopy (EIS), heating	Such tests are needed for characterizing the battery's fundamental operational parameters. Internal resistance and impedance affect the efficiency and power output of the battery, while heating tests are necessary to ensure the battery can manage its thermal output during operation. They also help to optimize battery design for better performance, longer life and higher safety.
Durability and life cycle testing	• Cycle life	The test determines how many charge and discharge cycles a battery can undergo before its capacity falls below a usable threshold, which provides insights into the long-term value.
Environmental testing	• Leakage • Electrolyte spillage	The tests are vital to identify the battery potential for corrosion, contamination and environmental damage, which are also highly critical for batteries that operate in sensitive environments or are used close to the human body.
Recycling and environmental impact	• Recycling • End-of-Life disposal • Environmental impact assessment	These ensure that batteries can be disposed of or recycled in an environmentally conscious manner for minimizing the ecological footprint of battery technologies.
Regulatory compliance and certification	• Product safety, conformity and certification • Transport safety and compliance • Safety data sheet (SDS) compliance	These are non-negotiable for market entry, and they ensure the consumers and industry partners that a battery product is compliant with the highest safety and quality standards.

testing for battery safety, performance and environmental responsibility during its life cycle.

As can be seen, by undergoing a comprehensive testing regime listed in the table, batteries can be ensured to be safe, reliable and environmentally responsible. Therefore, safety and abuse tests are conducted to evaluate responses to overcharges, short circuits, mechanical damage, temperature extremes and external threats. These tests ensure the battery operates safely, preventing failures like thermal runaway or fire. In addition, **performance tests** assess the **capacity, function and health** of the battery under various conditions. This includes evaluating behaviour during **temperature fluctuations, over-discharge and long-term charging**, along with measuring **voltage, resistance and electrochemical processes**.

Furthermore, **cycle life testing** determines the battery lifespan under repeated use, ensuring reliable performance over time. Moreover, **regulatory compliance** tests verify compliance with regulations relating to **shipping, handling and safety information** to ensure safe and responsible practices. Finally, **environmental/end-of-life** tests address the **environmental impact** of the battery throughout its life cycle, which includes evaluating **recycling practices, disposal methods and potential environmental risks**.

Table 5.24 presents a comprehensive perspective on the safety protocols of BSS, covering a range of considerations from regulatory compliance to physical safeguards and extending to the specific details of technology-specific features and operational procedures. While an in-depth discussion is reserved for Chapter 6, it is important to recognize that battery storage is often employed at the grid scale, constructed from interconnected battery modules, units and containers. These battery subunits specifically address the distinct demands of their applications. Therefore, the protection and safety of batteries are critical in an overall system.

It should be emphasized here that one of the key components not explicitly explained previously is the DC circuit breaker, which is crucial in BSS for complying with safety regulations by protecting against overcurrent and short circuits.

DC circuit breakers are specialized devices designed to interrupt direct current by overcoming the absence of a natural zero current point, unlike AC circuit breakers that take advantage of the alternating current's zero-crossing points. As a result, they are equipped with robust contacts and arc chutes to manage and extinguish the persistent arc generated during operation. In addition, blowout coils are included to stretch the arc

TABLE 5.24 Prioritized Safety and Protection Strategies for Grid-Scale Battery Systems

Key Areas	Details and Strategies	Sensors and Measures
Regulatory compliance	• Compliance with international standards and local regulations for design and operational safety. • Regular audits and certification processes are needed, reference to standards such as IEC, IEEE, UL.	Tracking systems, documentation management software.
Physical protection	• Structural integrity of modules and containers. • Protection against temperature and moisture. • Suitable enclosures, controlled environments, fire suppression systems (such as FM200, NOVEC 1230) are needed.	Temperature sensors, moisture and smoke detectors integrated with automated suppression systems.
Technology-specific protection	• For example: Li-ion has volatile electrolyte and flow batteries can cause electrolyte leakage. • Hence Li-ion needs BMS for thermal management, and • Flow batteries require leak detection and containment.	Thermal cameras, electrolyte level sensors and gas detectors.
Integrated and comprehensive system protection	• Module and unit level protection and temperature control at rack/container level. • Grid interaction and charge management at system level. • BMS for module-level monitoring, rack-level management systems for group protection. • Cooling systems, ventilation for thermal management.	Individual cell voltage monitors, current sensors, rack temperature sensors, infrared sensors and PTC thermistors.
Control systems safety	• Robust and redundant control systems. • Fail-safes to prevent operational errors. • Redundant control systems and emergency shutdown protocols.	Fault detection algorithms and automatic DC circuit breakers.
Capacity and performance monitoring	• To prevent overcharging or deep discharging. • Continuous SOC and SOH monitoring through BMS and predictive maintenance.	SOC estimation algorithms and impedance measurement.
Grid and load integration	• Protection against grid faults and load surges. • Secure communication protocols, anti-islanding. • Use of static switches, isolation transformers, advanced inverters with anti-islanding protection.	Grid-tie monitoring systems, power quality analysers, circuit breakers, isolation devices, surge protection.

(Continued)

TABLE 5.24 (*Continued*)   Prioritized Safety and Protection Strategies for Grid-Scale Battery Systems

Key Areas	Details and Strategies	Sensors and Measures
Cybersecurity and communication	• Protection against unauthorized access. • Robust cybersecurity measures. • Secure communication channels, encrypted data transmission, cybersecurity audits.	Firewall devices and intrusion detection systems.
Emergency response and maintenance protocols	• Defined emergency response procedures. • Regular maintenance and health checks. • Establishment of clear emergency response procedures, routine maintenance schedules, remote monitoring.	Remote diagnostics tools and emergency alert systems.
Application-specific characteristics	• Safety strategies based on application needs. • Considerations for renewable integration or frequency regulation. • Customizing safety and protection strategies.	Application-specific monitoring and control algorithms.
Risk and compliance management	• Compliance with safety regulations • Hazard identification and mitigation • Training for system safety management • Risk assessments, safety training for personnel, implementation of safety management systems	Risk management software and training simulation tools.

(to safely interrupt the DC and break the circuit, since the arc can persist after the contacts part) and energy storage mechanisms for rapid contact separation, these breakers are built to withstand the "continuous" nature of DC. The structure of these breakers also includes auxiliary contacts for signalling purposes. Higher voltage ratings and additional surge suppression are distinctive features that set DC circuit breakers apart from their AC counterparts, reflecting the unique demands of DC power systems in effectively managing and interrupting the flow of electricity to ensure safety and reliability, all contributing to their higher cost.

Therefore, DC circuit breakers can be considered a key component that supports the physical integrity of the system by preventing overheating and potential fires, which is particularly critical for technologies like lithium-ion batteries that are sensitive to current surges. Integral to both individual battery module protection and the broader system-wide safety framework, these breakers act in conjunction with the BMS to provide an immediate electrical fault response. As fail-safes within the BSS control systems, they enable safe operational shutdowns and contribute to the system's performance monitoring by signalling electrical anomalies. Furthermore, they ensure stable grid integration by guarding against load fluctuations and grid faults, indirectly supporting the system's cybersecurity by maintaining operational integrity against potential disruptions.

Therefore, DC circuit breakers collectively form an essential component of risk mitigation, maintaining the BSS within its SOA and implementing safety and protection strategies. As technology progresses, it is expected that WBG semiconductors will play an important role in DC circuit breakers as well, leading to devices that are more efficient, faster, smaller and potentially lower in cost as manufacturing processes mature.

Table 5.25 outlines the post-use phases of batteries, where each phase aims to maximize utility. Initially, used batteries undergo assessment for secondary roles with reduced performance demands, like stationary energy storage post-EV use. As capacity further declines, batteries may serve low-energy applications, possibly with some refurbishment. The end-of-life stage involves collection and recycling to extract valuable materials for new products, complying with environmental regulations. This process, though not yet fully implemented, envisages a comprehensive system to track battery history and condition, facilitating informed decisions on repurposing or recycling. Advancing a circular economy for batteries calls for collaboration among manufacturers, consumers, recyclers and regulators, leveraging technological innovations for battery health management and cost-effective business models for battery life extension.

Finally, it should be noted that selecting the right battery chemistry for specific uses requires a comprehensive evaluation of how battery life aspects respond to environmental and operational stresses. As known, this involves weighing performance, durability, safety, cost and environmental considerations against the unique challenges posed by each application's

TABLE 5.25　The Descriptions of Post Battery-Life

Life Stage	Description
Second life	After primary use, assessment and repurposing of used batteries for secondary applications with lower performance requirements, like stationary energy storage after use in EVs.
Third-life	Further use of batteries in applications with even lower energy and power requirements after their second-life capacity diminishes, which may require limited refurbishments.
End-of-life	The final stage involving the collection and recycling, where valuable materials are recovered from depleted batteries for reuse in new products, while considering compliance with environmental regulations.

conditions. An illustrative example is the battery testing and comparison project sponsored by ARENA in Australia, conducted by ITP Renewables. The project highlighted the complex interplay between technology, market dynamics and operational realities in the battery sector, and tested various residential and commercial battery packs under controlled conditions (in typical Australian conditions over a year), revealing significant insights:

- A total of 26 batteries tested, including lithium-titanate and sodium-nickel.

- Reliability varied across different technologies, with certain batteries outperforming others in terms of dependability.

- There were noted discrepancies between expected and actual performance, particularly concerning product integration and compatibility.

- A trend towards integrated battery-inverter systems and the necessity for compatible equipment from the same brand was observed.

- The incidence of battery failures and the need for replacements were higher than anticipated, affecting both new and established technologies.

- The instability of some manufacturers, including cases of insolvency, highlighted the critical need for choosing providers with proven reliability and the ability to honour warranties.

- The effort required for maintenance and the level of support from manufacturers were significantly underpredicted and varied widely.

BIBLIOGRAPHY

C. Chetri, A. Huynh and S. S. Williamson, "A Comprehensive Review of Active EV Battery Cell Voltage Balancing Systems: Current Issues and Prospective Solutions," *2022 IEEE 1st Industrial Electronics Society Annual On-Line Conference (ONCON)*, Kharagpur, India, 2022, pp. 1–6, https://doi.org/10.1109/ONCON56984.2022.10126563.

EPRI, *Common Functions for Smart Inverters: 4th Edition*. EPRI, Palo Alto, CA, 2016.

N. Ertugrul, "Battery Storage Technologies, Applications and Trend in Renewable Energy," *2016 IEEE International Conference on Sustainable Energy Technologies (ICSET)*, Hanoi, Vietnam, 2016, pp. 420–425, https://doi.org/10.1109/ICSET.2016.7811821.

N. Ertugrul and F. Castillo, "Maximizing PV Hosting Capacity and Community Level Battery Storage," *2019 29th Australasian Universities Power Engineering Conference (AUPEC)*, Nadi, Fiji, 2019, pp. 1–6, https://doi.org/10.1109/AUPEC48547.2019.211903.

N. Ertugrul and G. Haines, "Impacts of Thermal Issues/Weather Conditions on the Operation of Grid Scale Battery Storage Systems," *2019 IEEE 13th International Conference on Power Electronics and Drive Systems (PEDS)*, Toulouse, France, 2019, pp. 1–6, https://doi.org/10.1109/PEDS44367.2019.8998808.

N. Ertugrul, G. Bell, G. G. Haines and Q. Fang, "Opportunities for Battery Storage and Australian Energy Storage Knowledge Bank Test System for Microgrid Applications," *2016 Australasian Universities Power Engineering Conference (AUPEC)*, Brisbane, QLD, Australia, 2016, pp. 1–6, https://doi.org/10.1109/AUPEC.2016.7749304.

E. Fotouhi et al, Lithium-sulfur cell equivalent circuit network model parameterization and sensitivity analysis. *IEEE Transactions on Vehicular Technology*, 66(9), pp. 7711–7721, 2017.

G. Haines, N. Ertugrul, G. Bell and M. Jansen, "Microgrid System and Component Evaluation: Mobile Test Platform with Battery Storage," *2017 IEEE Innovative Smart Grid Technologies – Asia (ISGT-Asia)*, Auckland, New Zealand, 2017, pp. 1–5, https://doi.org/10.1109/ISGT-Asia.2017.8378466.

R. Khezri, A. Mahmoudi, N. Ertugrul, M. F. Shaaban and A. Bidram, "Battery Lifetime Modelling in Planning Studies of Microgrids: A Review," *2021 31st Australasian Universities Power Engineering Conference (AUPEC)*, Perth, Australia, 2021, pp. 1–6, https://doi.org/10.1109/AUPEC52110.2021.9597840.

S. Liu, J. Chen, C. Zhang, L. Jin and Q. Yang, Experimental study on lithium-ion cell characteristics at different discharge rates. *Journal of Energy Storage*, 45, pp. 103418, 2022.

G.L. Plett, *Battery Management Systems, Volume I, Battery Modeling*, Artech House, London, 2015.

J. T. Pukrushpan, Modeling and control of fuel cell systems and fuel processors, PhD Thesis, The University of Michigan, 2003.

E. Redondo-Iglesias, P. Venet and S. Pelissier, Efficiency degradation model of lithium-ion batteries for electric vehicles. *IEEE Transactions on Industry Applications*, 55(2), pp. 1932–1940, 2019. https://doi.org/10.1109/TIA.2018.2877166.

H. Wang, SA. Pourmousavi, WL. Soong, X. Zhang, N. Ertugrul and B. Xiong, Model-based nonlinear dynamic optimisation for the optimal flow rate of vanadium redox flow batteries. *Journal of Energy Storage*, 68, pp. 107741, 2023.

H. Wang, SA. Pourmousavi, WL. Soong, X. Zhang and N. Ertugrul, Battery and energy management system for vanadium redox flow battery: A critical review and recommendations. *Journal of Energy Storage*, 58, pp. 106384, 2023.

H. Wang, WL. Soong, SA. Pourmousavi, X. Zhang, N. Ertugrul and B. Xiong, Thermal dynamics assessment of vanadium redox flow batteries and thermal management by active temperature control. *Journal of Power Sources*, 570, pp. 233027, 2023.

CHAPTER 6

Microgrids with Distributed Energy Resources

Design, Operation and Case Studies

6.1 BACKGROUND

Traditional power networks primarily consist of power lines from the generation source to the load connection and current flows are unidirectional. Along the power lines, there are step-up transformers to increase the voltage for transmission across extensive high-voltage lines. Substations distributed along these lines feature protective circuit breakers and switchgear, stepping down the voltage through additional transformers to match the distribution level.

The current travels along distribution lines, reaching step-down transformers near end-users, ultimately delivering power to a range of consumers, from residential homes to large industrial complexes. The cable size in the power lines must be sufficient to handle the current level without overheating or experiencing excessive voltage drop, ensuring safety and reliable power delivery. However, other factors also influence cable selection, including voltage level, cable type and construction and cost-effectiveness. Therefore, it is important to emphasize that traditional power grids were

DOI: 10.1201/9781032692173-6

not designed to account for large-scale integration of renewable energy sources.

Figure 6.1a shows a one-way flow of power from the generator to the load in a traditional power grid. Due to the inherent resistance of power lines and transformers, voltage naturally drops at each stage. Traditional power grids are designed and managed to account for these expected voltage drops and maintain voltage within an acceptable range at the consumer end.

Figure 6.1b introduces the concept of a prosumer, a consumer who also generates electricity, typically through solar panels. As it is known, during peak sun hours, solar panels can generate more power than the prosumer consumes. This excess power can flow back into the grid, causing a voltage rise at the point of common coupling, PCC, the point where the prosumer connects to the grid. Note that although typical % voltage drops and rises are given in the figure, these values are highly dependent on the characteristics of the local network as well as the density and the nature of the loads and time of usage. However, they still indicate the potential voltage rise in the current power system due to every increasing local generation. It is likely that the future systems are most likely to integrate Battery Storage Systems (BSS) at the prosumer as well as community level. This will not only increase the level of power flow but will increase the time of zero power flow level (as shown in Chapter 1) and is likely to form the highest ratio as more BSS are accommodated at the prosumer and community level.

The voltage rise can be problematic for several reasons: safety (which can damage connected appliances), power quality (which can lead to malfunctions in connected devices) and solar inverter shutdowns (which can occur when voltage exceeds a defined range of solar PV inverters, but the shutdown may cause instability as well). Currently, several methods are employed to mitigate voltage rise from solar PV, including grid infrastructure upgrades, reactive power control on smart inverters and energy storage to avoid reverse current flow. However, one key conclusion under this grid structure is that the power demand from the conventional generator drops, assuming that there are other loads distributed along the power lines.

Before further discussions on Figure 6.1, it may be worth briefly explaining the concept of Distributed Energy Resources (DERs) first. DERs are small-scale power generation technologies strategically located near or within the point of use. Unlike traditional, centralized power plants that

Microgrids with Distributed Energy Resources ■ 427

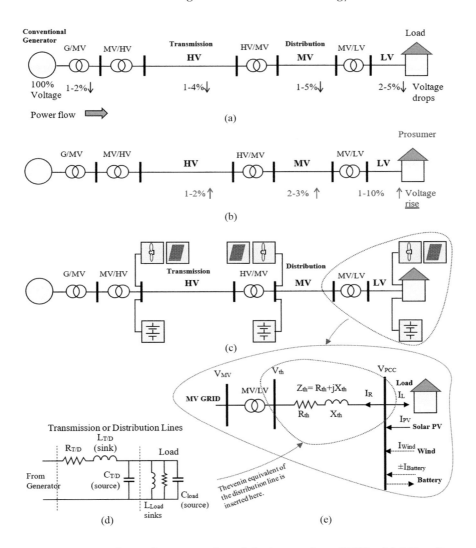

FIGURE 6.1 Traditional power grids and the integration of DERs: (a) A line diagram of a traditional power grid. (b) Introduction of a prosumer to the traditional grid. (c) Distribution of renewable energy (solar PV and wind) and battery storage to the traditional grid. (d) A typical electrical equivalent circuit of transmission and distribution power lines connected to an impedance load. (e) Application of the Thevenin equivalent circuit to a section of the traditional power grid that accept DERs. In the figure, $R_{T/D}$, $L_{T/D}$, and $C_{T/D}$ represent the resistance, inductance and capacitance, respectively, of the equivalent circuit of transmission or distribution power lines, along with common load model components that can be either resistive or reactive.

rely on large, remote facilities to generate electricity, DERs are designed to provide power closer to where it is consumed, utilizing technologies like solar PV panels, wind turbines, hydrogen fuel cells, microturbines, combined heat and power (CHP) systems and biomass generators. This distributed approach offers several advantages: increased reliability and security, reduced transmission losses, lower overall energy costs and environmental benefits. While current DER technologies primarily focus on renewables like solar and wind, advancements in microreactors (potentially in the 1–5 MW range) could also offer an additional energy option for future DER applications.

Therefore, Figure 6.1c expands on the concept of DERs and how Thevenin's equivalent circuit can be used to analyse voltage rise and reverse power capability based on acceptable voltage rise levels. Unlike a single reverse power generator on the prosumer side, this figure highlights the possibility of integrating various other DERs like solar PV, wind power and battery storage at different grid levels (transmission, distribution and load). It also emphasizes that voltage rise can be a critical issue, especially with unpredictable load patterns and the introduction of EVs and battery storage that can act as both generators and loads.

Figure 6.1d illustrates a general electrical equivalent circuit for transmission and distribution lines, highlighting the applicability of Thevenin's equivalent circuit for analysing voltage rise at the PCC between any voltage bus bar and a load. This analysis provides insights into voltage behaviour at the PCC, predicting potential voltage rise (or decay) and control issues when integrating DERs. Such integration is critical for smart inverter operation connected to the grid, performing various tasks as discussed in Chapter 5.

Thevenin's equivalent circuit simplifies the AC circuit analysis with variable loads by reducing the complex impedance networks of the power lines to a single voltage source V_{th} and a single impedance Z_{th}. This transformation involves replacing the linear, bilateral network (the direction of the current flow does not have any effect) of impedances and AC sources with an equivalent circuit comprising a single AC voltage source and a single series impedance. The initial step in this general approach involves identifying the circuit portion to replace with the Thevenin equivalent, particularly where the load connects. Following this, the load impedance is removed to isolate the network for analysis, and the open-circuit voltage across the load-connecting terminals is calculated. To find Z_{th}, all independent sources are deactivated (voltage sources shorted, current sources opened), and the total impedance seen at the open terminals is calculated. This process allows for the replacement of the original network with V_{th} in series with Z_{th}, upon which the load is reconnected, as shown in Figure 6.1e.

However, it is important to note that while Thevenin's theorem facilitates the representation of a complex network, it has limitations due to the assumption of a linear circuit. Real-world power system behaviour might slightly deviate with non-linear elements, and Thevenin's equivalent is primarily used for steady-state analysis, potentially not capturing transient voltage variations comprehensively. Despite these limitations, Thevenin's equivalent circuit remains a valuable tool for analysing voltage rise and effectively designing and managing grids to integrate DERs.

Note that DER integration, including solar PV and battery storage, usually happens at the Low Voltage (LV) and Medium Voltage (MV) levels, following the voltage regulation standards for these networks, but additional criteria for connecting DERs ensure grid stability and safety, such as limits on voltage rise due to the export of generated electricity. While DER integration typically involves connections at the LV and MV levels for rooftop solar PV and small-scale battery storage, wind energy, with its potential scale, often necessitates connections at the High Voltage (HV) level. This also applies to large-scale solar PV farms and BSS.

6.2 VOLTAGE REGULATION IN POWER GRID, ROLES OF INVERTER-BASED RESOURCES (IBRS) AND IMPACTS ON TRADING

Voltage regulation in power networks is needed to achieve the reliability and efficiency of electrical power delivery. Permissible voltage ranges vary across different parts of the power network, reflecting the diversity in national standards, grid infrastructure and operational practices. For example, in Australia, the Australian Standards AS60038 specifies standard voltages, including guidelines for voltage levels throughout the distribution and transmission network. Moreover, market operators and local distribution network service providers (DNSPs) impose specific requirements for integrating DERs based on the power grid's status and strength. Generally, voltage regulation levels are set at +10% to −6% for the LV network (230/400 V), defining a voltage range of 216–253 V for single-phase and 360–440 V for three-phase systems. At the MV level (11, 22 or 33 kV), the allowable variation is about ±6% and for HV networks, it is around ±10%, depending on DNSP regulations and network conditions.

Table 6.1 summarizes the various methods for voltage regulation at the PCC. Note that while each method has a role to play, as the power grid evolves to incorporate more inverter-based resources (IBRs), the dynamics of voltage regulation also change. Therefore, their effectiveness and suitability can vary significantly in an evolving power grid, which requires more dynamic/faster, flexible and intelligent solutions, highlighting the

TABLE 6.1 Comparative Analysis of "Voltage Regulation Methods" for Evolving Inverter-Dominant Power Systems

Regulation Methods	Pros	Cons	Applicability
Active power curtailment	Simple	Leads to energy wastage	Less desirable due to renewable energy loss
On-load tap-changing transformers, OLTC	Proven and robust	Wear, slower, designed unidirectional flow	Needs faster technologies
Volt-var control	Improves power factor and voltage stability	Requires careful coordination, potential control conflicts	Remains applicable with smart inverter integration
Smart inverters with reactive power control	Fast response; enhances grid flexibility	Complex control strategies, potential for conflicts	Highly suitable for dynamic voltage support
Static var compensators (SVC) and static synchronous compensators (STATCOM)	Fast reactive power compensation; improves stability	High cost, complex maintenance	Useful for managing fluctuations in any grid type
Battery energy storage systems, BESS	Provides voltage support and energy storage, quick response	High cost, storage and lifetime limitations	Increasingly important for energy shifting and regulation
Demand response, DR	Reduces peak load indirectly without infrastructure changes	Variable participation; delayed impact	Useful as part of broader voltage management strategy
Advanced distribution management systems, ADMS	Integrates multiple controls, optimizes grid performance	Significant technology and infrastructure investment	For managing diverse resources and stability

growing importance of smart inverters, BESS and advanced grid management systems.

As the table indicates, to mitigate voltage drops in the conventional power grid, either manual or on-load tap-changing transformers are utilized. However, the voltage regulation levels previously mentioned may result in undesirable voltage levels on these transformers due to strong reverse currents from DERs at various PCC locations. Note that this is a very critical issue when reverse power flow is at the transmission level as highlighted in the figure. Figure 6.2a shows the real-world power flow at the transmission level at midday due to the large-scale solar PV integration in the region.

A significant concern is also the frequency of tap changes in on-load tap-changing transformers (see Figure 6.2b), which is influenced by reverse current fluctuations arising not only from load demand but also from the intermittency of renewable energy sources at a renewable-rich power grid.

Note that while some voltage regulation and management methods listed in Table 6.1 for AC networks are not directly transferable to DC grids, the stability issues in DC grids are distinct but equally critical to address. The development of sophisticated control strategies and PE devices, inclusive of integrated protection schemes, is vital to accommodate the unique characteristics of DC systems for their reliable and efficient operation. As DC systems become more dominant, especially in microgrids, EV charging infrastructure and renewable energy integration, comprehending and managing these stability issues becomes essential to fully realize the advantages of DC technology. Table 6.2 summarizes the

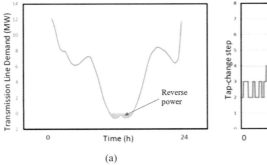
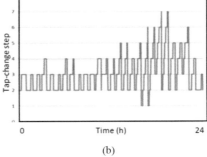

FIGURE 6.2 Correlation between transmission line demand and tap-changing activity in a day. A real-world scenario of reverse power flow at the transmission level was during midday due to large-scale solar PV integration in a region.

TABLE 6.2   Summary of the Future Impacts of IBRs on AC and DC Grid Energy Trading and Power System Stability

Aspect	Impact on AC Grid Energy Trading and Power System Stability	Key Features in AC Grid Context
Market dynamics	Faster markets favouring short-term ancillary services, increased DER participation.	Development of new market products and regulatory adaptations for DER integration.
Grid reliability and efficiency	Improved stability and efficiency through quick adjustments and optimized energy use.	Advanced coordination mechanisms; integration of AI and machine learning for real-time management.
Technological and regulatory	Seamless integration of DERs and evolution of regulatory frameworks	Regulatory reforms and technological advancements for effective DER management.
Energy trading innovations	Growth of peer-to-peer trading platforms, decentralized microgrids, adoption of highly dynamic pricing models.	Blockchain and secure technologies for trading; flexible pricing models to reflect real-time conditions.
	Impact on DC Grid Energy Trading	**Key Features in DC Grid Context**
Real-time energy markets	Facilitates instantaneous energy balancing, critical for DC grid stability	Shorter trading intervals, faster grid responses
Service-based trading	Shifts focus from mere energy sales to stability services	Emphasis on voltage control and power quality management
Decentralized and peer-to-peer models	Enables direct energy transactions between DC grid users	Use of blockchain for secure, transparent trading
Dynamic pricing	Encourages efficiency and flexibility in a fast-paced DC environment	Prices reflect real-time supply and demand conditions
Participatory grid	Requires active involvement from all generators for immediate response	Cooperative control and compensation mechanisms

potential future impacts of IBRs on energy trading and power system stability, outlining critical considerations that accompany this technological evolution towards the DC grid.

Note that the power quality in DC grids mentioned in the table is defined by maintaining voltage stability, minimizing voltage ripple, managing transients and surges, controlling harmonic distortion, mitigating EMI, regulating conducted and radiated emissions and ensuring proper grounding. These elements are essential for the reliable and efficient operation of electronic equipment and the overall integrity of the DC grid.

6.3 ANALYSIS OF REVERSE CURRENT FOR VOLTAGE REGULATION

The key reference in analysing reverse current is the voltage profile at the PCC; it is essential to maintain the minimum and maximum voltages within a range prescribed by standards.

To offer a mathematical framework for analysing the circuit given in Figure 6.1e, consider a simplified scenario in which only a solar PV system is connected to the final bus bar on the right side. The voltage at the PCC is defined as V_{PCC}. The complex current, containing both real and imaginary components and injected into the bus bar by the solar PV system, is represented as I_{PV}, while the load current is indicated by I_L. Therefore, the net reverse (feed-in) current, I_R, can be accurately defined by employing the complex conjugate of the net apparent power, S_R, which is exported to the grid at the PCC, alongside the complex conjugate of V_{PCC} at the PCC. In addition, this relationship can be simplified by assuming that V_{PCC} is approximately equal to V_{th}, which streamlines the reverse current equation. The following complex expressions contain these descriptions.

$$I_R = I_{PV} - I_L = \left(\frac{S_R}{V_{PCC}}\right)^* \left(\frac{P_R - jQ_R}{V_{th}}\right) \quad (6.1)$$

$$V_{PCC} = V_{th} + Z_{th}I_R V_{th} + \left(R_{th} + j\, X_{th}\right)\left(\frac{P_R - jQ_R}{V_{th}}\right) = V_{real} + jV_{imaginery} \quad (6.2)$$

Since $V_{real}^2 \gg V_{imaginery}^2$, the voltage increase at PCC can become,

$$\Delta V = V_{PCC} - V_{th} \approx \left(P_R R_{th} + Q_R X_{th}\right) \quad (6.3)$$

As it can be seen, this equation can be rearranged to define the amount of reverse active power that can ensure that the voltage increase limit is not exceeded,

$$P_{R(\max)} = \frac{(\Delta V_{\max}\, V_{th} - Q_R V_{th})}{R_{th}} \qquad (6.4)$$

Now, Figure 6.1e can be adapted as in Figure 6.3 to utilize the above equation when multiple DERs, such as solar PV, wind, BSS and EV (AC or DC coupled), are connected to the same PCC. It can then be observed that to maintain the voltage within the standard ranges as previously mentioned, the total reverse active and reactive power can be regulated. This regulation can be achieved by controlling the active and reactive power outputs of the DER inverters, while the load's active and reactive power may also vary. Note that, BSS and EV can be utilized to control the overall demand at PCC during charging.

The voltage rise limit and the amount by which voltage increases in a power system are also connected to droop control, which is a method used in power systems to stabilize voltage and frequency as discussed previously in Chapter 4. In the context of voltage control, as the "generator" power output increases, the system's voltage tends to rise. Droop settings can be configured to allow for a certain amount of voltage rise before the control system responds to bring the voltage back within the desired range. By

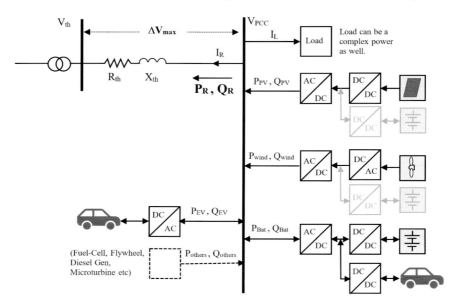

FIGURE 6.3 Schematic of a PCC with multiple DERs for calculation of maximum reverse active power and reactive power to stay within maximum voltage rise level of $\Delta V_{\max}$.

doing so, droop control contributes to maintaining the balance between generation and load while complying with voltage standards. Therefore, this is particularly relevant when integrating DERs, which can cause voltage fluctuations due to their variable output.

In addition, the DERs in the figure may represent larger IBRs, such as solar PV and wind farms or a large EV charging parking platform. Therefore, the network structure given in the figure is also suitable for use in a Virtual Power Plant (VPP) if the storage units are interconnected and sufficiently large enough to be dispatched as a combined power through a control system of the VPP. Such a VPP can provide power to the grid or absorb power from the grid, depending on demand.

Furthermore, the structure can provide Frequency Control Anciliary Services (FCAS), as it can be used by transmission system operators to maintain the frequency of the grid within the prescribed limits. For example, the BSS can rapidly discharge or charge to provide immediate power to the grid or absorb excess power, respectively, helping to regulate frequency. Similarly, EVs can be controlled to charge or stop charging to assist with frequency control.

6.4 POWER QUADRANTS, POWER FACTOR AND HARMONIC POWER

In the analysis and design of power systems, a fundamental understanding of various electrical parameters is essential. Therefore, a foundational education includes courses such as circuit theory, power systems engineering, power system stability and control, electrical protection, electrical machines and power electronics. This chapter does not aim to cover these fundamentals again but rather to provide a structured approach to help through detailed results in subsequent sections.

To achieve this, Table 6.3 provides essential definitions and equations that will be frequently referenced in the descriptions and results given. Note that the table is divided into two sections to address both sinusoidal and non-sinusoidal waveforms. Sinusoidal waveforms enable simplified descriptions due to the direct relationship between their maximum and RMS values, a simplification that is further utilized in a balanced three-phase system. However, the integration of PE converters into a three-phase system can result in imbalances and waveform distortions. Therefore, non-sinusoidal waveforms, characterized by significant harmonic content, require the use of specialized meters for measurement and analytical tools to consider waveform complexity. These "true" quantities

TABLE 6.3   Fundamental Definitions and Equations for Power System Analysis in Single-Phase and Three-Phase AC Circuits and in PE Converters

Term	Definition	Single-Phase	Three-Phase
Sinusoidal Waveforms			
RMS (effective) voltage in V	It is equal to the DC which, flowing through a resistance R, delivers the same power to R as the periodic current does.	$V_{rms} = V_{peak}/\sqrt{2}$	$V_{rms(line)} = \sqrt{3} V_{rms}$
RMS current in A	Same as above but for current waveform	$I_{rms} = I_{peak}/\sqrt{2}$	$I_{rms(line)} = I_{peak(line)}/\sqrt{2}$
Complex power, S	The total power in an AC circuit	$\underline{S} = P \pm jQ$	$\underline{S} = P_{total} \pm jQ_{total}$
Real (active) power in W	The average power that performs work or generates heat.	$P = V_{rms} I_{rms} \cos\phi$	$P_{total} = \sqrt{3} V_{rms(line)} I_{rms(line)} \cos\phi$
Reactive power, in var	The power that oscillates back and forth due to the reactive components in the circuit.	$Q = V_{rms} I_{rms} \sin\phi$	$Q_{total} = \sqrt{3} V_{rms(line)} I_{rms(line)} \cos\phi$
Apparent power in VA	The product of the RMS voltage and current.	$S = V_{rms} I_{rms}$	$S_{total} = \sqrt{3} V_{rms(line)} I_{rms(line)}$
Power factor, pf	The ratio of real power to apparent power, indicating the efficiency of power usage.	$pf = \cos\phi = S/P$	$pf = \cos\phi = S_{total}/P_{total}$
Instantaneous power in W	Power at any instant in an AC circuit.	$p(t) = v(t)\, i(t)$	$p(t) = 3[v(t)\, i(t)]$
Average power, in W	The average of instantaneous power over a complete cycle.	$P_{ave} = \frac{1}{T}\int_0^T v(t) i(t) dt$	$P_{ave} = \frac{1}{T}\int_0^T 3v(t) i(t) dt$
Total Harmonic Distortion, THD, in %.	The ratio of the sum of all rms magnitudes of harmonic (H_n) content of a waveform to the fundamental content H_1.	$THD\% = 100 \frac{\sqrt{\sum_{n=2}^{\infty} H_n^2}}{H_1}$	$THD\% = 100 \frac{\sqrt{\sum_{n=2}^{\infty} H_n^2}}{H_1}$
Non-Sinusoidal Waveforms			
Average value of voltage; V_{ave}, V_{mean} or V_{dc}	It is the arithmetic mean of the voltage over a specified time interval	$V_{ave} = \frac{1}{T}\int_0^T v(t) dt$	$V_{ave} = \frac{1}{T}\int_0^T v(t) dt$
Average value of current, I_{ave}, I_{mean} or I_{dc}	It is the arithmetic mean of the current over a specified time interval	$I_{ave} = \frac{1}{T}\int_0^T i(t) dt$	$I_{ave} = \frac{1}{T}\int_0^T i(t) dt$
True RMS voltage	It is calculated as the square root of the average of the squares of the instantaneous value	$V_{rms} = \sqrt{\frac{1}{T}\int_0^T i(t)^2 dt}$	Same as in single phase, but for phase and line voltages
True RMS current	Same as above	$I_{rms} = \sqrt{\frac{1}{T}\int_0^T i(t)^2 dt}$	Same as in single phase, but for phase and line currents
Total Harmonic Distortion, THD	As described in sinusoidal waveforms. It can be calculated for current, voltage or power functions.	$THD\% = 100 \frac{\sqrt{\sum_{n=2}^{\infty} H_n^2}}{H_1}$	$THD\% = 100 \frac{\sqrt{\sum_{n=2}^{\infty} H_n^2}}{H_1}$
Power factor, pf	For nonlinear loads in PE converters, it is defined using the total harmonic distortions.	$pf = \dfrac{1}{\sqrt{1 + THD_I^2}}$	$pf = \dfrac{1}{\sqrt{1 + THD_I^2}}$

Notes: ϕ is known as the phase angle of sinusoidal current waveform commonly reference to the voltage waveform. T is the period of the waveform and the integration should be carried out in an integer number of periods. In AC grids, THD is primarily defined for current waveforms reference to sinusoidal supply voltage. In pf calculations in sinusoidal waveforms, the angle ϕ is positive (lagging pf) in inductive loads, and the angle ϕ is negative (leading pf) in capacitive loads where the current leads the voltage. The rms values can also be estimated using harmonics contents of waveforms, say for current: $I_{rms} = \sqrt{I_{dc}^2 + I_{rms(1)}^2 + I_{rms(2)}^2 + \cdots + I_{rms(n)}^2}$

are required for analysing the performance and power quality of circuits generating non-sinusoidal waveforms. In addition, remember that the active power does the useful work, the reactive power only represents oscillating energy.

It is important to note that the terms average, mean and DC value of a voltage or current waveform are often used interchangeably. However, their meanings may differ based on the context and the specific waveform under consideration. The definitions provided for non-sinusoidal waveforms assume that voltage and current waveforms are not purely sinusoidal since their average value over one complete cycle is zero. Note that for rectified waveforms of sinusoidal waveforms or waveforms with a DC bias, the average voltage or current is the DC component of the signal.

Note also that in AC circuits, the RMS value of voltage or current represents the equivalent DC value that would deliver the same power to a resistive load, effectively expressing the power capacity of an AC waveform due to its time-varying characteristic. Therefore, in DC applications, such as batteries supplying DC power, the RMS and mean values are effectively identical, given that the output is constant and does not vary over time. The true RMS value of a voltage or current waveform is a measure of its equivalent DC power-carrying capability. As shown in the table, it is calculated as the square root of the average of the squares of the instantaneous values. True RMS meters, commonly used in power electronics converters, are designed to accurately measure the RMS value of any waveform, regardless of its shape.

As is known, power represents the rate at which energy is transferred or converted over time. For DC systems, where voltage and current are stable over short intervals, calculating power is straightforward—simply the product of voltage and current. In contrast, AC systems feature voltage and currents that oscillate between positive and negative values. The transfer of real, or active, power occurs when voltage and current waveforms coincide temporally. The optimal transfer, where all power is utilized effectively, occurs when these waveforms are not only identical in shape but also perfectly synchronized, a state referred to as being "in-phase".

When there is a misalignment of voltage and current waveforms, known as a phase difference, the instantaneous power fluctuates between positive and negative, leading to a bidirectional flow of energy. The most extreme case of misalignment results in all the power that was sent forward in the grid being returned, netting zero productive power flow—this is known as pure reactive power. Although reactive power does not perform any real work, it is necessary for the functioning of certain components in the system, such as inductors and capacitors, which require this energy to maintain electric and magnetic fields.

Most loads on the power grid consume a combination of active and reactive power. As given in Table 6.3, pf quantifies the proportion of power that is active relative to the total or apparent power, with the equation $pf = P/S$, where P is active power and S is apparent power, calculated as the product of voltage and current magnitudes ($S = V\,I$). Apparent power represents the total power flow without considering the phase difference and is always a positive value, indicating the upper limit of useful power for given electrical parameters.

It is also known that in power electronics, inverters play a crucial role in converting DC into AC, which is commonly used in DERs. However, the output waveforms of inverters are not perfect sine waves due to the switching actions inherent in their operation. These imperfections introduce harmonics into the power system, which are multiples of the fundamental frequency of the AC signal. Despite these non-idealities, for the purposes of power system analysis and design, it is common practice to focus on the fundamental components of the inverter's output—that is, the primary sine wave at the intended frequency, assuming that the inverter is well-designed and the Total Harmonic Distortion (THD) is kept low. This means that the fundamental frequency dominates and the higher-order harmonics are relatively insignificant. This assumption allows us to simplify analyses and discussions around power quality by focusing on the fundamental components of voltage and current waveforms, which are essential for calculating active, reactive and apparent power as well as for assessing the pf in the system.

Figure 6.4 illustrates three different power scenarios, indicating the relationship between voltage, current and resultant power over time. In the case on the left, voltage and current are in phase ($pf = 1$), indicating that all power is active with no reverse flow. In the middle figure, the current lags the voltage by 45° ($pf = 0.7$), resulting in a predominant forward power flow but with some reverse reactive power. In the waveforms on the right, current and voltage are 90° out of phase ($pf = 0$), where there is an equal forward and reverse power flow, indicating pure reactive power without any net transfer of active power.

Note that when the voltage and current waveforms are out of alignment, the instantaneous power starts alternating between positive and negative power, and power starts flowing back and forward. The worst-case

Microgrids with Distributed Energy Resources ■ 439

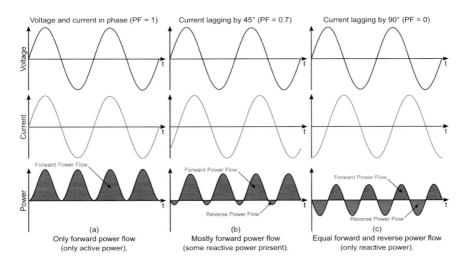

FIGURE 6.4 Instantaneous power flow for different voltage–current alignments, for the same apparent power. From left to right: $pf=1$, $pf=0.7$ and $pf=0$.

alignment causes all the power that flowed forward to flow back, resulting in a net zero power flow. This is a reactive power flow and provides no real/active/useful power despite the presence of voltage and current (assuming no losses in the system). Despite not delivering any useful power, reactive power still requires the power system and generators to supply the current component of the reactive power, which increases the load on the power system and reduces the amount of useful power the system can deliver. Therefore, as illustrated in Figure 6.3, reactive power will still be required in AC networks due to the reactive loads (such as an electric motor) and power line equivalent circuits (as illustrated in Figure 6.1d).

Understanding power quadrants is crucial in the power grid, especially when it involves uni- or bidirectional inverters or converters, such as those used in EVs, BSS and other DERs. As it is given in Figure 6.5c power quadrants illustrate the direction of power flow and the nature of power (active or reactive) in AC systems, providing insight into how power electronics devices operate under different conditions. Note that when discussing power quadrants (hence pf), it is important to note that while an inverter output may not be a perfect sine wave, the fundamental components are usually considered in the analysis, with the assumption that the level of THD is low enough not to significantly impact the overall power system

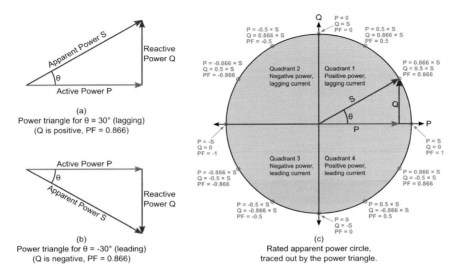

FIGURE 6.5 Power triangles and the circle of constant apparent power showing four-quadrant operation at a PCC.

performance. This simplification is key to making the complex behaviours of power systems more tractable and manageable in practical applications.

Knowledge of power quadrants is valuable for designing, operating and optimizing modern power systems with advanced power electronics. It aids in the effective integration of renewable energy, enhances grid stability, and enables innovative energy storage solutions, contributing to a more flexible and resilient power grid.

For a constant apparent power (e.g., the rated apparent power of a generator), the combinations of active and reactive power trace out a circle in the power quadrant diagram known as the "constant apparent power circle". This characteristic is visible from the apparent power definitions. In fact, the circle represents the operating area of a device, with the edge of the circle representing the rated operating limit. For a given apparent power, increasing reactive power will reduce the amount of active power that can be supplied. Hence it is desirable to control and minimize reactive power to maximize the active power that can be supplied by the power system.

In the ideal/best case, all power is useful, and the real power and apparent power are equal. This gives a power factor of 1. When there is reactive power in a system, the power factor is less than 1. In the worst case of all reactive power (no active power), the power factor will be 0, and $S=Q$.

The power factor can also be negative, which has the same meaning as a positive power factor, except that the active power is flowing in the opposite direction (−P).

The power triangles given in Figures 6.5a and b provide a graphical way of visualizing the relationship between active (real) power, reactive power and apparent power. Positive active power represents power flow out of a generator and into a load, and negative active power is power flow out of a load (e.g., a house with solar PV). Positive reactive power represents reactive power "consumption" by a load, where the current waveform is lagging behind the voltage waveform, and is typically caused by inductive loads (such as motor and transformer windings and any other coil structures). Negative reactive power represents reactive power being "supplied" by a load, where the current waveform is leading the voltage waveform, and is typically caused by capacitive loads (such as *pf* correction equipment, capacitors in PE converters and cable systems).

The concept of a four-quadrant operation when discussing inverters, refers to the ability of the inverter to handle bidirectional power flow both in terms of active and reactive power. This is essential for applications like BSS, where the inverter must be capable of charging (absorbing power) and discharging (delivering power), as well as providing or consuming reactive power for grid support functions.

Figure 6.6 is given as an operational capability chart for the PQ quadrants, indicating the boundaries within which DERs can function in terms of real and reactive power. Each quadrant represents a unique mode of operation: the production or consumption of real power and the generation or absorption of reactive power. As previously discussed, this quadrant representation also indicates the multifunctional capabilities of PE devices, such as bidirectional inverters, which are capable of operating across all four quadrants to provide dynamic grid support and battery management. The figure also demonstrates the flexibility available to DERs for VPP and FCAS operations, allowing them to act as a singular, controllable entity at the PCC within the power grid.

In the figure, BSS curves reveal the extent to which, for a specified level of active power, reactive power can either be absorbed or delivered to the grid, in both positive and negative directions. The curve's rectangular shape indicates that the capability for reactive power remains constant across a range of active power outputs until reaching the physical limits of the battery inverter. In addition, it indicates that the maximum real power the DER can absorb from the grid for charging batteries can change.

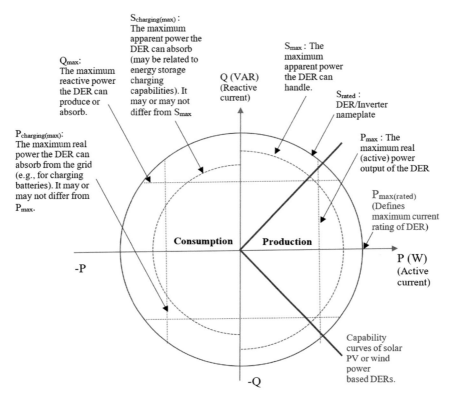

FIGURE 6.6 Operational capability chart for a DER (battery) showing maximum real and reactive power limits in consumption and production quadrants reference to the rated name plate data, and other DERs (solar PV and wind power).

For solar PV and wind power, the linear relationship between active and reactive power shown by the lines indicates that the generation of reactive power is directly proportional to the active power output. This characteristic is common for fixed-speed wind turbines lacking intricate reactive power control mechanisms. In the case of solar PV power, the slope of the line defines the amount of reactive power the solar inverter is able to generate at various levels of active power.

Figure 6.7 shows load-sharing principles among multiple DERs with varying control characteristics. It indicates how different control strategies, such as constant and droop, impact the distribution of both active and reactive power among DERs. This understanding is valuable for ensuring stability and efficiency in power systems that integrate diverse renewable energy sources and technologies, as it directly influences grid dynamics

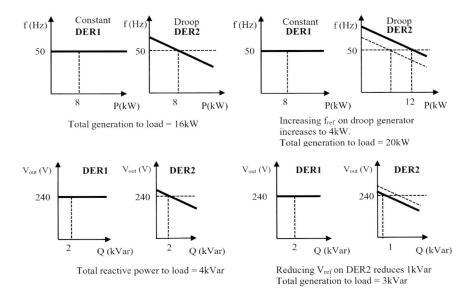

FIGURE 6.7 Load sharing between DERs with constant and droop control: The interaction of active power with frequency and reactive power with voltage in a multi-DER system.

and the interaction between various generation units. By adjusting control settings such as frequency and voltage references, it is possible to balance the load and maintain the desired power quality and supply levels that can be associated with the load demand or limit voltage increase, reflecting the complex orchestration required for power systems management as the number of DERs increase.

As illustrated, the figure demonstrates the load-sharing behaviour of two DERs employing distinct control strategies—constant control for DER1 and droop control for DER2—across two different sets of graphs: frequency versus active power and voltage versus reactive power. In the top graph, DER1 consistently delivers 8 kW, maintaining constant frequency control. Initially, DER2 also contributes 8 kW, operating under droop control, resulting in a combined delivery of 16 kW to the load. On the right, DER1 continues at a constant frequency, while DER2 has increased its active power output to 12 kW by raising the frequency reference, bringing the total supplied active power to 20 kW.

The lower diagrams in Figure 6.7 illustrate DER1 providing a constant voltage output while delivering 2 kVar of reactive power. Concurrently, DER2 under droop control also delivers 2 kVar, accompanied by a slight voltage reduction. The bottom right graph shows DER1 maintaining voltage while DER2, upon lowering its voltage reference, reduces its contribution to 1 kVar, resulting in a net reactive power of 3 kVar to the load.

6.4.1 Harmonic Power

The management of four-quadrant power and harmonics is an important aspect of PE converters, playing an essential role in the integration and efficient functioning of DERs and BSS across both AC and DC networks. Harmonics, defined as voltage or current components at frequencies that are integer multiples of the system's fundamental frequency, can be produced by non-linear loads and the switching actions of PE devices, including inverters. In four-quadrant inverters, controlling harmonics is vital to maintaining power quality, a task commonly achieved through the use of filters or control algorithms designed to modulate the inverter's switching activity and minimize harmonic injection into the grid, as mentioned in Chapter 4.

Non-linear loads introduce an additional power component, often named distortion power or harmonic power, due to their non-sinusoidal current draw. This phenomenon adds complexity to the power system without contributing to active or reactive power. The power tetrahedron model provides a means to visualize and explain the power interactions in systems with non-linear loads, dividing total power into its active, reactive and distortion components. This model can be used for the analysis, design and optimization of systems with a substantial presence of non-linear loads.

As shown in Figure 6.8, the power tetrahedron, first described by Constantin Budeanu, represents the different components of power in a three-dimensional form. The third axis represents distortion power, which occurs from the harmonic content of the load due to the time distortion of the current waveform, which neither performs work nor generates magnetic fields. The vertical plane on the right side is divided into fundamental reactive power, $Q1$, associated with the fundamental frequency necessary for magnetic field creation in inductive loads, and harmonic reactive power, D, related to the harmonic currents from non-linear loads.

The active and reactive powers can be determined using a data logging system employing a method devised by Budeanu in 1927. This method separates voltage and current into frequencies, calculating active and reactive power at each frequency up to 2,500 Hz or the 50th harmonic, as will be further explored in later case studies. The sums of these active and reactive powers constitute the total active power and reactive power.

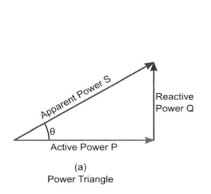 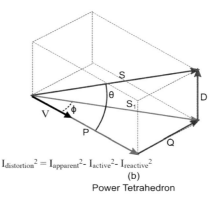

FIGURE 6.8 Power tetrahedron reference to the power triangle, relating the active, reactive and distortion power to the apparent power under non-sinusoidal conditions.

In scenarios with sinusoidal voltage and current waveforms, the power triangle illustrates the relationship between active, reactive and apparent power, described by the equation $S = \sqrt{P^2 + Q^2}$. However, this relationship does not hold for non-sinusoidal or distorted waveforms, leading to a discrepancy in the terms. Budeanu's solution introduced distortion (or void) power, $D = \sqrt{S^2 - P^2 - Q^2}$, transitioning from a power triangle to a power tetrahedron.

In conclusion, while distortion power modifies the relationship between P, Q and S for non-sinusoidal waveforms, rendering reactive power less meaningful than in sinusoidal cases, a zero reactive power no longer indicates a power factor of 1. Therefore, attempting to reduce reactive power to zero using capacitors or inductors may not effectively improve the power factor, underscoring that reactive power becomes a less relevant parameter in enhancing the power factor of a system with non-sinusoidal waveforms.

6.5 GENERAL DESIGN CRITERIA OF BATTERY STORAGE SYSTEMS

Table 6.4 summarizes the diverse applications of battery storage at different system levels (generation, transmission and distribution) reflecting a transformation in the traditional power grid, which is also evolving at a greater pace. As is known, historically, power generation plants were primarily focused on electricity production. However, with battery storage integration, these plants are also adopting batteries for fast-response frequency regulation, black start capabilities and spinning reserve, enhancing

TABLE 6.4 Battery Storage Applications in Power Systems

Generation Level	Transmission Level	Distribution Level
Fast-response frequency regulation	Dynamic line rating support Dynamic stability support	Energy storage for utilities
Black start	Reducing interconnection cost	Facilitating high PV penetration in embedded microgrids
Spinning reserve	Voltage support of long radial circuit	Energy arbitrage
Backup and mission-critical power	Transmission deferral	Ramp-rate control of PV inputs
Power plant hybridization	Minimize reverse power flow	Peak demand management
Ramp rate management	Voltage support	Power quality improvement
Peak demand management		Distribution deferral
		Non-spinning reserves
		Microgrid support
		Demand response programs
		Islanding
		Peak shaving, load and time shifting
		Power reduction in curtailment events
		Voltage support
		Loss reduction

Emerging Applications and Practices

Renewable integration (wind and solar)
Asset deferral
Reactive power control
Mitigating intermittency (firming)
Electric vehicle (EV) charging infrastructure support
Frequency regulation in ancillary services market

grid stability and resilience. Batteries are increasingly providing backup power for critical infrastructure as well.

At the distribution level, battery storage is forming decentralized generation and storage, enabling bidirectional power flow. This is apparent in scenarios such as embedded microgrids and islanding operations. Moreover, emerging load types, such as EVs, contribute to the system not only as consumers but also as potential generators when linked to a bidirectional connection point. Therefore, battery storage is critical in load management, enabling energy consumption shifts through programs like demand response and supporting the integration of intermittent renewable energy sources.

Examining the table, it can be seen that at the generation level, battery storage supports functions like frequency regulation, black start and spinning reserve, traditionally managed by conventional power plants. This characteristic feature of battery storage supports the way for a more responsive grid that can handle a higher share of renewable sources while reducing variability and intermittency. Battery storage applications at the transmission level also continue to develop and provide essential grid stability support. The most significant changes, however, are evident at the distribution level, where battery storage is transforming the network by enabling high renewable integration, managing peak demand and improving power quality, which is a transformation that also empowers consumers and facilitates microgrid and decentralization concepts.

Figure 6.9 illustrates the characteristic time scale of various energy storage technologies on a logarithmic scale, ranging from ms to weeks, demonstrating their suitability for an array of applications within the power grid. Note that these durations are approximate and subject to variation depending on the technology, manufacturer and operational conditions. The exact values depend upon the design and size of the energy storage system as well as the requirements of the grid in question. For example, PE manages harmonics in the ms range, while magnetic storage types exhibit a comparable response range. Supercapacitors are designed to release energy within seconds, making them optimal for managing short-term power surges. Flywheels and batteries are utilized for frequency regulation, with flywheels being typically effective for up to 15 minutes. Load levelling involves a variety of batteries, including Li-ion, which can provide power for 1–4 hours. Compressed air energy storage and pumped hydro storage are configured for daily operation, typically ranging from 10 to 24 hours or longer. At the highest end of the scale, hydroelectric variations enable storage durations of several weeks, even across seasonal periods.

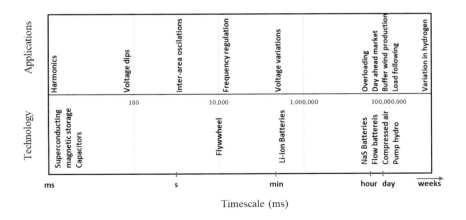

FIGURE 6.9 Correlation between energy storage technologies and their operational timescales with corresponding applications.

Note that the time between cycles is also very critical because it relates directly to the efficiency and longevity of the storage technology. Technologies designed for frequent cycling need to be robust and durable to withstand the rapid charge and discharge, whereas those used less frequently can be optimized for energy density and overall capacity rather than speed of response.

Table 6.5 also summarizes the critical technical characteristics of battery storage systems when utilized in various energy and power applications. The characteristics include backup time, the number of cycles per year and storage response time. These parameters are also needed for evaluating the performance and suitability of battery storage solutions for specific applications. For example, energy storage applications tend to require a higher power capacity and longer backup times. In contrast, power applications require rapid response to transient events and are typically involved in improving the quality and reliability of the power supply.

From the table, it can also be concluded that BSS offer a range of energy-related benefits, including the minimization of costs and emissions, enhancement of overall efficiency and reduction of peak demand which helps emission reduction. They also contribute to maximizing the life and reliability of both the energy storage system and power generators, ensuring a secure energy supply. In terms of power-related advantages, these systems are important for stabilizing the power grid by providing

TABLE 6.5 Technical Characteristics of Battery Storage for Energy and Power Applications

Application Type	Common Storage Applications	Power (MW)	Backup Time	Cycles/ Year	Storage Response Time
	Spinning reserve	~100	Hours	20–50	Sec to min
Energy	Load levelling	~100	Hours	250	Minutes
	Black start	~100	Hours	Seldom	<1 min
	Investment deferral	~100	Hours	>100	Minutes
	Power regulation with intermittent sources	<10	Minutes	1000s	<1 minute
Power	Integration of non-predictable sources	~10	Minutes	Frequent	<1 minute
	Power quality	<1	Minutes	<100	10 sec–1 min
	Line stability	~100	Seconds	100	~ cycles
	Power oscillation damping	<1	Seconds	100	~ cycles

frequency and voltage support and increasing the grid's inertia, hence enhancing the quality and reliability of the power supply.

The implementation of BSS in grid-scale applications is a complex task. It includes a number of components and activities that go beyond the battery unit itself and contribute to the final cost. From the major architecture that connects with the electrical grid to the specific control and communication systems necessary for operation, each component plays an important role in both operational and electrical efficiency. For example, a range of engineering, procurement and construction, EPC activities are needed for integrating the system. Moreover, legal requirements and permits are necessary for the BSS to function according to its primary design. Table 6.6 classifies all relevant aspects of BSS in grid-scale applications and provides a detailed description of each. These factors must be considered in system development, as was done and included for the microgrid test platform described in Section 6.7.

The costs associated with building and deploying a BSS can vary greatly due to factors such as differing project requirements, evolving grid demands and the availability of technologies and services. Despite these variations, it is possible to outline a general cost distribution for a utility-scale BSS. Typically, battery costs may represent approximately 40%–60% of the total

TABLE 6.6 Classification and Description of Components and Tasks Associated with the BSS in Grid-Scale Applications

Category	Components	Descriptions
Power conversion system, PCS	Bi-directional converter	For charging and discharging of the battery, capable of operating in four quadrants.
	Battery unit	The primary energy storage components
	Isolation transformer	Provides galvanic isolation to ensure safety, while mitigating noise and harmonics, and to match voltage levels between the grid and the battery system.
Balance of plant, BoP	Battery management system BMS	Manages the operational parameters of the battery to ensure safety and longevity.

(*Continued*)

TABLE 6.6 (*Continued*) Classification and Description of Components and Tasks Associated with the BSS in Grid-Scale Applications

Category	Components	Descriptions
(All the supporting components and auxiliary systems of a power plant)	Cooling system	Maintains battery temperature within safe limits
	Protection hardware	Includes devices that protect against overcurrent, overvoltage, undervoltage conditions and fire risks.
	AC switches and DC switches	To safely connect or disconnect the system from the AC grid or loads and to manage the continuous flow of DC from the batteries for safe isolation and protection.
	Enclosures	Houses the system's electrical components, providing protection from environmental factors.
	Bus bars	To distribute power efficiently and safely
	Alarm systems	To alert system malfunctions and safety breaches.
Engineering procurement construction	Design, purchase and integration	Involves designing the system to meet specific criteria and integrating with existing infrastructure; Covers acquiring all necessary components ensuring technical and quality standards; and system integration.
Grid interconnection	Location selection; Interconnection point specifics; Safety protocols; Noise management; Lightning protection; Grounding; and Communication and protection requirements.	Selecting a site based on strategic, technical, and regulatory considerations; Physical and technical specifications for connecting to the grid; Protocols to protect personnel and equipment during operation and maintenance; Ensuring the system operates within local noise regulations; To prevent damage from lightning; To ensure proper electrical grounding; To define the technical requirements for system communication and protection from the T/D providers.

(*Continued*)

TABLE 6.6 (*Continued*)   Classification and Description of Components and Tasks Associated with the BSS in Grid-Scale Applications

Category	Components	Descriptions
Legal and compliance	Regulatory approvals; Land use and zoning compliance; Environmental impact assessments; Compliance with building and safety codes; Interconnection agreements; Insurance; Financing agreements; and power purchase agreements, PPAs.	Necessary legal permissions from relevant authorities to construct and operate the system; To comply with local land use regulations and zoning restrictions; To evaluate the potential environmental impacts and the development of mitigation strategies; To follow codes that govern the construction and operation of such facilities; To detail legal contracts and terms of connecting and operating in relation to the grid; To cover and to protect against risks associated with the system's construction and operation; To detail the terms of financing, including loans and investor funds; and outlining the terms in an agreement under which the stored power will be sold.

system cost, with PE converters accounting for about 10%–15%. The balance for plant and installation might cover 15%–30% of the costs, while additional soft costs related to grid connection and legal aspects could contribute another 5%–20%. Operation and maintenance expenses are often estimated to be around 2.5% of the capital costs annually. It is important to note that these percentages are subject to change based on the specific details of a project and prevailing market conditions.

In contrast, smaller BSS units, such as those used in residential and small-scale industrial systems, may have different cost ratios. This difference arises due to factors like design standardization, mass production efficiencies and uniformity in grid connection procedures. As grid-scale BSS development begins to benefit from similar economies of scale, costs are expected to decrease significantly. The move towards decentralized and microgrid configurations that integrate multiple DERs may further drive the adoption of standardized, low-cost units, influencing the overall system cost.

Within the next decade, cybersecurity-related costs are anticipated to initially increase due to the growing complexity of hybrid power grids. However, the trend towards decentralization may eventually lead to a reduction in these costs. Building a BSS demands a multidisciplinary approach, requiring expertise across software, communication, civil, electrical and mechanical disciplines.

As the power grid evolves, with reduced system inertia hence a changing RoCoF, and as the penetration of IBRs increases, there will likely be a higher demand for software control and PE converters capable of responding to these new conditions. This is likely to increase software and related costs constitute a significant portion of the controlled hardware associated with batteries, converters, protection and monitoring. Such systems provide numerous functions, including ancillary services capability, reactive support, black start capability, ramp rate control, voltage isolation and stepping up, power quality control, time response, power management, thermal management, safety and protection, as well as frequency control, battery management and data logging.

Two critical components of BSS are the battery and the bidirectional inverter, which are interconnected and need to be designed together. Defining the maximum voltage level of the battery for the DC link, and consequently for the inverter, involves the inverter's voltage range, battery technology and system efficiency.

In large-scale BSS applications, using multiple inverters in a three-phase system offers several advantages, specifically in enhancing system reliability. If one inverter fails, others can continue operating, hence reducing the risk of complete system downtime. Moreover, scalability is easier with multiple inverters, as adding more units can meet growing demand. This modular approach provides flexibility in system design and expansion.

Note that the size of inverters is influenced by practical factors related to power electronics. Three-level neutral-point clamped converters, for example, are used in BSS applications for their higher output voltage and better harmonic performance, which reduce filter requirements. Despite their complexity, these converters can handle higher voltage levels, making them suitable for high-power applications.

H-bridge converters and modular multilevel converters are also used in BSS, enabling direct connections to the medium voltage grid without step-up transformers. This simplification and the high efficiency and low harmonic content of multilevel converters make them effective for grid-connected energy storage applications, though they require more semiconductor switches, increasing costs. Note that multiple inverters can operate more efficiently at partial loads compared to a single large inverter. Inverters can be turned on or off depending on the load, optimizing the system's overall efficiency. Transformerless topologies using two-level voltage source converters are still used for a direct connection to the medium voltage grid.

The input DC voltage levels and power ratings for designing voltage and current configurations of batteries can vary widely. For example, integrating second-life EV batteries into large-scale stationary BSS applications involves connecting battery cells in series and parallel. Lithium-based systems, common in BSS, often have a wide range of voltages depending on the series connection of cells.

Power conversion systems in BSS, responsible for bidirectional power flow during charging and discharging, must effectively handle voltage levels. Therefore, inverter selection should account for battery cell voltage variability, which can change by up to 40% between fully charged and discharged states.

High voltage levels are preferred in PE converters to minimize semiconductor current stress as well as to reduce losses. International norms consider the border between low and medium voltage at 1.5 kV. Note that higher efficiency requires higher-rated AC and DC voltages for efficient DC-AC conversion.

The physical layout, whether containerized units or dedicated buildings with modular setups with multiple battery and inverter strings, is critical for efficient, safe and reliable grid operation.

Power ratings for BSS range from small residential systems to large-scale utility grid applications, designed according to project needs, battery technology and integration methods, whether AC-coupled or DC-coupled with solar PV systems.

For large-scale BSS, power ratings of inverters and corresponding battery voltage and configurations are variable and highly dependent on the project's specific requirements. Applications can range significantly in scale, from 1–100 MW BSS. BSS setups using 1 MWh and 2 MWh Li-Ion batteries coupled with a 500 KVA inverter, indicate the flexibility in system design. Furthermore, configurations like a BESS with 4 MWh Li-Ion batteries and a 2 MVA power conversion system also demonstrate the scalability of such systems to support the evolving demands of smart grids and large-scale commercial solar PV systems. Note that the choice of battery technology affects performance characteristics such as energy density, specific power, charge/discharge efficiency, cycle durability and self-discharge rates.

In summary, selecting the appropriate battery voltage, power and energy capacity for a grid-scale application, as well as identifying the correct inverter ratings, involves a detailed analysis of several factors. Firstly, BSS should meet the specific needs of the application, such as peak demand shaving, load levelling, renewable energy integration and backup power. A high-level overview of the selection criteria may involve:

- The choice of battery voltage is influenced by the application's power and energy requirements, the type of battery technology and the overall design architecture. Higher voltages are often preferred in large-scale applications to reduce current and minimize losses. The selected battery voltage must be compatible with the bidirectional inverter.

- The definition of application also defines the power in kW or MW and energy in kWh and MWh. Note that charging and discharging C-ratings are probably the most critical parameters when considering the applications listed in Table 6.4. Duration and Depth of Discharge (DOD) should be known. Note that energy capacity is a product of power capacity and discharge duration in hours or min, which is also related to the frequency of operation.

- The energy capacity of the battery can be overrated to account for degradation over time, ensuring the system meets its performance requirements throughout its lifespan. In addition, considering the DOD and the impact on battery life, systems should be designed not to fully utilize the total capacity, extending the battery life.

- Inverter ratings must match or exceed the system's maximum power output to handle peak loads, which also defines the fault current level in hybrid grids. Inverters should be chosen for high efficiency (and also ideally with flat-efficiency characteristics as in WBG devices) and a suitable operational range for the expected battery voltage and system configuration. Inverters should also meet relevant grid interconnection standards for safety and power quality.

- In addition, inverters are slightly overrated compared to the battery system's nominal power to accommodate peak loads, enhance reliability and provide headroom for future expansion. This approach helps manage inefficiencies and ensures the inverter can handle transient conditions without overstressing.

6.5.1 Reference Applications in BSS Design: LDC, RoCoF and Virtual Inertia

6.5.1.1 Using Load Duration Curves, LDC for BSS

Load duration curves are generated by ranking the power (usually active power) levels of a system from highest to lowest, providing a clear picture of the frequency and duration of different loads. These curves are particularly useful in BSS design for several reasons. They help in recognizing the highest power demands and the aggregate energy requirement over a period, facilitating the scheduling of charge and discharge cycles of the BSS in response to variable or consistent energy demands. They are also useful for reliability planning during peak demand periods and help to avoid the excessive costs associated with overcapacity.

Designing a BSS to accommodate the full spectrum of a Load Duration Curve (LDC) results in significant costs, as it requires a system with sufficient capacity to handle peak loads that may occur infrequently. As the adoption of renewable energy grows, this strategy could result in oversized and underutilized systems. Therefore, to optimize costs, the BSS design should be based on an in-depth analysis of the LDC, with an emphasis on meeting the most critical load segments. This commonly involves designing a system that addresses substantial, rather than absolute, peak demands, which is more cost-effective.

The trend in decreasing costs for battery components, compared to other system components, suggests that future BSS projects could be more feasible even with broader LDC coverage. The integration of second-life EV batteries into large-scale stationary BSS applications is also anticipated to lower costs. Furthermore, advancements in PE technology are expected to improve round-trip efficiency, hence enhancing the cost-effectiveness of BSS solutions.

In addition, with the rise of negative pricing due to reverse power flow, particularly from high solar PV penetration, a well-designed BSS is required to meet energy needs while remaining cost-effective and sustainable in the long run.

Figure 6.10 displays three LDCs for distinct settings—a mining site, a residential area and during the summer season in South Australia. These curves illustrate the power demand over different time scales, reflecting the variability and consistency in electricity usage across sectors. The mining site's LDC suggests a high, consistent power demand typical for industrial operations with heavy machinery. The summer season LDC for South Australia likely represents a sharp peak in demand, indicative of increased

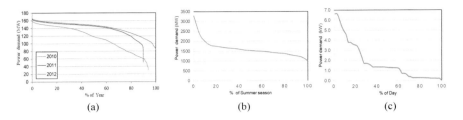

FIGURE 6.10 LDCs for different settings demonstrating power demand patterns: (a) in a mining site over three consecutive years, (b) in South Australia in a summer season, and (c) in a household.

use of cooling systems during the hottest part of the year. In contrast, the residential LDC shows a more moderate and varied demand pattern, with a baseline level of consumption and peaks reflecting typical household activity.

While future electrification practices, efficiency improvements and energy and demand management activities are likely to transform the LDCs of typical load sites, the current landscape indicates a wide array of distinct LDC characteristics across various load types. Table 6.7 presents such load types, highlighting their typical peak demand characteristics and LDC shapes, along with predicted shifts resulting from trends such as increased EV usage and the adoption of smart technologies. Therefore, it can be indicated that BSS solutions will be application-specific dictated by the unique LDC of each load type for optimizing energy storage and management.

Note that the desirable shape of an LDC for integrating BSS depends on the specific objectives of the energy storage system as well as the characteristics of the load and the broader energy system it is part of while providing cost savings. The specific shape and characteristics of the LDC for each type of load reflect the usage pattern and are critical for BSS design because they dictate how much energy storage is needed and how it will be used.

LDCs give insight into the daily or seasonal load patterns of a system, which is critical when sizing a battery to ensure it can handle peak demands. In addition, by showing the periods when energy demand is at its highest, LDCs help in determining the required capacity of the battery to meet these demands without frequent cycling, which can reduce battery life. Furthermore, they assist in understanding for how long the battery needs to discharge to meet the load requirements and how often these

TABLE 6.7 Existing Characteristics and Future Predictions of LDCs across Various Loads

Load Type	Peak Demand Characteristics	Typical LDC Shape	Future Predictions
Residential	Individual homes: Morning and evening peaks. Residential complexes: Similar patterns with additional variability due to shared facilities.	Double-peaked	Increased use of home energy management systems might flatten peaks. EV charging could introduce new peaks unless managed via smart charging.
Commercial	Buildings: Peaks during work hours on weekdays. Retail outlets: Peaks during business hours, weekends or shopping seasons.	Double-peaked with low night/ weekend load	Flexible work arrangements and online shopping might reduce peaks. Energy efficiency and solar adoption could further modify demand profiles.
Industrial	Continuous or shift-based peaks. Manufacturing plants: Constant if 24/7 or variable with shifts.	Flat during operations or variable	Advanced technologies may lead to more energy-efficient operations and potentially shift energy use to off-peak times.
Public services	Predictable and aligned with schedules. Street lighting: Increases after dusk, decreases before dawn. Water treatment: Consumption-based peaks.	Predictable or constant	Advances in LED and smart lighting can reduce energy consumption for street lighting. Water and utility demand might become more efficient with better infrastructure.
Transportation	EV charging, railways, airports: Peaks aligned with travel patterns.	Variable with travel patterns	EV adoption will significantly increase electricity demand. Smart charging and V2G technologies could alter demand.
Hospitality services	Hotels and restaurants: Peaks during check-in/out and meal times.	Variable with service times	Energy management systems could optimize demand, while renewable energy sources might offset some grid demand.

(*Continued*)

TABLE 6.7 (*Continued*)   Existing Characteristics and Future Predictions of LDCs across Various Loads

Load Type	Peak Demand Characteristics	Typical LDC Shape	Future Predictions
Healthcare	Hospitals: Constant demand for critical operations.	Flat, consistent demand	On-site renewable energy generation and energy storage systems might potentially smooth the overall demand curve.
Education	Schools and universities: Peaks during class hours.	Double-peaked with off-time dips	Remote and hybrid learning models might lead to less pronounced peaks in energy demand at educational institutions.
Entertainment	Cinemas, theatres: Peaks during events.	Peaks with events	Digital streaming might reduce demand peaks for traditional entertainment venues, while energy efficiency improvements could lower baseline consumption.
Agriculture	Variable, with seasonal peaks for planting/harvesting.	Seasonal variability	Technological advances in farming could lead to more predictable and efficient energy use, potentially flattening peaks.
Mining	Continuous high load or shifts with breaks.	Constant or peaked	Energy efficiency measures and shifts towards renewable energy sources could lead to different energy use patterns.
Data centres	Constant demand to support 24/7 operations.	Flat, 24/7 operation	Increased focus on energy efficiency and use of renewable energy might not alter the flat demand curve but could reduce the energy consumption.

discharges will occur, which impacts the battery's depth of discharge and, consequently, its overall lifespan and specifications.

Figure 6.11 presents an analysis of power demand characteristics in a microgrid with reverse power flow and the utilization of an LDC for BSS design. Figure 6.11a provides a data file showing active, reactive and apparent power, required for understanding the site's demand profile, including periods of reverse power flow when solar PV generation surpasses local demand. Figure 6.11b graphically demonstrates the fluctuations in power throughout a day, which informs the sizing of a BSS to manage excess generation and compensate for demand shortfalls. Figure 6.11c illustrates the LDC, highlighting where the BSS could provide the most benefit, particularly noting the reverse power segments indicating surplus generation. Finally, Figure 6.11d provides the relationship between demand, generation capacity, network capability and BSS operation. It emphasizes that the network's capacity must always surpass the load demand, which includes BSS and prosumer contributions, and details the optimal BSS capacity to ensure complete demand coverage, at the 50% level.

6.5.1.2 Using BSS to Alter the Rate of Change of Frequency, RoCoF and Virtual Inertia

As is known, traditional generation units utilize governor controls to respond to imbalances and stabilize frequency, which is primarily defined by the inertia of rotating machinery, contributing to the overall power system inertia. In a hybrid real power system, however, stability involves highly complex interactions of multiple quantities, including resonance, rotor angle (transient and small disturbance), voltage (large and small disturbance), frequency (short and long term) and converter-driven (fast and slow interaction) stability.

A critical indicator of grid stability is the rate of change of frequency, RoCoF, measured in Hz/s. It reflects how quickly the power grid's frequency changes over time. It can be calculated as the ratio of $\Delta f/\Delta t$, where Δf is the change in frequency from the nominal grid frequency within a time duration Δt.

Figure 6.12a illustrates the frequency response of a power system over time during a decelerating RoCoF event. It marks the start of the event and the subsequent application of virtual inertia. Although the components of RoCoF can be highly complex stability-related elements, this figure highlights the principle role of a BSS in providing virtual inertia, hence the changing characteristics of the frequency response of the power grid.

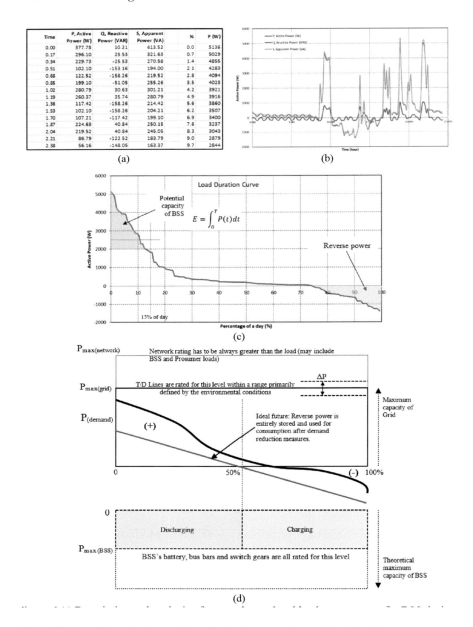

FIGURE 6.11 Description and analysis of power demand and load management for BSS design in solar PV-enhanced grid networks: (a) a sample power demand data set, (b) active, reactive and apparent power fluctuations over time, (c) LDC with potential BSS capacity overlay, and (d) Network capacity and BSS operation schematic with the ideal BSS scenario (50% level).

Microgrids with Distributed Energy Resources ■ 461

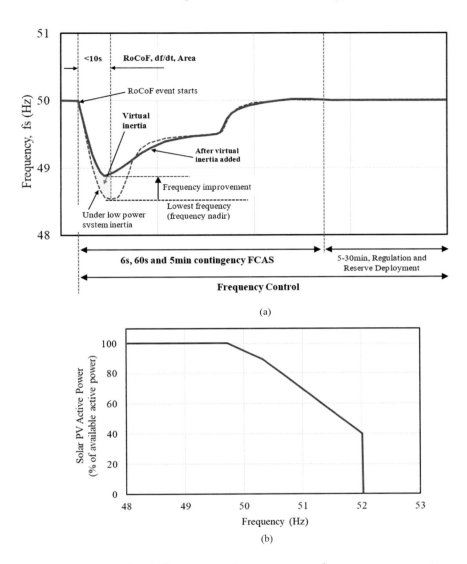

FIGURE 6.12 Graphical illustration of power system frequency response to a major loss of generation (dashed lines) and RoCoF mitigation through virtual inertia from battery storage systems, solid lines (a) and a typical solar PV active power versus frequency characteristics in a specific grid (b).

The figure indicates that BSS can simply inject active power to reduce the RoCoF, providing a buffer for other system responses. The curve labelled "after virtual inertia added" indicates improved frequency stability, allowing the system more time to deploy additional reserves or implement load shedding to avoid reaching the frequency nadir (the lowest frequency reached during an event).

It is important to note that as the integration of renewable energy and BSS continues to grow, grid codes are being defined to address critical RoCoF values. Due to the unique characteristics of local power grids, various codes are established in different parts of the world, such as those by ENTSO-E in Europe, NERC in North America and AEMO in Australia. It is worth emphasizing that as grids evolve, becoming hybridized and dominated by IBRs, the power system inertia is reduced as a result of decommissioned conventional synchronous generators. This requires further revisions to grid codes, imposing stricter requirements for fast frequency response from renewables and the development of advanced techniques for managing variable power sources.

For example, a typical active power versus frequency response curve for a specific region is shown in Figure 6.12b. This figure illustrates that in the case of a positive frequency deviation (above the rated frequency), active power should be reduced. Conversely, when the frequency is reduced, active power should be increased. This can be achieved via DERs and BSS. As illustrated in the figure, in the case of a frequency rise, active power should be reduced (which may be obtained from a solar PV system or a wind farm via a PE converter). If the frequency is above 50.2 Hz, for example, the solar PV system has to reduce active power by 10% of the nominal value within a defined time. When the threshold limit is exceeded, active power must be restored with a gradient of 40% per Hz.

It should also be noted that reactive power control may also be needed in conjunction with active power control. This control can be implemented by the control systems of intermittent renewable energy sources during large voltage deviations and under fault conditions.

Note that the virtual inertia highlighted in the figure, achieved through control algorithms, DERs, BSS and power electronics, can simply emulate the inertia response of traditional power systems. This emulation adjusts the RoCoF to prevent the network frequency from falling outside the stability range. However, the unpredictability of RoCoF events, especially when driven by low system inertia and external events, requires rapid recharging of the battery post-FCAS event response.

Therefore, a high C-Rate for battery charging is also critical in RoCoF events, as is ensuring the battery has sufficient capacity to deliver significant active power with the support of correctly rated PE converters. The battery's C-rating, a measure of its discharge rate capability and its response time is essential for batteries in FCAS applications, to discharge at rates sufficient to counter frequency drops within specific contingency time frames, such as 6 or 60 seconds.

Note that a higher inertia system requires more energy to change its frequency, affecting the amount of power needed to correct a deviation, and vice versa. Therefore, DERs in microgrids, especially those using PE converters and high C-rating battery-based BSS, can simulate inertia very quickly and at the right level well beyond mimicking the behaviour of traditional synchronous generators. As it can be observed, aggregations of DERs, managed as a VPP at the community or even household level, can be managed as a VPP to coordinate and provide a substantial active power response to a RoCoF event, leveraging the distributed nature of resources for a more resilient and flexible frequency support, although it can be costly.

However, for BSS design and battery selection, calculating the necessary active power to counteract RoCoF involves highly complex interactions as the time constant of the network and power imbalances need to be identified accurately since all power network components at the PCC have an impact.

The model in Figure 6.13 represents a block diagram representation of a power grid with only one simplified rotating machine as $P_{generation}$ but with all other possible major components. It is assumed that the model does not represent local oscillations that could trigger some frequency-sensitive

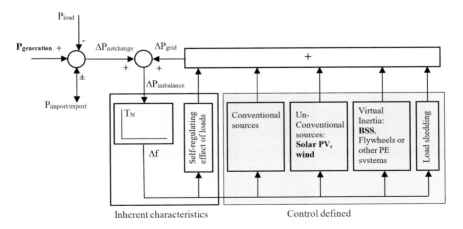

FIGURE 6.13 Block diagram of a power system model, illustrating the interactions between components that influence grid frequency stability and active powers.

relays in the grid, as this phenomenon will be located only in some parts of the grid, hence it is considered negligible.

Note that in a conventional power grid, inertia is linked to the rotating masses of synchronous machines connected to the grid. When such a grid experiences an increase in electrical load, these synchronous generators are naturally loaded up to their rated values, contributing additional power to maintain grid stability. However, when unconventional sources (such as solar PV and wind turbine systems) function as generators in the power network, they do not inherently provide additional inertia or power during increased electrical load. This is because they are designed to deliver maximum active power only when their primary fuels, solar irradiance and wind, are readily available.

In addition, as unconventional sources become more dominant in the grid without energy storage and firming, they can significantly impact load shedding. This is because their variability makes it difficult to precisely match power generation with fluctuating demand, hence increasing the risk of imbalances. For example, during a sunny day, excess solar generation could surpass demand, leading to over-frequency and potentially triggering load shedding, or a sudden drop in strong winds powering wind turbines can cause a significant shortfall in generation. Furthermore, the overall inertia of the grid might decrease for two reasons: the decommissioning of conventional power stations and the increasing ratio of renewable energy sources relative to total generation.

As known, firming the grid can be achieved through additional dispatchable generation resources, such as natural gas plants, which can be ramped up or down quickly to compensate for the variability of renewable energy sources. However, this approach can be costly and highly critical resource-dependent. Hydrogen fuel cells and pumped hydro storage may also offer solutions for firming a grid when available.

As illustrated in the figure, both virtual (synthetic) inertia sources and load shedding can contribute to changing the active power change on the grid. By injecting active power rapidly in response to frequency deviations, virtual inertia sources can help achieve grid stability without relying on load shedding. Load shedding, by reducing overall demand, can also reduce the required active power, mitigating the imbalance and helping to restore frequency. However, there is a key distinction between these approaches. BSS acting as virtual inertia sources offer a proactive approach to grid stability by providing a rapid active power response, but load shedding is a reactive measure used to avoid complete system collapse during critical situations.

The major grid components and their principal interactions are illustrated in Figure 6.13. The second node in the block diagram essentially acts as an adder, combining two key terms to calculate the net change in active power: $\Delta P_{net} = \Delta P_{grid} + (-\Delta P_{imbalance})$. This calculation directly influences the overall active power balance.

- ΔP_{grid} defines an increase in active power injected into the grid from additional power generation sources, as indicated.
- $\Delta P_{imbalance}$ reflects the change in active power consumption on the grid. This term can be positive or negative:
  - A positive $\Delta P_{imbalance}$ indicates a decrease in active power consumption by the loads. This could be due to consumer behaviour adjustments (using less power during peak hours) or demand-side management programs incentivizing reduced consumption.
  - A negative $\Delta P_{imbalance}$ indicates an increase in active power consumption by the loads. This is less common in the context of grid stability but could happen due to unexpected surges in demand or large industrial loads coming online.

As indicated in the figure, the calculated net change in active power (ΔP_{net}) influences the intrinsic mechanisms in the model. These mechanisms represent inherent characteristics of certain loads that can automatically adjust their power consumption in response to frequency changes. Here are two examples:

- Power shortage (frequency drop):
  - ΔP_{grid} would likely be positive, indicating an increase in power generation to address the shortage.
  - $\Delta P_{imbalance}$ might be slightly positive (consumers reducing consumption) or zero.
- Excess power (frequency rise):
  - ΔP_{grid} would likely be negative or zero, indicating reduced generation or maintaining existing levels.
  - $\Delta P_{imbalance}$ would be positive, signifying a decrease in power consumption to reduce the excess power and bring the frequency down.

In summary, these explanations emphasize the importance of managing both power generation and consumption to maintain grid frequency stability. By considering both ΔP_{grid} (injection) and $\Delta P_{imbalance}$ (consumption changes), the net change in active power (ΔP_{net}) provides a valuable indicator for grid operators. This allows them to take corrective actions and utilize various resources, including BSS, to ensure a balanced and stable power system. It is important to note that this also highlights the centralized role of BSS control within a large grid.

The amount of active power (which can define the BSS capabilities in a network) required to counteract a frequency deviation can be obtained from the power system's frequency response equation given below,

$$\text{RoCoF} = \frac{df}{dt} \quad \text{RoCoF}_{max} = \frac{df}{dt_{max}} = \frac{\Delta P_{imbalance}}{P_{load}} \frac{f_{nominal}}{T_{N,min}} \quad (6.5)$$

The equation identifies the amount of active power that can provide an insight into the amount of active power that can be injected into the power grid to alter the time constant of the network. It should be noted here that as the grid transformation and evolution continue, both $\Delta P_{imbalance}$ and P_{load} will change. At given values of $\Delta P_{imbalance}$ and P_{load} values the minimum time constant of the network T_N will occur when RoCoF is maximum.

Note that the time constant of the network T_N is a direct consequence of the equation of motion and the inertia of the power system (represented by the inertia constant), which reflects how quickly the system responds to changes in power imbalances based on its inherent resistance to changes in speed due to its rotating masses.

Note also that while not explicitly shown in the above figure and discussions, droop control plays a critical role in power generation ΔP_{grid} and indirectly influences some load behaviour $\Delta P_{imbalance}$, which ultimately contributes to the net change in active power ΔP_{net} at the second node. This can then trigger adjustments in the intrinsic mechanisms to maintain grid stability. The BSS described in Figure 6.13 can represent entire battery storage applications that alter the value of $\Delta P_{netchange}$, including community-level BSS applications and EVs.

As known, wind energy is a major renewable source, utilized both onshore and offshore. Table 6.8 presents a heuristic evaluation of the benefits of integrating BSS into wind farms with highly diverse layouts, compared to scenarios without BSS. The table clearly demonstrates that optimizing wind farms as a collective unit, considering various battery capacities leads to more effective battery sizing and storage utilization.

TABLE 6.8 Evaluation of the Impact of BSS Integration on Wind Farm Performance Metrics

Reference	No BSS	Single BSS at Substation	One BSS per Wind Turbine	Multiple BSS in Optimized Locations
Power quality	Low	Medium	High	High
Reliability as generator	Medium	Medium	High	High
System cost	Medium	High	Very High	Medium
Intermittency	High	Low	Low	Medium
Dispatchability	Very limited	Medium	Medium	High
Response time	Low	High	High	High
Reactive power control/rate	Limited	4 quad/ high	4 quad/ high	4 quad/ medium
Backup capacity	Very low	Medium	High	High
Power consumption	0.1%	>0.1%	>0.1%	>0.1%
Power/Energy	Power	Power/ Energy	Power/ Energy	Power/Energy
Availability	95%	>95%	>95%	>95%
Utilization of battery	None	Low	Medium	High

The optimization should focus on improving reliability, efficiency and cost-effectiveness. The primary goal of BSS has to be to provide dispatchable energy and backup capacity. It is important to note that creating a similar table for solar PV systems is feasible, although the process tends to be simpler. This is due to the more uniform structure of solar PV farms and the predictability of their daily and seasonal energy output, which simplifies BSS sizing.

In conclusion, the explanations given above highlight the key differences between the traditional, centralized power grid and the ever-evolving, complex hybrid grid and microgrid that accommodates BSS. Variable renewable energy sources, alongside increasing demand and ageing infrastructure, create a delicate balance that is becoming increasingly difficult to maintain. A decentralized, smarter and more responsive grid will be the answer.

Energy management at every level, from individual homes to entire communities, is critical. As discussed in Chapter 1, smart appliances and smart houses empower consumers to become active participants, adjusting consumption based on real-time needs and grid conditions.

Furthermore, DERs bring power generation closer to consumption points, reducing reliance on long-distance transmission and associated losses that are becoming bidirectional. This distributed approach, coupled with BSS, provides local resilience and flexibility.

The potential of DC grids also emerges as a future-proof solution. Their inherent compatibility with a wide range of DC electrical loads eliminates the energy losses associated with AC conversion, further enhancing efficiency. Additionally, DC grids seamlessly integrate with renewable energy sources and BSS, creating a more streamlined and responsive power system.

The decentralized approach can both reduce demand and significantly minimize the complex interactions described above within the power grid, regardless of whether it is AC or DC. However, the battery sizing concept remains the same. It can be defined as using an energy capacity-based approach based on LDCs and a short-term power rating-based approach to consider generation-demand balance.

6.6 MICROGRIDS

The electrical grid system operates across a hierarchy of interconnected networks, each with unique characteristics and functions. Macrogrids, the large-scale national, multinational or regional grids, deliver electricity from suppliers to consumers across vast areas. They integrate a variety of power sources while maintaining high standards for reliability and stability. Microgrids are localized networks that can operate independently from the macrogrid. They enhance reliability, reduce emissions and support critical services with a mix of renewable and conventional energy sources. Nanogrids, the smallest of the three, typically serve single buildings or small clusters. They optimize energy use and provide backup power through simpler systems, usually focused on a single type of DER.

The decentralization of the electrical grid system involves integrating microgrids and nanogrids into the existing macrogrid infrastructure. This approach enables a shift towards localized generation and consumption, resulting in a more resilient and efficient power system that accommodates a wider range of energy sources and consumption patterns.

Microgrids offer a range of benefits categorized into three main areas, as listed in Table 6.9. By providing local control and optimization of energy resources, microgrids deliver significant technical, environmental and economic benefits.

However, microgrids also face limitations. These include technical complexities in control and management, grid integration issues, the dependence on weather for renewable sources, and limitations on scale.

TABLE 6.9 Microgrid Benefits across Technical, Environmental and Economic Categories

Category	Benefit	Brief Description
Technical	Enhanced reliability and resilience	Ensures power continuity for essential functions during main grid failures.
	Improved power quality	Address issues like voltage sags and swells, improving supply for sensitive devices.
	Energy efficiency	Reduce transmission and distribution losses and can be finely tuned to local energy demands.
	Scalability and flexibility	Can be adapted to include various energy sources and respond to changing demands.
	Peak load management	Help in managing and reducing the peak demand on the central grid.
Environmental	Increased renewable energy integration	They facilitate the use of renewables.
	Reduced greenhouse gas emissions	Efficient local generation lowers carbon emissions.
	Reduced land use and habitat impact	Decrease the need for large power plants and transmission systems, preserving habitats.
	Support for electrification of transportation	Can support the infrastructure for EVs, promoting cleaner transportation.
Economic	Cost savings	Lower energy costs and reduced transmission fees.
	Energy market participation	Can provide grid services and energy trading.
	Deferred infrastructure investment	Can offset costly grid infrastructure upgrades.
	Local economic development	Offer job creation and business opportunities in local communities.
	Energy independence	Offer a self-sufficient energy solution for remote areas, reducing external energy sources.

Environmentally, reliance on non-renewable generation methods and resource consumption for construction and maintenance can diminish their green advantages. Economically, high initial investments, regulatory and financial barriers, fluctuating cost-effectiveness, and market participation restrictions pose significant obstacles. Socially, public acceptance, regulatory limitations and disparities in accessibility and benefits require addressing.

To mitigate these limitations, a multi-directional approach is necessary. Investment in advanced control technologies and grid management systems is required to handle the complexity of operations. Enhancing predictive analytics and energy storage solutions can help address the variability of renewable sources. Economically, innovative financing models, subsidies and incentives could lower initial costs and encourage investment. Regulatory reforms should aim to streamline integration processes and support market entry for microgrids. Finally, public engagement and inclusive policies can promote broader acceptance and ensure equitable distribution of microgrid benefits. Addressing these issues through a combination of technology, policy and community engagement can maximize the potential of microgrids to transform energy systems.

It is highly critical to indicate that each microgrid should be designed to achieve specific objectives, including improving reliability, enabling cost efficiencies and facilitating the integration of renewable energy sources. The design and operational strategies should meet the user demands and align with the specific features of the geographical area served.

As illustrated in Figure 6.14, microgrids can be classified based on various criteria, such as the type of electricity they utilize, their power source or key attributes. These classifications cover a broad spectrum of sectors, each with unique energy demand profiles and operational requirements.

For example, community microgrids typically include residential and small businesses, with fluctuating energy patterns daily and seasonally. Campus or institutional microgrids serve educational, research, and

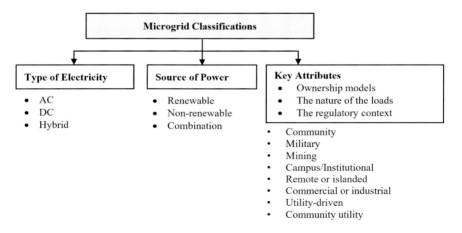

FIGURE 6.14 Microgrid classifications based on electrical type and key attributes.

healthcare facilities, demanding stable power with peaks tied to institutional schedules. Military microgrids, on the other hand, require continuous operation for mission-critical activities, with a focus on reliability and security. Remote or islanded microgrids, detached from central networks, rely on local renewable resources and emphasize energy conservation, usually due to elevated generation costs. Commercial or industrial microgrids address the uniform and considerable energy demands of business and industrial processes, prioritizing cost reduction as well as power quality improvement. Additionally, utility-driven or community utility microgrids, which are often more extensive, are designed to enhance overall grid resilience and manage peak loads among a diverse set of residential, commercial, and industrial users, mirroring the varied energy consumption within a utility's coverage area.

Finally, as an emerging application, microgrids at mining sites are designed to sustain the constant and substantial energy demands inherent to time-sensitive operations. Such microgrids must deliver a stable and continuous baseload power for heavy machinery and operational processes, alongside dynamic management of occasional peak demands. Considering the remoteness of most mining locations, microgrids in mining often combine traditional generation with renewable sources like solar and wind, supplemented by energy storage systems, to ensure resilience and reliability away from the main grid. Therefore, since efficiency is key, smart energy management systems are also integrated into the microgrid control. As mines expand, these microgrids are usually designed to be scalable. In addition, to maintain critical operations during any supply constraints, load shedding strategies are in place, prioritizing essential services and enabling uninterrupted mining activity. As previously mentioned, hybrid microgrids can be viewed as a stepping stone towards more DC-centric systems, as the adoption of DC sources and loads escalates.

Figure 6.15 shows a schematic of a microgrid connected to a conventional power grid (discussed previously in Figure 6.1) through a PCC. This network configuration, while relevant to general network structures accommodating DERs and microgrids, highlights the major case study to be discussed in the following section of this chapter. The grid side in this section of the macrogrid includes voltage regulation and communication control for the 11 kV–0.415 V network over a distance of 5 km. On the microgrid side, there are 35 customers in a small community with a total of 22 kW of rooftop solar PV systems installed.

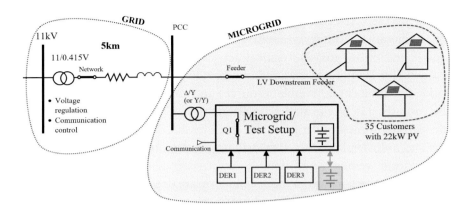

FIGURE 6.15 A schematic diagram of grid-to-microgrid interaction.

The microgrid test setup, including a BSS with highly reconfigurable connection options, will be covered comprehensively later. Therefore, as an expanded version of the power grid studied previously, this diagram serves as a basis for discussing the dynamic interactions between the grid and the microgrid.

Microgrid operation within a power grid can be highly complex, as illustrated in Figure 6.13 and already discussed. Some of the complexities are primarily associated with switching processes, connection, disconnection and reconnection issues between the main grid and the microgrid, as well as within the microgrid structure itself that contains DERs. Figure 6.16,

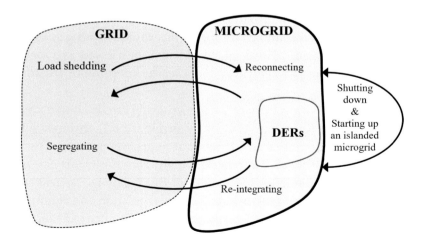

FIGURE 6.16 Interactions between the grid and microgrid.

which is an overview of Figure 6.15, summarizes and illustrates these interactions. The key aspects of microgrid operation can be described as

- Load Shedding: This is a last-resort measure in the main grid to avoid overloading the system, which could cause widespread outages or damage to infrastructure. It can cover the area of a microgrid, although inconvenient for those affected. Reconnecting a microgrid to the main grid after load shedding (or any other reasons) involves several steps to ensure safety and stability for both systems.

    - The microgrid's generation sources must be stable and synchronized with the main grid in terms of frequency and voltage, with all protection systems checked.

    - The microgrid's load needs to be managed to avoid instability. This can involve gradually increasing the load to match the generation capacity.

    - If the microgrid is part of a larger utility system, communication and coordination with the control centre are required.

    - Reconnection may be staged, starting with critical loads and followed by less critical ones. Close monitoring is required after connection, including post-event analysis of load shedding.

- Segregation: This refers to isolating a section of the grid (microgrid) or specific DERs from the main grid. Segregation ensures:

    - Reliable power supply to critical infrastructure or areas vulnerable to grid outages.

    - More effective management of energy production, consumption and storage within microgrids and for DERs.

    - Protection of workers and prevention of wider outages during grid maintenance or faults.

    - Compliance with legal or regulatory frameworks requiring isolation for data privacy, energy sovereignty or local energy trading schemes.

    - Reduced energy costs by managing local generation and consumption.

- Minimized reliance on the main grid, especially during peak price periods, leading to increased energy efficiency and cost savings.
- Accommodation of variable renewable energy sources like solar and wind without affecting the main grid's stability.
* Re-integration: This involves seamlessly reconnecting the microgrid to the main grid, requiring careful management of electrical properties and control systems to maintain grid stability and reliability.
* Shutting down an islanded microgrid: This is a carefully sequenced process to ensure safety and minimize impact:
  - Inform relevant personnel and potentially affected customers.
  - Gradually reduce the microgrid's load, prioritizing non-critical systems and equipment.
  - Systematically decrease DER output to match the reduced load and prevent energy overproduction.
  - Manage battery storage systems (if available) to maintain a safe state of charge.
  - Physically isolate the microgrid from any interconnected grids.
  - Issue the shutdown command through the control system, turning off DERs.
  - Confirm safe shutdown and security of all systems.
* Starting up an islanded microgrid: This requires ensuring the readiness and safety of all components:
  - Inspect DERs, control systems and protective devices for maintenance activities or hazards.
  - Activate control and monitoring systems to oversee the start-up sequence.
  - Check battery storage systems (if available) for sufficient charge level.
  - Start DERs one by one, carefully monitoring each for stable operation before reconnecting loads.

- Connect essential services first, followed by less critical loads, to maintain system stability and balance.

- Closely monitor performance to address any issues promptly and ensure a smooth transition back to normal operations.

In conclusion, the descriptions given above provide a structured approach to manage the transition between grid connection and disconnection for microgrids, whether for isolation or resuming normal operations.

The following subsection will present a highly detailed case study. This case study utilizes a state-of-the-art BSS design and explores its autonomous operation with detailed real data captured seasonally and including grid faults. The data encompasses every conceivable parameter, ranging from temperature and solar irradiance to those critical for engineering analysis purposes. This analysis not only helps the reader to understand the complex operation of a microgrid under real-world conditions and scenarios but also provides valuable insights for designing future systems.

6.7 A DETAILED CASE STUDY: AUTONOMOUS MICROGRIDS WITH BSS AND DERS OPTIONS

Building on the exploration of microgrid characteristics and benefits in the previous section, this case study indicates the autonomous operation of a real-world microgrid. Figure 6.15 provides a basic microgrid structure with DERs and its connection to the main grid. In this section, a specific setup includes several prosumer households (having both generation and consumption capabilities) with solar PV systems. This "Microgrid/Test Setup", connected to the PCC, forms the core of this case study.

The key aspect of this setup is its grid-scale BSS. Designed from the ground up, the system incorporates all the battery storage applications listed in Table 6.4 and offers flexible DER integration options. Most importantly, it includes autonomous operation capabilities, a feature likely to be dominant in future decentralized microgrid structures.

The following subsections will provide detailed information on this setup to guide future design studies and share operational results and characteristics. Data for this analysis was obtained from a real-world operational environment, containing seasonal changes. A highly advanced data logging system, designed with high bandwidth and accurate sensors, was integrated during the setup phase to capture comprehensive power quality level data, from steady state to transients that is important to highlight the dynamic waveforms during transition phases as well as in the case of network or the setup related faults.

The results will include not only electrical data but also environmental data to understand the impacts of external and internal heat sources, critical factors in battery storage applications. The entire range of grid-microgrid interactions will be explored, including load shedding, segregation, reconnection, re-integration, shutdown, start-up and islanded microgrid operation. The setup can also be considered a model for both community-level and remote area microgrids, with additional capabilities such as EV fast charging. In the subsections, the complete electrical and mechanical system design will be provided with all associated details.

Figure 6.17 shows the main components of containerized autonomous microgrid test systems designed for flexibility and real-world deployment. The system prioritizes safety, functionality and adaptability for various grid connection scenarios.

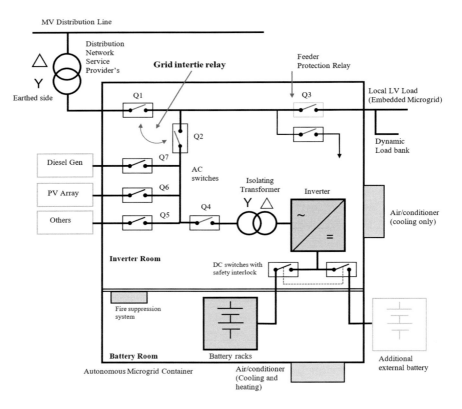

FIGURE 6.17 Single-line diagram of the autonomous microgrid test system.

Firstly, it is housed in a containerized structure for ease of transport and deployment. This allows for rapid setup and operation in diverse locations, whether within a large grid or a remote, isolated area. It also features two thermally and fully isolated rooms. This separation isolates the BSS for independent control of cooling and cooling/heating requirements in each room optimizes system efficiency, and helps contain potential fire risks within the BSS room.

In addition, a proper grounding system is accommodated to protect personnel and equipment by providing a low-impedance path for fault currents, preventing electrical shock and equipment damage. Grounding also helps maintain stable voltage levels and reduces electromagnetic interference, ensuring reliable system operation.

If the containerized system is parked in an open environment, a lightning protection system is also needed to safeguard the microgrid from transient voltages caused by lightning strikes, which have a potential to damage sensitive electronic components within the system.

Furthermore, the design incorporates wind resistance considerations to ensure the container's stability in high wind conditions. This is critical for safe operation, especially in exposed locations.

The test system also includes protection systems to safeguard against electrical faults, overloads and short circuits. Fire protection and suppression systems are in place to minimize fire risks and potential damage.

An advanced data logging system with local storage and remote access capabilities allows for comprehensive monitoring and analysis of system performance from any location.

Finally, the battery storage capacity is strategically chosen to provide a reasonable impact on the three-phase local network. This balance allows for effective operation while maintaining the flexibility for potential upgrades to higher power and capacity ratings in the future.

The following subsections will explain the technical specifics of this microgrid test system, serving as the primary case study. This will include a detailed explanation of the single-line diagram, BSS configuration, grid connection options and the functionalities of the data logging system used to capture unique data sets presented after a detailed analysis.

6.7.1 Mechanical Design Features of a Custom-Built Container

The utilization of any BSS needs to carefully consider all associated system components to be able to achieve the primary aim of a target application. Although the technology may look simple, as it basically involves a single battery type with AC and DC switches and PE bidirectional smart converter, a complete functional BSS requires a significant number of auxiliary components that need to be sized and designed all around a specific storage technology. The key components of BSS were listed in Table 6.6 which were all applied during the development, commissioning, deployment and operational phases of the BSS solution.

As well known, containerized solutions are commonly used in grid-scale BSS applications. Firstly, such a design is preferred since it offers mobility and scalability. Containerized BSS systems are housed in prefabricated, transportable containers. This allows for easy deployment and relocation to areas where energy storage is most needed. Additionally, multiple containers can be combined to scale up the BSS capacity as required.

In addition, since the containerized BSS is prefabricated and pretested, it can be deployed much faster compared to traditional, site-built BSS solutions. This reduces overall project lead time, allows for quicker response to grid needs. In addition, they offer a standardized approach to BSS design and construction, simplifying the process and reducing costs. The compact nature of containers minimizes the required footprint compared to traditional customized solutions. Furthermore, containers can be designed with features for thermal insulation and climate control, and they can be equipped with security measures to deter unauthorized access and a level of physical protection for the BSS equipment. Finally, containerized BSS systems can be pre-integrated with power conversion equipment and control systems, making them easier to connect to the grid. This flexibility allows for seamless integration with various grid connection scenarios.

While commercially available shipping containers offer a highly rigid and cost-effective solution, they often require significant thermal insulation work on the entire inner surfaces. In this case study, a custom-built container, shelter solution is implemented to achieve the best possible thermal insulation. This approach primarily aims to eliminate external heat sources, by using proper insulation structure, hence minimizing the impact of external heat sources on the BSS, as it is highly critical in hot climates. In addition, such custom design has allowed the strategic integration of temperature sensors into the walls, ceiling and floor structure of the container. This allows measuring only the outside temperature of the container surfaces and monitoring how the container's orientation affects internal temperatures, hence cooling. The results have been obtained during a very hot day of operation and will be demonstrated later. Figure 6.18 illustrates the top view and the 3D view of the container design.

Microgrids with Distributed Energy Resources ■ 479

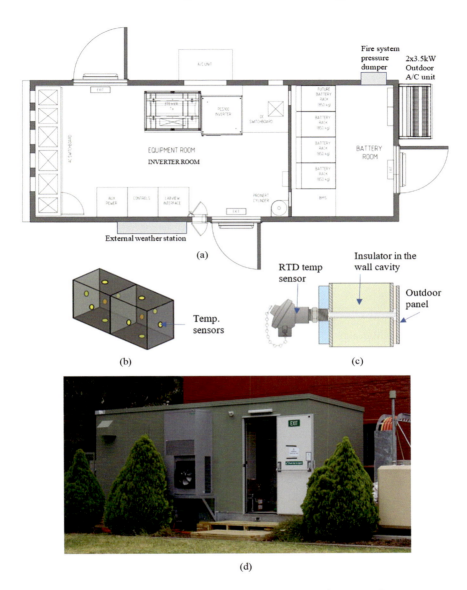

FIGURE 6.18 Containerized design details: (a) top view of the unit showing major components and their locations; (b) locations of ten platinum RTD temperature sensors: four on external walls, two on the external roof, two on the floor frame and two for ambient temperature; (c) an RTD temperature sensor assembly for measuring external surface temperatures; (d) the final system assembly parked next to a diesel generator for nonlinear load testing.

Table 6.10 summarizes the considerations and characteristic features used in the design of the custom-built container.

TABLE 6.10 The Custom-Built Container Design Considerations and Features

Category	Description
General specifications	• Designed according to relevant loading codes (dead, live, wind, and snow loads), steel structure codes, concrete structure codes, and high-strength structural bolting codes. • Dimensions: $L=6.95$ m, $W=2.85$ m, footprint • $L=7.2$ m, $W=2.85$ m, $H=3$ m, excluding the A/C units • $L=7.85$ m, $W=4.85$ m with the A/C units and the door openings
Structure	• Steel base frame, 18 mm CFC Flooring, 100 mm EPS external walls, 100 mm EPS Roof, 50 mm EPS dividing wall • A low centre of gravity shelter with most of the weight fixed directly to base frame
Weight distribution and foundation	• Weights of major components: 1,700 kg of isolation transformer, 850 kg inverters, 200 kg of DC switchboard, 2,550 kg of three battery racks, 200 kg of control system, 200 kg of auxiliary power systems and 800 kg of AC switchboard. • Total weight: 7,800 kg without battery modules fitted, and approximately 10 tonnes with modules fitted. extra space is provided in the battery room to house a future fourth battery rack (with an extra weight of 850 kg) • Lift reaction force considered for frequent movement • 8 × reinforced heavy-duty foundation
Lifting and transport	• Requires removal of battery modules before transport for safety. • Eight footing pads included in design. • Maximum deflection of 24.4 mm during lifting
Wind load	• Designed for a regional wind standard of 56 m/s wind speed (used to define panel thickness) • Deployable to any Class C (Cyclonic) region wind zone
Anchoring	• Tie down at each anchor point (minimum 2 per side or end) • Uplift at each anchor point: 17.7 kN • With suitable support points for lifting by a crane for loading/unloading to a truck

(Continued)

TABLE 6.10 (*Continued*) The Custom-Built Container Design Considerations and Features

Category	Description
Slab footing system	• Requires a nominal 200 mm thick slab foundation • M16 chemset anchors with 160 mm minimum embedment into 32 MPa concrete slab
Electrical components	• The wall behind the main switchboard is reinforced with a penetration hood and flat plate to compensate for penetrations required for electrical components. • Locations of DC and AC switchboards are optimized to minimize bus bar losses. • Grounding connections are established according to relevant electrical codes. • Grounding points are readily accessible for all electrical equipment.
Air conditioning	• A/C outdoor units for the battery room are located on a double-stack frame next to the door. (Note: Justification for two A/C units with different temperature ranges will be provided later - related to lithium battery requirements for both heating and cooling) • Internal battery room A/C units (21 kg each) are mounted on the ceiling level. • A separate 320 kg wall-mounted A/C unit (WMF0661) are removed for transport and reinstalled on-site.
Access and safety	• Three separate doors are included for safe access to the container. • Fire suppression system is installed inside the battery room, while pro inert cylinder placed inside the equipment room). • A fire system pressure dumper is included in the battery room near the ceiling level. • Exit lights are strategically placed.
Monitoring and protection	• A weather monitoring unit is attached to a metal bar attached to the container roof. • Lightning protection system with two air terminals, assembled on a long metal pipe structure that is removable during transportation.
Power connections	• Dedicated outlets are located at the floor level behind the AC switchgear to avoid water, humidity while providing easy access for connecting DERs and other major power inlet and outlet connections.

6.7.2 Electrical Design and Characteristic Features

Figure 6.19a presents a single-line electrical diagram of the microgrid test platform. This diagram highlights the major components and includes voltage and current measurement points, which will be discussed in

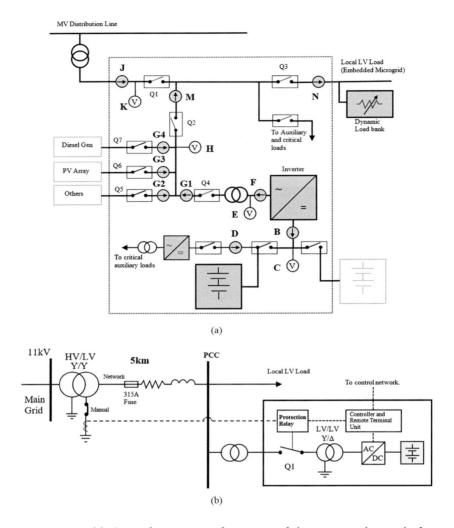

FIGURE 6.19 (a) General system configuration of the microgrid test platform single-line diagram, showing electrical measurement locations for voltage (shown as V sign inside the circle) and current (shown as an arrow inside the circle) with locations designated by the 14 node labels: B, C, D, E, F, G1, G2, G3, G4, H, J, K, M, N. Note that except the DC nodes of B, C and D, the rest of the nodes are related to three phase AC measurements; (b) structure of communication between microgrid platforms and remote locations.

more detail later. This serves as a simplified overview. A more detailed circuit diagram is provided in Figure 6.20. This comprehensive diagram highlights the increased complexity of the system when all safety and protection circuits, as well as auxiliary power supplies, are included.

It is important to emphasize that most microgrid platforms utilizing multiple DERs, including a BSS, can adopt a similar circuit topology to ensure safe operation. Even larger systems will require these same core components, but with higher voltage and current busbars, and multiple, higher-power inverter and battery racks. Note that signal conditioning, control level and communication level wiring are not included in the figure. These are excluded because they are highly variable due to both product brand and their rapid evolution.

While the figures exclude signal conditioning, control level and communication level wiring due to their variability, Figure 6.19b highlights the critical role of communication between microgrid platforms and remote locations. This communication, facilitated by the grid interface relay (via Q1 and Q2), enables four key functions:

- Communication allows for coordinated operation between the microgrid and the main grid, ensuring regulatory standards are followed.

- The relay detects faults within the microgrid and isolates them from the main grid, preventing them from spreading and causing wider outages. Conversely, it can also protect the microgrid from faults originating on the main grid (as will be shown later).

- The relay synchronizes the frequency and voltage of the microgrid's power sources with the main grid, allowing for seamless power exchange between the two systems.

- By continuously monitoring various system parameters, the relay provides valuable data for optimal grid management.

Note that there are many types of protection relays designed for specific fault scenarios. Some, like ground fault relays, residual current devices (RCDs), ground fault circuit interrupters (GFCIs), zero sequence current relays and directional relays, require a reference to earth for effective operation and utilize earth current in their detection. The ease of ground fault detection depends on the system's grounding type. Solidly grounded systems offer simpler detection, while ungrounded or high-impedance grounded systems might require special relays like ground fault overvoltage relays (detecting voltage rise on healthy phases during a fault). However, not all protection relays rely on earth

484 ■ Reinventing the Power Grid

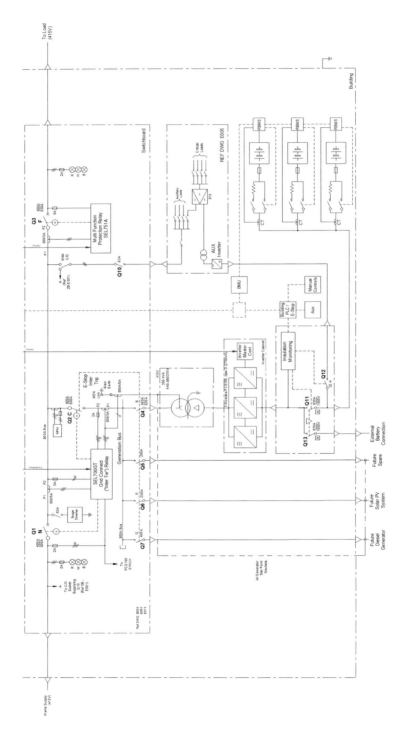

FIGURE 6.20 Single-line diagram of the microgrid test system with BSS, power converters, AC and DC switch gears, the control system and the data logging system.

current. Differential relays (protecting against phase-to-phase faults) and overcurrent relays can function without it, comparing currents entering and leaving a circuit section.

It is also important to note that while inverter-related voltage and current measurements are typically embedded within the PE circuit topology, commercial devices often lack integrated high bandwidth and accurate power quality sensors and data capture and analysis tools. These additional functionalities require separate equipment which is accommodated in the system described here to demonstrate the possible operating modes and all possible measured quantities to understand both steady-state and dynamic behaviours of the core system component.

In microgrid testing, a dynamic load bank is critical for mimicking real-world fluctuations in power demand. It allows designers and users to simulate load shedding scenarios and step load increases. This helps assessing the microgrid's ability to respond effectively to these changes, ensuring it can maintain stable voltage, frequency and power delivery. By testing under dynamic load conditions during the commissioning phase, potential weaknesses can be identified. This information can then be used to optimize the microgrid's design and control systems for reliable and efficient operation.

Therefore, Table 6.11 summarizes the external resistive practical load bank configurations suitable for testing the microgrid setup reported in the case study. The table provides the group number to identify different configurations within the load bank, resistances per load string and the number of resistors per string. Finally, active power values at minimum and maximum voltages are given to show the power dissipation at the

TABLE 6.11 External Resistive Practical Load Bank Configurations Suitable for Testing the Setup

	Group 1	Group 2	Group 3	Group 4	Group 5	Group 6	Group 7
Ohms per load string	100	100	50	50	25	25	25
Number of resistor strings in circuit	1	2	3	4	5	6	6
Total resistance (ohms)	100	50	25	16	10	7.1	5.6
Power (kW) @ min battery voltage	3.4	6.7	13.5	20.2	33.6	47.1	60.6
Power (kW) @ max battery voltage	7.2	14.5	28.9	43.4	72.3	101.2	130.1
Power (kW) @ nominal DC battery voltage	5	10	20	30	50	70	90

minimum and maximum DC voltages of the battery system. Note that the minimum and maximum DC voltage values in the battery rack designed are 580 and 850 V, respectively.

Table 6.12 is provided to summarize the electrical characteristics and features of the microgrid test unit and see Table 6.13 for BSS packing details. Note that when the sensors are simultaneously activated, the IG55 inert gas suppression system is activated after a countdown sequence that is controlled by the fire panel. In addition, a partitioned room structure with separate air conditioning units (and redundant units in the battery room) for effective, reliable cooling and safety measures for fire extinguishing. Precision air conditioning units for the inverter room (Unifilair) and battery room air conditioner (Mitsubishi) are used. Note also that the Uniflair unit in the inverter room with an operating set point of 30°C and two 3.5 kW DX A/C units in the battery room operate independently from each other to maintain the room temperature between 20°C–25°C. Extra cooling capacity is available in the battery room for a fourth battery rack, which can also be considered as a backup unit in high-temperature climates.

Note that after complete system integration, the following additional tasks may be required. Some of these tasks can be time-consuming and incur extra costs depending on site specifications, complexity and application integration (Figure 6.21).

- Control integration via SCADA system
- Integration with the local network's voltage regulator
- Enabling further remote access for engineering
- Commissioning phase to ensure the system functions as designed

As previously mentioned, the described setup offers autonomous operation capabilities, which can be defined remotely or through the local user interface (see Figure 6.22d). This allows autonomously managing voltage levels, peak loads and state of charge. In addition, multiple parties may be involved in some cases, which could extend design and testing times. While there is still some uncertainty surrounding the immediate future of grid-scale BSS assets due to cost and access to full value, a year-long trial can provide valuable real-world benefits.

Note that peak shaving, also known as peak lopping, is an energy management method to curtail grid demand during high-consumption

TABLE 6.12 Key Electrical Characteristics of the Microgrid Test Unit Given in Figure 6.20

Group	Characteristics and Features
Power conversion and battery storage	• Capable of connecting to a three-phase, 400 V, 50 Hz grid, including unbalanced 230 V single-phase networks due to per-phase inverter modules. • 3× ABB PCS100 D (250–820 Vdc, nominal 750 Vdc), bi-directional and smart inverter modules (total 270 kVA), expandable to 360 kVA (4 units), (see also Figure 6.22f). • 4-Qadrant operation (in voltage or current source mode) of the inverters with a wide range of battery storage applications. • 3× LG Chem R800 91 kWh racks using JH3 cell modules (total 273 kWh), expandable to 364 kWh; 270kW/270kWh BSS, space for a fourth rack, hence expandable to 350kW/350kWh. See Table 6.13 for BSS packing details. See also Figure 6.22h. • Isolating transformer: 370 kVA, star/delta; 415 V (with taps at 400 V/440 V); secondary current of 577 A; efficiency: 98.8%. See also Figure 6.22d. • AC auxiliary supply with a static transfer switch and a 5 kVA Schäfer inverter. • Two potential AC input sources for auxiliaries, (line side/load side) controlled by a manual transfer switch provide further flexibility for independent operation of the auxiliary inverter when the system is running in island mode.
Protection and control	• SEL700GT relay is used to control the "Q1" breaker facing the network, and the "Q2" main breaker for the generation bus. It provides all the standard grid-tie passive anti-islanding protections such us under/over voltage, under/over frequency, sustained (10min) low-level over-voltage, etc. in addition to over current, earth fault, reverse power (if required). It also provides a synch-check function which provides an enabling window for the inverter to close back onto network supply after a segregation event. • SEL751A feeder relay is used to protect loads connected at the load terminal, and this relay operates the "Q3" (Load) breaker. This relay provides the typical protections required for feeders, such as positive and negative sequence over current, earth fault, reclosing, loss of phase etc. in addition to under/over voltage and under/over frequency. • See also Figure 6.22e.

(*Continued*)

TABLE 6.12 (*Continued*) Key Electrical Characteristics of the Microgrid Test Unit Given in Figure 6.20

Group	Characteristics and Features
Busbars, switchgears, safety and alarms	• All generator bus circuit breakers are moulded case breakers fitted with a standard thermal-magnetic trip unit. The AC generation bus (800 A) (see Figure 6.22b) allows connection of multiple AC sources, including an internal inverter battery connection (with a 630 A nominal rating), an AC coupled solar PV system (200 A), a diesel generator (400 A) and one spare AC connection (200 A) for other potential sources (such as fuel cell or flywheel), (see Figure 6.22a for the switchboard). • A separate "load" bus/switchgear (800 A) accommodates embedded local grids or external loads for dynamic testing. • Switchgear is motorized for AC grid connection and 800 A/1,000 V, 24 kA DC battery busbar (for internal and external links), (see Figure 6.22 g for location). • The DC switchboard (SB) comprises three DC four-pole circuit breakers interlocked with two 800 A circuit battery breakers for internal/external battery configurations with a key trapped in the "ON" breaker. • The AC SB designed to allow direct connections with LV fault levels up to 36 kA. • Main DC breaker and its SB selected to allow for the fault level of the installed battery set and also to provide a margin of additional capacity to cater for the characteristics of alternate battery sets that may be connected externally. • The voltage, frequency, current and active/reactive power for protection, control and alarms are measured by the protection relays, the PaDECS® control system and a power meter (see Figure 6.22e). • With multiple indication lamps on the switchboard and around the shelter to help local operation and minimize any associated operational risks. • A surge protection device (TDX100M – 277/480, rated to 40 kA 8/20 μs) is installed at the network interface and monitored by the PLC and with alarm. • A Safety PLC (located in the inverter room) manages all critical safety functions, including monitoring of conditions in the separate battery compartment. • With an APAC fire panel (linked to the Safety PLC) connected to one smoke detector in the inverter room and two smoke detectors in the battery room. PLC activates an active alarm in the alarm log.

TABLE 6.13 Battery Packing System Components

Component	Unit Photo	Descriptions
Cell		BSS specialized cell.
Module		Battery module with 142P configurations containing the cell, LG JH3 cell modules
Battery protection unit (BPU)		Rack BMS, circuit breaker, contactor, pre-charge circuit, and sensors (voltage and current).
System controller		Bank management, bank BMS control, interface with power conversion system (PCS) master controller and bank BMS. PCS typically includes inverters, converters and other electronics.
Rack		1 Rack = showing only 2 modules + 1 BPU LG Chem R800 91 kWh racks, using JH3 cell modules.

periods, typically the late afternoons and early evenings in residential areas, or throughout the day during heatwaves due to increased air conditioning use. The approach often involves the utilization of alternative power sources such as solar panels and wind turbines, ideally coupled with energy storage systems, to meet demand, as well as reducing consumption by shutting off non-essential systems.

The BSS are a critical component in this strategy as they store energy when demand is low or when renewable sources generate surplus power, then release it during peak demand times. This process not only flattens the load profile, leading to reduced strain on the grid and lower peak demand charges levied by utilities, but also offers cost savings, improves power system efficiency, and potentially diminishes the need for further investment in generation capacity or network upgrades.

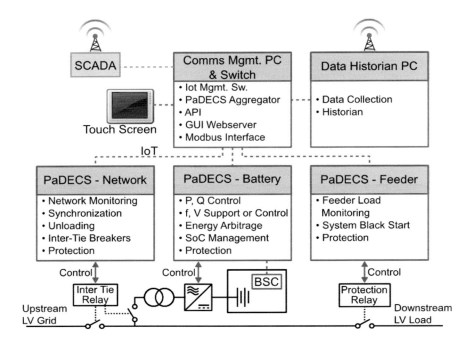

FIGURE 6.21 Distributed IoT-based control system of the test unit and DNSP Integration.

In current grid practices, advanced energy management systems are essential for dynamically optimizing power utilization in response to fluctuating energy prices and consumption patterns. Peak shaving offers two primary benefits: it reduces electricity bills and reduces grid load, thereby decreasing overall energy use. In addition, this method supports grid stability by preventing brownouts or load shedding, functions that are becoming increasingly critical as the industry is moving towards dynamic electricity pricing models. It is also noteworthy that peak shaving is an integral component of the VPP concept.

6.8 DATA CAPTURE AND POWER QUALITY ANALYSERS

Microgrid research often prioritizes understanding core functions like power conversion, control strategies and grid interaction, primarily influenced by major components like inverters, converters and batteries. Including auxiliary systems and real-world operating cycles significantly increases model complexity, making it challenging to isolate and analyse the specific effects of different control strategies or DER integration methods.

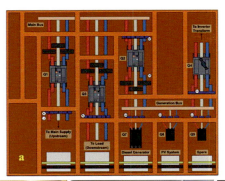

FIGURE 6.22 Infrastructure components of the microgrid: (a) AC switchboard busbar arrangement, (b) AC switchboard, (c) three-phase AC, Q1 switch (d) touch screen GUI, (e) isolation transformer unit, (f) control and relay units, (g) smart inverter modules, (h) DC switchboard, (i) battery racks, (j) HV feeder with voltage regulator at the zone substation, (k) The IoT unit, (l) the microgrid test platform, (m) LV grid connection transformer at the PCC.

However, unlike typical microgrid research models, the data logging system designed for this test setup does consider air conditioning systems, auxiliary systems, operating cycles and losses of all components as they significantly affect system efficiency and component wear.

To ensure a comprehensive evaluation of the microgrid test platform with DERs, a high-performance data logging system has been built. This system captures and analyses both electrical parameters (as shown in Figure 6.19a) and temperature data (as in Figure 6.18b).

The system features simultaneous high-speed sampling across multiple channels, enabling detailed data capture for various microgrid configuration modes (as shown in Figure 6.23 and see Table 6.14 for the detailed descriptions). Precise time synchronization via GPS with minimal uncertainty guarantees accurate data correlation, and redundancy with automatic failsafe ensures continuous operation.

Modular, custom-built hardware and software allow for future expansion and adaptation to evolving test requirements. Furthermore, the system integrates with an external weather station, capturing environmental data such as wind speed, direction and solar irradiance. This comprehensive data collection allows for in-depth analysis of the microgrid's performance under diverse operating conditions and environmental influences, as will be demonstrated in the following subsection.

Tables 6.15 and 6.16 summarize the sensors used in the test setup and the capabilities of the data logging and analysis system. As illustrated in the previous circuit diagrams and listed in the table, voltage and current transducers are distributed throughout the entire power network, including the DC side and three-phase system. Temperature probes are embedded throughout the container, and an external weather station with a pyranometer, wind speed and direction, pressure, temperature, humidity and rain level is also included. As shown in the simplified system diagram with sensor groups and software architecture in Figure 6.24, LabVIEW-based data acquisition software is developed and integrated with associated hardware with GPS-locked frequency source 100–500 kHz sample rate high-speed waveform recorder, IEC 61000-4-30 power quality recording and IEEE C37.118.1 phasor measurements.

In the test setup, a plug-in power analyser (see Figure 6.24a) has been developed as a Phasor Measurement Unit (PMU) () reference to the relevant IEEE standards to produce measurements of synchrophasors, frequency and RoCoF. The plug-in analyser receives raw voltage and current waveforms that have been acquired by the data acquisition hardware and

Microgrids with Distributed Energy Resources ▪ **493**

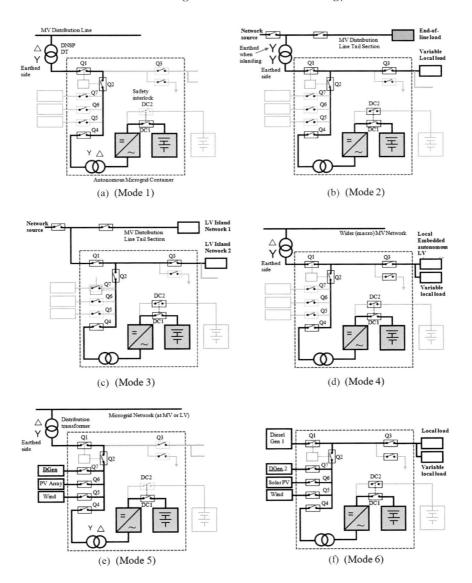

FIGURE 6.23 The principle connection diagrams of AESKB test system illustrating operational modes for specific tests in microgrid applications: (a) parallel to mains only, no islanding, (b) parallel to mains with islanding of an MV feeder 'tail' section, (c) parallel to LV network mains only and carrying an LV 'tail' section only, (d) parallel to mains with an embedded LV microgrid load, (e) parallel to mains with embedded LV microgrid plus PV array and/or diesel generator, (f) parallel to isolated microgrid.

TABLE 6.14   Descriptions of the Potential Core Tests as Illustrated in Figure 6.23

Modes	Descriptions of Potential Tests
1. Parallel to mains only, no islanding	• Typically used to provide network support via a step-up transformer; peak shaving above a set point; reactive power support capabilities, voltage support; demonstrating arbitrage; and showing stability at maximum continuous charge/discharge rates.
2. Parallel to mains with islanding of an MV feeder 'tail' section, via a step-up transformer	• Off-line demonstrations of inverter grid-forming of islanded MV microgrid and maximum step load capability; inverter behaviour at top of charge with excess microgrid generation; confirmation and test of realistic "induced fault" scenarios to test adequacy of protection schemes applied to the "normal parallel to mains" and "MV island" scenarios. • On-line demonstration to form an island (MV microgrid) supply from standstill; ability to maintain stable supply to MV microgrid.
3. Parallel to LV network mains with islanding and carrying an LV 'tail' section only	• Off-line demonstrations of inverter grid-forming of islanded LV microgrid and maximum step load capabilities; demonstration of inverter behaviour at top of charge with excess microgrid generation; confirmation and test of protection relay set-points relevant to the "normal parallel to mains" and "LV island" scenarios. • On-line confirmation of ability to form an island (LV microgrid) supply from 'standstill' and ability to maintain stable supply to MV microgrid. Demonstrates seamless segregation and smooth reintegration with network.
4. Parallel to mains with an embedded LV microgrid load (connected directly to the grid battery)	• Off-line demonstration of inverter's grid-forming capability, maximum step load capability and forming microgrid supply following loss of normal supply. • Demonstrate seamless segregation and smooth reintegration with network, with the battery supplying the embedded LV load during disconnection with network. • Demonstrate adequacy of protection arrangements through simulated faults. • Trial of control schemes for grid support (peak shaving, voltage support). • Maximizing PV array yield and output of a trial, when connected to a weak feeder.

(Continued)

TABLE 6.14 (*Continued*) Descriptions of the Potential Core Tests as Illustrated in Figure 6.23

Modes	Descriptions of Potential Tests
5. Parallel to Mains with embedded LV microgrid plus PV array and/or Isolated Diesel Generator (or other sources) connected directly to the grid battery and controlled.	• Seamlessly disconnect from the network supply following failure of that supply (voltage levels or supply frequency out-of-bounds) • Demonstrate inverter capability to "form" a grid supply (to the embedded LV network), follow the microgrid load using a frequency droop response, operate a connected diesel generator as a set point slave, and curtail any controllable PV connected either directly at the grid battery or distributed around the microgrid. • Demonstrate grid battery ability to carry out "smooth reintegration" once stable network supply becomes available again. • PV curtailment to maintain load on diesel generator and reduction of step loads on diesel generator. • Ramp rate control for loads on the microgrid.
6. Parallel to isolated microgrid. (Similar to 'PTM', but a network is a microgrid.)	• Synchronize grid battery to an existing microgrid supply at LV when the voltage and frequency of that supply is within bounds. • Safely disconnect from the microgrid supply following failure of that supply (voltage levels or supply frequency out-of-bounds) • Read a remote power meter indicating total load values (P, Q, pf, etc.) as would apply at the existing DG power station; to curtail PV inverter output to a (moving) target; to maintain total load on an in-service DG to be just less than a pre-set critical threshold level, while connected load on the microgrid is increasing beyond that level. • Individual energy storage system components (inverters, batteries, PV, novel generation devices). • Simulation of load profile and PV output profile.

software. Testing and verification is also made using the simulated waveforms defined by these standards. The developed PMU plug-in does not provide real-time communication capabilities and processes waveforms that already include latency (typically 2s when waveforms are acquired in blocks of 1s). This power analyser is only concerned with the M-class performance, which is intended for higher accuracy logging applications.

It should be noted that PMUs provide high-resolution, real-time data on voltage, current and frequency in power grids. They convert these waveforms into phasors, simplifying analysis and enabling comparisons across

TABLE 6.15 The Sensors and Characteristic Features Used in the Data Logging System

Sensors
- 36 high bandwidth voltage and current transducers, capable of capturing detailed power quality events (see Figure 6.19a for the locations)
  - 6× voltage (14 kHz), located on the main bus and generation bus (H and J)
  - 4× voltage (500 kHz), located on the inverter AC side and DC bus (C and E)
  - 10× current (100 kHz), located on the generation bus and Aux DC line (D, G2, G3 and G3)
  - 16× current (200 kHz), located on the main bus, generation bus, inverter AC side and DC bus (B, F, G1, M, K, and N)
- 10× platinum RTD temperature sensors located in the container walls, roof and floor (6× external skin, 2× floor, 2× ambient)
- External weather station, providing wind speed and direction, rain, pressure, temperature, humidity and solar radiation.
- GPS synchronized to universal time (UTC), time uncertainty < 100 ns.
- Simultaneous sampling at high speed (8×500 kHz, 32×100 kHz).

Characteristic Features
- Hot standby redundant with auto start and auto recovery for greater reliability.
- Distributed and expandable design, capable of extending into adjacent battery enclosures or other network assets.
- Continuous IEC 61000-4-30 power quality and IEEE C37.118.1 (M class PMU) analysis, including the following measurements:
  - RMS, DC voltage and current measurements, including fundamentals
  - Frequency, phase, RoCoF
  - Active, reactive and apparent powers
  - Energy (net, positive, inductive, reactive, capacitive)
  - Power factor and cos ϕ
  - THD, harmonics and interharmonics (typically to the 50th harmonic, but analysis up to the 5,000th harmonic is possible)
  - Zero, positive and negative sequence components
  - Battery SoC, total voltage, cell voltage and temperatures
  - Event waveforms
  - Total 7,613 channels recorded, 537 million values per day
- High-speed waveform recorder, triggered by power system events (dip, swell, interruption, inrush, frequency and RoCoF). For example, on 6 January 2018, 2099 events have been captured.
- Further analysis can provide:
  - Inverter and battery losses and efficiency
  - Battery capacity, for full discharge or during peak shaving cycles
  - Auxiliary power losses, including in response to environmental variations
  - Detailed view of protection system operation and fault analysis
  - Insight into rooftop PV generation and reverse flows, impact of momentary clouds on PV generation.
  - Insight into microgrid islanding operation, including anti-islanding performance, segregation and reintegration
  - Power factor control and operation around a circle of constant apparent power with varying real and reactive power

(*Continued*)

TABLE 6.15 (*Continued*)   The Sensors and Characteristic Features Used in the Data Logging System

- Environmental response of the container, including skin temperatures in response to solar radiation and wind, and the required auxiliary cooling system power.
- Response of the battery energy storage system to upstream grid faults.
- Operation and performance of automatic voltage support and peak load lopping functions of the control system.
- Long-term trends in daily demand, energy consumption and reverse power flow (negative) demand from rooftop PV generation, including the impact of changing seasons.

grid locations. In large grids with numerous potential PCCs, PMUs synchronize with a high-precision time source (such as GPS) for accurately aligned measurements. PMUs provide the high-frequency grid frequency measurements needed for accurate RoCoF estimation, unlike slower traditional SCADA systems. Real-time PMU data helps operators understand how quickly and where grid frequency is changing, which is crucial for determining the appropriate FCAS actions, such as injecting/absorbing active power (as discussed previously).

Before performing a comprehensive operational study on a microgrid with BSS and DERs, a testing matrix is needed. This ensures the safe, reliable and efficient integration of the BSS working alongside DERs with the power grid. The testing process validates the system's functionality before connection, guaranteeing it meets grid requirements and operates as designed. It complements the commissioning phase, which focuses on verifying the design against standards and ensuring proper function at the PCC.

In addition, reference standards should be specified based on the region where the BSS will be connected to the grid. A breakdown of some major regions and their relevant standards include:

- North America: IEEE 1547, UL 1974 and NFPA 855
- Europe: E.ON Netz, National Grid ESO, E-Netz Berlin and VDE-AR-N 4105
- Asia: China Electricity Council (CEC), Japan Electrical Safety & Environment Institute (JESI), Grid Code for Power System Interconnection of Distributed Energy Resources (DER) in South Korea
- Australia: AS/NZS 4777.2, AS/NZS 4859, IEEE 1547, and ARENA Best Practice Guidelines for Grid-Scale Battery Storage

TABLE 6.16 The Key Specifications of the Current and Voltage Sensors

Voltage and Current Transducers	Accuracy and bandwidth
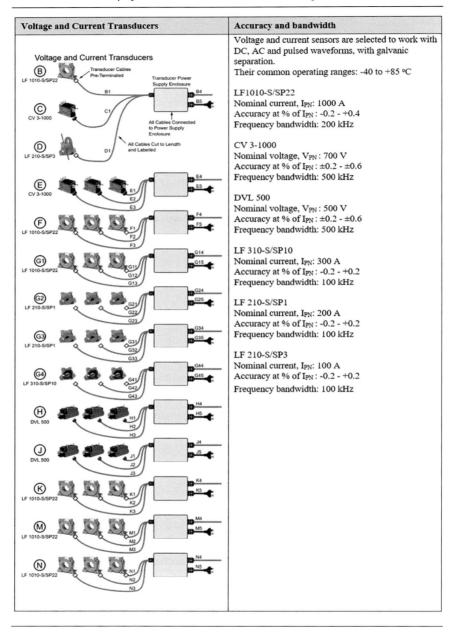	Voltage and current sensors are selected to work with DC, AC and pulsed waveforms, with galvanic separation. Their common operating ranges: -40 to +85 °C LF1010-S/SP22 Nominal current, I_{PN}: 1000 A Accuracy at % of I_{PN}: -0.2 - +0.4 Frequency bandwidth: 200 kHz CV 3-1000 Nominal voltage, V_{PN}: 700 V Accuracy at % of I_{PN}: ±0.2 - ±0.6 Frequency bandwidth: 500 kHz DVL 500 Nominal voltage, V_{PN}: 500 V Accuracy at % of I_{PN}: ±0.2 - ±0.6 Frequency bandwidth: 500 kHz LF 310-S/SP10 Nominal current, I_{PN}: 300 A Accuracy at % of I_{PN}: -0.2 - +0.2 Frequency bandwidth: 100 kHz LF 210-S/SP1 Nominal current, I_{PN}: 200 A Accuracy at % of I_{PN}: -0.2 - +0.2 Frequency bandwidth: 100 kHz LF 210-S/SP3 Nominal current, I_{PN}: 100 A Accuracy at % of I_{PN}: -0.2 - +0.2 Frequency bandwidth: 100 kHz

Microgrids with Distributed Energy Resources ■ **499**

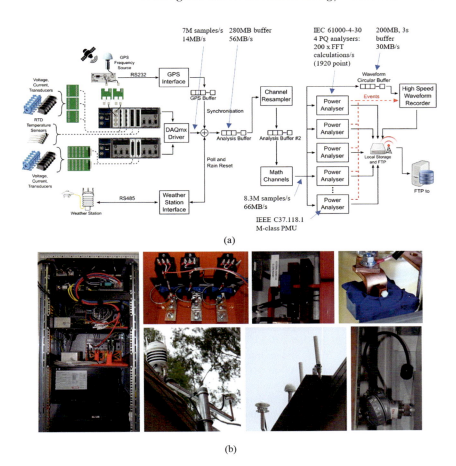

FIGURE 6.24 (a) Simplified system diagram with sensor groups and software architecture; (b) the key data logging system hardware components, left to right: the complete data logging rack system with NI DAQ modules, high bandwidth three phase voltage sensors, a Hall-Effect current sensor assembly for AC circuit, a DC current sensor assembled on a DC busbar, the weather station assembly, the communication antennas and temperature sensor.

Table 6.17 provides a general guide for such a testing matrix. Note that some additional tests may also be required depending on the specific system configuration and functionalities of the BSS and DERs. These include environmental testing, ageing and degradation, cybersecurity testing, grid frequency response (for weak grids), voltage regulation (in high solar PV generation), black start capability (for islanded systems) and performance during faults (specifically under fault ride-through cases of wind and solar PV generation systems).

TABLE 6.17 Grid-Scale Battery Testing Matrix (BSS with DERs)

Test Category	Test Description	Pass/Fail Criteria
Electrical Performance		
Active power control	Evaluate ability to deliver/absorb active power at various setpoints.	Meets accuracy and response time
Reactive power control	Evaluate ability to deliver/absorb reactive power at various setpoints.	Meets accuracy and response time
Ramp rate control	Evaluate ability to change active/reactive power at various ramp rates.	Meets accuracy and response time
Round-trip Efficiency	Measure at various charge/discharge rates.	Meets min. efficiency requirements.
Harmonic distortion	Measure current and voltage harmonic content at various operating conditions.	Meets harmonic distortion limits
Protection & Safety		
Overcharge/discharge protection	Verify system disconnects or limits power when battery reaches critical SoC	System disconnects or limits power
Overcurrent protection	Verify system interrupts current flow in case of fault conditions.	System interrupts current flow within specified timeframes.
Ground fault protection	Verify system detects and interrupts ground faults.	System detects and interrupts ground faults within specified timeframes.
Islanding protection	See the dedicated section below	
Battery Management		
Cell voltage monitoring	Verify the BMS accurately measures the voltage of each battery cell	Meets predefined tolerance range during discharge and charge cycles.
Temperature monitoring	Verify the BMS accurately measures battery temperatures including individual cells, module level and ambient	Falls within a reasonable tolerance range during charging, discharging and idle.
Cell balancing	Verify the BMS can balance cell voltages	Difference between any two cells remains within the range over time.
SoC estimation	Verify the BMS accurately estimates the remaining battery capacity.	Meets the actual battery capacity during discharge and charge cycles.

(*Continued*)

TABLE 6.17 (*Continued*)   Grid-Scale Battery Testing Matrix (BSS with DERs)

Test Category	Test Description	Pass/Fail Criteria
SoH monitoring	Verify the BMS can monitor battery health and detect potential degradation.	Meets health parameters (eg. increased internal resistance, reduced capacity)
Safety interlocks	Verify the BMS triggers system shutdown, overcharge, over-temperature.	Performs safety actions listed
Communication and Control		
Communication interface	Verify communication between BSS, DERs, and grid operator systems.	Bidirectional communication established and data transmitted/received error-free.
Control system functionality	Verify control system operates as designed based on setpoints and grid signals.	System responds correctly to control commands and grid signals.
Data acquisition	Verify system collects and transmits operational data accurately.	Data acquisition system functions correctly and transmits accurate data.
Islanding Protection		
Loss of mains (islanding detection)	Verify system detects loss of grid connection and disconnects from the grid.	System detects islanding condition and disconnects from the grid within specified timeframe.
Anti-islanding protection	Verify system prevents energizing the grid during island mode operation.	System does not energize the grid during island mode.
DER Integration		
DER power management	Verify BSS interacts correctly with DERs to optimize system operation.	BSS and DERs coordinate effectively to achieve setpoints and grid support functions.
DER communication compatibility	Verify communication between BSS and DERs functions properly.	BSS and DERs exchange data seamlessly using the defined communication protocol.

Chapter 7 will provide a comprehensive set of read data obtained from the hardware and software platform explained in Sections 6.7 and 6.8.

BIBLIOGRAPHY

N. Ertugrul, "Battery Storage Technologies, Applications and Trend in Renewable Energy," *2016 IEEE International Conference on Sustainable Energy Technologies (ICSET)*, Hanoi, Vietnam, 2016, pp. 420–425, https://doi.org/10.1109/ICSET.2016.7811821.

N. Ertugrul and F. Castillo, "Maximizing PV Hosting Capacity and Community Level Battery Storage," *2019 29th Australasian Universities Power Engineering Conference (AUPEC)*, Nadi, Fiji, 2019, pp. 1–6, https://doi.org/10.1109/AUPEC48547.2019.211903.

N. Ertugrul and G. Haines, "Impacts of Thermal Issues/Weather Conditions on the Operation of Grid Scale Battery Storage Systems," *2019 IEEE 13th International Conference on Power Electronics and Drive Systems (PEDS)*, Toulouse, France, 2019, pp. 1–6, https://doi.org/10.1109/PEDS44367.2019.8998808.

N. Ertugrul and W. K. Wong, "Power System Inertia in High-Renewable Penetration Power Systems and the Emerging Role of Battery Energy Storage," *2020 International Conference on Smart Grids and Energy Systems (SGES)*, Perth, Australia, 2020, pp. 133–138, https://doi.org/10.1109/SGES51519.2020.00031.

N. Ertugrul, G. Bell, G. G. Haines and Q. Fang, "Opportunities for Battery Storage and Australian Energy Storage Knowledge Bank Test System for Microgrid Applications," *2016 Australasian Universities Power Engineering Conference (AUPEC)*, Brisbane, QLD, Australia, 2016, pp. 1–6, https://doi.org/10.1109/AUPEC.2016.7749304.

N. Ertugrul, C. E. McDonald and J. Makestas, "Home Energy Management System for Demand-Based Tariff Towards Smart Applicances in Smart Grids," *2017 IEEE 12th International Conference on Power Electronics and Drive Systems (PEDS)*, Honolulu, HI, USA, 2017, pp. 511–517, https://doi.org/10.1109/PEDS.2017.8289156.

A. Firlit, *Power Theory with Non-sinusoidal Waveforms, Annex 3, Handbook of Power Quality*, edited by A. Baggini, John Wiley & Sons, Ltd, New York, 2008.

S. Hashemi et al, Efficient control of active transformers for increasing the PV hosting capacity of LV grids. *IEEE Transactions on Industrial Informatics* 13(1), pp. 1–1, 2016. https://doi.org/10.1109/TII.2016.2619065.

G. Haines, N. Ertugrul, G. Bell and M. Jansen, "Microgrid System and Component Evaluation: Mobile Test Platform with Battery Storage," *2017 IEEE Innovative Smart Grid Technologies - Asia* (ISGT-Asia), Auckland, New Zealand, 2017, pp. 1–5, https://doi.org/10.1109/ISGT-Asia.2017.8378466.

R. Khezri, A. Mahmoudi, M. H. Khooban and N. Ertugrul, "Optimal Sizing of Grid-tied Residential Microgrids Under Real-Time Pricing," *2021 IEEE Energy Conversion Congress and Exposition (ECCE)*, Vancouver, BC, Canada, 2021, pp. 771–776, doi: 10.1109/ECCE47101.2021.9595828.

R. Khezri, A. Mahmoudi, N. Ertugrul, M. F. Shaaban and A. Bidram, "Battery Lifetime Modelling in Planning Studies of Microgrids: A Review," *2021 31st Australasian Universities Power Engineering Conference (AUPEC)*, Perth, Australia, 2021, pp. 1–6, https://doi.org/10.1109/AUPEC52110.2021.9597840.

Report on "Frequency Stability Evaluation Criteria for the Synchronous Zone of Continental Europe: Requirements and impacting factors", Prepared by RG-CE System Protection & Dynamics Sub Group, by European Network of Transmission System Operators for Electricity, ENTSOE, March 2016. https://doi.org/10.1109/AUPEC.2016.7749358.

E. Troester, New German Grid Codes for Connecting PV Systems to the Medium Voltage Power Grid, 2nd International Workshop on Concentrating Photovoltaic Power Plants: Optical Design, Production, Grid Connection, 2009.

L. Wang, N. Ertugrul and M. Kolhe, "Evaluation of Dead Beat Current Controllers for Grid Connected Converters," *IEEE PES Innovative Smart Grid Technologies*, Tianjin, China, 2012, pp. 1–7, https://doi.org/10.1109/ISGT-Asia.2012.6303109.

Y. Yao and N. Ertugrul, "An Overview of Hierarchical Control Strategies for Microgrids," *2019 29th Australasian Universities Power Engineering Conference (AUPEC)*, Nadi, Fiji, 2019, pp. 1–6, https://doi.org/10.1109/AUPEC48547.2019.211804.

Y. Yao, N. Ertugrul and S. A. Pourmousavi, "Power Sharing and Voltage Regulation in Islanded DC Microgrids with Centralized Double-Layer Hierarchical Control," *2021 31st Australasian Universities Power Engineering Conference (AUPEC)*, Perth, Australia, 2021, pp. 1–6, https://doi.org/10.1109/AUPEC52110.2021.9597804.

B. Young, N. Ertugrul and H. G. Chew, "Overview of Optimal Energy Management for Nanogrids (End-Users with Renewables and Storage)," *2016 Australasian Universities Power Engineering Conference (AUPEC)*, Brisbane, QLD, Australia, 2016, pp. 1–6.

CHAPTER 7

Comprehensive Results and Analysis of an Autonomous Microgrid with BSS

7.1 BASIC OPERATION OF A BATTERY STORAGE SYSTEM (BSS) AND BATTERY

As previously explained, although a BSS utilizes batteries for storage, additional components are needed for its function within an electrical network. These components manage power flow and ensure safe operation. For example, isolation transformers, protection devices, cooling systems and a high-level control system are included, which will be covered in the comprehensive testing phases.

This section details the testing procedures and results for a BSS integrated with the microgrid test platform. Figure 6.19 illustrated the BSS, including a bidirectional inverter that facilitates AC/DC power conversion for both charging and discharging the battery. The figure also highlighted the locations of voltage and current sensors. These sensors measure the primary electrical quantities used to analyse the microgrid's operation under various distributed energy resource (DER) and load configurations. Due to the bidirectional nature of the power flow in microgrids, Table 7.1

TABLE 7.1 Typical Operating Scenarios and the Polarity of Active Power at Each Location

Scenario	\multicolumn{6}{c}{Active Power Polarity}
	Grid (K)	GenBus Tot. (M)	GenBus Inv. (G1)	Inverter AC (F)	Inverter DC (B)	Battery DC (Calculated)	Load (N)
Battery charging from grid	+	−	−	−	+	+	
Battery exporting to grid	−	+	+	+	−	−	
Battery islanded, supplying load		+	+	+	−	−	+
Battery islanded, absorbing from load (and/or peak shaving)		−	−	−	+	+	−
Grid supplying load (battery off)	+						+
Grid absorbing from load (reverse power flow)	−						−
Grid and battery supplying load	+	+	+	+	−	−	+

is given to summarize the possible polarity of the power measured at each node in the system for different scenarios.

The first tests connected the test platform to a low-voltage (LV) grid to perform a battery charging test. Figure 7.1 demonstrates system operation during a 20 kW charge test, which served as a simple demonstration of basic system functionality. The inverter was initially controlled to turn on with zero power consumption, followed by gradual increases in charging power delivered to the battery. The 20 kW operating point represents approximately 7.4% of the inverter's rated power.

Figure 7.1a displays the DC power delivered to the battery (calculated from nodes B and D). At the test's beginning, the inverter was instructed to draw zero power from the grid. However, since the inverter requires power to operate, it initially drew power from the battery, resulting in brief dips of negative battery power. Shortly after, the inverter received instructions to draw power from the grid and charge the battery, with power levels increasing in steps up to 20 kW. The actual DC power delivered to the battery was lower than the 20 kW consumed by the inverter from the grid due to inverter losses.

Note that in Figure 7.1b, the battery voltage (measured at node C) rises at an approximately constant rate throughout the constant positive power charging period, as expected. Figure 7.1c shows the inverter's AC terminal voltage for each phase, measured between the transformer and the inverter (node E). A small voltage drop occurs once charging starts and the inverter begins to load the grid connection.

506 ▪ Reinventing the Power Grid

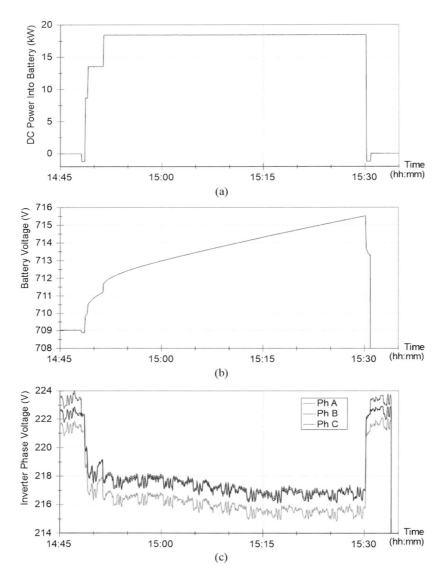

FIGURE 7.1 System operation during a 20kW battery charge demand from the inverter, which represents 7.4% of the inverter's rated power: (a) DC power into the battery for about 43 minutes, (b) battery terminal voltage and (c) inverter phase voltages (RMS) during charging.

The following test involved connecting the previously mentioned load bank to the Q3 LV network line and testing the full output power of the inverter and battery. Figure 7.2 shows the results, in which the test begins

Results and Analysis of an Autonomous Microgrid ▪ 507

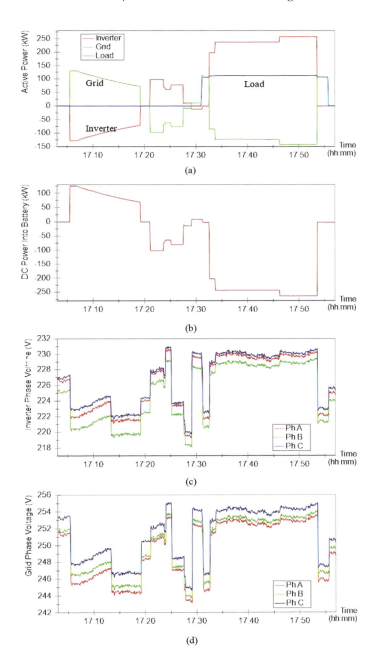

FIGURE 7.2 BSS testing under the full output power of the inverter and battery: (a) total active power at the inverter (node F), grid (node K) and load (node N), (b) DC power into battery, (c) inverter terminal voltage (RMS) during the full power discharge test, d) Grid voltage (RMS) during the full power discharge test.

power from the grid, and after the load bank is running, the inverter is enabled to operate at just under full power. As illustrated in Figure 7.2a, the inverter outputs 258 kW, with 113 kW sent to the load and 143 kW exported to the grid. The missing 2 kW represents losses in the isolation transformer between the inverter and both the grid and load connections.

During the test, the peak DC power (calculated from nodes B and D) was measured at −264 kW, indicating power flowing out of the battery (see Figure 7.2b). Due to inverter and transformer losses, not all battery power reaches the grid. This lost power, dissipated as heat, necessitates the operation of air conditioning systems for the battery, inverter and transformer.

Figure 7.2c shows the inverter's AC terminal voltage for each phase, measured at node E. Similar to Figure 7.2c, it demonstrates a voltage drop as the inverter supplies power to the grid by discharging the battery. Conversely, when charging the battery from the grid (and load), the voltage increases. Figure 7.2c shows the grid voltage (measured at node J), exhibiting a similar pattern but with a different voltage level due to the isolation transformer.

For the full power discharge test described above, the total battery voltage, individual cell voltage and SoC were also measured. Figure 7.3a shows both the battery voltage and the SoC. The battery voltage was directly measured at the inverter DC terminals (node C), but SoC was calculated by the battery management system (BMS) in each battery rack.

As known, SoC estimation relies on a complex model for accurate battery measurement. While battery voltage loosely correlates with SoC, it is not an accurate standalone indicator. For example, when charging begins, the battery voltage might increase by nearly 10 V instantly, but due to the electrochemical reactions, the SoC cannot immediately reflect this change. Figure 7.3b displays the cell voltages for the battery, plotting maximum, average and minimum values over time. Each battery module individually measures its cell voltages, and the data are collected by the BMS of each battery rack.

As known, a large, high-power battery consists of numerous LV cells. The cell voltage is determined by the battery chemistry (lithium ion in this case) and shares the same nominal voltage with other batteries of the same type. Each cell in the utilized battery has a nominal voltage of 3.7 V. These cells are connected in series, with individual voltages adding up to a level suitable for a grid-connected energy storage system. Each battery module uses 14 cells in series, resulting in a nominal module voltage of 51.8 V. Each battery rack then uses a stack of 14 modules connected in series to produce

Results and Analysis of an Autonomous Microgrid with BSS ■ 509

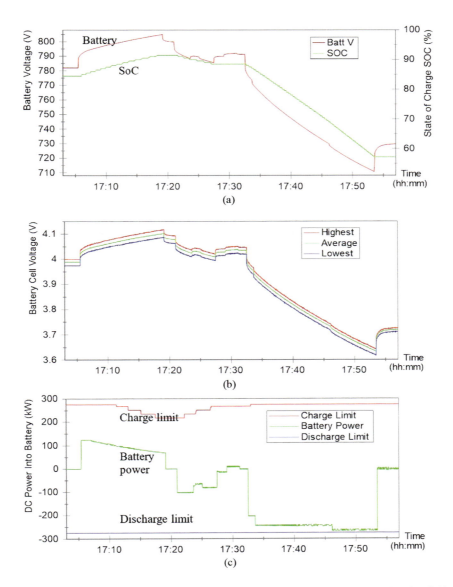

FIGURE 7.3 The battery voltage and the SoC during BSS testing under the full output power of the inverter and battery: (a) the battery voltage and the SoC, (b) battery cell voltages, (c) battery power, charge and discharge limits from the battery section controller (BSC).

by charging the battery for 15 minutes to just over 91% state of charge (SoC). The full power discharge (negative power) test starts shortly after 5:30 PM by turning on the 100 kW load bank. Initially, this only consumes a nominal voltage of 725 V. In total, each rack has 196 lithium-ion cells connected in series.

Despite having the same total voltage (sum of 196 cells), each battery rack stores 91 kWh of energy (126 Ah at 725 V). Three racks are connected in parallel to provide a total energy storage capacity of 273 kWh. Each battery rack is rated for "1C", meaning a single 91 kWh rack can discharge its energy in 1 hour (or longer) or operate at a power of up to 91 kW. In the context of energy capacity, a lower C-rating signifies a lower power rating, while a higher C-rating indicates a higher power rating. For example, a rating exceeding "1C" means a battery can discharge its capacity in less than an hour.

The C-rate of a battery is a simplified representation of its general power capability and does not reflect the actual power rating under all conditions and states of charge. For example, a battery nearing 99% SoC cannot be charged at full power and requires a significantly slower rate. Figure 7.3c shows the battery power for the same period as Figure 7.2b, but with the addition of instantaneous charge and discharge power limits. These limits are determined in real time by each rack's BMS and the overarching battery section controller managing all battery racks. Charging starts above 100 kW but automatically reduces as the SoC increases. This decrease in charge power is controlled by the high-level control system, which coordinates the inverter's power commands based on the battery's SoC. As the SoC rises (approaching 90%), the battery's charge power limit drops and remains so until the SoC decreases.

Due to these power limitations, a grid-connected battery experiences reduced effectiveness if the SoC is too high or too low. Grid applications like energy arbitrage, voltage support, reactive power support and peak shaving necessitate a SoC that allows for high-power charging and discharging at any time.

The efficiency of a device describes two key aspects: How effectively the device utilizes the input electrical energy for its intended output and the amount of energy lost as heat, which defines the cooling requirements. This is especially critical for batteries and power electronic converters due to their sensitivity to heat. Inverter efficiency represents the instantaneous power loss with respect to the power flow through the inverter (when charging or discharging). Figure 7.4a shows the inverter efficiency for the 20 kW charge test, and Figure 7.4b shows the inverter efficiency for the full power charging and discharging conditions. As shown in Figure 7.4a and b, when the inverter runs at light loads, it demonstrates poor efficiency.

Results and Analysis of an Autonomous Microgrid with BSS ■ 511

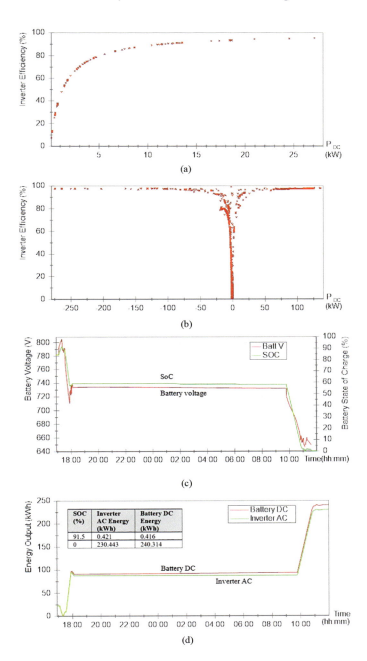

FIGURE 7.4 Inverter efficiency and energy capacity curves: (a) inverter efficiency during the 20 kW charge test, (b) inverter efficiency during the full power discharge test, (c) battery voltage and SoC characteristics during the full discharge test, and (d) battery and inverter accumulated energy during the full discharge test.

This is a typical behaviour for all Si device-based inverters and also highlights the importance of correctly sizing the inverter for the application to minimize losses. Note that when wide bandgap (WBG) devices are used in inverters, the efficiency characteristics will have a flat trend, similar to the high efficiency operating region as in the efficiency figures given, which is between 92% and 97.5% efficiency and above.

The capacity of an energy storage system is measured using an energy meter while fully discharging the battery. Figure 7.4c shows the battery voltage and SoC from the start of the full power discharge test (91.5% SoC) to the end of the full discharge test (0% SoC) on the following day. Figure 7.4d 20 shows the energy meter plots for the inverter AC connection and the battery/inverter DC connection, obtained over the same time. The table inside the figure summarizes that the energy delivered by the inverter's AC connection was lower than the energy delivered by the battery. This is because of inverter losses, but also because of the auxiliary inverter's standby current (which caused a small drop in SoC overnight). This test indicates that 91.5% SoC corresponds to 240 kWh of energy in the battery or 230 kWh that can be delivered by the inverter.

7.2 FOUR-QUADRANT POWER, POWER FACTOR CONTROL AND INVERTER IDLE REACTIVE POWER WAVEFORMS

Power definitions were covered in Chapters 4 and 5. As discussed, most loads connected to a power system are not ideal, resulting in reactive power flow and a power factor less than 1. While desirable, not all devices have built-in power factor correction. This leads to net reactive power flows through the power system, increasing system load and reducing available active power. Typically, this is addressed at the network/grid level by adding power factor correcting devices such as capacitors, inductors, synchronous condensers, static var compensators (SVCs) and static synchronous compensators (STATCOMs).

As demonstrated in this case study, a BSS with a smart inverter can also provide power factor control functions similar to a STATCOM. The inverter operates as a voltage source in all four quadrants, allowing it to not only charge or discharge the battery but also supply or consume reactive power.

Figure 6.19a illustrates the system's design, permitting the inverter to supply reactive power to the LV grid and/or downstream load connections. When the inverter supplies the same amount of reactive power consumed by the downstream load/feeder, the reactive power cancels out, leaving only real, active power flow at the LV grid connection. Consequently, the power factor of the load can be controlled and improved (brought closer to 1) even if the load itself lacks power factor control capability.

If the downstream load in Figure 6.19a is a 100 kW resistive load, it can only absorb active power and naturally has a high-power factor (minimal reactive power). Therefore, in such tests, any reactive power flows occur only between the inverter and the grid.

To demonstrate reactive power control and four-quadrant operation, the inverter was set to operate at eight different points around a 100 kVA unit power circle. The 100 kVA apparent power represents 37% of the inverter's rated apparent power. Four-quadrant operation indicates that the inverter can operate with positive or negative reactive power while charging or discharging the battery.

Active, reactive, apparent power and power factor were measured at node M in Figure 6.19a, representing the total at the generation bus. The polarity at node M is defined relative to the inverter (and other embedded generators) operating as a generator. As shown in Table 7.1, positive power indicates power flow from the generation bus to the grid/load, while negative power indicates flow from the grid/load to the generation bus. Positive reactive power represents the supply of reactive power to the grid/load. Conversely, negative reactive power represents reactive power consumption from the grid/load.

The four-quadrant power test or P-Q unit circle test was performed in two parts in this section. The first part involved operation at five points (A to E) and operation at zero active and reactive power (F). In Figure 7.5a, the apparent power for the five points (A to E) is kept approximately constant, while the value of real and reactive power varies for each point. Figure 7.5b displays the power factor for the same period. Because apparent power is held constant, the power factor plot follows the same shape as the active power plot.

Plotting reactive power versus active power provides a more insight for this test. Figure 7.6 displays all measured active and reactive power points from both test parts, corresponding to the time periods marked in Figure 7.5a and c. The figure clearly shows the eight operating points and their location on the 100 kVA unit apparent power circle. For better clarity, corresponding operation points and the actual 100 kVA circle traced by these points are annotated.

One key observation from Figure 7.6 is the spread of points in each direction. The measured points for each operating point have minimal

514 ■ Reinventing the Power Grid

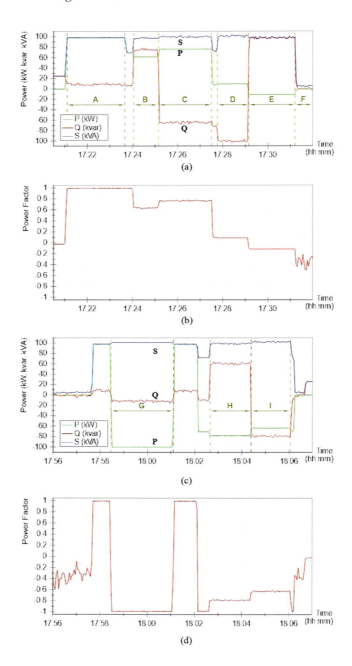

FIGURE 7.5 Four-quadrant power and power factor control tests: (a) operation in quadrants 1 and 4, (b) corresponding power factor during operation in quadrants 1 and 4, (c) operation quadrants 2 and 3, and (d) power factor during operation quadrants 2 and 3.

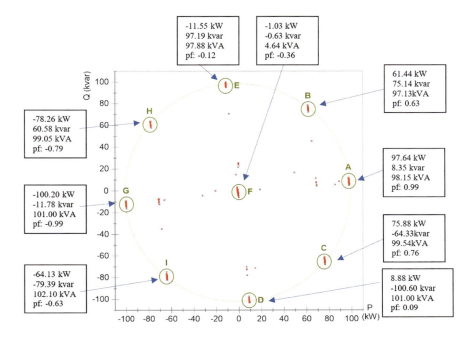

FIGURE 7.6 P-Q plots showing operation around the P-Q unit circle with 100 kVA apparent power, and average values of key parameters for each operating point.

horizontal spread, indicating good precision in the inverter's active power control. However, the noticeable vertical spread suggests the inverter's reactive power control has less precision compared to real power control. While the inverter can still operate in all these regions, the reactive power will fluctuate more than the real power.

For each operating point, the average values of active power, reactive power, apparent power and power factor over the corresponding region are listed next to each region. In region F (zero P and Q), the apparent power appears significantly higher than the averages of active and reactive power. This occurs because both active and reactive power can fluctuate around zero (being positive or negative momentarily), but apparent power is always positive and produces a non-zero average when both active power and reactive power are varying.

During the P-Q unit circle test (Figure 7.5a), high-speed waveforms were captured at 5:23:58 PM (hours:minutes:seconds). This corresponds to the transition into region/operating point "B" just before 5:24 PM in the figure. The transition involves a shift from a power factor slightly below 1 to a power factor above 0.63. The waveforms show the change after active power reduction and the simultaneous increase in reactive power to maintain 100 kVA apparent power.

Figure 7.7a displays the inverter's phase A voltage and current waveforms. The voltage appears sinusoidal but "fuzzy" due to the inverter's switching noise. To generate a sine wave, the inverter rapidly switches the battery's DC voltage (around 800 V) between positive and negative values using pulse width modulation (PWM). While a filter smooths the voltage into a sine wave, some noise persists. However, this noise has minimal impact on active power and power factor, which remain essentially equivalent to those of the cleaner generation bus and grid connection.

Figure 7.7b shows the phase voltage and current for the generation bus. The voltage is measured at node H, after the transformer, which filters out any remaining noise from the inverter's voltage waveform. In Figure 7.7a and b, the current waveform appears sinusoidal. This represents the same current waveform measured on either side of the transformer. Just before 58.93 seconds, the inverter's set point changes. This shift causes the phase of the current waveform to change relative to the voltage waveform. The phase shift is evident in both figures, but clearer in Figure 7.7b. In this case, the current waveform lags behind the voltage waveform, indicating an increase in reactive power (positive).

The increase in reactive power is more clearly visible in the instantaneous power waveforms for the inverter and generation bus, shown in Figure 7.7c and d, respectively. Figure 7.7d provides a clearer view as it lacks the inverter switching noise.

Before the transition, only real power flows. In a single phase, this pulsates at twice the power system frequency. In a three-phase system, each phase's power pulsates such that the total instantaneous power flow across all phases sums to a constant value. After the transition to reactive power, the minimum power dips below zero, and power starts to flow in the reverse direction. The active power measurement can be obtained from the waveform by simply averaging the pulsating power.

Note that at the very start of Figure 7.5a and at the very end of Figure 7.5c, the inverter's real power is zero; however, the reactive power and apparent power are not. At this point in time, the inverter was electrically connected to the grid, but was disabled. This means the inverter was powered on, but was not running, was not performing high frequency switching to produce a voltage and was not trying to control the flow of power into its AC connection.

Results and Analysis of an Autonomous Microgrid with BSS ■ 517

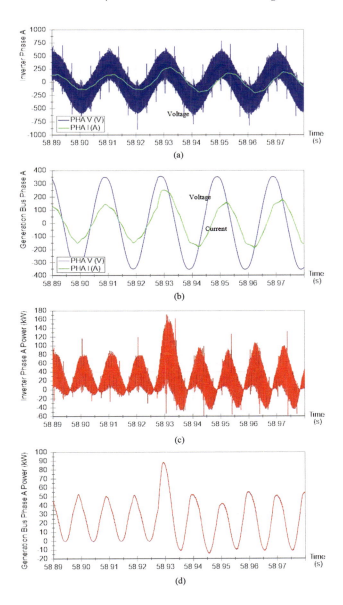

FIGURE 7.7 Reactive power waveforms as reactive power is increased: (a) inverter phase A voltage and current waveforms, (b) generation bus phase A voltage and current waveforms, (c) inverter phase A instantaneous power waveform, (d) generation bus phase A instantaneous power waveform.

This is an unwanted side effect of this inverter deployed in this configuration. Ideally, when disabled, no current should be flowing between the inverter's AC connection and the grid, and hence, no power should be flowing. In this case of an idle or disabled inverter, the current flow results in unwanted reactive power consumption and an apparent power flow just under 10% of the inverter's rated apparent power (270 kVA).

With the inverter not running, Figure 7.8a and b has a similar shape. In both figures, the inverter current is out of phase with the voltage (in this case, lagging), resulting in the consumption of reactive power. Because the current is lagging by 90°, all the apparent power is reactive, with no active power flowing. A clearer view can be seen in the instantaneous power waveforms at the inverter connection and generation bus, shown in Figure 7.8c and d respectively. Both waveforms show equal parts of positive and negative power flow. When averaged, this power cancels out and results in no net power flow.

The power waveform has a slightly different shape in Figure 7.8d. Figure 7.8c has ripples in the peaks and troughs, indicating harmonic power flow (power flowing at a multiple of the power system frequency, in this case third harmonic). Figure 7.8d shows that the harmonic power flow is lower, with smaller ripples in the power waveform. The two power waveforms are different because of the transformer that sits between the generation bus and the inverter.

7.3 MICROGRID ANTI-ISLANDING, ISLANDING, SEGREGATION AND REINTEGRATION WITH A BSS

As is well known, a microgrid can operate in three modes: connected to the grid, disconnected (islanded) or dynamically switching between the two. When a grid-connected microgrid disconnects dynamically, it enters island mode. In the event of a grid outage (e.g., blackout), safety regulations require grid-connected inverters to shut down. This anti-islanding protection prevents inverters from maintaining or forming an independent electrical supply. In other words, the inverter must not provide power or voltage in such scenarios.

The following method was used to test anti-islanding:

- The load bank is activated, drawing power from the grid.

- The inverter is then enabled and set to supply most of the power consumed by the load. This significantly reduces the load on the grid, but the inverter remains connected and synchronized.

Results and Analysis of an Autonomous Microgrid with BSS ▪ 519

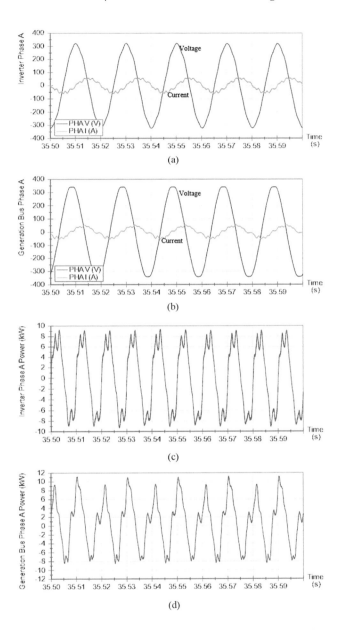

FIGURE 7.8 Inverter idle reactive power waveforms: (a) Inverter phase A voltage and current waveforms, (b) Generation Bus phase A voltage and current waveforms, (c) Inverter phase A instantaneous power waveform, (d) Generation Bus phase A instantaneous power waveform.

- The upstream grid connection is manually tripped, disconnecting the inverter and load from the grid.

Figure 7.9a shows the 10 kW active power for the inverter (measured at node F), the grid (measured at node K) and the load (measured at node N). The inverter supplies most of the load power for roughly 1.5 minutes. After the grid supply is tripped, all power within the microgrid system ceases, including the inverter power, as intended.

Figure 7.9b illustrates the positive sequence voltage magnitude measured by the phasor measurement unit (PMU) of the embedded data logging system. For this test, the PMU was configured to capture fast changes by taking 25 measurements per second. The inverter voltage was measured at node E, and the grid voltage was measured at node J. The positive sequence represents the nominal operating voltage of the three-phase system. Following the supply disconnection, the voltage within the microgrid drops in less than a second.

The anti-islanding test was repeated at 110 kW using the same method as the 10 kW test. Figure 7.9c displays the power within the microgrid, with the inverter supplying most of the load power for approximately 4 minutes. All power in the microgrid system ceases after tripping the grid supply. Figure 7.9d shows the voltage measured by the PMU. Once again, the voltages at nodes J and E drop in less than a second after the grid supply is disconnected.

As previously stated, the autonomous microgrid test platform is capable of intentional islanding operation. The system can perform smooth segregation and reintegration, where the high-level control system coordinates the dynamic disconnection and reconnection of both embedded generators and loads from the grid. "Smooth" refers to the fact that generators and loads inside the microgrid continue to operate normally during the transition. Loads experience no outages or interruptions and do not even notice that the grid has been disconnected.

Referring to the single-line diagram and sensor layout given previously, the system starts operation in grid-connected mode. Circuit breakers Q1 (to main grid), Q2 (to generation bus/inverter) and Q3 (to downstream load) are all closed, and power can flow through them. Upon receiving a command to segregate and form an island, the control system coordinates the transition of all devices to this new operating mode. This includes a change in protection settings and a change in the inverter's operating mode.

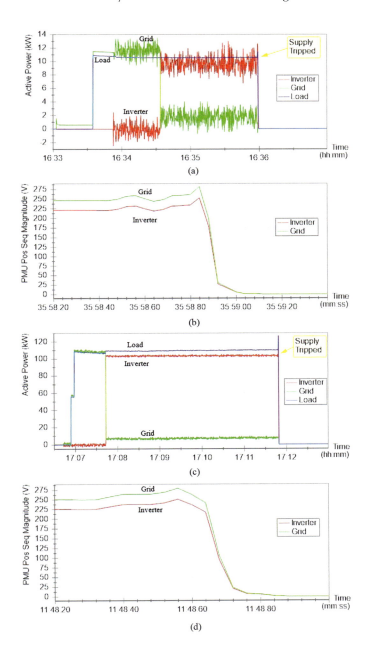

FIGURE 7.9 Microgrid anti-islanding: (a) Total active power at the inverter (node F), grid (node K) and load (node N) connections during the 10kW anti-islanding test, (b) Voltage magnitude during the 10kW anti-islanding test, (c) Total active power at the inverter (node F), grid (node K) and load (node N) connections during the 110kW anti-islanding test at Thebarton SA, (d) Voltage magnitude during the 110kW anti-islanding test.

The inverter transitions to providing the voltage and frequency reference for the microgrid. This is done using a voltage and frequency droop curve, where voltage and frequency decrease in response to the power being delivered by the inverter. Figure 7.10a shows the frequency of the grid connection (measured on the grid side of Q1, at node J) and the load (measured at the generation bus, at node H). When Q1 is closed, the frequency at both locations is the same. After segregation (Q1 opens), the generation bus frequency changes independently of the grid. Figure 7.10a highlights the frequency stability of the inverter when operating in island mode. After reintegration (Q1 closes), the frequency at both locations is the same again.

Figure 7.10b shows the power flowing through the microgrid during the islanding test. Before segregation, the inverter provided almost all the power used by the load. After segregation, the grid power drops to zero, but the load power remains constant. Just after reintegration, there is a small peak in power exported from the inverter to the grid, lasting less than a second. Throughout the process, the load power is constant and operates continuously and seamlessly during both transitions.

The electrical power measurements are made by measuring the voltage and current (of each phase) at each node in the system. In grid-connected mode, the voltage of the grid, load and generation bus is the same because these buses are all directly connected. However, this is not the case in island mode. In the electrical diagram given previously, there are two voltage nodes: node J located on the grid connection and node H located on the generation bus. Depending on the state of the grid connection (Q1) and generation bus connection (Q2), the voltage at the load connection (at node N) may be the same as either node J, node H, both nodes J and H, or neither (case where both Q1 and Q2 are open, and there is no voltage at the load).

The power at the load (node N) is measured twice, with a different voltage reference for each measurement. Figure 7.10c shows the same power plot as Figure 7.10b, but with the load power measurement that uses the grid voltage (node J) as its reference. Because the grid voltage is disconnected from the microgrid during islanding mode, the resulting voltage and current waveforms drift out of phase and no longer produce a meaningful result. In this case, the plot appears to show power flowing back and forth (this is not possible, though, because the load used can only absorb power and not generate power).

Results and Analysis of an Autonomous Microgrid with BSS ■ 523

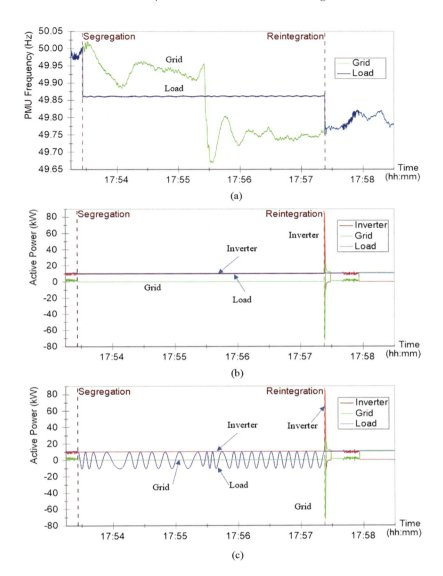

FIGURE 7.10 Microgrid islanding: (a) grid and generation bus frequency during the microgrid islanding test, (b) power during the microgrid islanding test, and (c) incorrect voltage reference for load power calculation during the microgrid islanding test.

Figure 7.11a–d show the moment of microgrid segregation in detail. These data are measured using the data logging system's PMU. As previously indicated, a PMU uses advanced analysis techniques to provide more information than conventional electrical measurement techniques.

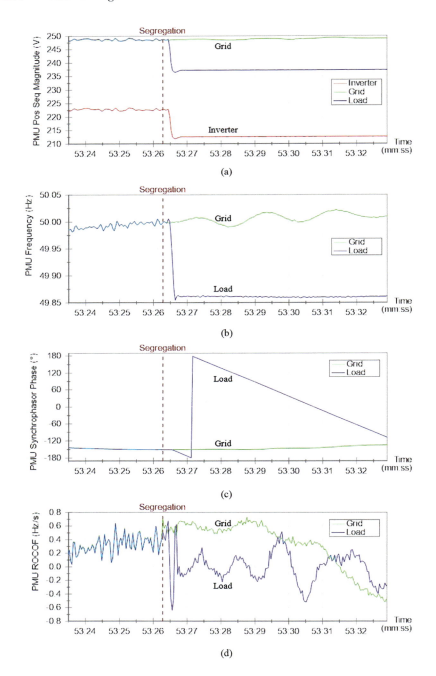

FIGURE 7.11 Microgrid segregation: (a) voltage magnitude during segregation, (b) voltage frequency during segregation, (c) voltage phase during segregation, and (d) RoCoF during segregation.

After segregation, the grid and load voltages separate (Figure 7.11a). Both the inverter and load voltage drop because of the voltage droop curve used by the inverter in island mode. The inverter voltage is lower than the load voltage due to the isolation transformer. Frequency also drops (Figure 7.11b), due to the frequency droop curve of the inverter.

Figure 7.11c shows the phase of the voltages at the grid connection and the generation bus (corresponding to the voltage at the load). Both voltages are synchronized before segregation, having the same phase angle. Because the frequency is near, but not exactly 50 Hz, the phase angle slowly changes. After segregation, the drop in inverter frequency (which sets the generation bus and load frequency) results in a relative phase angle that changes more quickly and is no longer synchronized with the grid.

Figure 7.11d shows the rate of change of frequency (RoCoF). Again, both the grid and load voltages have the same value until segregation. Just after segregation, there is a small impulse/negative peak in the RoCoF waveform, showing the sudden change in frequency.

Note that the PMU uses a GPS-locked timing reference to provide power system measurements that are synchronized to Universal Coordinated Time (UTC). The resulting measurements are called synchrophasors because they produce magnitude and phase measurements (the components of a phasor) with respect to the ideal power system frequency of exactly 50 Hz (synchronized to UTC via the GPS receiver). For a voltage waveform, the synchrophasor magnitude is just the voltage. The synchrophasor phase is the relative phase angle difference between the voltage waveform and an ideal 50 Hz sine wave (aligned to UTC). For an exact 50 Hz waveform, the phase will be constant. For a waveform not at 50 Hz, the phase angle will slowly change. For example, a 49.9 Hz waveform will have a synchrophasor phase that changes at a constant rate and repeats every 10 seconds (the phase angle is wrapped into a range of −180° to 180°).

The key benefit of synchrophasors is the ability to see fast changes in the phase angle and to look at the phase angle alignment of different parts of the power system. Frequency (rate of change of phase) and RoCoF are derived from the phase angle measurements. They can provide deeper insight into the operation and stability of a power system.

Figure 7.12a–d show the moment of microgrid reintegration in detail. These data are also measured using the data logging system's PMU.

Before reintegration can occur, the voltage source inverter must synchronize with the grid. Synchronization requires the inverter's voltage waveform to match the voltage level of the grid and be in phase with the

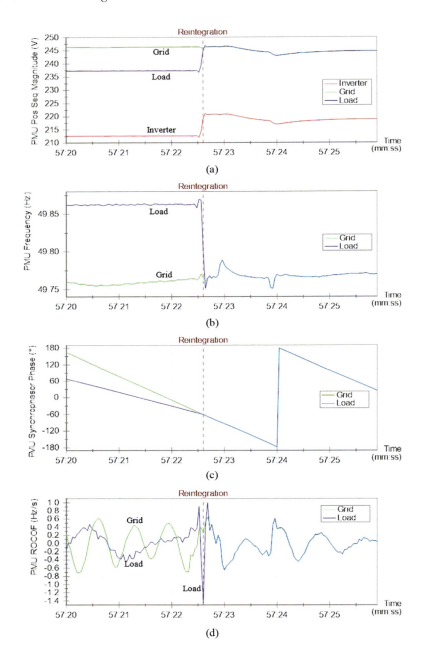

FIGURE 7.12 Microgrid reintegration: (a) PMU's positive voltage sequence magnitude during reintegration, (b) PMU frequency during reintegration, (c) PMU synchrophasor phase during reintegration, and (d) PMU RoCoF during reintegration

grid waveform. If the connection to the grid (Q1) is closed without synchronization, very large currents and power may briefly flow between the inverter and the grid, potentially damaging components of the microgrid. Normal and desired power flow between the grid and the microgrid is also impossible if the voltages are not synchronized.

Figure 7.12b shows the frequency of the grid and the load (set by the inverter). Before reintegration, the two frequencies are different. Because of this, the phase angle between them is constantly changing. The grid and load voltages are repeatedly coming into and going out of phase. After receiving a command to reintegrate, the system waits until the grid and inverter voltages are in phase. Then, Q1 can be closed to connect the generation bus and load back to the grid.

At the point of reintegration, the generation bus voltage increases to match the grid voltage (Figure 7.12a), and the generation bus frequency drops to match the grid frequency (Figure 7.12b). The RoCoF plot in Figure 7.12d shows an impulse/negative peak in the generation bus frequency, caused by the sudden drop in frequency required to match the grid frequency. The RoCoF measurement is very sensitive to small changes and is useful for detecting events and changes in the power system.

7.4 POWER QUALITY TESTS WITH NONLINEAR LOAD IN BSS

Power quality analysis becomes increasingly complex for non-ideal power system conditions, particularly with the rise of nonlinear loads like battery storage and variable speed drives (VSDs). This section explores power quality issues in a microgrid powered by a BSS supplying VSDs. While the system functioned continuously, it exhibited less than ideal power quality due to rapidly fluctuating power, non-ideal power factor and current harmonics.

Power quality, which involves voltage and current characteristics, directly impacts power system reliability, stability and efficiency. Ideally, sinusoidal and in-phase voltage and current waveforms maximize usable power flow. As is known, power factor measures how efficiently voltage and current contribute to real (active) power. Linear loads (resistance, capacitance, inductance) maintain sinusoidal waveforms, with the phase angle affecting power factor. However, nonlinear loads (primarily semiconductor-based devices) distort voltage and current waveforms, causing harmonics and noise. These distortions reduce real power flow, increase heating losses in components, may affect voltage regulation and system stability and may cause other devices to malfunction or age prematurely.

Therefore, there is a need to understand the impacts of nonlinear loads at the LV distribution level (where most customers connect), which is important to ensure that these loads coexist without harming each other. Note also that as nonlinear loads become more prevalent, there is a need to understand their impact on power systems. Therefore, the knowledge gained here can be used to optimize new technologies to achieve higher power quality, minimize interference and ensure efficient energy delivery.

Note that the VSDs are nonlinear loads, consisting of a rectifier and inverter that convert AC to DC and back to AC at varying frequencies. To capture detailed system behaviour, the test platform utilized high-bandwidth AC measurements at nodes E and F in Figure 6.19a for the islanded test with VSD loads in the same network line as the dynamic loads.

To investigate power quality issues, the diesel generator and load bank were disconnected for the islanded test. This left only six VSD-controlled wind tunnel fans, with a total power rating under 800 kW, operating at low speeds (to achieve low power consumption, worst-case power quality and low efficiency). With the diesel generator and load disconnected, the inverter was started in islanding mode. Then, the VSDs were turned on and run at low power for just under 10 minutes.

Figure 7.13a shows the power aggregated over 3 seconds, where each point represents approximately 150 electrical cycles. The power analyser module that produces these measurements uses a method based on IEC 61000-4-30. Here, the incoming voltage and current waveforms are dynamically resampled to achieve a sample rate locked to the power system frequency (i.e., the number of sample points in one cycle is kept constant, and the sample rate varies with the power system frequency). The baseline analysis is performed at a rate of 5 Hz (nominal) or ten electrical cycles per reporting time/measurement. The 3-second aggregation averages the measurements over each 3-second period.

Viewing the 3-second aggregated data, the microgrid island appears to be functioning smoothly. However, this is not the case. Using the underlying higher rate data, a very different picture emerges.

Figure 7.13b shows the power, sampled at 5 Hz or every ten electrical cycles. When recording at this faster rate, it is clear that the delivered power is not smooth or constant. Instead, it varies quite rapidly and erratically. Figure 6.37b includes a breakdown of the power components: apparent power (S), active power (P), reactive power (Q) and distortion power (D). In the plot, the apparent power is larger than the active power, with the difference mainly coming from the distortion power (as the reactive power

Results and Analysis of an Autonomous Microgrid with BSS ■ 529

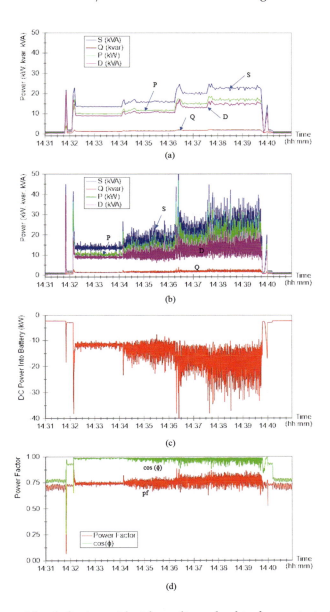

FIGURE 7.13 Islanded microgrid with nonlinear load to demonstrate distortion power: (a) power during islanded operation of the wind tunnel fans loads aggregated over 3 second intervals (approx. 150 cycles per point), (b) power during islanded operation of the wind tunnel fans, sampled at 5 Hz (10 cycles per point), (c) battery power during islanded operation of the wind tunnel fans, sampled at 5 Hz (10 cycles per point), and (d) power factor and fundamental power factor $\cos(\varphi)$ during islanded operation of the wind tunnel fans.

remains quite low). Note that, as discussed previously, non-sinusoidal or distorted waveforms form a power tetrahedron instead of a power triangle in the sinusoidal waveforms.

Note also that although distortion power confirms the relationship between P, Q and S for non-sinusoidal waveforms, the resulting reactive power no longer has much meaning as it does for the sinusoidal case. A reactive power of zero no longer means a power factor of 1. Therefore, trying to reduce the reactive power to zero using capacitors or inductors may not improve the power factor. Hence, reactive power is no longer a useful parameter for improving the power factor of a system with non-sinusoidal waveforms.

Figure 7.13d shows the power factor and the fundamental power factor (pf_1). pf_1 uses the fundamental active and apparent powers (P_1 and S_1), which are calculated by extracting the 50 Hz sinusoidal component in the waveforms. In Figure 7.13d, pf_1 shows that most of the fundamental power is active/useful power. However, the overall power factor is much lower, resulting from the non-fundamental components of apparent power, which represent instantaneous energy flow at a frequency other than 50 Hz. This is a problem for most devices, which can only use the power delivered at the fundamental frequency. For example, higher frequency power applied to a motor does not provide mechanical power and is instead wasted as heat.

To better describe a power system with non-sinusoidal operation, different techniques are required for more meaningful analysis. The IEEE standard 1,459 provides some guidance. Despite the data logger's real-time analysis using an IEC 61000-4-30 method, the data set contains enough information to perform additional IEEE 1459 non-sinusoidal analysis.

Firstly, the voltage and currents are broken down into their fundamental and harmonic components. Because the inverter is being analysed as a complete three-phase system, effective values of voltage, current and apparent power are calculated as they are better suited for unbalanced and non-sinusoidal systems.

Figure 7.14a and b shows that the fundamental effective voltage is close to the effective voltage, with a low non-fundamental voltage component indicating a small harmonic component. Figure 7.14c shows the effective current components, with a much larger non-fundamental component, indicating the presence of significantly more harmonics compared to the voltage.

Results and Analysis of an Autonomous Microgrid with BSS ■ 531

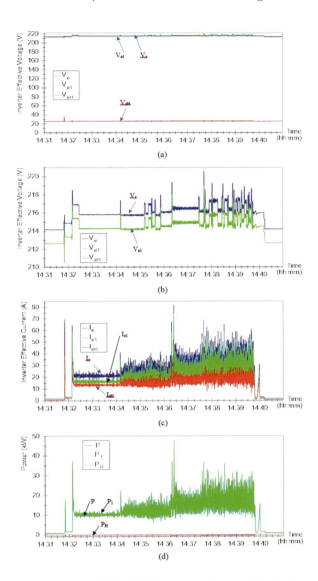

FIGURE 7.14 Non-sinusoidal load behaviour during islanded operation: (a) effective voltage components during islanded operation of the wind tunnel fans. V_e is the effective voltage of all three phases, V_{e1} is the effective fundamental voltage, and V_{eH} is the effective nonfundamental voltage, (b) closer view of the effective voltage and effective fundamental voltage, (c) effective current components during islanded operation of the wind tunnel fans. I_e is the effective current of all three phases, I_{e1} is the effective fundamental current, and I_{eH} is the effective nonfundamental current, and (d) power components during islanded operation of the wind tunnel fans. P is the total active power, P_1 is the fundamental active power, and P_H is the nonfundamental active power.

The active power can also be broken down into fundamental and non-fundamental components. Figure 7.14d shows that almost all active power is fundamental active power, with very little harmonic power flowing. Harmonic power is active power (net energy transfer) that occurs at a frequency other than the fundamental frequency. Despite being actual energy (not energy that flows back and forth), most power devices cannot utilize it, and it is wasted as heat.

The power triangle relationship can also be applied using the IEEE 1459 method, but the result is non-active power, N ($N = \sqrt{S_e^2 - P^2}$). This represents all fundamental and non-fundamental, non-active components. In the special case of only pure sinusoidal waveforms, the value of N is equal to the traditional reactive power (Q). The plot of N in Figure 7.15a should be contrasted with the plot of Q in Figure 7.13b.

The effective apparent power (Figure 7.15b) can also be decomposed into fundamental and non-fundamental components. The non-fundamental apparent power (S_{eN}) can be further decomposed into three components:

- Effective current distortion power, D_{eI}, produced by current harmonics and the fundamental voltage component.
- Effective voltage distortion power, D_{eV}, produced by voltage harmonics and the fundamental current component.
- Effective harmonic apparent power, S_{eH}, produced by voltage harmonics and current harmonics.

Figure 7.15c shows these three components, each produced from different combinations of the effective fundamental and non-fundamental voltage and current. This immediately provides insight into the source of the lower power factor and high non-active apparent power. It shows that the non-fundamental apparent power is dominated by current distortion power produced by the current harmonics of the VSDs. In comparison, there is a much smaller amount of voltage distortion power and even lower harmonic apparent power.

The effective harmonic apparent power can be used with active harmonic power to find the effective harmonic distortion power, $D\left(D_{eH} = \sqrt{S_{eH}^2 - P_H^2}\right.$. In this case, the active harmonic power was very low, resulting in the effective harmonic apparent power being dominated by harmonic distortion power, as shown in Figure 7.15d.

Results and Analysis of an Autonomous Microgrid with BSS ■ 533

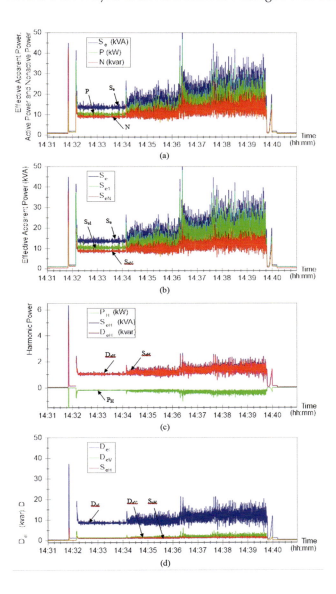

FIGURE 7.15 Power variations during islanded operation with nonlinear loads: (a) nonactive power, derived from the effective apparent power and the active power, (b) apparent power components. S_e is the effective apparent power, S_{e1} is the effective fundamental apparent power and S_{eN} is the effective nonactive apparent power, (c) nonfundamental apparent power components. D_{eI} is the effective current distortion power, D_{eV} is the effective voltage distortion power, and S_{eH} is the effective harmonic apparent power, and (d) effective harmonic distortion power D_{eH}, effective harmonic apparent power S_{eH}, and harmonic active power P_H.

IEEE 1459 uses the same definition of power factor and $\cos\phi$ fundamental power factor, resulting in the same plot as shown in Figure 6.37d. In addition to power factor, the harmonic pollution factor is also defined (Figure 7.16a) as the effective non-active apparent power divided by the effective fundamental apparent power (S_{eN}/S_{e1}). Ideally, this value should be as low as possible.

The fundamental power factor $\cos\phi$ is useful for examining how efficiently power flows in each phase at the fundamental frequency. In a three-phase system, the instantaneous power must also flow in the correct phase order. This power pulsates at twice the system frequency in each phase, with each phase pulsating at evenly spaced times and in the correct sequence (phase A, B, C, then repeating back to phase A). Ideally, the total power flow of all phases adds up to a constant value. If only some of the power flows in the correct phase order, a load may not be able to utilize all the power, even if the power factor of individual phases is high.

The technique of symmetrical components can decompose the three-phase system into three sequence components. The positive sequence component represents the portion of a signal with the correct phase order/sequence. The other two components are the negative sequence (reverse phase order) and the zero sequence (power pulsating at the same time with no phase order). Using the fundamental positive sequence active power component (P_1^+) and the fundamental positive sequence apparent power component (S_1^+), the fundamental positive sequence power factor can be evaluated: $PF_1^+ = P_1^+/S_1^+$. This calculation is similar to $\cos\phi$ but further isolates the fundamental quantities to the power-producing positive sequence component. Figure 7.16b shows PF_1^+ compared to $\cos\phi$.

Load unbalance represents the portion of fundamental apparent power that is not a part of the fundamental positive sequence. This represents apparent power flow that results in the three-phase system not being balanced (the phases are not operating identically). Figure 7.16c shows the load unbalance in red. For the case of the VSDs with fan loads, there is some unbalance, particularly as the fans increase in speed.

By dividing the load unbalance by the fundamental positive sequence apparent power, the load unbalance factor can be produced (Figure 7.16d).

For more context, the load unbalance factor can be compared to the IEC 61000-4-30 method of unbalance, which compares the negative and zero sequence components to the positive sequence. In a balanced system, both the negative and zero sequence components of a signal are zero. In the wind tunnel test, there were no zero sequence components in the

Results and Analysis of an Autonomous Microgrid with BSS ■ 535

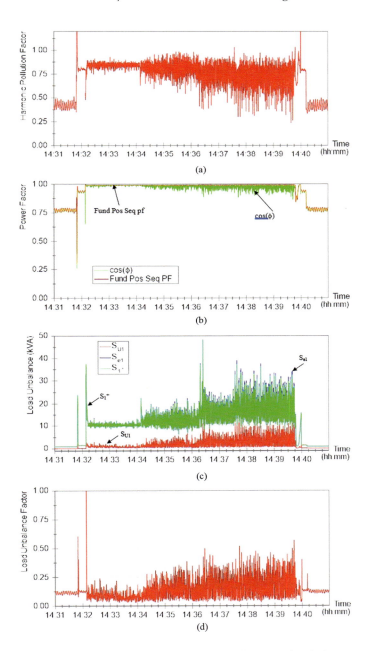

FIGURE 7.16 Harmonics and load unbalances during islanded operation of nonlinear loads (a) harmonic pollution factor, (b) fundamental positive sequence power factor, (c) load unbalance S_{U1}^+ plotted alongside the effective fundamental apparent power S_{e1} and fundamental positive sequence apparent power S_1^+, and (d) load unbalance factor.

voltage and current, leaving only positive and negative sequences. The unbalance calculated from these values is shown below in Figure 7.17a. This shows that the voltage is balanced, but the current has a large amount of unbalance.

Figure 7.17b summarizes the power in the microgrid island with nonlinear VSD load. The fundamental positive sequence active power (P_1^+) represents the total useful, real power flowing from the inverter to the VSDs at the 50 Hz power system fundamental frequency and in the correct phase sequence. This represents the usable active power for a typical three-phase load. For contrast, the effective apparent power (S_e) shows the total apparent power of the three-phase system and represents the actual loading on the power system: in this case, the islanded microgrid's inverter.

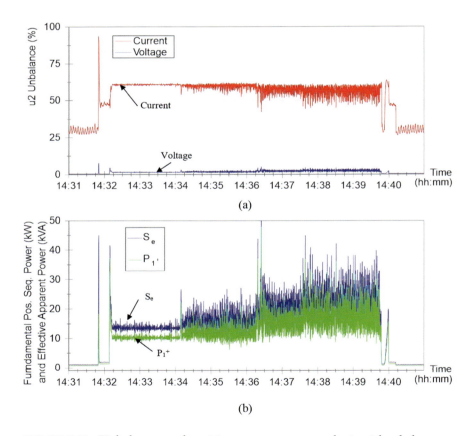

FIGURE 7.17 Unbalances and positive sequence powers during islanded operation of the nonlinear loads: (a) u2 unbalance for voltage and current, and (b) fundamental positive sequence active power P_1^+ and effective apparent power S_e.

7.5 HARMONIC COMPONENTS AND THD WITH NONLINEAR LOAD IN BSS

As stated in Chapter 4, when dealing with non-ideal and distorted signals, THD is used to define the level of distortion of the waveform. It is calculated as the ratio of the non-fundamental voltage (or current) to the fundamental voltage (or current): $THD_V = V_H/V_1$ and $THD_I = I_H/I_1$. There are two methods for calculating THD, arising from the methods of calculating the non-fundamental term:

- Sum method, used by IEC 61000-4-7, $THD_V = \dfrac{V_H}{V_1} = \sqrt{\dfrac{\sum_{h=2}^{h\max} V_h^2}{V_1^2}}$,

  but the summation also includes the DC term V_0.

- Difference method, used by IEEE standard 1459, $THD_V = \dfrac{V_H}{V_1} = \sqrt{\dfrac{V^2 - V_1^2}{V_1^2}}$.

The sum method simply compares the sum of all harmonics (excluding the fundamental) to the fundamental. The difference method takes the total root mean square (RMS) value minus the fundamental and compares it to the fundamental. Both methods then express the result as a percentage. There are two main differences between these approaches:

- The sum method only includes harmonic components (multiples of the 50 Hz system frequency) and does not include components in between these frequencies (no interharmonic components). The sum method also only includes harmonic components up to a maximum harmonic, typically the 50th (which is 2,500 Hz for a 50 Hz power system).

- In contrast, the difference method captures everything that is not the fundamental, including interharmonics and components above the 50th harmonic. It is also simpler to calculate because it does not require the waveform to be deconstructed into frequency components.

Figure 7.18a and b shows the THD for voltage and current, respectively, calculated using the difference method. The voltage THD is sitting at

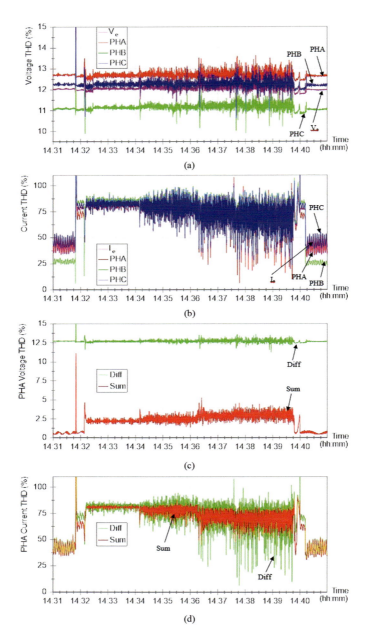

FIGURE 7.18 THD in sum and difference methods and their comparison during islanded operation with a nonlinear load: (a) voltage THD (difference method), (b) current THD (difference method), (c) comparison of the THD sum and difference methods for the same phase A voltage, and (d) comparison for the same phase A current.

about 12.5%, but the current THD is over 75%. This indicates a lot of current distortion, and the current waveform is more distorted than the voltage waveform.

Figure 7.18c and d shows comparisons between the sum and difference method, for voltage and current, respectively. The voltage THD values are very different, with the sum method showing a much lower THD below 5%, compared to 12.5% for the difference method. This discrepancy is caused by non-harmonic components in the voltage waveform that are not captured by the sum method. This is expected, considering that the inverter produces significant switching noise at a frequency above the 50th harmonic and at frequencies not synchronized to the power system frequency (hence not registering as harmonics but interharmonics).

Note that although the voltage waveform appears to have high distortion, this is only the case at the inverter terminals. The isolation transformer filters the voltage waveform; hence, the grid only sees a clean voltage with very low distortion.

Figure 7.18d shows that both THD methods produce a similar result for current. This suggests that most of the distortion is from harmonics of the fundamental power system frequency (multiples of 50 Hz). The difference method has a wider spread of values and more noise for current, indicating that it is also detecting interharmonic and noise components.

Figure 7.19a shows all voltage harmonics, from the zeroth (DC) to the 50th (2,500 Hz). These are calculated as subgroups, using the method in IEC 61000-4-7. This method breaks the signal into 501 frequency components, each separated by 5 Hz and each 5 Hz wide. The harmonic subgroup is calculated from three frequency bins (15 Hz total width), centred on each harmonic. The subgroup allows detection of harmonics that are not sitting exactly on a multiple of the power system frequency, preventing these components from being accidentally excluded.

Figure 7.19b shows the voltage interharmonics, which represent anything that is not a harmonic and consequently not captured by the harmonics subgroups. Using the method in IEC 61000-4-7, the interharmonics are calculated from all the frequency bins between the harmonic subgroups. In this case, seven bins are used with a width of 35 Hz, centred between harmonic subgroups. The zeroth interharmonic subgroup represents components between 7.5 and 42.5 Hz, and the first represents components between 57.5 and 92.5 Hz, all the way up to the 49th interharmonic which represents components between 2,457.5 and 2,492.5 Hz.

540 ■ Reinventing the Power Grid

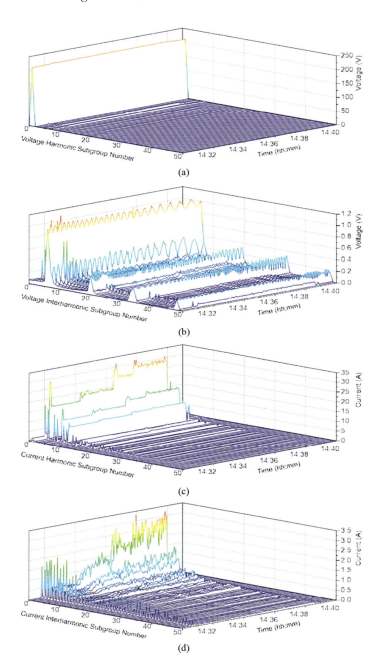

FIGURE 7.19 Interharmonics due to the inverter operations, during islanded operation of the nonlinear load: (a) all voltage harmonic subgroups, (b) all voltage interharmonic subgroups, (c) all current harmonic subgroups, and (d) all current interharmonic subgroups.

Recording harmonic and interharmonic data consume significant space. In this case, recording both requires 101 channels of data per input channel. Three voltages and three currents require a total of 606 channels to capture this information. This is impractical for long-term data logging; hence, measures like THD are preferred, where the harmonic information can be condensed into one or two channels per waveform channel. For this test of the microgrid island with VSD loads, the embedded data logger was recording harmonics aggregated over 3-second intervals (not at 5 Hz).

The voltage harmonics in Figure 7.19a show the fundamental being the largest component, with all other harmonics being very small. Therefore, the sum method of THD produces a low result. Figure 7.19a also shows minimal interharmonics, with all having a magnitude of approximately 1 V or less. In this case, there is minimal difference between the sum and difference methods.

The current harmonics in Figure 7.19c show a large fundamental and large fifth and seventh harmonic components. All other harmonics are small. Figure 7.19d shows that a large number of interharmonics are present, with a magnitude below 10% of the fundamental. The largest two interharmonics are the first two components, representing current with a frequency just below and just above the 50 Hz microgrid frequency. These interharmonics are expected, considering that the VSDs produce a variable frequency that is not synchronized to the microgrid's 50 Hz frequency and operates well below 50 Hz when the fans run at low speed. The presence of interharmonics explains the small difference between the THD sum and difference methods, but because the interharmonics are small compared to the harmonics, both THD methods achieve a similar result.

7.6 VOLTAGE WAVEFORMS AND RMS MEASUREMENTS

The embedded data logging system includes three methods of RMS measurement, used to determine the magnitude of the voltage and current waveforms. Two are performed in the time domain: RMS (TD) and fundamental RMS (TD).

- RMS (TD) measurement: This is the conventional RMS measurement (square root of the average of all squared waveform data points) performed over each waveform block of ten electrical cycles (approximately 200 ms in length in 50 Hz supply).

- Fundamental RMS (TD) measurement: This measurement first decomposes the waveform into two fundamental components: one in phase with the reference voltage (inverter's phase A is the reference) and the other 90° out of phase (the fundamental phasor). The RMS of the phasor's instantaneous magnitude is then measured over the waveform block of ten cycles.

The third method (labelled "RMS" in the figures below) is the RMS of all frequency components used for harmonic analysis. This RMS calculation is performed from the zeroth (DC or 0 Hz) bin up to the 500th bin (2,500 Hz or 50th harmonic).

Figure 7.20a shows the three RMS values for the inverter's phase A current waveform. The RMS (TD) and RMS measurements are effectively the same, and the fundamental RMS (TD) measurement is lower. This result is expected due to the presence of large current harmonics captured by both the RMS and RMS (TD) measurements.

Figure 7.20b shows the three RMS values for the inverter's phase A voltage waveform. Unlike the current measurement, the RMS (TD) and RMS values are different, but the RMS and fundamental RMS (TD) measurements are similar. This results from the limited bandwidth (2,500 Hz) of the RMS measurement. As shown in Figures 7.19a and 7.20b, there are no significant voltage harmonic or interharmonic components other than the fundamental. The fundamental RMS (TD) and RMS measurements only see the fundamental voltage; hence, they both return a similar value. The discrepancy between RMS (TD) and RMS can be explained by the presence of voltage components beyond 2,500 Hz.

The embedded data logger can record high bandwidth waveforms directly from the voltage and current transducers. For the inverter, these waveforms have a 500 kHz sample rate and use high bandwidth transducers: 500 kHz for voltage and 200 kHz for current. This allows for capturing high frequency components beyond the typical 2.5 kHz upper limit of conventional analysis.

The voltage waveform for the islanded microgrid was analysed with the VSDs powered off. These waveforms have the same set of three RMS values as shown at the very end of Figure 7.20b.

Figure 7.20c shows the noisy and distorted voltage waveform of the inverter over a 100 ms period. Despite the large amount of noise, the fundamental shape (a sine wave) is visible. As is known, the inverter produces a sine wave by very quickly switching the battery DC voltage (~800 V) between a positive and negative value, using PWM. The inverter includes a

Results and Analysis of an Autonomous Microgrid with BSS ■ 543

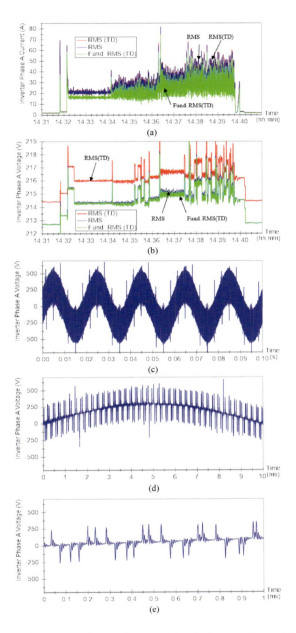

FIGURE 7.20 Voltage waveforms during islanded operation of the nonlinear load: (a) comparison of current RMS measurements, (b) comparison of current RMS measurements, (c) inverter voltage waveform, shown for a 100 ms period (5 electrical cycles), (d) inverter voltage waveform, shown for a 10 ms period (half an electrical cycle), and (e) inverter voltage waveform, shown for a 1 ms period.

filter to smooth out the voltage into a sine wave; however, some noise still gets through, as seen in Figure 7.20c.

Note that this high-frequency noise is filtered out by the isolation transformer. Hence, the voltage seen by the grid and load is very close to a sine wave. The examples below look specifically at the inverter output terminals and the quality of the voltage waveform when there is no further filtering beyond the inverter's internal sine wave filters.

Figure 7.20d and e observe the voltage waveform, showing the inverter's switching transients. The high-speed waveform recorder has a high enough bandwidth to capture these transients, including the ringing/oscillations that occur after each switching event. These transients and associated ringing are common to power electronic devices and depend on the inverter's power module design and internal filtering.

More information about the inverter's voltage can be found in the frequency domain. Figure 7.21a displays the full inverter voltage spectrum from 0 Hz to 250 kHz, with a resolution of 0.5 Hz (the size of each frequency bin). The spectrum is produced from the fast Fourier transform (FFT) of 1 million waveform data points spanning a 2-second period. The amplitude is displayed using a logarithmic scale, and each frequency component's magnitude is an RMS value (e.g., the 50 Hz frequency bin has an amplitude matching the fundamental RMS value).

The largest component is the 50 Hz fundamental frequency. The next largest components come from a very broad line spectrum of inverter switching components. Figure 7.21b shows the inverter spectrum between 0 and 10 kHz. The 4 kHz PWM switching frequency of the inverter can be clearly seen, along with harmonics of the switching frequency appearing at 8 kHz and beyond. There are no inverter switching components below 3.5 kHz.

Figure 7.21c shows the inverter spectrum between 0 and 500 Hz, corresponding to the region of the zeroth–tenth harmonic. The spectrum shows the presence of a fifth harmonic, a seventh harmonic and interharmonic components below the seventh harmonic (the sixth interharmonic region). Although the seventh harmonic appears to be larger than 1 V, this is actually an interharmonic component at 348 Hz. The true seventh harmonic at 350 Hz has a magnitude of 0.3 V. However, when using ten-cycle analysis (per IEC 61000-4-7), this 348 Hz component would be included in the harmonic bin (which has a width of 5 Hz). The other interharmonic at 336 Hz would be included in the sixth interharmonic bin.

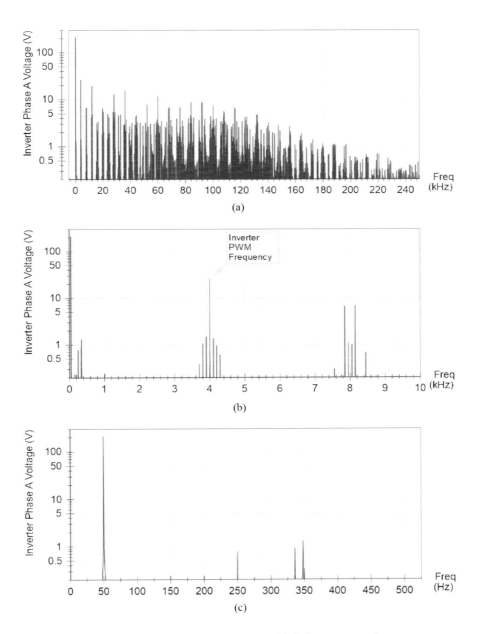

FIGURE 7.21 Analysis of voltage spectrums: (a) full inverter voltage spectrum (RMS), from 0 to 250 kHz (up to the 5000th harmonic), (b) inverter voltage spectrum (RMS), from 0 to 10 kHz (up to the 200th harmonic), and (c) inverter voltage spectrum (RMS), from 0 to 500 Hz (up to the 10th harmonic).

Because the spectrum below 2.5 kHz is sparsely populated (excluding inverter switching noise), it is not efficient to record all spectral content (or record the waveform at a very high rate) for any length of time. The techniques for harmonic and interharmonic analysis in IEC 61000-4-7 provide a more efficient way of capturing these data, relying on the fact that most distortion components will be harmonics of the fundamental. Recording harmonics still consumes a large amount of storage space; hence, measures like THD provide a good compromise between effectively measuring waveform distortion and the volume of data produced.

To investigate the cause of the RMS discrepancies in Figure 7.20b (where RMS (TD) is larger than RMS), detailed analysis was performed using the high-bandwidth voltage waveform. Figure 7.22a shows the fundamental voltage, obtained by filtering the high-bandwidth waveform to exclude any components beyond 100 Hz (the second harmonic). This is another method for extracting the fundamental component, although it is typically done with a very sharp filter that keeps the 50 Hz component and rejects the 100 Hz component. Such a filter is very difficult to design and use in a real-time measuring device, often being too computationally intensive to function in real time.

Figure 7.22b reproduces the RMS voltage, fundamental RMS voltage (using the filtered waveform) and the non-fundamental RMS voltage. The RMS and fundamental RMS values were produced using a sliding window RMS calculation with a window width of ten cycles. This method produces an RMS measurement every 2 μs, but each value is produced from 200 ms of data around that point. The non-fundamental voltage is produced using the IEEE 1459 method. Figure 7.22c shows the corresponding values produced in real time by the data logging system (except V_H, which was added during offline analysis).

Figure 7.22b shows a much larger non-fundamental voltage (V_H) of 73 V compared to the end of Figure 7.22c, which shows only 27 V. The fundamental RMS voltage is similar at 212.8 V. The RMS voltage in Figure 7.22b is 225 V, higher than the 214.5 V shown at the end of Figure 7.22c.

These differences are caused by the limited bandwidth of the power quality analyser module used by the data logging system. All waveforms are decimated (filtered and then have their sample rate reduced) as part of the resampling process that synchronizes the new sample rate to the power system frequency. For the data in this report, the nominal sample rate used for real-time analysis was 9,600 Hz, which can only capture frequency components below 4,800 Hz. Some of the PWM switching noise is

Results and Analysis of an Autonomous Microgrid with BSS ■ 547

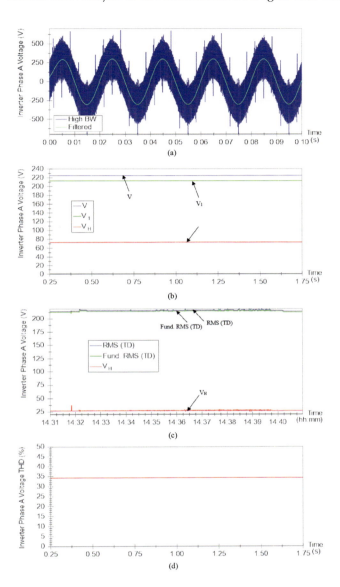

FIGURE 7.22 Further analysis of inverter voltages for RMS and THD: (a) inverter voltage waveforms, showing both the original high bandwidth waveform and a filtered waveform to obtain the fundamental, (b) inverter RMS voltage V, fundamental RMS voltage V_1, and nonfundamental RMS voltage V_H, all produced using a sliding RMS method of 10 cycle width, (c) RMS, fundamental RMS measured by the data logging system, and nonfundamental RMS calculated from these values, during islanded operation of the wind tunnel fans, and (d) inverter voltage total harmonic distortion THDV, produced by the THD difference method using V, V_1, and V_H.

captured by the power quality analyser module; however, the broad spectral components (from 8 kHz onward, see Figure 7.21a) are not captured. Consequently, the RMS and non-fundamental RMS values of the waveforms are much higher.

Using the non-fundamental RMS measurement of the high-bandwidth voltage waveform, the true total harmonic distortion can be calculated. Figure 7.22d shows this measurement, with a value just over 34%. In contrast, the corresponding real-time analysis THD values at the end of Figure 7.18c show the THD sum method producing a value less than 1% THD and the THD difference method producing a value of 12.6%.

These results highlight that high-frequency switching noise and non-sinusoidal waveforms should be assumed when power electronic devices like inverters are used. The specific implementation of standard measurements like RMS and THD is critical for providing accurate and valid results. This is crucial when comparing these measurements and using them to characterize the underlying behaviour of the power system.

7.7 COMMUNITY-LEVEL OPERATIONS: DAILY CYCLES OF BSS WITH HIGH PV PENETRATION, P-Q CONTROL AND FAULT RESPONSE

The presented results are obtained from an 8-month microgrid test platform deployment in a community, including a summer season. This test platform was initially designed for peak shaving and deferral in a small town experiencing increased power demand due to air conditioning loads during hot weather. The town is located on the fringe of the regional grid and has significant rooftop solar PV installations, functioning as a prosumer.

Instantaneous power (active or reactive) measurements are taken at the point of common coupling/connection (PCC), where Kirchhoff's current law dictates that the sum of powers is always zero ($P_{in}(t) = P_{out}(t)$). Since three-phase power lines connect to the PCC, the direction of power flow determines both power and energy balance for the network. Figure 7.23 summarizes the potential reasons for the direction of the power flow at the PCC, excluding unrealistic scenarios (e.g., three simultaneous inward or outward flows). In addition, the figure summarizes the reasons for the four-quadrant power control from grid, BSS and prosumer viewpoints.

It should be noted that the power line rating ($P_{max(line)}$) defines the transformer rating. If the load (P_{load}) exceeds $P_{max(line)}$, either a line upgrade (often including a step-down transformer) or a DER like a BSS is necessary.

Results and Analysis of an Autonomous Microgrid with BSS ▪ 549

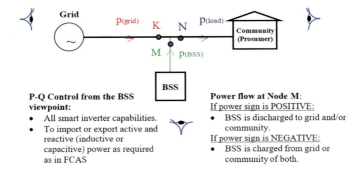

FIGURE 7.23 General description of power flow and purpose of four-quadrant control at the PCC from the viewpoints of grid, BSS and prosumer. Refer to Table 7.1 for the directions of the power flows.

Ideally, BSS is coupled to the grid near load centres with numerous prosumers. This configuration reduces transmission and distribution (*T/D*) losses while increasing *T/D* line capacity.

A sample daily power characteristic at node K and corresponding load duration curve (LDC) is given in Figure 7.24. Note that measured solar irradiance for the same day is also illustrated in the measured power figure.

The positive and negative sections of the LDC based on the measurements at node K indicate the energy used by the load (BSS + community) and the amount of energy exported to the grid that may be due to BSS and/or prosumer (community). The duration of reverse power and the amount of reverse energy depends upon the status of BSS and community as well as the seasonal impacts.

7.7.1 Analysis of Four-Quadrant Control with BSS in Daily Cycles

A smart inverter with four-quadrant control (*P-Q* control) offers significant benefits for BSS. It allows accurate power management (active, reactive and frequency) at the source of voltage and frequency variations,

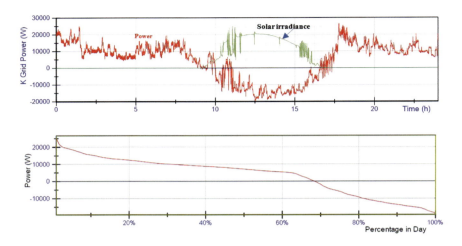

FIGURE 7.24 A typical daily power characteristic at Node K (reference to the solar irradiance of the day) and corresponding LDC.

near prosumers. This enables faster response and targeted solutions. Four-quadrant control also contributes to grid stability by managing reactive power flow for services like voltage regulation, frequency control and reserves. In addition, it minimizes power losses, reduces reliance on the grid for reactive power and improves power quality by mitigating harmonic distortion. These benefits extend to the distribution system, increasing line capacity and enabling integration of new loads such as EVs. Furthermore, in autonomous microgrids, BSS with four-quadrant control acts like a smart substation, combining functionalities of conventional voltage regulators, reactive power compensators and tap-changing transformers, while offering a wider range of control options. Figure 7.25 illustrates a set of results obtained in the network structure given in Figure 7.23.

Figure 7.25a illustrates P and Q data points based on real-time measurements of three-phase voltage and current at the PCC. Sampling time was 3 seconds. By controlling P and Q values, voltage and frequency can be regulated as highlighted by three hypothetical data points (1, 2 and 3). Note that, the BSS can quickly shift between these points. For example, if a large voltage regulation is required, operating point can be changed from 1 to 2 (e.g., 1 to 2) very fast, from positive maximum Q value to negative maximum Q value, hence offering greater flexibility compared to conventional generators.

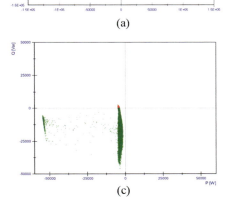
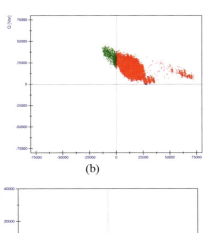
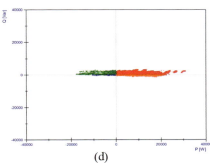

FIGURE 7.25 A selected set of P-Q control results in a community level microgrid in a 24-hour cycle (14th October 2018): (a) a typical 4-quadrant power data, (b) P-Q data in Node K (total duration of operation in three quadrants, Q1: 23.22h; Q2: 0.7425 h; Q4: 0.0058h), (c) P-Q data in Node M (total duration of operation in two quadrants only, Q2: 0.0691h; Q3: 23.94h), and (d) P-Q data in Node N (total duration of operation in all quadrants, Q1: 19.27h; Q2: 1.54h; Q3 : 0.086h; Q4: 3.0633h).

The battery and smart inverter enable the grid, BSS and prosumers to operate in both "generator" and "load" modes, influencing power flow direction. Refer to Figure 6.5 for details on operating quadrants. Note also that total durations of operations in each quadrant in Figure 7.25 are given, which indicates the duration of power quality events and the response time of smart inverter to each state of operation. The operating data points in different quadrants are shown in different colours.

7.7.2 Longer-Term Operation of the BSS in Grid

As stated previously, the microgrid system features an IoT high-level control system with a SCADA interface, enabling remote access and control when integrated into a power distribution system. This allows operators to optimize energy production and consumption within the microgrid and utilize all capabilities of the BSS inverter under various operating modes and fault response.

Most tests focused on exporting power to reduce the load on the 11 kV feeder supplying the community. This operation brought the feeder load close to zero, demonstrating the microgrid's ability to function in near-grid forming mode. During this 40-minute demonstration, the grid battery supplied nearly all the community's power requirements.

Power set points for active power were manually controlled, while reactive power was automatically managed by the PaDECS system to maintain voltage levels at the 11 kV feeder.

Due to limitations of the pole-mounted transformer connected to the grid battery, it's unlikely for export power to exceed 200 kW in this specific location. Figure 7.26 illustrates the overall grid battery operation during and after the discharge test. Following the demonstration, the battery intelligently transitioned to a "smart/autonomous" mode, where it recharged itself and then returned to an idle state.

Figure 7.27 shows a mode when no load is connected to the system by the distribution system operator and while the battery was supporting itself only when islanded.

The voltage plots in Figure 7.28 illustrate grid stability over a 4-month period. While two grid interruptions occurred during this time, the inverter and generation bus voltage experienced a few additional brief drops. It is important to note that these additional interruptions were limited to the generation stage and did not impact power delivery to downstream customers. They also occurred during the commissioning phase, before the system was connected to any customers. The last figure in Figure 7.28 shows the battery voltage over 4 months and can provide an approximate view of the batteries SoC and depth of discharge.

At the bottom of Figure 7.28, the depth of discharge of the battery during its autonomous operation is given, which was calculated by integrating the battery power (10-minute aggregated measurements). Because integration is very susceptible to small offsets, the battery power measurements were first "AC-coupled" by removing a −281 W offset in the battery power measurements. This offset may appear large but is quite small considering that the battery power transducers can measure up to 1 MW of power (up to 1,000 VDC, and up to 1,000 A), with the offset representing about 0.02% of the measurement range. The battery power measurements use a convention of positive power for charging and negative power for discharging.

Results and Analysis of an Autonomous Microgrid with BSS ■ 553

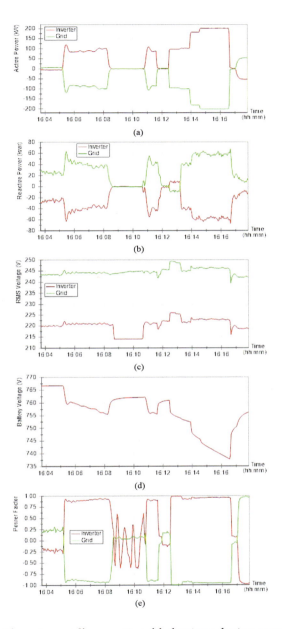

FIGURE 7.26 The microgrid's operational behaviour during power export scenarios, under a discharge test of approximately 35 minutes, representing key parameters related to power control (up to 100 kW and 200 kW) across various components: (a) active power flow on the grid and the inverter lines, (b) reactive power flow on the grid and the inverter lines, (c) RMS voltages measured on the grid and the inverter lines, (d) the voltage of the BSS DC side, and (e) the power factors on the grid and the inverter lines.

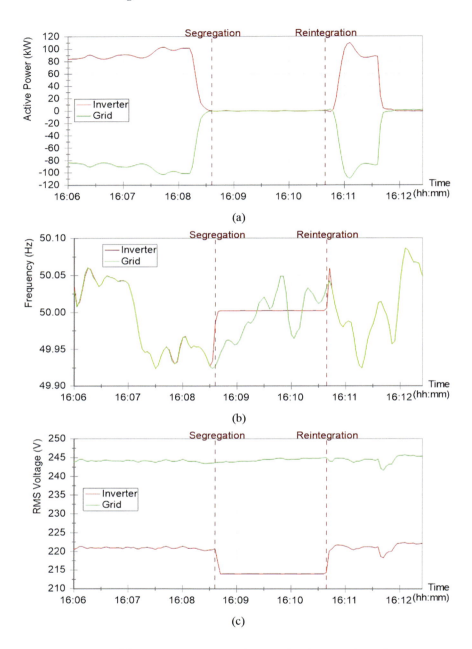

FIGURE 7.27 Grid battery operation under an islanding mode and its segregation and integration: (a) active power on the grid and the inverter lines, (b) frequency on the grid and the inverter lines, (c) RMS voltages on the grid and the inverter lines.

Results and Analysis of an Autonomous Microgrid with BSS ■ 555

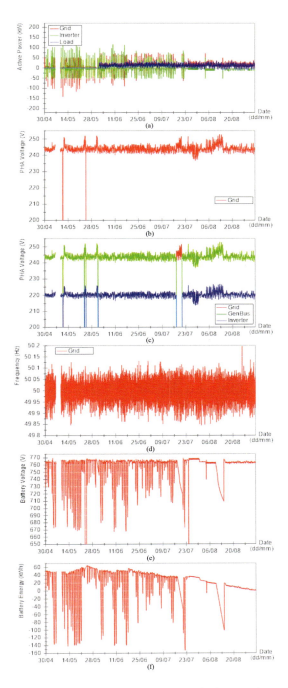

FIGURE 7.28 Grid battery operation over a period of 4 months and power quality results including DC battery voltage.

This translates to negative energy when the battery has discharged and positive energy when the battery has charged. The battery energy plot represents the net energy that is cycling through the battery.

Figure 7.29a–d shows the autonomous operation of the battery during the first week of July (winter month) and in a day. Note that the date markers are at midnight local time. Evening peak lopping/grid support can be seen by the positive inverter power and negative grid power. This operation is shortly followed by recharging of the grid battery (negative inverter power).

Similarly, in Figure 7.30, the grid battery operation at the end of August and start of September is given. Note that the grid battery is largely idle, with occasional charging top-ups. Note that in the middle of the days, the downstream feeder has reverse (negative) power flow, which is partially absorbed by the grid battery (note the difference between the minimum grid and load power).

7.7.3 Response of the BSS to a Fault

Figure 7.31 shows the autonomous microgrid's response to a fault event (two lines to ground fault) that occurred between 4:37 PM and 9:45 PM. The voltage plots (grid and generation bus) cover a period from 9:30 AM on 1 day to 9:30 AM the next day for context. It is observed that during the fault, phases A and C experienced a significant voltage sag (drop) until 7:27 PM, followed by a complete outage for all phases. At 8:06 PM, the microgrid's generation bus voltage recovered (black start), demonstrating the system's islanding capability; and the grid voltage returned at 9:45 PM; and the microgrid successfully reintegrated at 9:46 PM.

It is highly important to note that during this fault, the microgrid was not supplying power to any downstream loads. In addition, the voltage comparison between the grid and generation bus indicates that the microgrid automatically disconnected from the grid (opened Q1 in Figure 6.17) upon detecting the fault. Furthermore, the black start of the microgrid's generation bus occurred approximately 3 hours and 29 minutes after the initial fault or 39 minutes and 30 seconds after the complete outage. This highlights the system's ability to respond to certain two-line-to-ground faults (where the conventional protection system might fail and has failed).

Figure 7.32 is given to show the frequency of the grid and the inverter and before and after reintegration and the grid voltage asymmetric components. Note the symmetric components during the fault (for clarity: the zero-sequence component remains at approximately zero).

Results and Analysis of an Autonomous Microgrid with BSS ■ 557

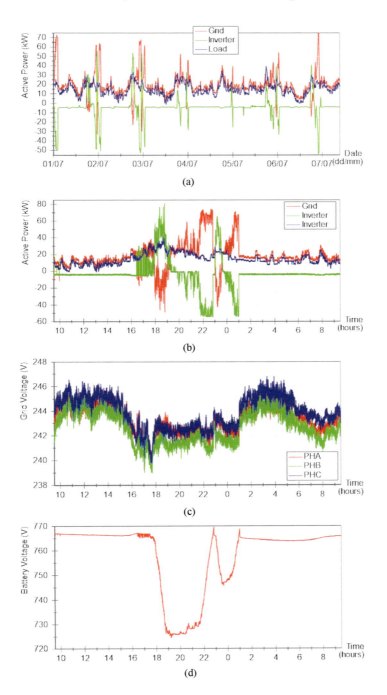

FIGURE 7.29 The autonomous operation of the battery during a week (a), and the detailed views of the grid battery operation in a day (b, c, d).

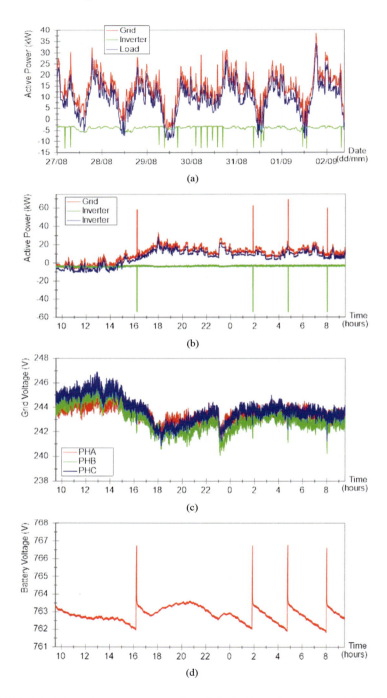

FIGURE 7.30 The autonomous operation of the battery: during a week in a Spring period (a), detailed views of the grid battery operation within a day (b, c, d)

Results and Analysis of an Autonomous Microgrid with BSS ■ 559

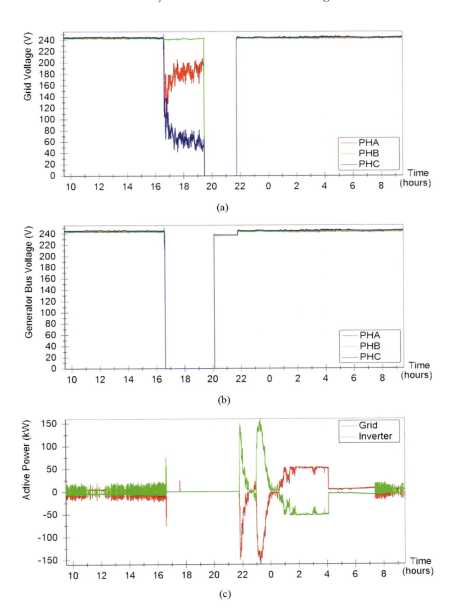

FIGURE 7.31 The grid and the generation phase voltages and the active power during a fault.

560 ■ Reinventing the Power Grid

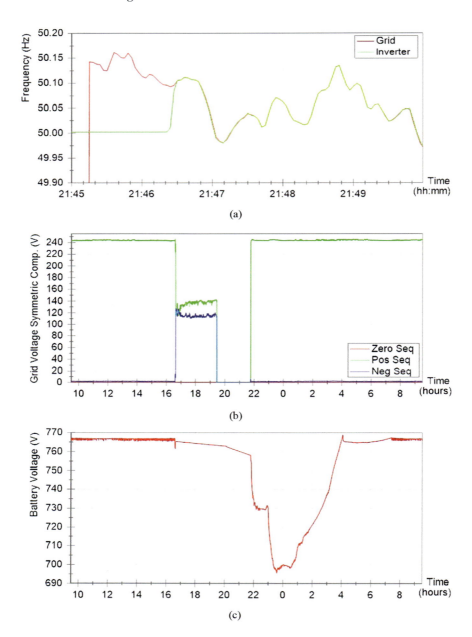

FIGURE 7.32 The frequency of the grid and the inverter (a) and before and after reintegration and the grid voltage asymmetric components (b), and the battery voltage over the day (c).

7.7.4 Inverter Efficiency Duration Curve and Thermal Management

Figure 7.33 presents the efficiency duration curve for the BSS in the microgrid test system, including the inverter and isolation transformer. The curve is plotted over a 24-hour period, similar to the way a LDC is obtained, but charging and discharging periods are separated which can be identified from the direction of current of the BSS. Note that the inverter efficiency varies significantly throughout the day due to two main factors: Fluctuations in charging and discharging current levels and also variation in inverter switch losses depending on the current levels.

As highlighted in the figure, the system operates at an efficiency below 50% for about 8.4 hours per day, during charging cycles. This lower efficiency translates to higher energy losses, impacting the overall performance and lifespan of the BSS. In addition, the highest efficiency occurs during discharging periods when the current injected into the PCC is significantly higher but for shorter durations.

As discussed in Chapter 5, lithium-ion batteries require precise temperature control due to their limited operational temperature range. This control is typically achieved using dedicated air conditioning systems, which can lead to significant energy losses often overlooked in the literature.

Figure 7.34 illustrates the impacts of thermal management on BSS, which contains a data set captured on a hot summer day, highlighting several critical parameters:

- BSS wall/ceiling and floor temperatures: These directly correlate with solar radiation, outside temperature and wind speeds, which also influence battery room and equipment room temperatures.

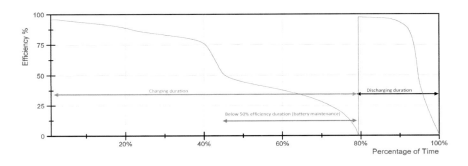

FIGURE 7.33 A typical efficiency duration curve of the inverter through the isolation in a complete battery cycle in a day.

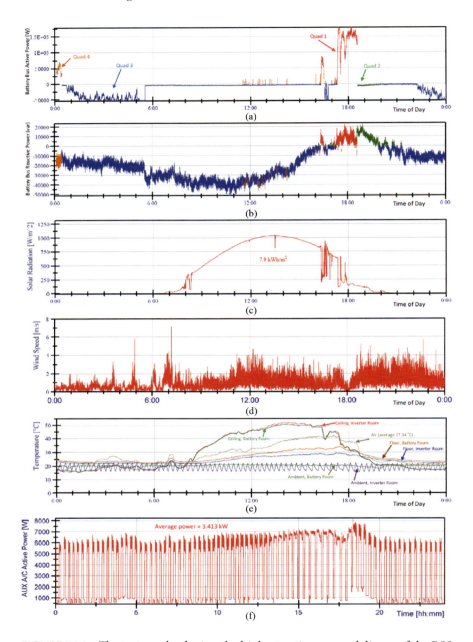

FIGURE 7.34 The test results during the highest active power delivery of the BSS and under the highest ambient temperature, from top: battery-bus active and reactive powers, solar radiation, wind speed, BSS temperature measurements and auxiliary power consumed by the air conditioning system in the battery room.

- Solar radiation: The data suggest a mostly clear sky with some cloud cover between 4 and 6 PM. This change is reflected in the decrease of ceiling surface temperatures and the absence of solar PV generation.

- Power delivery: In response to the changing load demand, the autonomous BSS delivers approximately 150 kW of active power to the grid during the day.

- Ambient temperature: The average temperature of 27.34°C represents a typical hot day (up to 50°C max).

As it is illustrated, the combination of the heatwave and internal heat generation forces the A/C system to operate at full capacity for 24 hours, consuming approximately 82 kWh—a significant portion of the total battery capacity (270 kWh).

It can be concluded that various strategies can be accommodated to improve thermal management. For example, effective insulation plays a critical role in reducing A/C power requirements and operational costs. Ideally, battery and inverter rooms should be physically separated for optimal temperature control. In addition, painting the BSS rooftop with heat-reflective colours can further minimize heat absorption. Furthermore, the data suggest that floor temperatures remain relatively stable despite ambient temperature variations. Therefore, where feasible, housing BSS units in an underground structure is recommended, particularly in hot climates.

7.8 MICROGRID WITH FLOW BATTERY IN A REMOTE CAMPUS

DER-powered microgrids offer an effective solution for fringe-of-grid and remote communities, including campuses, as they address reliability, cost, fuel security and environmental impact. Studies show that simply utilizing entire rooftops in remote areas for solar PV installations can generate two to three times more energy than the local demand, even without additional energy management or efficiency improvements.

For instance, Figure 7.35a shows a typical household's daily peak demand curve. Figure 7.35b focuses on the air conditioner's energy consumption and how much solar PV could potentially offset it. Data analysis suggests a total daily energy requirement of around 37 kWh. By strategically reducing non-critical loads, daily consumption can be significantly lowered, influencing the sizing of the community/campus BSS and the

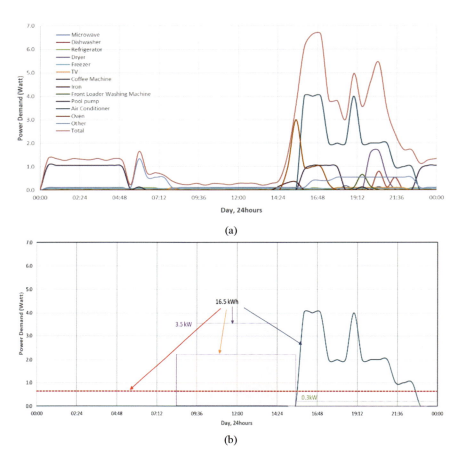

FIGURE 7.35 The maximum daily demand curve of a typical urban house (a), and the air conditioner's power demand (b) highlighting the energy windows.

corresponding solar PV farm. Additionally, the power and energy demand of EV fast chargers can also be factored into BSS sizing calculations.

However, fringe-of-grid and remote sites, including campuses, face several critical issues when it comes to reliable and cost-effective power generation. These challenges can be effectively addressed by microgrids powered by DERs.

Figure 7.36 shows the microgrid system layout in a remote campus, which involves a solar PV farm, a Li-ion and a flow battery system, all AC coupled to the local distribution power network using inverters and transformers.

Results and Analysis of an Autonomous Microgrid with BSS ■ 565

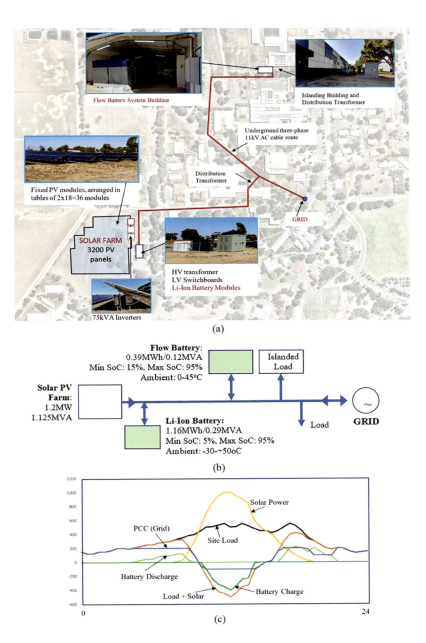

FIGURE 7.36 The microgrid with flow battery for a remote campus: (a) major microgrid infrastructures, where PV module structures are connected to the HV transformer site via 120 mm² bare copper earth conductors, (b) systematic microgrid system layout with specs, (c) a typical daily DER and load profiles under load following control.

566 ▪ Reinventing the Power Grid

The campus microgrid utilizes two BSS: Li-ion and flow batteries. While the Li-ion BSS has twice the discharge capacity and three times the storage capacity of the flow battery, it also has a higher impact on grid imports.

Figure 7.37a highlights an anomaly observed in one flow battery cell. Despite a constant SoC reading of 27%, this bypass level did not prevent overall system operation. Notably, the microgrid functioned effectively during eight summer days (exceeding 40°C), which is due to the flow batteries' air-conditioned enclosure, while the Li-ion battery packs were exposed to the outdoor environment.

As expected, solar generation peaks during midday, significantly reducing daytime grid imports. During the afternoon and evening, when solar generation drops, both the Li-ion and flow batteries discharge to meet the campus's load (see Figure 7.37b). This generation and demand shift are also due to the fact that the microgrid prioritizes solar power during peak periods, reducing reliance on the grid. These can be observed in Figure 7.37c and d visually wherein pre- and post-microgrid grid import patterns are considered in a weekly

As shown in Figure 7.36c, the microgrid successfully shifted the campus' peak grid demand from mid-afternoon to early mornings or evenings. This is because the microgrid prioritizes solar power during peak periods, reducing reliance on the grid. Figure 7.37c and d visually demonstrates this change in pre- and post-microgrid grid import patterns in weekly and yearly cycles.

This campus microgrid has delivered the following results: 41% reduction in energy consumption, 49% reduction in peak annual demand, 12% reduction in anytime actual demand and, in addition, significant energy cost savings due to tariff reductions from the high-voltage (HV) connection.

In addition, the microgrid offers two control options: load following and remote operator:

- The load following mode automatically manages peak shaving and battery charging based on solar generation and load demand. Batteries may be charged from the grid when necessary to maintain acceptable SoC levels, particularly for the flow battery with a narrower SoC range (15%–95%). Figure 7.37b and c indicates typical profiles under load following control.

Results and Analysis of an Autonomous Microgrid with BSS ▪ 567

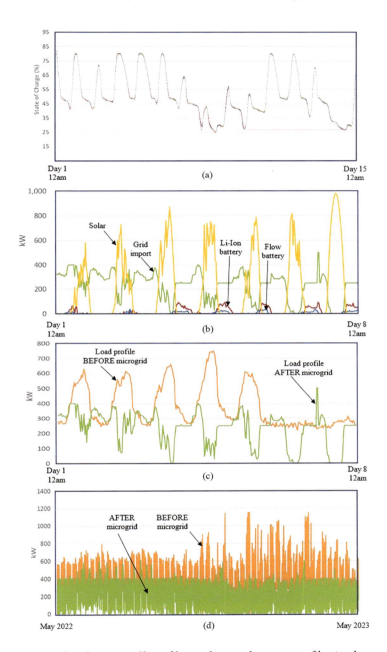

FIGURE 7.37 Flow battery cell profiles and typical power profiles in the campus microgrid: (a) SoC of the 13 flow battery cells during 2 weeks period, (b) weekly generation and load profiles during March (autumn season), (c) weekly demand profiles before (typical) and after (measured) microgrid, and (d) annual demand profiles before and after the microgrid.

- Remote operator mode allows an external operator to directly control energy resources and batteries to maximize revenue within the electricity market. This could involve prioritizing battery charging during low or negative wholesale electricity prices and reserving power for grid stability events.

The microgrid has limited islanding capability, allowing only a critical research building to disconnect from the grid and rely solely on the flow battery during outages or planned disconnections.

Furthermore, the chosen control method is ultimately subject to infrastructure and network constraints. These constraints primarily concern battery SoC and grid operator limitations. For example, if battery SoC falls below minimum levels due to extended cloudy periods or the flow battery's parasitic load exceeding solar generation, grid charging becomes necessary. Additionally, network constraints like target power factor, export power limits and battery charge/discharge limitations must be considered.

Finally, Figure 7.38 shows LDCs for three diesel generators that used in a remote site. Key observations are that the total power rating of the three generators is 1,052 kW. However, the total peak power demand is about 446 kW.

The capacity factors of the generators are all relatively low, ranging from 15% to 21%, which means that the generators are not producing power at their maximum potential for a large portion of the year. This suggest that the diesel generators at this remote site are likely oversized for the

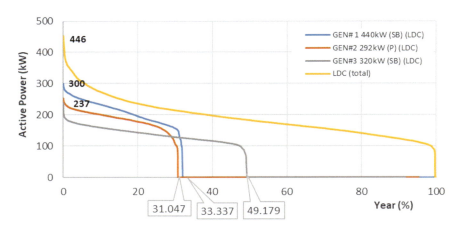

FIGURE 7.38 LDCs for remote diesel generators: High capacity but low utilization.

actual power needs, which is usually a common practice specifically in residential settings. This can lead to further inefficiencies, due to wasted fuels (since generators run at low capacity) and higher maintenance costs (since they frequently cycle).

Integrating renewable energy sources with the diesel generators could help to improve efficiency and reduce reliance on fossil fuels, since such sources typically have very high capacity factors especially when integrated with BSSs. Note also that, if the peak demands are infrequent and short-lived, no diesel generator option may still be sufficient.

In summary, there is potential to improve the efficiency of the power generation systems at remote sites by integrating renewable energy sources.

Index

AC
    coupled 552
    in data centres 35, 36
    machines 275, 276, 279
active load 313
active power 302, 303, 436, 437, 439, 440, 442
    balance 463, 465
    curtailment 430
active stall regulation 222–224
active thermal management 359, 362, 363
active yaw control 159, 161, 229, 230
actual
    energy capacity 341
    state of charge 341, 342
    wind speed 220
adaptive control 286
aerodynamic
    efficiency 158, 161, 174
    loads 229, 230
    thrust forces 226, 227
    torque 226, 227
AEMO 462
air density with altitude 175, 176
airfoil 220, 221
air-mass ratio 95–96
air-source heat pump 41
albedo 89
alkaline fuel cells 408
all electric vehicle 19, 20
altitude correction for air density 175, 176
ambient temperature 359, 360, 364, 367, 368
American wind energy association (AWEA) 179
amorphous silicon (a-Si) 101

amplitude modulation index 295, 296, 298
analog control circuits 144
analog-to-digital converters (ADCS) 144
ancillary services 273, 302–304
anemometer 162, 170, 171, 206, 211
angle of attack 165, 220–223
angle stability 246
anode 328, 331, 332, 333, 340, 364, 374, 375, 406–411
anolyte 332, 333
anti-reflective coating 116
anti-islanding protection 302, 313
apparent power 280, 436–438, 440, 444, 445, 456, 461, 496
    circle 440
appliance landscape 30, 40
applications of battery storage 446
area control error (ACE) 68
AS/NZS 78
aspect ratio 166, 221
asynchronous (induction) generators 172
atmospheric losses 144, 146
autonomous microgrid 476
    with BSS 504–569
average
    power 436, 437
    value 436, 437
    voltage 336, 339, 344, 354
axial momentum theory 181–188
    tilt 90
azimuth 90, 91

back electromotive force (EMF) 281, 283
back EMF constant 201, 202
back feeding *see* reverse power

571

back-to-back PWM converters 254, 256, 272–274, 287
balance of plant, BoP 449–452
band-gap 103
bandwidth 252, 310
batteries 16, 35, 53, 60–63, 68, 79, 136, 138, 139, 328
   balancing 337, 340, 393–395
   case configuration 356, 357
   cell 331
   cell encapsulation 356, 357
   cells electrical equivalent circuits 349–352
   characteristic curves of 378, 379
   characteristics 378, 379
   charge-discharge cycle 360, 36
   chargers 138
   charge voltage 339
   charging 380, 381
   comparison of 369–377
   configurations 352, 353, 356, 358, 359
   C-rate 336, 337, 381, 386, 387
   cut-off voltage 336, 339, 342, 382, 383
   depth of discharge (DoD) 337, 338, 342, 377, 380
   discharging 380–382
   efficiency 343–345, 378, 379, 391, 392
   energy 328–330, 336–338, 341–345, 349, 353, 354, 370–372, 374–376, 378, 381, 382, 386, 390–393, 395–398, 401, 413–416, 418
   energy density 336–338, 370–376, 390
   energy storage systems 328–330, 358, 359, 366, 367, 381, 382, 393, 397, 401, 402, 405
   -EVs 56
   intercell connections 356
   life 346–348
   life metrics 346
   management 393–397, 449, 450, 452
   management systems 340, 343, 345, 349, 357, 358, 365, 368, 378, 393–397, 401, 402, 405, 420, 421
   nominal voltage 336, 339, 343, 354
   open circuit voltage 336, 339, 340, 380, 391, 392
   operational parameters 336–338, 372, 373
   packaging configurations 352–358
   parallel connection 352, 353, 396
   parameters 336–338
   peak voltage 339
   performance 332, 338, 342, 346, 349, 357, 360, 361, 364, 373, 378, 379, 381–383, 385–389, 391, 395, 396, 401, 412, 413, 418
   power density 336, 337, 361
   protection 416–422
   protection and safety 486, 488
   rack/unit/bank configurations 354–359, 397, 401
   selection 369–373, 377, 378, 390
   series connection 352, 353, 396
   series-parallel connection 352, 353, 396
   sizing 369–373, 377, 378
   specific energy 336, 337, 374, 375, 390
   specific power 337, 338, 374, 375, 390
   state of charge (SoC) 336–342, 345, 378, 379, 380, 382, 391–393, 395, 396, 401, 402, 418
   state of discharge (SoD) 337, 338, 380
   state of health (SoH) 337, 338, 340, 341, 358, 373, 395–397, 400, 401
   storage 247, 249, 250, 253–256, 272, 274, 304, 317, 318, 324, 328–330
   storage applications in power systems 446
   storage systems 426
   technologies 334, 335, 374–377
   temperature control 359–368
   terminals and contacts 356, 357
   test classification 417, 418
   testing methods 416–419
   thermal management 356–368, 398, 400, 420
   voltage 336, 338–343, 353, 354, 373, 378, 380–383, 385–389, 391–393, 395, 396, 401, 402, 412, 413, 415, 416, 418
   voltage levels 338–340, 342, 354, 395
beginning of life (BOL) 346, 347
Bernoulli's Equation 182
BESS *see* battery, energy storage systems

Index

Betz limit 182, 186, 187, 190, 192, 208, 231
Betz's Law 182
bidirectional
    converter 313, 317, 318, 320
    DC/DC converter 136
    inverters 452–454
    power flow 441, 453
bifacial solar PV cells 89, 116, 117, 121
bioenergy 86
biomass 87
blackbox testing 108
blackouts 8, 9, 59, 61, 62, 68
    and consequences 8
black start capability 273, 302, 303
blade
    adapter 169
    design 165–168, 221
    element momentum (BEM) 188, 195
    pitch 195, 196, 223–227, 231, 232, 233, 235
    pitch angle 195, 196, 226, 227
    regulation 219
blocking diodes 128, 129
BMS 340, 343, 345, 357, 358, 365, 368, 378, 393–397, 401, 402, 405, 420, 421
BOL 346, 347
boost (step-up) converter 253, 254, 256, 265–268
brake system 170
brownouts 62
brushless permanent magnet generators (BPMGs) 172, 173, 198, 201–205, 226–228
brushless permanent magnet synchronous generators (PMSGs) 172, 219, 226
BSS
    grid-scale 452–454
    in grid-scale applications 449–452
    integration impact of 466, 467
buck-boost converter 137, 253, 254, 256, 268–271
buck converter 137, 253, 254, 256, 262–265, 268, 271
Budeanu, Constantin 444, 445
building-integrated photovoltaics (BIPV) 85, 121, 149

busbars 450, 483, 488
bypass diode 116, 118, 127–136

cadmium telluride (CDTE) 101
calendar life 346–348
camber 221
capacity factor 154, 155, 207, 239, 240, 568
capital cost 41
cell balancing 393–402
central inverter 145
charge 336, 338
    efficiency 344, 345
    speed 335, 336, 373, 375, 376
charging methods 380, 381
CHP 408, 409
Clarke transformation 275–279, 290, 294
clamped converters 452
clean energy sources 87
closed-circuit voltage 336, 339
cloud-edge effect 48
coefficient of performance (CoP) 43–45, 328
cogging torque 199
collective pitch control 234
communications 2–5, 17, 27
community microgrids 470
comparison of
    horizontal-axis wind turbines (HAWTs) 158
    vertical-axis wind turbines (VAWTs) 158
complex models 111
complex power 436, 437
components of electricity prices 7
components of power grid 5
computational fluid dynamics (CFD) 195, 218, 220, 364
concentrated solar power (CSP) 85
concentration loss 413
concentrator photovoltaics 101
conservation of
    mass 182
    of momentum 182
constant
    current (CC) 108, 138, 142
    current charging 380, 381
    current discharge 380–382, 386, 387

574 ■ Index

constant (*cont.*)
    power discharge 380–382
    resistance discharge 380, 381
    voltage 108, 138
    voltage charging 380, 381
containerized flow battery systems 359
continuity equation 182
control
    strategies 205, 231, 232
    system diagram 233
converter control 235
cooling methods 359, 362–365, 367
Cooper's method 93
COP *see* coefficient of performance (COP)
copper indium gallium selenide 101
cost of cooling 39
Coulombic efficiency 344, 345, 392
C-rates 336, 337, 381, 386, 387
current limit control 259, 260
current sensors 496, 498, 499
current source inverter (CSI) 299
current-voltage characteristics 102,
        108–115, 123–125, 131–137,
        140, 147
curtailment 59, 68, 69, 70
    event 69
cut-in speed 199, 208, 231, 232
cut-off voltage 336, 339, 342, 382, 383
cut-out speed 208, 232
cybersecurity 18, 73, 77, 79, 246
    for power systems 73, 79
cycle life 346–348, 370, 371, 372, 375, 376,
        377, 385

daily cycles 549
damping coefficient 197
dark I-V characteristic 109, 110
data centres 1, 35–39
    energy consumption 38
data logging 475, 477, 492, 496–502
DC
    circuit breaker 419, 421, 422
    circuits 436, 437
    in data centres 35, 36
    -link capacitor 257, 272, 274, 297, 318
    powered air conditioners 40
DC/DC converters 136–145, 247, 249, 250,
        253–255, 272, 317, 320

dead beat current controllers 326
decreased inertia 18, 59, 60, 72
delta
    connection 319
    windings 319
demand
    management 30, 34
    response services 62
    response systems 9, 68, 79
density ratios 176
DER
    integration 495, 500, 501
    technologies 427, 428, 468, 470,
        472, 474
design guidelines 147, 148
deterministic loads 229, 230
diagonal shading 130–133
differential
    evolution (DE) 142
    power processing (DPP) 255
    relays 483
diffuse irradiances 89
digital control techniques 253
diode(s) 101–103, 108, 110, 112–115, 118,
        127–130, 132, 135, 136, 248, 252,
        254, 255, 258, 262, 263, 265–268,
        283, 284, 317, 320, 324, 325
diode-clamped 317
diode saturation current 114, 115
directional relays 483
direct (beam) irradiances 89
direct methanol fuel cells 408, 409
direct torque control (DTC) 286
discontinuous load current 259
dispatchable generation resources 464
distributed energy resources (DERs) 5, 6,
        24, 246, 299–305, 315, 328,
        330, 427
distribution
    lines 425, 428, 436, 437
    network service providers 429
    system operators (DSOs) 12, 13
distortion (or void) power 444, 445
distributed resistance models 111
    energy resources 427
DoD *see* battery depth of discharge
Doppler shift 210
droop

coefficients 309, 310, 312
control 301, 307–310, 312, 435, 442, 443, 465
settings 435
driving levels in automotive 25
dual-axis tracking systems 106, 107
duty cycle 253, 262–265, 267

e-mobility 22, 25, 27, 39, 77
empirical models 111
encapsulation 356, 357
end-of-charge voltage 339
end-of-life 346–348, 397, 422
energy
capacity 328–330, 332, 333, 341, 342, 353, 357, 369, 371, 372, 375, 376, 377, 397, 399, 410
conservation 182
consumption 1, 13, 14, 28, 35–38, 43, 4
efficiency 343–345, 367, 378, 383, 391, 392, 399, 400, 407, 413, 414
efficiency ratio (EER) 44
factor (EF) 44
management systems (EMS) 9, 12
storage systems 91, 250, 298, 313, 315
storage technologies 334, 335, 374, 375, 377, 390, 391
engineering, procurement, and construction (EPC) 449, 450
equation of motion 196, 197, 199, 228
equivalent circuit
models 108, 110–112, 349–352, 411, 412
of solar PV cells 107–117
parameters 108
excess power 465
expert systems 286

FACT 322
failure in time (FIT) 325
failures in solar PV systems 149–151
faradaic efficiency 344, 345
Farada's Law 200
fault current management 273
fault levels 59, 63, 64
fault ride-through capability 273, 301
FCAS 435, 462, 463
feathering 225
feed-in tariffs 120

field-oriented control (FOC) 277, 280, 286
field-weakening 281, 282
fill factor 124
filling factor 147
filter inductor 309
filters 254, 257, 259, 298, 308–312, 317, 318
fire protection 477, 481, 488
firming the grid 464
flat efficiency 52, 250
floating life 346–348
flow battery 317, 318, 321, 322, 332–335, 350, 359, 366, 374–377, 397, 406, 408–411
comparison of 334, 335, 374–376, 390, 391, 408–411, 563, 565–568
electrical characteristics 411, 412
equivalent circuit 410
vanadium redox 366
fluid mechanics 181, 182, 188
flux-oriented vector control 287
flyback 254, 255
flywheels 375, 390, 391
energy storage systems 317, 318
FOC see field-oriented control
forced air cooling 364, 365
Fourier analysis 202
four-quadrant
control 313, 549
inverters 441
operation 313–315, 440, 441
freewheeling diode 268
frequency control ancillary services (FCAS) 57, 60–63
frequency modulation 259, 261
frequency nadir 59, 61, 62
frequency variation 57–61, 63
fuel cell EVs 56
full-bridge 254, 255, 317, 318, 321, 322
full-shading 130, 134, 136
fundamental frequency 436, 438, 444

GaN devices 265, 268, 270
gas storage heater 41
general dynamic wake (GDW) 195
generation and demand variability 47, 50, 51
generator-side converter 273, 280, 287, 290, 312

Index

generator torque 197
geographic smoothing 94
geostrophic wind speed 179
global warming 13, 15
GPS-locked timing 525
gravitational loads 227
green hydrogen 86, 406
grid
    -following 56, 302
    -forming 56, 302
    impedance 291, 308–311
    interconnection 63, 70
    inverters 298, 300–303, 313, 320
    modernization 6, 12, 18, 22, 71, 77, 79, 82
    reliability 6, 8, 18, 72
    -side inverters 291
    -supporting inverters 300, 302
    -tied 56, 57
    transformation 18, 22
ground fault circuit interrupters 483
grounding 450, 477, 480
ground-sourced heat pump 41

half-bridge 254, 255, 317, 318, 321, 322
harmonics 51, 57, 261, 308, 317, 436–438, 444, 452, 496
    analysis 274
    cancellation 319
    components 537
    distortion power 532–533
    power 444
    pollution factor 534
h-bridge 251, 317, 320, 322
    converters 452
    topologies 251, 317
heating methods 359, 362, 363
heating seasonal performance factor (HSPF) 44
heat pumps 17, 27, 28, 40–45
Hellmann exponent 177, 178
high
    current regions 109
    voltage regions 109
high-frequency
    isolation transformer 317, 322
    transformer 254
high-pole-count generators 173

historical turning points 2–4
hill climbing 330
horizontal-axis wind turbines (HAWTs) 153, 157, 158, 160–162, 181–183, 193, 219, 222
horizontal shading 130, 133, 134
hot spots 129, 130, 364, 403
hot-wire sensors 206
hour angle 92–94
hub 157, 160, 162, 163, 168–170, 174, 195, 200, 205, 206, 226–228, 230, 234, 235
HVDC
    submarine cables 95
    transmission 238, 274, 298
hybrid AC grid 9, 19, 20, 22
hybrid electric vehicle 19–22
hybrid microgrids 471
hydrodynamic 225, 226
hydroelectric 87
hydrogen production methods 406, 407
hydrokinetic turbines 188
hydropower 86
hysteresis 380, 391
    voltage 383

ideal diode 108, 110, 113
IEC 78
IEEE 78, 79
impedance 291, 308–311, 39, 139, 140
    matching 139, 140
incremental conductance 140–143
induction cooktops 46, 47
induction factor 185, 186, 190
infant mortality failures 151
infinite bus 272, 273
inline torque transducer 200, 201, 206
inner rotor designs 173
insolation patterns 106
instantaneous
    electric heater 41
    gas Heater 41
    power 438, 439, 441, 516, 517, 548, 549
    power efficiency 344, 345
interharmonics 437, 496, 537–541
internal resistance 338, 343, 344, 349–352, 364, 373, 378, 382, 383, 386, 396, 412, 417, 418

Index

intermediate (hybrid) steps of transitions 19, 20
Intermittency 47, 48, 50, 56, 57, 67
internal resistive losses 110
International Electrotechnical Commission (IEC) 179, 211
International Standard Atmosphere (ISA) 176
internet of everything (IoE) 247
internet of things (IoT) 247
inverters 246, 247, 250, 251, 254–256, 258, 261–263, 272–274, 277, 283, 284, 286, 291–303, 305–310, 312–320, 322, 441, 452–454, 517–533, 542–548, 550, 552, 553, 556–559
   -based resources (IBR) 429, 430, 432
   characteristics 298, 299
   -dominated infrastructure 298
   idle reactive power 517
   ratings 452, 454
   technologies 300, 301
Ion exchange membrane 332
irradiance 89, 92–98, 103, 114, 115, 124–126, 131–135, 140–144, 146
islanded microgrid 472–475, 494, 495, 496, 500
islanding 446, 472–474, 488, 493, 494, 496, 500, 501
islanding protection 500, 501
ISO 78
isolated microgrid 495
ITU 78

Jean Meeus' method 93
joule heating 359, 360

Kalman Filter 234
knee points 109, 110, 132–134

latitude 90–94, 104
LCOE 14, 15, 88, 91, 107, 120
leakage
   current 110, 112
   resistance 415
LEDs 101–103, 138
levelized cost of energy (LCOE) 14, 88, 91, 107, 120
LiDAR 210, 211, 235

lift
   and drag forces 220
   coefficient 222
lift-based wind turbine 221
lift-to-drag ratio 166, 220, 221
light-induced
   current 103, 112
   power degradation (LID) 151
lithium-based battery criteria 369–372
lithium-ion batteries 328, 330–332, 334–340, 342–345, 349–354, 356–366, 368–373, 374–383, 385–393, 395–402, 410, 418–420, 422, 565, 566
   average voltage 339
   cell 331
   charge cut-off voltage 339
   charge efficiency 344, 345
   efficiency 344, 345
   end-of-charge voltage 339
   midpoint voltage 339
   nominal voltage 336, 339, 343, 354
   open-circuit voltage (OCV) 336, 339, 340, 380, 391, 392
   over-charge voltage 339, 340
   over-discharge voltage 339, 340
   peak voltage 339, 340
   theoretical voltage 339
   voltage levels 338–340, 342, 354, 395
   working voltage 339
lithium iron phosphate 331, 339, 372, 374–377, 380, 388, 391
lithium manganese oxide 374, 380
lithium plating 375
lithium titanate 375–377
load
   angle 291–293, 309, 310
   bank 485, 506, 507, 509, 518, 528
   duration curves (LDCs) 65, 66, 455–460, 466, 549
   levelling 315
   shedding 59, 62, 67–70, 430, 435, 446, 460, 461, 464, 465, 473, 485, 490
   test 109
   types 138
   unbalance factor 534, 535
logarithmic wind profile 177–179
long axis (pitch) control 165

loss ownership 12
low voltage 429, 471, 472, 482, 485, 486, 488, 494, 495
    ride-through (LVRT) 281, 282, 301

macrogrids 468
magnitude control 291–293
maintenance charging 381
major components of wind turbines 162, 163
maximum power
    locus 235, 236
    point 109, 120, 140
    point tracking 136–145, 149, 150, 204, 205, 235
    point tracking methods 141, 142
    transfer theorem 139
maximum torque per ampere (MTPA) control 289
mean time between failures (MTBF) 325
mechanical losses 199, 200
medium voltage (MV) 429, 471, 494
meteorological masts 210, 211
microgrids 246, 300, 303, 315, 327, 425–435, 441, 443, 446, 447, 451–453, 454, 455, 459, 463, 467, 468–476, 482, 484, 485, 487, 488, 491, 493–495, 496–502
    benefits 468, 469
    electrical design 481
    at mining sites 471
    operation 472–475, 492
    test unit 486–488
microinverters 119, 135
midlife failures 151
minimum usable SOC 341, 342
mining electrification 22, 27, 30–34, 77
mobility 2–5, 17, 27
model predictive control (MPC) 286
modulation
    index 295, 296, 298
    techniques 251, 261, 295, 298
molar flow rate 413
molten carbonate fuel cells 408
monocrystalline silicon 101, 120
motor control 246, 272, 275, 286, 298, 315
MTPA control 289
multi-junction solar cells 101

multi-level
    inverter 251, 317, 318, 321, 322
    converters 452

nacelle 157, 160, 162–164, 168–171, 200, 226–228, 230, 234, 235
    and hub 170
nanogrid 300, 468
National Renewable Energy Laboratory (NREL) 179
natural gas 86, 87, 88
negative impacts of AC 9–11
negative sequence 534, 536
NERC 462
net reverse (feed-in) current 433
network disturbances 58
neural network control 286
Newton's Second Law 182
nickel cobalt aluminium 372, 374
nickel manganese cobalt 372, 374
nominal operating cell temperature (NOCT) 125, 126
non-linear loads 436, 437, 444, 445, 527–533
non-sinusoidal waveforms 436, 437, 445
nonsynchronous converters 268
north-south phase shift 93, 94

off-grid systems 302
offshore wind
    energy 86, 153
    farms 154–156, 162, 163, 165, 172, 173, 179, 211, 225, 226, 228, 229, 237–239
    turbines 154–156, 162, 163, 165, 172, 173, 211, 225, 226, 229, 238
ohmic resistance 351–352, 383, 412
one-sun of insolation 126
on-load tap-changing transformers 429, 430, 431
onshore wind farms 154–156, 162, 163, 167, 237–239
on-state resistance 249
open-circuit voltage (OCV) 336, 339, 340, 342, 380, 391, 392, 412
    point 120
    potential 412

Index

voltage 108–110, 112–115, 117, 120, 123–125, 139, 142, 147, 149, 301, 336, 339
operational
    loads 227, 229
    system losses 44, 45
optimal torque locus 236
orbital plane 90, 91
oscillatory instability 59
outer rotor generators 173, 174, 198
out of operation 231, 232, 234
output filters 308
overcurrent conditions 299
over-charge voltage 339–342
overcharging 339, 340, 342, 359, 360, 393, 395, 401, 420
over-discharge voltage 339, 340
overexcited condition 292
over-generation 72

panel tilt 104, 106, 145
parallel
    connection 352, 353, 396, 416
    resistance 110, 112, 114, 124, 130
park transformation 275, 276–279, 282, 289, 290, 293, 294
passive
    components 249, 250, 253, 268
    load 313
    stall regulation 222–224
    thermal management 359, 362, 366, 396
peak shaving 232
PE building blocks 252, 253
PEM fuel cells 407, 408, 414
performance ratio (PR) 146, 147, 271
perovskite solar cells 101, 103, 111
perturb and observe 142
phase angles 292
phase-locked loop (PLL) 293
phasors 274–276, 280, 283, 284, 292, 293
    diagram 283, 291, 292
    measurement units 492, 496, 497
phosphoric acid fuel cells 408
photocurrent 108, 110, 112, 113, 117
photodiodes 108
photovoltaic 85, 101
pitch
    angle 195, 196, 223–228, 231–235, 287, 289, 290
    control 165, 197, 222–225, 231–235
    regulation 222–224
    system 170, 171, 234
pitot tubes 206
P-N junction 101–103, 112, 124
point of common coupling 273, 426–429, 433–435, 441, 454, 462, 471, 472, 475, 482, 483, 499, 500
polarization resistance 351–352
polar moment of inertia 197
polycrystalline silicon 101, 117, 118, 120
polymer electrolyte membrane (PEM) fuel cells 406–408, 410, 413, 414
post battery-life 422
power
    capacity 86, 328–330, 333, 371, 372, 377, 397, 410, 86
    coefficient 185, 186, 190, 192, 194, 195, 196, 205, 206, 208, 231, 232
    conversion 449, 450, 452, 453, 490
    density 249, 252, 268, 322, 336, 337, 361, 370–372, 374–376, 390
    electronics 4, 17–21, 24, 25, 29, 37–40, 45–48, 52–57, 63, 79, 82, 116, 137–139, 246–254, 256–263, 265–270, 272, 274, 280–287, 290–304, 307–315, 317–326
    factor 272, 273, 285, 289, 295, 304–306, 314, 315, 432, 436–441, 445, 488, 495, 512, 513, 515, 529–530, 532, 534, 553, 568
    factor correction 313, 314, 439, 441, 512
    in the wind 174, 175, 177, 182, 184, 185
    law profile 177–179
    optimizers 119, 135
    quadrants 439–442
    quality 9, 10, 50, 51, 57, 73, 77, 432, 488, 490, 492, 496
    ratings 117–119
    semiconductor devices 299
    sharing 308–310, 312
    system inertia 8, 58–60, 63, 72
    system security 32, 39, 73–75
    system stability 52, 58, 74, 429, 432, 438, 461, 462, 464, 468, 473, 474, 485, 490

P-Q control 550, 551
P-Q unit circle 513, 515
predictability and forecasting 72
pre-industrial era 1
pressure ratios 176
prioritization 13, 15
prismatic cells 356
probability density functions (PDFS) 216–218
proportional-integral-derivative (PID) 286
protection 416–422, 477, 481, 486, 488
    circuits 306
PV cell types 101
P-V characteristics of PV panels 112, 115, 123–125, 131–136, 140
PV panel configurations 118–120, 144, 145
pumped hydro storage 390, 391
pumped hydroelectric storage 328, 329
pulse width modulation 253, 258–261, 263, 272, 277, 279, 287, 290, 295–297, 317
pure DC grid 19–21
PWM converter/rectifier 256, 272, 273, 283–287, 289, 290
pyranometer 97–99
pyrheliometer 98

q-axis current 280, 287, 289, 290, 294, 295, 297, 312
quadrant control 313
quantum dot solar cells 101

radar chart 43, 46
Ragone
    efficiency 344, 345
    plot 378, 389, 390
rapid ramping 72
rated power 204, 205, 231, 232, 235
rated torque 236
rate of change of frequency 60–62, 523–526
rate of change of power 47, 48, 50
Rayleigh PDFs 216, 217
reactive power 273, 274, 278, 280, 282, 289, 290, 292–295, 297, 301–306, 309, 310, 312–314, 432, 436, 437, 512, 513, 515–518, 528, 530, 532, 550, 553, 562

rectifiers 250, 251, 256, 258, 272–274, 283–287, 320
reference-frame 275, 276, 279
reference height 177–179
reflected irradiances 89
regenerative braking 272
regulations 18, 21, 22, 27, 30, 77, 78, 81
reinforcement learning 286
relative wind speed 220
reliability 15, 24, 35, 39, 51, 52, 56, 64, 71–73, 79
renewable integration 6, 53–55, 63, 64, 68, 70, 72, 75, 79
resistance water heating 328, 329
resonant converters 254, 255, 322
reverse flow duration 66, 67
reverse power 51, 64, 65, 66, 67, 71, 73
RF solid state cooking ovens 40, 47
ripple 259–261
RoCoF see rate of change of frequency
rolling blackouts see load shedding
rooftop solar array 145
root mean square (RMS) 201, 202, 204, 274, 276, 541–548
rotor
    blades 157–160, 162–165, 166, 167, 169, 170, 171, 173, 174, 178, 180, 181–183, 188, 189, 190, 191, 192, 194, 195, 196, 201, 205, 206, 208, 220, 224, 226, 227, 231–236
    plane 182–185, 190, 191
    speed 195–201, 204, 231, 232, 234–236
roughness coefficient see Hellmann exponent
roughness length 177–180
round-trip efficiency 344, 345, 392
running cost 41

SAE International 78
safety 416–422, 477, 481, 486, 488
saliency 288
sampling instant 141, 143
scattering 48
Schottky diodes 128
seasonal energy efficiency ratio (SEER) 44
seasonal phase-shift see north-south phase shift

Index ■ 581

second
  life 422
  order model 351–352
self-discharge 347, 376, 388, 393, 415
sensor integration 401
sensors in battery systems 401–405
series
  connection 352, 353, 396, 416
  output impedance 308
  resistance 110, 112, 114, 124, 147
series-parallel connected solar PV cells 118–120, 130, 135, 145, 352, 353, 396
service life 346–348
shadow effect in turbine types 161, 162
shading 110, 118, 119, 127–136, 141, 145–147, 150
shading scenarios 130
shelf life 346, 347, 388
short circuit current 108–115, 117, 123–125, 131, 147
shunt resistance 110–112, 114, 147, 149
single-axis horizontal tracking (SAHT) 104–106
single-axis vertical tracking (SAVT) 105, 106
single diode model 110, 111, 113, 115
smart
  appliances 27–30
  energy meters 9, 12, 42
  houses 22, 27, 28, 30, 77
  inverter(s) 12, 56, 57, 63, 74, 300–304
  inverter functions 304–306
SOC *see* state of charge (SOC)
SOD *see* battery state of discharge (SOD)
SOH estimation methods for batteries 396, 397
soiling 109, 125, 146, 147, 150
solar
  altitude angle 92, 94, 95
  and moon calculator 93, 94
  array configuration 118, 119
  declination angle 92, 93
  drying 85
  eclipse 48–50
  electric heater 41

  energy 85, 86, 88, 90–92, 94, 96, 102–104, 120, 122, 136, 137, 139, 144, 147–149
  farms 86, 98, 118–121, 125, 135, 145
  gas heater 41
  geometry 92
  heat pump 41
  irradiation 95, 97–99
  noon 92, 99, 104, 106
  ovens 85
  pumps 85
  PV 5, 14, 15, 40, 42, 43, 48–50, 52–54, 60, 63, 65, 68, 69, 71, 72, 79, 107, 115, 137, 139, 144, 147–149
  PV cells 85, 89, 102, 103, 107–118, 120, 121, 127–131, 135–137, 139, 144
  PV integration 5, 72
  PV panel sizes 119
  PV system inverters 254, 274, 297, 306, 307, 315
  radiation 88, 89, 562
  resources 88
  simulator 122, 123
  spectrum 96, 97, 122
  trackers 104–106, 116
  water heaters 85
  zenith angle 90, 92, 94–96
solid
  electrolyte 333–335, 340, 350, 351
  electrolyte interphase 385
  -state batteries 333–335, 374–377
  -state RF cooking ovens 47
solidity 191, 192
solid oxide fuel cells 408
solid-state transformers 248, 251, 317, 321, 322
source impedance 139, 140
space vector modulation 295
spatial diversity *see* geographic smoothing
speed and pressure in air and water 188, 189
stall 192, 220, 222–225, 236
  -regulated turbines 222
  torque 236
standards 13, 58, 73, 77–80
  and regulations 77
  spectrum 96, 122
  test conditions (STC) 122–124, 146, 147

state of
　charge (SOC) 336–338, 339, 340–342,
　　345, 378–380, 382, 391–393, 395,
　　396, 401, 402, 418, 508–512
　discharge (SOD) 337, 338, 380
　health (SOH) 337, 338, 340, 341, 358,
　　373, 395–397, 400, 401
step-down converter 251, 253, 254, 256,
　　262–265, 268, 271, 272
step-up converter 254, 256, 265–268
super capacitors 334, 335, 374–376, 390,
　　415, 416
superconducting magnetic energy storage
　　390, 391
supervisory
　control 234
　control and data acquisition
　　(SCADA) 9
　systems 358, 359
supply/demand imbalance 62, 67–70, 73
surface roughness 178, 180, 221
swept area 165, 175, 191, 195, 208
switch control 258–261
switched
　inductor method 398, 399
　mode rectifier (SMR) 256, 258,
　　283–286
switching frequency 249, 250, 254, 260,
　　263, 317, 318, 322
swing equation 289
synchrophasor 525
synchronizing inverters 313
synchronous buck-boost converter
　　268–271

tandem cells 103
temperature
　for air density 175, 176
　on I-V Characteristics 124, 125
temporal and spatial diversity *see*
　geographic smoothing
the beginning of life 346, 347
the components of the modern wind
　　turbines 162–164
the end of life 346–348
thermal
　control in batteries 359–368, 398,
　　400, 420

efficiency 344, 345
losses 125, 146, 344
management 561–563
modelling 368
thermocouples 402
thin-film panels 110, 111, 114, 120, 121
third-life 422
three-blade HAWT design 193, 219
three-phase inverter bridge 317, 318, 320
thrust force 184, 185, 226, 227, 232
tilt angle 90, 104, 105
tilted single-axis tracking (TSAT) 106
time-of-use pricing 12
time ratio control 258, 259
tip speed ratio (TSR) 189–192, 194–196,
　　231, 232
torque transducers 200, 201
total harmonic distortion (THD) 437
　current 537–539
　voltage 537–539
tower and foundation 167, 168
tower shadow effect 161, 230
tracking in solar PV panels 104
transient instability 59
transmission and distribution charges 13
transmission/distribution losses 12
transmission losses 120, 145, 146
transmission system operators (TSOs)
　　12, 13
trickle charging *see* maintenance
　charging
triplen harmonics 319
TSR *see* tip speed ratio (TSR)
two diode model 111
two-level inverters 317
two-mass system 228
types of solar irradiances 88, 89

ultracapacitors 334–336, 376
ultra-lift luo 254, 255
ultra-WBG 248
under excited condition 293
unidirectional
　converter 250
　power flow 272, 313
uniform shading 134
uninterrupted power supplies 298, 299
unity power factor 289, 314

upwind and downwind turbines 159, 160, 161, 219
usable energy capacity 341, 342

variability 47, 48, 50, 51, 65, 67
variable speed motor drives 272, 298, 315
variable-speed wind turbines 174, 232
Varshni equation 115
vector control 276, 277, 286, 287, 289–295, 297, 298, 302, 303, 308, 312, 313
vehicle-to-grid (V2G) applications 320
velocity ratio 186, 190
vertical partial shading 130, 134, 135
vertical-axis wind turbines (VAWTs) 157, 158, 161, 162, 219
virtual
    generators 6, 52
    inertia 60–63
    power plant (VPP) 63
viscous friction coefficient 197
voltage
    control 259, 260, 293
    efficiency 344, 345, 414
    limit control 259, 260
    magnitude 520
    quality 57
voltage source
    converters (VSC) 258, 261, 295, 309
    inverter (VSI) 291, 299, 317, 318

wake effect 238
Warburg impedance 352, 415
water heating 40–43, 45
wave energy 4, 14, 47, 50, 53
WBG devices 247–250, 252, 268, 270, 271, 315, 322, 324
wear-out failures 151, 324, 325
weather vane 160
Weibull PDFs 216–218
wide bandgap devices 246–250
wind
    levels 209, 210
    map 209, 210
    power density rose 215, 216
    profiles 177–181
    quality 209
    quality 209, 210
    regimes 209
    resource assessment 179, 181, 209–213, 216, 217
    resources 209
    roses 215, 216
    shear exponent 177
wind energy 153, 157, 160, 164, 166, 167–170, 174, 175, 176–183, 186, 188–192, 194–198, 200, 201, 204–207, 208–210, 211–219, 220–232, 233–239
    offshore 86
wind farm
    incident categories 241–243
    layouts 237, 238
    offshore 154–156, 162, 163, 165, 172, 173, 179, 211, 225, 226, 228, 229, 237–239
    onshore 154–156, 162, 163, 165, 168, 237–239
wind generation 4, 14, 15, 47, 50, 53, 59, 60, 63, 67, 69, 79
winding
    Inductance 202–204
    Resistance 202–204
wind turbine
    blades 157–160, 162–171, 173, 174, 178, 180–183, 188–192, 194–196, 201, 205, 206, 208, 220, 224, 226, 227, 231–236
    design 155–192, 193–208, 209, 210, 211–219, 220–232, 233–239
    failure categories 241–243
    rotor 157–166, 169, 170, 171, 173, 174, 180–185, 189–193, 197, 200, 204, 205, 206, 219, 220, 222–226, 227, 228, 231–236
    sensors 162, 163, 170, 171, 206, 211, 233–235
    sizes 154, 156, 157
    standards 240, 241
    systems 153–198, 200–243, 254, 272, 273, 281, 286, 287, 315
wind system inverters 306, 307
wind vane 162, 170, 171, 211, 235
wireless
    charging system 318–320
    power transfer (WPT)
        systems 320

working voltage 339
wound rotor synchronous generators (WRSGs) 172

yaw
    angle 193, 194, 226, 227, 229–231, 234, 235
    control 234
    mechanism 157, 159, 160, 161
    misalignment 229, 230

zenith 90–92, 95, 96
zero-current switching (ZCS) resonant converters 254, 255
zero power flow duration 66, 67
zero sequence 534, 536, 559
zero-voltage switching (ZVS) resonant converters 254, 255

For Product Safety Concerns and Information please contact our
EU representative GPSR@taylorandfrancis.com Taylor & Francis
Verlag GmbH, Kaufingerstraße 24, 80331 München, Germany